Art from the Ashes

ART from the Ashes

A Holocaust Anthology

EDITED BY LAWRENCE L. LANGER

New York
OXFORD UNIVERSITY PRESS 1995
Oxford

R

Oxford University Press

Oxford New York Toronto
Delhi Bombay Calcutta Madras Karachi
Kuala Lumpur Singapore Hong Kong Tokyo
Nairobi Dar es Salaam Cape Town
Melbourne Auckland Madrid

and associated companies in
Berlin Ibadan

Published by Oxford University Press, Inc.,
200 Madison Avenue, New York, New York 10016

Library of Congress Cataloging-in-Publication Data
Art from the ashes : a Holocaust anthology / edited by Lawrence L. Langer.
p. cm.
ISBN 0-19-507559-5
ISBN 0-19-507732-6 (pbk.)
1. Holocaust, Jewish (1939–1945)—Personal narratives.
2. Holocaust, Jewish (1939–1945)—Literary collections.
I. Langer, Lawrence L.
D804.3.A78 1995 940.53'18'092—dc20 94-11446

9 8 7 6 5 4 3 2 1

Printed in the United States of America
on acid-free paper

Avraham Tory, "Memoir." Reprinted by permission of the publishers from *Surviving the Holocaust: The Kovno Ghetto Diary* by Avraham Tory. Translated by Jerzy Michalowicz. Cambridge, Mass.: Harvard University Press. Copyright © 1990 by the President and Fellows of Harvard College.

Ida Fink, "The Key Game" and "Spring Morning," from *A Scrap of Time*. Translated by Madeline G. Levine. Copyright © 1987 by Random House, Inc. Reprinted by permission of Pantheon Books, a division of Random House, Inc., and Peter Owen Publishers, London.

Isaiah Spiegel, "Bread." Translated by David H. and Roslyn Hirsch. Appeared originally in *Midstream*, December 1984. Reprinted by permission of *Midstream*.

Bernard Gotfryd, "The Last Morning," from *Anton, the Dove Fancier*. Copyright © 1990 by Bernard Gotfryd. Appeared originally in *Midstream*. Reprinted by permission of Simon & Schuster, Inc.

Sara Nomberg-Przytyk, "The Verdict" and "Friendly Meetings," from *Auschwitz: True Tales from a Grotesque Land*. Translated by Rosyln Hirsch. Edited by Eli Pfefferkorn and David H. Hirsch. Copyright © 1985 by The University of North Carolina Press. Reprinted by permission of the publisher.

Aharon Appelfeld, *Tzili: The Story of a Life*. Translated by Dalya Bilu. Copyright © 1983 by Aharon Appelfeld. Reprinted with the permission of Wylie, Aitken & Stone, Inc.

Tadeusz Borowski, "This Way for the Gas, Ladies and Gentlemen," from *This Way for the Gas, Ladies and Gentlemen*. Translated by Barbara Vedder. Translation copyright © 1967 by Penguin Books Ltd. Original text copyright © 1959 by Maria Borowski. Used by permission of Viking Penguin, a division of Penguin Books USA, Inc., and Penguin Books Ltd.

Arnošt Lustig, "Infinity," from *Street of Lost Brothers*. Translated by Vera Borkovec. Copyright © 1990 by Arnošt Lustig. Reprinted by permission of Northwestern University Press.

Adolf Rudnicki, "The Crystal Stream," from *Ascent to Heaven*. Translated by H. C. Stevens. Copyright © 1951 by Roy Publishers A.N.

Alexander Kluge, "Lieutenant Boulanger," from *Attendance List for a Funeral* Translated by Leila Vennewitz. Copyright © 1966 by Alexander Kluge. Reprinted by permission of Alexander Kluge.

Jakov Lind, "Resurrection," from *Soul of Wood*. Translated by Ralph Manheim. Copyright © 1964 by Jakov Lind. Reprinted by permission of Hill and Wang, a division of Farrar, Straus & Giroux, Inc., and Reed Consumer Books Ltd.

Pierre Gascar, "The Season of the Dead," from *Beasts and Men*. Translated by Jean Stewart. Reprinted by permission of Georges Borchardt, Inc. Copyright © by Editions Gallimard. Translation copyright © 1956 by Little, Brown and Company.

For Sandy

who in an age of fragments keeps me whole

Contents

II

Journals and Diaries 153

III

Fiction 235

IV

Drama 471

V

Poetry 553

VI

Art from the Ashes

On Writing and Reading Holocaust Literature

We may never know what the Holocaust *was* for those who endured it, but we do know what has been said about it and, as this volume will show, the varied ways writers have chosen to say it. If the Holocaust has ceased to seem an event and become instead a theme for prose narrative, fiction, or verse, this is not to diminish its importance, but to alter the route by which we approach it. Our vision of it may never be complete, but the composite portrait offered by these texts does much to rescue it from obscurity and to light up its dreadful features with the deciphering rays of language.

Language, of course, has its limitations; this is one of the first truths we hear about in Holocaust writing. Reflecting on rumors that children under the age of ten were being slaughtered in the Lodz ghetto, Warsaw diarist Abraham Lewin admits: "It is hard for the tongue to utter such words, for the mind to comprehend their meaning, to write them down on paper." Then a few lines later, turning to his own dilemma in the Warsaw ghetto, he adds:

> [P]erhaps because the disaster is so great there is nothing to be gained by expressing in words everything that we feel. Only if we were capable of tearing out by the force of our pent-up anguish the greatest of all mountains, a Mount Everest, and with all our hatred and strength hurling it down on the heads of the German murderers of our young and old—this would be the only fitting reaction on our part. Words are beyond us now. Our hearts are empty and made of stone. (May 25, 1942)

If we read carefully, we realize that Lewin is rebuking not the poverty of language, as posterity will judge it, but its value as a weapon against the current enemy bent on destroying him and his fellow victims. Two months later, Lewin would witness the deportation of about 300,000 Warsaw Jews (including his wife) to their death in Treblinka, and would sense even more firmly the gulf between his "useless" diary entries and Jewish doom. If Lewin's sense of futility has carried over to our own time—some still believe that it is impossible to write about the Holocaust and that silence is the only fitting response—perhaps one reason is a failure to understand the difference between language as "ought" and language as "is." Since a vast body of Holocaust literature already exists, the issue of whether it "should" or "can" be written is pointless. The question we need to address, dispensing with excessive solemnity, is how words help us to imagine what reason rejects—a reality that makes the frail spirit cringe.

Our expectation is that language should play an active role in the affairs of men, that a link should exist between what we say and what we do, or what is done to us. But the more Lewin and other diarists, poets, and commentators wrote, the less effect they seem to have had on the incidents consuming them. This explains the impulse to abandon words, to nullify their power to rule over or portray human events. But even as he conceded the vanity of his efforts, Lewin went on writing, leaving behind a record of scenes that nothing *but* language could have captured for the future. In such moments of crisis, the pen must have seemed a brittle instrument indeed; a half century later, we realize that little else could so keenly conjure up that vanished time.

The path to Holocaust literature is strewn with clichés bred by the conflict between "ought" and "is." We still hear that writers needed a decade or more before they could begin to imagine the horrors of the catastrophe, in spite of the evidence that major poems by Abraham Sutzkever and Paul Celan and Itzhak Katzenelson were written *during* the disaster, and that classics such as Tadeusz Borowski's short stories, Charlotte Delbo's memoir *None of Us Will Return* (1965), and Ilse Aichinger's novel *Herod's Children* (1948) were all finished within a few years of the war's end. We need to believe that one *ought not* to be able to write about that defilement of human dignity, as if the act of writing would swell the trespass or soil the sanctity of the ordeal. Yet fifty years after the havoc, we have such an abundance of texts that Holocaust literature has grown into a genre of its own, needing neither excuse nor vindication.

It does, however, require some context because, unlike most other literature, it draws on experiences so foreign to the ordinary reader that one might be inclined to mistake its vision for an alien world of fantasy. Between the uttering tongue and the comprehending mind of Lewin's words lies the intervening medium of the page (which for Lewin was the intervening fact of total chaos), and what appears there so threatens all systems of value that we have cherished for generations and millennia that we shrink instinctively from its implications—and, hence, often from its truths. Consider, for example, the impact of Holocaust literature on our Romantic heritage: infinitude of spirit dwindles to the defeat of the body, physical despair; the inviolable self ebbs into the violated self, defenseless against the fury of power; the idea of the future as a dream of unbounded possibility and automatic progress subsides into a nightmare of violence and annihilation, an abrupt end to everything we consider human.

Holocaust literature is a literature of the moment, seeking vainly to unite with the stream of time. Examples of its villains and victims defy precedent and analogy; Edmund and Iago seem mild in comparison with Nazi savagery, and the choices available to traditional heroes bear little resemblance to the situations into which Jews were plunged. In a classic chapter in *The Survivor: An Anatomy of Life in the Death Camps* (1976), Terrence Des Pres uses the image of excremental assault to describe the humiliation the Germans inflicted on their prey. If we can understand that he was speaking in literal, not metaphorical, terms, we will have come a long way in entering the world of Holocaust literature.

The most vivid example of excremental assault I know of emerged from an interview with a woman who described her escape from a German deportation roundup. With her sister-in-law, she fled to the village square of her Polish town and found the public outhouse. Together, they pulled up the floorboards and climbed down into the cesspool beneath. There they remained for three days and nights, without food or water, up to their necks in liquid filth. They were soon joined by several other women. On the second night, someone led the Germans to their hiding place; they shot into the darkness, and the woman standing next to her was fatally struck. The following day, the "action" was temporarily suspended, and the witness and her sister-in-law sneaked out and returned home.

No amount of polishing can cleanse our memories of the lingering stench; its fetid odor clings to the woman's nostrils today. Holocaust literature cannot escape the taint of decay that extends far beyond the motif of a single testimony. The Enlightenment belief in human beings as reasonable creatures, pursuing their own good and the good of the community, must now share its influence with a darker and more demeaning heritage. Whether or not we can decree a treaty between the contending views remains an area of intense dispute. We are still wrestling with the loss of stature that a disaster like the Holocaust imposes on our ideal of civilization.

Much of the Holocaust writing included here uses the power of expression to tell the story of men and women who matched their waning strength against the ruthless enemy resolved to destroy them. Myth and tradition are of little use in consoling them or us. The encounter is unequal from the start. Camus's Sisyphus may provoke the gods by the nobility of his defiance as he pushes his rock once more up the hill, but the Jews who dragged their stones up the endless steps of Mauthausen quarry knew a more tangible version of despair. The issue is not *how* they will prevail, but, *if* they endure, what they will survive with, and what they will have survived. Our pursuit of these questions may usher in meager answers, but the history of the Holocaust itself leads to a spiritual universe more haggard than the one we inhabited before its arrival.

Among the leading ideas we are forced to surrender as we read through these pages is the comforting notion that suffering has meaning—that it strengthens, ennobles, or redeems the human soul. The anguish of the victims in ghettos and camps, in hiding or flight, in brave if futile rebellion, can be neither soothed nor diminished by a vocabulary of consolation. The challenge of this literature is to discover and *accept* the twisted features of the unfamiliar without searching for words, like "suffering", to shape what we see into more congenial façades. Incidents of resistance and uprising, courageous as they undeniably were, existed within a larger framework of loss: an axiom of this universe, so different from the world of independent striving we appear to thrive on today, is that one's own survival *almost never could be severed from someone else's death.*

Examples abound in this literature, but one will have to suffice. A group working outside the barbed-wire fences of the Sobibor deathcamp overpowered and killed the Ukrainian guards and escaped into the surrounding woods. The SS summoned all the remaining prisoners in the camp to roll call, and randomly chose every tenth one to be sent to the gas chamber. With what language or spirit do we admire the heroic daring of the escapees, once we have learned the effect of their flight? Does the life of the few "redeem" the death of the others? Or in celebrating one, do we defame the other? Neither query nor language begins to grasp the complexity of situations like these, which elude common definition and cast us into a morass of moral confusion.

The way out is not to try to jostle the confusion back into an unwarranted clarity, but to find our bearing by using landmarks native to this uncertain terrain. For example, one of the chroniclers of the Lodz ghetto describes his visit to a mica-splitting shop, where Jews "who are not fit to work at anything else" (a sarcastic jibe at their German taskmasters, and perhaps at the internal ghetto administration, too) sit at tables and with trembling fingers perform their lung-destroying work. They are, he says, "mostly 'by-gone' people—ancient men, old women, people with 'obituary' faces." When we begin to grasp the impact of that image—that the purpose of Jewish daily existence under the Germans was *wasting away,* not living (though this entry is undated, it probably was made in early 1941, before the Final Solution had begun)—we will see how easily different words can create different eyes for meeting this literature. Those accustomed to reading obituary pages are thrust into a confrontation with "obituary faces." These unsettling visages of death-in-life help to define the frontiers of the realm we are about to invade.

The most compelling Holocaust writers reject the temptation to squeeze their themes into familiar premises: content and form, language and style, character and moral growth, suffering and spiritual identity, the tragic nature of existence—in short, all those literary ideas that normally sustain and nourish the creative effort. Just as the Holocaust experience crushed the structures of self that usually favored survival, forcing victims to find new means for staying alive, so its literature sabotages the reader's hopes for a durable affirmation lurking in the dusk of atrocity. Reading and writing about the Holocaust is an experience in *un*learning; both parties are forced into the Dantean gesture of abandoning all safe props as they enter and, without benefit of Virgil, make their uneasy way through its vague domain.

I have deliberately excluded from this collection the extensive (and very familiar) Holocaust literature that depends on these safe props for its impact. Anne Frank, for example, was a very talented young writer, but she has been ill-served by fervent admirers who have refurbished her work with their own sentimental provocations. Those who would convert death in Auschwitz or Bergen-Belsen into a triumph of love over hate feed deep and obscure needs in themselves having little to do with the truth. In addition, they pander to a hungry popular clamor for reassurance that mass murder had its redeeming features. The best Holocaust literature gazes into the depths without flinching. If its pages are seared with the heat of a nether world where, unlike Dante's, pain has no link to sin and hope no bond with virtue, this is only to confirm the dismissal of safe props that such an encounter requires.

We are left with an art that is rich in its unsparing demands on our sacred beliefs. In an odd sense, it invites us to share the aura surrounding the new arrival at Auschwitz or Treblinka—disoriented, hesitant, fearful, hoping for the best—until he or she grew acquainted with the worst, and then had to find a vacant chamber in the imagination for the unthinkable. No one can be blamed for not wanting to dwell there for long, but the actual victims had little choice. If reading about it becomes a venture in strengthening consciousness rather than the ordeal of survival that they endured, that venture nonetheless gives us access to the central event of our time, and perhaps of the modern era.

Western civilization has always prided itself on achieving the thinkable. When Hitler and his cohorts corrupted this vision by making the morally and physically *unthinkable* thinkable, and then practical, possible, and finally *real,* they not only stained the idea of civilization, but infected the vaunted sources of its pride. Holocaust literature plays a vital role in raising questions about the integrity of language and identity and the dominion of history itself that if left unchallenged would plunge us back into the moral innocence that legend ascribes to the Garden of Eden. Its unsettling contours help us to face the estrangement of the world we live in from the one we long to inhabit—or the one we nostalgically yearn to regain.

The need to wrest meaning from such estrangement (and the atrocities that lie at its root) continues to goad us, since nothing less than a renewal of the Golden Age seems able to satisfy our desires. And certainly, it would be possible to assemble a body of literature that mistakes survival for renewal and celebrates the triumph of human continuity over the disruptive forces of the Holocaust. But that is as far from my purpose here as it is from the truth of the event. The dubious feat of wresting meaning from the murder of 5 to 6 million innocent men, women, and children I leave to more hopeful souls. The majority of the texts I draw on refuse that temptation, though many address the painful irony of the annihilation of European Jewry in a world where nature still flourished and love prevailed. Like much tragic literature, these texts elicit pity and terror, but since most of the victims they describe lacked agency in their fate, both the texts and their characters are barred from the consolations of tragedy. We are left to make new space for this episode in history and art, circling a rough-hewn reality that declares its monumental presence while balking our efforts to chisel its form.

I have chosen *Art from the Ashes* as the title of this anthology not to proclaim a phoenix reborn from the mutilation of mass murder, redeeming that time of grief, but to suggest a symbiotic bond linking art and ashes in a seamless kinship. Whatever "beauty" Holocaust art achieves is soiled by the misery of its theme. The early sections of this collection, drawn from nonfiction accounts of the ordeal, introduce us to the landscape of disaster as it was seen (with one or two exceptions) by those who occupied it. Many of the reports are contemporaneous, vivid in authentic detail but retaining the uncertainty of vision that governed their authors (some of whom did not survive their efforts at documentation). As they shift between hope and despair, seizing on the tiniest rumor to verify their yearning for rescue, we are given glimpses of the dual world of promise and doom that nurtured Holocaust victims then and continues to haunt its survivors today.

Such testimony forms a ballast for the chorus of fictional, dramatic, and poetic voices that succeeds it. As we move from the literal to the literary, we begin to understand those commentators who insist that the impact of Holocaust reality exceeds the force of any imaginative work that might seek to capture it. Readers of the selections in this volume can reach their own conclusions on the issue. Although I expect responses to vary, this will merely be a tribute to the fertile challenge of the subject to both writer *and* reader. Because Holocaust facts often seem so unimaginable, they assume the features of fiction; on the other hand, wary of promoting disbelief, Holocaust fiction clings to its moorings in the grim truths of the event. My aim in this anthology is to give readers a chance to encounter the variety and complexity of both—the facts, and the fictions about them.

My main principles of selection are the artistic quality, intellectual rigor, and physical integrity of the texts. Of course, standards of critical taste differ, and I have chosen works that appeal to mine. Some of these works are familiar; others remain less popular or have been ignored. I believe that everything I include deserves to be preserved for its ability to engage our attention and stimulate reflection, and, most of all, to liberate responses on the deepest levels of psychological, mental, emotional, and aesthetic concerns. Except in the case of diaries and journals, where including the whole text would have been impossible, I have tried to avoid excerpts. But even a dozen poems by Abraham Sutzkever or Nelly Sachs, or a single short story by Tadeusz Borowski or Ida Fink, cannot convey the enormous achievements of these artists in distilling the complex anguish of the event into a few perfectly finished lines or pages. One must turn to the full journals and diaries, the collected poems and stories, to appreciate the distinction of this writing. My hope is that this anthology will deepen among old readers and inaugurate among new ones an ongoing exploration of the precious lode of Holocaust literature, whose extensive veins are no closer to exhaustion than are the memories of the catastrophe these writings portray.

I

The Way It Was

When Lina Wertmüller's film *Seven Beauties* first appeared, some historians objected that no camp had ever been located in the area to which she assigned it, and that no woman had ever served as the commandant of a concentration camp. Such distortions, they argued, warped the candor of her vision and thus thwarted the film's efforts to capture the reality of the Third Reich. Supporters insisted that the accuracy of detail was less urgent than fidelity to the inner tensions of the victims, who were driven to find *any* means to keep themselves alive. The charges against the film were not frivolous, since versions of the Holocaust experience, whether verbal or visual, that wander too far from the truth threaten to trivialize its legacy as the gravest moment in modern history. But its champions were credible, too, because art limited by the rigors of historical precision denies to artists the freedom to experiment with style, tone, character, and narrative form, which is the lifeblood of their craft.

Holocaust fact and Holocaust fiction are Siamese twins, joined at birth and severed at their peril. There is no way we can prize one without appraising both. The opening section of this collection is devoted to Holocaust fact, but its essays gain meaning as we proceed through the subsequent readings. Although accounts of "the way it was" have a raw power of their own, sometimes, paradoxically, they stun us into disbelief and even denial, for they speak of deeds that need to be eased into our imagination with less naked force. The following excerpt from Jankiel Wiernik's "One Year in Treblinka," which is included in this section, makes our instincts clamor that he must be exaggerating:

> One of the Germans, a man named Sepp, was a vile and savage beast, who took special delight in torturing children. When he pushed women around and they begged him to stop because they had children with them, he would frequently snatch a child from the woman's arms and either tear the child in half or grab it by the legs, smash its head against a wall and throw the body away. Such incidents were by no means isolated.

Tragic scenes of this kind occurred all the time. In the presence of such sudden rage and silent grief, we seem intruders in an alien sphere. Are the pictures too graphic, the terror too cruel? If scenes like these did indeed occur "all the time"—and we have many other witnesses to sustain Wiernik's claim—then how from the refuge of our privileged lives can we enter sympathetically into the grisly Nazi world that was the daily ordeal of their victims? Painful as it may seem, we must begin with the unembellished truth, stripping off our civilized "lendings" just as Lear shed his garments on the heath. Only after that can the process of clothing ourselves with more suitable apparel begin. And then we will have to face the question that critics of *King Lear* still dispute today: Is this truly a form of healing or just a necessary illusion?

This section begins with some vivid reports of what happened and ends with efforts by Auschwitz survivors, such as Elie Wiesel, Primo Levi, Jean Améry, and Charlotte Delbo, to assess the meaning of these events. Delbo's distinction between "ordinary memory" and "deep memory" helps us to establish a bridge between the Holocaust experience and the so-called normal years following liberation. The link is difficult to describe, still more complex to maintain. Looking back on the nightmarish episode when he was tortured by the SS, Améry is left with a sense of "foreignness in the world that cannot be compensated by any sort of subsequent human communication." He confirms the feeling of estrangement shared by many survivors of Hitler's deathcamps: "Whoever has succumbed to torture can no longer feel at home in the world." This is the operation of deep memory, and its stains are indelible.

But life goes on, and, as Delbo suggests, slowly one reenters the rhythms of daily existence and relearns its routines—eating with a knife and fork, using a toothbrush, trying to smile. But her Auschwitz self, she argues, is different from the postcamp self who writes about the experience today. No one reading Wiernik's account of what he endured before his escape from Treblinka can fail to grasp Delbo's plight when she speaks of her double existence: Auschwitz and the pursuit of happiness in a free society do not and never will coalesce. Wiernik cannot charm his children and grandchildren with tales of his adventures in the camp. Common memory is faithful to a more amiable past.

Many years ago a well-known American poet, who had just returned from a visit to Auschwitz, suggested to me that he thought the camp should be razed and an amusement park built on the site. I was tempted to reply with Romeo's romantic riposte: "He jests at scars, who never felt a wound." What would this change, I wondered, for those who had been there or those like myself who were committed to exploring their ordeal? Yet the idea was not merely fanciful; deeply ingrained in the post-Auschwitz sensibility is a need for some form of cosmetic surgery to cover the scars of our disfigured era. The essays in this section brandish the ache that those scars barely conceal. They conjure a vision of our culture that threatens to strangle its spiritual roots.

For example, one of the many ironies of the postwar period is that the memories of guilt remained more silent than the memories of grief. We might have expected a surge of remorse and confession to calm some troubled consciences—but this did not happen. The selection by Christopher R. Browning, describing the exploits of a German police battalion assigned to the mass murder of Jews in Poland, is the only work included here that does not derive from the testimony of a participant. It is based on the research of an American historian who had to reconstruct the episode despite the decades of silence by the men who had taken part, sometimes reluctantly, in the killings. In the few brief reports in this section that present testimony in the voices of the perpetrators themselves, mostly taken from judicial proceedings after the war, the witnesses remain morally neutral, furnishing little evidence of regret. The difference between those who were forced to die against their will and those who agreed to carry out the slaughter, though free to refuse, is stark and unsettling. It challenges platitudes about both human nature and inhuman nature, as well as placid categories of good and evil that do little to help us assess the roles of victims and killers in these unprecedented conditions.

We do have memoirs from Rudolf Höss, commandant of Auschwitz, volumes of testimony from Adolf Eichmann, and weeks of interviews with Franz Stangl, commandant of Treblinka, but it would be pointless to include them. They are so embroidered with denials, lies, and a dense confusion about the nature of moral conduct that they shed little light on the ethical issues of choice, consequence, and responsibility, to say nothing of the diabolic goad that drove such men. Indeed, under the scrutiny of the dispassionate inquirer, their words erode the absolute value of such categories, making them provisional and introducing the possibility of revised notions of character and destiny that infiltrate many and perhaps most of the writings in this volume. Höss, Eichmann, and Stangl did not perceive what they were doing as we perceive it, or as we would *expect them* to perceive it, at least after their defeat. Partly as a result of this, we are left with a history that continues to frustrate our attempts to reduce it to some kind of intellectual and moral order.

When men and women became random ciphers in a chaotic equation that followed no predictable patterns, the rules for living, and for thinking about existence, changed. The cherished axiom making the individual the agent of his or her own fate was displaced by a new, unflattering principle that intentionally reduced human beings to states of helplessness and fear and imposed on them the role of victim. In "Torture," his account of the torture chamber of Fort Breendonk, Améry identifies the source of this unique status: "[T]he name of it was power, dominion over spirit and flesh, orgy of unchecked expansion." He questions the value of nostrums like "the 'moral' power of resistance to physical pain," a formula that supports dignity as long as it is not tested, or remains safely enshrined in the temple of literary vision. When such pieties crumbled and victims abandoned "living" in an effort to simply stay alive, desperate gestures of survival poured into the vacuum, replacing familiar tactics of will and choice.

Such gestures were radically limited by the Germans, who paradoxically retained for themselves the moral "freedom" of will and choice—to do evil, rather than good. In "One Year in Treblinka," Wiernik offers several vivid descriptions of the "victim", but one will suffice to define the issue. He speaks of a "poor wretch" called the Scheissmeister (shit master), one more invention of the malicious SS mentality to mock and degrade the Jews: "He was dressed like a cantor and even had to grow a goatee. He wore a large alarm clock on a string around his neck. No one was permitted to remain in the latrine longer than three minutes, and it was his duty to time everyone who used it. The name of this poor wretch was Julian." Julian's survival depended on his compliance in this role, though as Wiernik tells us, the human price was severe: "Julian was a poised and quiet man, but when they began their horseplay with him, he wept bitterly. He wept also while he worked on the fire grates [where the corpses of the gassed were burned]. His garb, his appearance and the task he had to perform provoked the German fiends to abuse him all the more and to amuse themselves at his expense." In this situation, what could Julian do to defend his dignity, without provoking immediate and fatal reprisal? Martyrs may be made of sterner stuff, but with rare exceptions their heroic exploits are the fruit of a concentrated spiritual zeal unavailable to ordinary men and women.

Perhaps in acknowledgment of this, and to forestall any judgment or criticism of the victim that might ensue (indeed, his essay suggests that it had already begun), Elie Wiesel in "A Plea for the Dead" tries to substitute a different kind of nobility for the doomed. Lapsing into the language of Camus, he defends those less fortunate than himself: "Knowing themselves abandoned, excluded, rejected by the rest of humanity, their walk to death, as haughty as it was submissive, became an act of lucidity, of protest, and not of acceptance and weakness." Wiesel's portrait of the prototypical victim clashes with Wiernik's sketch of the hapless Julian, who defers not through weakness, but through the absence of meaningful sources of defiance. One suspects that most camp inmates struggled to survive in an obscure terrain somewhere between haughty protest and abject fear.

Wiesel's words mask the deeper origin of their injured tone: his justifiable exasperation at the inaction and indifference of the free world, both before and during the ordeal of the Jews. He identifies the intertwined challenge to Holocaust witness and audience: How shall they tell what happened, and how shall we hear it? He seems to call for a kind of empathy beyond reason and comprehension and, perhaps most important of all, beyond reassurance. As he walks a tightrope between defiance and despair, his shifting positions from his precarious perch remind us of the uncertainty of *our* footing, too. If his conclusion undermines many of his own premises, that is but to reenact the effect of the Holocaust experience on *any* system of values that glorifies the human spirit. The depth of its lesson, he ends—if lesson there was—is that "our strength is only illusory, and that in each of us is a victim who is afraid, who is cold, who is hungry. Who is also ashamed." This is the legacy we must somehow absorb.

It is also an image of man that ravages our memories—and our hopes. The option of "not knowing" remains, to tempt us to free our future from this tainted past. But as Primo Levi cautions us in "Shame," he and his fellow prisoners were denied this "screen of willed ignorance"; they were not able "not to see." Discreet avoidance, properly named, is a form of willed ignorance, and Levi makes its price so high that, after reading his words, we are forced to wonder who would agree to pay it:

> It was not possible for us nor did we want to become islands; the just among us, neither more nor less numerous than any other human group, felt remorse, shame, and pain for the misdeeds that others and not they had committed, and in which they felt involved, because they sensed that what had happened around them and in their presence, and in them, was irrevocable. Never again could it be cleansed; it would prove that man, the human species—we, in short—had the potential to construct an infinite enormity of pain, and that pain is the only force created from nothing, without cost and without effort. It is enough not to see, not to listen, not to act.

If the human species is differentiated from the beasts by the marvel of consciousness, then we enact our humanity and the very authenticity of our being by straining to "know" through awareness the "unthinkable" experience of others. In the case of the Holocaust, this requires sharing in the infinite enormity of pain that consumed its victims, both the living and the dead. The mute voices of the latter resound through the pages that follow by their very absence, though paradoxically when we examine the mise-en-scène we see how they dominate the landscape with their silent presence: the corpses of Treblinka, Lidice, and Józefów; the "dead" for whom Wiesel pleads; those who "saw the Gorgon," in Levi's words, and did not return to tell about it; the tortured, like Améry, who saw the Gorgon and did return, leaving a portion of the living self behind; or one of Delbo's fellow survivors who, as Delbo tells us in *Mesure de nos jours* (1971), "died in Auschwitz though no one sees it." This final paradox verifies the dismal truth that our century is awash in the murder of others, and that only through "willed ignorance" can we separate our lives from their deaths. To be human today, it appears, requires us to grasp the fact, if not the meaning, of the inhumanity that some men, in the name of history, have decreed as part of our stern heritage.

In the text included here, Levi speaks of an "atavistic anguish," reflected in the second verse of Genesis: "inscribed in everyone of the 'tohu-bohu' [without form and void] of a deserted and empty universe crushed under the spirit of God but from which the spirit of man is absent: not yet born, or already extinguished." Perhaps we can view Holocaust writing as another kind of second creation, the spirit of man reborn this time not from the flood but from the flames, chastened but resolved to renew through language and art the seared world of its ordeal by fire. The following selections give us a glimpse into some of the efforts to commemorate that charred universe.

1. Jankiel Wiernik

Jankiel Wiernik was born in Poland, which at the time was still part of Russia, in 1890. He left home at the age of twenty, joined the Bund (socialist Jewish organization), and was subsequently arrested and sent to Siberia. He served in the tsarist army, and then settled in Warsaw. He was deported from the ghetto by the Germans to the deathcamp at Treblinka in August 1942.

Because of his carpenter's trade, Wiernik was not led directly to the gas chambers, but joined the so-called work Jews in the camp. He played a vital role in the camp uprising in August 1943, because his trade allowed him to move between the "lower camp," where the planning took place, and the isolated "upper camp," or *Totenlager,* where the gas chambers were located. He therefore was able to serve as a contact to coordinate the unified breakout attempt.

Wiernik survived the escape, made his way to Warsaw, and secured help in finding a hiding place in the countryside and in obtaining false identity papers. Members of the Jewish underground prevailed on Wiernik to write an account of his experiences in Treblinka, and this was subsequently published in Polish in a small edition in May 1944. A courier managed to smuggle out copies to Jewish members of the Polish government-in-exile in London. Thus details of mass murder by gassing in Treblinka were available a full year before the war's end. Wiernik's pamphlet was soon translated into Yiddish and English, gaining an even wider potential audience for his narrative of atrocity.

Between 750,000 and 1 million people were murdered in Treblinka, most of them Jews. A fictionalized, though not fully reliable, version of the uprising and breakout may be found in Jean-François Steiner's *Treblinka* (1968). A briefer account, based on testimony of surviving participants, appears in Gitta Sereny's *Into That Darkness: An Examination of Conscience* (1983). More detailed essays by survivors are in *The Death Camp Treblinka: A Documentary,* from which the following selection, "One Year in Treblinka," is taken.

After the war, Wiernik emigrated to Israel, where he died in 1972.

One Year in Treblinka

Chapter I

Dear Reader:

It is for your sake that I continue to hang on to my miserable existence, though it has lost all attraction for me. How can I breathe freely and enjoy all that which nature has created?

Time and again I wake up in the middle of the night moaning pitifully. Ghastly nightmares break up the sleep I need so badly. I see thousands of skeletons extending their bony arms towards me, as if begging for mercy and life, but I, drenched with sweat, am unable to help. And then I jump up, rub my eyes and actually rejoice that it was all only a dream. My life is embittered. Phantoms of death haunt me, specters of children, little children, nothing but children.

I sacrificed all those nearest and dearest to me. I myself took them to the execution site. I built their death chambers for them.

Today I am a homeless old man without a roof over my head, without a family, without any next of kin. I keep talking to myself. I answer my own questions. I am a nomad. It is with a sense of fear that I pass through places of human habitation. I have a feeling that all my experiences are etched upon my face. Whenever I look at my reflection in a stream or a pool of water, fear and surprise twist my face into an ugly grimace. Do I look like a human being? No, definitely not. Disheveled, untidy, destroyed. It seems as if I were carrying the load of a hundred centuries on my shoulders. The load is wearisome, very wearisome, but for the time being I must bear it. I want to bear it and bear it I must. I, who witnessed the doom of three generations, must keep on living for the sake of the future. The whole world must be told of the infamy of those barbarians, so that centuries and generations to come may execrate them. And, it is I who shall make it happen. No imagination, no matter how daring, could possibly conceive of anything like what I have seen and experienced. Nor could any pen, no matter how facile, describe it properly. I intend to present everything accurately so that all the world may know what western *Kultur* was like. I suffered as I led millions of human beings to their doom; therefore let many millions of other human beings know about it. That is what I am living for. That is my one aim in life. In quiet loneliness I go over the whole ground once again, and I am presenting it with faithful accuracy. Quiet and loneliness are my trusted friends and nothing but the chirping of birds accompanies my labors and meditations. Those dear birds! They still love me; otherwise they would not chirp away so cheerfully and would not become accustomed to me so easily. I love them as I love all of God's creatures. Perhaps the birds will restore my peace of mind. Perhaps some day I shall know how to laugh again.

Perhaps this will happen once I have accomplished my work and after the fetters which now bind us have fallen away.

Chapter II

It happened in Warsaw on August 23, 1942, at the time of the ghetto blockade. I had been visiting my neighbors and was never able to return to my own home. We heard the crack of rifle fire from every direction, but had no inkling of the whole truth. Our terror was intensified by the entry of German Scharführers* and of Ukrainian militiamen who shouted in menacing tones: "Everybody out!"

In the street one of the leaders arranged the people in ranks, without any distinction as to age or sex, performing his task with glee, a satisfied smile on his face. Agile and quick of movement, he was here, there and everywhere. He looked us over appraisingly, his eyes darting up and down the ranks. With a sadistic sneer he contemplated the great accomplishment of his mighty Fatherland which, at one stroke, would chop off the head of the loathsome serpent.

I looked at him. He was the vilest of them all. Human life meant nothing to him, and to inflict death and untold torture was his supreme delight. Because of his "heroic deeds," he subsequently was promoted to the rank of Untersturmführer.† His name was Franz. He had a dog named Barry, about which I shall speak later.

I was standing on line directly opposite the house on Wolynska Street where I lived. From there, we were taken to Zamenhof Street. The Ukrainians divided our possessions among themselves before our very eyes. They fought over our things, opened up all bundles and assorted their contents.

Despite the large number of people in the street, a dead silence hung like a pall over the crowd. We had been seized with mute despair—or was it resignation? And still we did not know the whole truth. They photographed us as if we were prehistoric animals. Part of the crowd seemed pleased, and I myself hoped to be able to return home, thinking that we were merely being put through some identification procedure.

*Staff sergeant.

†Lieutenant. Kurt Franz was arrested in 1959 and, after a trial, sentenced by a German court to a life term. He was quietly released from prison, at the age of seventy-nine, in 1993.

At a word of command we got under way. And then, to our dismay, we came face to face with stark reality. There were railroad cars, empty railroad cars, waiting to receive us. It was a bright, hot summer day. It seemed to us that the sun itself rebelled against this injustice. What had our wives, children and mothers done to deserve this? Why all this? The beautiful, bright, radiant sun disappeared behind the clouds as if loath to look down upon our suffering and degradation.

Next came the command to board the train. As many as 80 persons were crowded into each car, with no way of escape. I was wearing only a pair of pants, a shirt and a pair of slippers. I had left at home a packed knapsack and a pair of boots which I had prepared because of rumors that we would be taken to the Ukraine and put to work there. Our train was shunted from one siding to another. Since I was familiar with this railroad junction I realized that our train was not moving out of the station. We were able to hear their shouts and raucous laughter.

The air in the cars was becoming stiflingly hot and oppressive. It was difficult for us to breathe. Despair descended on us like a pall. I saw all of my companions in misery, but my mind was still unable to grasp the fate that lay in store for us. I had thought in terms of suffering, homelessness and hunger, but I still did not realize that the hangman's ruthless arm was threatening all of us, our children, our very existence.

Amidst untold agonies, we finally reached Malkinia, where our train stopped for the night. Ukrainian guards came into our car and demanded our valuables. Everyone who had any surrendered them just to gain a little longer lease on life. Unfortunately, I had nothing of value because I had left my home unexpectedly and because I had been unemployed and had gradually sold all my possessions in order to keep going. The next morning our train started to move again. We saw a train passing by filled with tattered, half-naked, starved people. They were trying to say something to us, but we could not understand what they were saying.

As the day was unusually hot and sultry, we suffered greatly from thirst. Looking out of the window, I saw peasants peddling bottles of water at 100 zlotys apiece. I had only ten zlotys on me plus two-, five- and ten-zloty coins in silver, with Marshal Pilsudski's effigy on them, which I had saved as souvenirs. And so I had to do without the water. But others bought the water and bread too, at the price of 500 zlotys for one kilogram of dark bread.

Until noon I suffered greatly from thirst. Then a German, who subsequently became a Hauptsturmführer,* entered our car and picked out ten men to get water for us all. At last I was able to quench my thirst to some extent. An order came to remove all dead bodies, if there were any, but there were none.

At 4 P.M. the train started to move again and within a few minutes, we pulled into the Treblinka camp. Only when we arrived there did the full truth dawn on us in all its horror. Ukrainians armed with rifles and ma-

*Captain.

chine guns were stationed on the roofs of the barracks. The camp yard was littered with corpses, some still in their clothes and others stark naked, their faces distorted with terror, black and swollen, the eyes wide open, with tongues protruding, skulls crushed, bodies mangled. And blood everywhere—the blood of innocent people, the blood of our children, of our brothers and sisters, our fathers and mothers.

Helpless, we intuitively felt that we would not be able to escape our destiny and would also become victims of our executioners. But what could be done about it? If only all this were just a nightmare! But no, it was stark reality. We were faced with what was termed "resettlement," but actually meant removal into the great beyond under untold tortures. We were ordered to get off the train and leave whatever packages we had in the cars.

Chapter III

They took us into the camp yard, which was flanked by barracks on either side. There were two large posters with big signs bearing instructions to surrender all gold, silver, diamonds, cash and other valuables under penalty of death. All the while Ukrainian guards stood on the roofs of the barracks, their machine guns at the ready.

The women and children were ordered to move to the left, and the men were told to line up at the right and squat on the ground. Some distance away from us a group of men was busy piling up our bundles, which they had taken from the trains. I managed to mingle with this group and began to work along with them. It was then that I received the first blow with a whip from a German whom we called Frankenstein. The women and children were ordered to undress, but I never found out what had become of the men. I never saw them again.

Late in the afternoon another train arrived from Miedzyrzec (Mezrich), but 80 per cent of its human cargo consisted of corpses. We had to carry them out of the train, under the whiplashes of the guards. At last we completed our gruesome chore. I asked one of my fellow workers what it meant. He merely replied that whoever you talk to today will not live to see tomorrow.

We waited in fear and suspense. After a while we were ordered to form a semi-circle. The Scharführer Franz walked up to us, accompanied by his dog and a Ukrainian guard armed with a machine gun. We were about 500 persons. We stood in mute suspense. About 100 of us were picked from the group, lined up five abreast, marched away some distance and ordered to kneel. I was one of those picked out. All of a sudden there was a roar of machine guns and the air was rent with the moans and screams of the victims. I never saw any of these people again. Under a rain of blows from whips and rifle butts the rest of us were driven into the barracks, which were dark and had no floors. I sat down on the sandy ground and dropped off to sleep.

The next morning we were awakened by loud shouts to get up. We jumped up at once and went out into the yard amid the yells of our Ukrainian guards. The Scharführer continued to beat us with whips and rifle butts at every step as we were being lined up. We stood for quite some time without receiving any orders, but the beatings continued. Day was just breaking and I thought that nature itself would come to our aid and send down streaks of lightning to strike our tormentors. But the sun merely obeyed the law of nature; it rose in shining splendor and its rays fell on our tortured bodies and aching hearts.

I was jolted from my thoughts by the command: "Attention!" A group of Scharführers and Ukrainian guards, headed by Untersturmführer Franz with his dog Barry stood before us. Franz announced that he was about to give a command. At a signal from him, they began to torture us anew, blows falling thick and fast. Our faces and bodies were cruelly torn, but we all had to keep standing erect, because if one so much as stooped over but a little, he would be shot because he would be considered unfit for work.

When our tormentors had satisfied their thirst for blood, we were divided into groups. I was put with a group that was assigned to handle the corpses. The work was very hard, because we had to drag each corpse, in teams of two, for a distance of approximately 300 meters. Sometimes we tied ropes around the dead bodies to pull them to their graves.

Suddenly, I saw a live woman in the distance. She was entirely nude; she was young and beautiful, but there was a demented look in her eyes. She was saying something to us, but we could not understand what she was saying and could not help her. She had wrapped herself in a bed sheet under which she was hiding a little child, and she was frantically looking for shelter. Just then one of the Germans saw her, ordered her to get into a ditch and shot her and the child. It was the first shooting I had ever seen.

I looked at the ditches around me. The dimensions of each ditch were 50 by 25 by 10 meters. I stood over one of them, intending to throw in one of the corpses, when suddenly a German came up from behind and wanted to shoot me. I turned around and asked him what I had done, whereupon he told me that I had attempted to climb into the ditch without having been told to do so. I explained that I had only wanted to throw the corpse in.

Next to nearly every one of us there was either a German with a whip or a Ukrainian armed with a gun. As we worked, we would be hit over the head. Some distance away there was an excavator which dug out the ditches.

We had to carry or drag the corpses on the run, since the slightest infraction of the rules meant a severe beating. The corpses had been lying around for quite some time and decomposition had already set in, making the air foul with the stench of decay. Already worms were crawling all over the bodies. It often happened that an arm or a leg fell off when we tied straps around them in order to drag the bodies away. Thus we worked from dawn to sunset, without food or water, on what some day would be our own graves. During the day it was very hot and we were tortured by thirst.

When we returned to our barracks at night, each of us looked for the men we had met the day before but, alas, we could not find them because they were no longer among the living. Those who worked at assorting the bundles fell victim far more frequently than the others. Because they were starved, they pilfered food from the packages taken from the trains, and when they were caught, they were marched to the nearest open ditch and their miserable existence was cut short by a quick bullet. The entire yard was littered with parcels, valises, clothing and knapsacks which had been discarded by the victims before they met their doom. As I worked, I noticed that some of the workers had red or yellow patches on their pants. I had no idea what this meant. They occupied a part of our barrack marked off by a partition. They were 50 men and one woman. I spent four days working with the corpses and living under these appalling conditions.

Chapter IV

One Friday, I believe it was August 28, 1942, we returned from work. Everything went off in accordance with routine, "Attention! Headgear off! Headgear on!" and a speech by Franz. He appointed a headman from among us and several bosses (kapos), who were to drive us to work. In his talk, Franz told us that if we worked hard, we would get everything we needed. If not, he would find ways and means of dealing with us. A German proved his skill, Franz said, by his ability to master any situation. Thus, the Germans carried out the deportations in such a way that the Jews pushed into the trains of their own free will, without thinking of what might be in store for them. All of Franz's talk was spiced profusely with his usual invective.

On August 29 there was the usual reveille, but this time it was in Polish. We got up quickly and went out into the yard. Since we slept in our clothes, we did not have to get dressed; accordingly, we were able to obey the order quickly and to form ranks. The commands were given in the Polish language, and by and large we were treated politely. Once again, Franz delivered a speech in which he said that from now on everybody was going to be put to work at his own occupation.

The first to be called were specialists in the building trades; I reported as a master construction worker. All those in this group were separated from the others. There were 15 of us in our construction group, to which three Ukrainians were assigned as guards. One of them, an older soldier by the name of Kostenko, did not look too menacing. The second, Andreyev, a typical "guard," was of medium size, stout, with a round red face, a kind, quiet individual. The third one, Nikolay, was short, skinny, mean, with evil eyes, a sadistic type. There were also two other Ukrainians, armed with rifles, who were to stand guard over us.

We were marched to the woods and were ordered to dismantle the barbed wire fences and cut timber. Kostenko and Andreyev were very gentle. Nikolay, however, used the whip freely. Truth to tell, there were no real specialists among those who had been picked for the construction gang. They had simply reported as "carpenters" because they did not want to be put to work handling corpses. They were continuously whipped and humiliated.

At noon we stopped working and returned to the barracks for our meal, which consisted of soup, groats and some moldy bread. Under normal conditions, a meal like ours would have been considered unfit for human consumption, but, starved and tired as we were, we ate it all. At 1 P.M. our guards came with the Ukrainians to take us back to work, at which we remained until evening, when we returned to the barracks. Then came the usual routine, commands, and so forth.

On that particular day there were many Germans around, and we were about 700. Franz was there, too, with his dog. All of a sudden he asked, with a smile on his face, whether any of us knew German. Approximately 50 men stepped forward. He ordered them all out and form a separate group, smiling all the while to allay our suspicions. The men who admitted knowing German were taken away and never came back. Their names did not appear on the list of survivors and no pen will ever be able to describe the tortures under which they died. Again, a few days went by. We worked at the same assignment and lived under the same conditions. All this time I was working with one of my colleagues and fate was strangely kind to us. Perhaps it was because we were both specialists in our trade, or because we had been destined to witness the sufferings of our brethren, to look at their tortured corpses, and to live to tell the tale. Our bosses gave me and my colleague boxes for lime. Andreyev supervised us. Our guard considered our work satisfactory. He showed us considerable kindness and even gave each one of us a piece of bread, which was quite a treat since we were practically starving to death. Some people who had been spared from another form of death, which I shall discuss later on, would become yellow and swollen from hunger and finally drop dead. Our group of workers grew; additional workers arrived. The foundations were dug for some sort of building. No one knew what kind of a building this would be. There was in the courtyard one wooden building surrounded by a tall fence. The function of this building was a secret.

A few days later a German architect arrived with an assistant and the construction work got under way. There was a shortage of bricklayers, although many pretended to be skilled laborers in order to avoid being ordered to handle corpses. Most of these men, however, had been killed off. Once, while doing some bricklaying, I noticed a man I had known in Warsaw. His name was Razanowicz. He had a black eye from which I inferred that he would be shot by evening. An engineer from Warsaw by the name of Ebert and his son were also working with us, but within a short time they, too, were put to death. Fate spared me nothing. A few days later I learned the purpose of the building behind the fence, and the discovery left me shuddering with terror.

The next job for my colleague and myself was to cut and process lumber. It was hard for the two of us. I had not done such work in 25 years, and my colleague was a cabinetmaker by trade and not very adept with an axe, but, with my help, he managed to hold on to the job. I am a carpenter by trade, but for many years I had functioned only as a member of the examining board of the Warsaw Chamber of Artisans. Meanwhile, eight more indescribable days of hard existence went by. No new transports were arriving. Finally, on the eighth day, a new transport arrived from Warsaw.

Chapter V

Camp Treblinka was divided into two sections. In Camp No. 1 there was a railroad siding and a platform for unloading the human cargo, and also a wide open space, where the baggage of the new arrivals was piled up. Jews from foreign countries brought considerable luggage with them. Camp No. 1 also contained what was called the *Lazarett* (infirmary), a long building measuring 30 by 2 meters. Two men were working there. They wore white aprons and had red crosses on their sleeves; they posed as doctors. They selected from the transports the elderly and the ill, and made them sit on a long bench facing an open ditch. Behind the bench, Germans and Ukrainians were lined up and they shot the victims in the neck. The corpses toppled right into the ditch. After a number of corpses had accumulated, they were piled up and set on fire.

The barracks housing the Germans and Ukrainians were located some distance away, and so were the camp offices, the barracks of the Jewish workers, workshops, stables, pigsties, a food storage house and an arsenal. The camp cars were parked in the yard. To the casual observer the camp presented a rather innocuous appearance and made the impression of a genuine labor camp.

Camp No. 2 was entirely different. It contained a barrack for the workers, 30 by 10 meters, a laundry, a small laboratory, quarters for 17 women, a guard station and a well. In addition there were 13 chambers in which inmates were gassed. All of these buildings were surrounded by a barbed wire fence. Beyond this enclosure, there was a ditch of 3 by 3 meters and, along the outer rim of the ditch, another barbed wire fence. Both of these enclosures were about 3 meters high, and there were steel wire entanglements between them. Ukrainians stood on guard along the wire enclosure. The entire camp (Camps 1 and 2) was surrounded by a barbed wire fence 4 meters high, camouflaged by saplings. Four watchtowers stood in the camp yard, each of them four stories high; there were also six one-storied observation towers. Fifty meters beyond the last outer enclosure there were tank traps.

When I arrived at the camp, three gas chambers were already in operation; another ten were added while I was there. A gas chamber measured 5 by 5 meters and was about 1.90 meters high. The outlet on the roof had a hermetic cap. The chamber was equipped with a gas pipe inlet and a baked tile floor slanting towards the platform. The brick building which housed the gas chambers was separated from Camp No. 1 by a wooden wall. This wooden wall and the brick wall of the building together formed a corridor which was 80 centimeters taller than the building. The chambers were connected with the corridor by a hermetically fitted iron door leading into each of the chambers. On the side of Camp No. 2 the chambers were connected by a platform 4 meters wide, which ran alongside all three chambers. The platform was about 80 centimeters above ground level. There was also a hermetically fitted wooden door on this side.

Each chamber had a door facing Camp No. 2 (1.80 by 2.50 meters), which could be opened only from the outside by lifting it with iron supports and was closed by iron hooks set into the sash frames, and by wooden bolts. The victims were led into the chambers through the doors leading from the corridor, while the remains of the gassed victims were dragged out through the doors facing Camp No. 2. The power plant operated alongside these chambers, supplying Camps 1 and 2 with electric current. A motor taken from a dismantled Soviet tank stood in the power plant. This motor was used to pump the gas which was let into the chambers by connecting the motor with the inflow pipes. The speed with which death overcame the helpless victims depended on the quantity of combustion gas admitted into the chamber at one time.

The machinery of the gas chambers was operated by two Ukrainians. One of them, Ivan, was tall, and though his eyes seemed kind and gentle, he was a sadist. He enjoyed torturing his victims. He would often pounce upon us while we were working; he would nail our ears to the walls or make us lie down on the floor and whip us brutally. While he did this, his face showed sadistic satisfaction and he laughed and joked. He finished off the victims according to his mood at the moment. The other Ukrainian was called Nicholas. He had a pale face and the same mentality as Ivan.

The day I first saw men, women and children being led into the house of death I almost went insane. I tore at my hair and shed bitter tears of despair. I suffered most when I looked at the children, accompanied by their mothers or walking alone, entirely ignorant of the fact that within a few minutes their lives would be snuffed out amidst horrible tortures. Their eyes glittered with fear and still more, perhaps, with amazement. It seemed as if the question, "What is this? What's it all about?" was frozen on their lips. But seeing the stony expressions on the faces of their elders, they matched their behavior to the occasion. They either stood motionless or pressed tightly against each other or against their parents, and tensely awaited their horrible end.

Suddenly, the entrance door flew open and out came Ivan, holding a heavy gas pipe, and Nicholas, brandishing a saber. At a given signal, they would begin admitting the victims, beating them savagely as they moved into the chamber. The screams of the women, the weeping of the children, cries of despair and misery, the pleas for mercy, for God's vengeance ring in my ears to this day, making it impossible for me to forget the misery I saw.

Between 450 and 500 persons were crowded into a chamber measuring 25 square meters. Parents carried their children in their arms in the vain hope that this would save their children from death. On the way to their doom, they were pushed and beaten with rifle butts and with Ivan's gas pipe. Dogs were set upon them, barking, biting and tearing at them. To escape the blows and the dogs, the crowd rushed to its death, pushing into the chamber, the stronger ones shoving the weaker ones ahead of them. The bedlam lasted only a short while, for soon the doors were slammed shut. The chamber was filled, the motor turned on and connected with the inflow pipes and, within 25 minutes at the most, all lay stretched out dead or, to be more accurate, were standing up dead. Since there was not an inch of free space, they just leaned against each other.

They no longer shouted, because the thread of their lives had been cut off. They had no more needs or desires. Even in death, mothers held their children tightly in their arms. There were no more friends or foes. There was no more jealousy. All were equal. There was no longer any beauty or ugliness, for they all were yellow from the gas. There were no longer any rich or poor, for they all were equal before God's throne. And why all this? I keep asking myself that question. My life is hard, very hard. But I must live on to tell the world about all this barbarism.

As soon as the gassing was over, Ivan and Nicholas inspected the results, moved over to the other side, opened the door leading to the platform, and proceeded to heave out the corpses. It was our task to carry the corpses to the ditches. We were dead tired from working all day at the construction site, but we had no recourse and had no choice but to obey. We could have refused, but that would have meant a whipping or death in the same manner or even worse; so we obeyed without grumbling.

We worked under the supervision of a Hauptmann (captain), a medium-sized, bespectacled man whose name I do not know. He whipped us and shouted at us. He beat me, too, without a stop. When I gave him a questioning look, he stopped beating me for a moment and said, "If you weren't the carpenter around here, you would be killed." I looked around and saw that almost all the other workers were sharing my fate. A pack of dogs, along with Germans and Ukrainians, had been let loose on us. Almost one-fourth of the workers were killed. The rest of us tossed their bodies into the ditches without further ado. Fortunately for me, when the Hauptmann left, the Unterscharführer relieved me from this work.

Between 10,000 and 12,000 people were gassed each day. We built a narrow-gauge track and drove the corpses to the ditches on a rolling platform.

One evening, after a hard day's work, we were marched to Camp No. 2 instead of Camp No. 1. The picture here was entirely different; I shall never forget it. My blood froze in my veins. The yard was littered with thousands of corpses, the bodies of the most recent victims. Germans and Ukrainians were barking orders and brutally beating the workers with rifle butts and canes. The faces of the workers were bloody, their eyes blackened and their clothes had been shredded by dogs. Their overseers stood near them.

A one-storied watchtower stood at the entrance of Camp No. 2. It was ascended by means of ladders, and these ladders were used to torture some of the victims. Legs were placed between the rungs and the overseer held the victim's head down in such a way that the poor devil couldn't move while he was beaten savagely, the minimum punishment being 25 lashes. I saw that scene for the first time in the evening. The moon and the reflector lights shed an eerie light upon that appalling massacre of the living, as well as upon the corpses that were strewn all over the place. The moans of the tortured mingling with the swishing of the whips made an infernal noise.

When I arrived at Camp No. 2 there was only one barrack there. The bunks had not yet been finished, and there was a canteen in the yard. I saw there a number of people I had known in Warsaw, but they had changed so much that it was difficult to recognize them. They had been beaten, starved and mistreated. I did not see them for very long, because new faces and new friends kept arriving on the scene. It was a continuous coming and going, and death without end. I learned to look at every living person as a prospective corpse. I appraised him with my eyes and figured out his weight, who was going to carry him to his grave and how badly his bearer would be beaten while dragging his body to the ditch. It was terrible, but true nonetheless. Would you believe that a human being, living under such conditions, could actually smile and make jokes at times? One can get used to anything.

Chapter VI

The German system is one of the most efficient in the world. It has authorities upon authorities. Departments and subdepartments. And, most important, there is always the right man in the right place. Whenever ruthless determination and a complete destruction of "vicious and subversive elements" are needed, good patriots can always be found who will carry out any command. Men can always be found who are ready to destroy and kill their fellow men. I never saw them show any compassion or regret. They never showed any pity for the innocent victims. They were robots who performed their tasks as soon as some higher-up pressed a button. Such human hyenas always find a wide field for activity in times of war and revolution. To them the road of evil is easy and more pleasant than any other. But a firm and just order, aided by education, good examples and wise discipline could check these evil tendencies.

Vicious types lurk in disreputable places where they carry on their subversive activities. Today, all ethics have become superfluous. The more vicious and depraved one is, the higher the position he will occupy. Advancement depends on how much one has destroyed, or how many one has killed. People whose hands drip with the blood of innocent victims receive adulation and there is no need for them to wash their hands. On the contrary, these are held aloft so that the world may pay them honor. The dirtier one's conscience and hands, the higher the glory their owner will achieve.

Another amazing character trait of the Germans is their ability to discover, among the populace of other nations, hundreds of depraved types like themselves, and to use them for their own ends. In camps for Jews, there is a need for Jewish executioners, spies and stool pigeons. The Germans managed to find them, to find such gangrenous creatures as Moyshke from near Sochaczew, Itzik Kobyla from Warsaw, Chaskel the thief, and Kuba, a thief and a pimp, both of them born and bred in Warsaw.

Chapter VII

The new construction job between Camp No. 1 and Camp No. 2, on which I had been working, was completed in a very short time. It turned out that we were building ten additional gas chambers, more spacious than the old ones, 7 by 7 meters or about 50 square meters. As many as 1,000 to 1,200 persons could be crowded into one gas chamber. The building was laid out according to the corridor system, with five chambers on each side of the corridor. Each chamber had two doors, one door leading into the corridor through which the victims were admitted; the other door, facing the camp, was used for the removal of the corpses. The construction of both doors was the same as that of the doors in the old chambers. The building, when viewed from Camp No. 1, showed five wide concrete steps with bowls of flowers on either side. Next came a long corridor. There was a Star of David on top of the roof facing the camp, so that the building looked like an old-fashioned synagogue. When the construction was finished, the Hauptsturmführer said to his subordinates, "The Jew-town has been completed at last."

The work on these gas chambers lasted five weeks, which to us seemed like centuries. We had to work from dawn to dusk under the ceaseless threat of beatings from whips and rifle butts. One of the guards, Woronkov, tortured us savagely, killing some of the workers each day. Although our physical suffering surpassed the imagination of normal human beings, our spiritual agonies were far worse. New transports of victims arrived each day. They were immediately ordered to disrobe and were led to the three old gas chambers, passing us on the way. Many of us saw our children, wives and other loved ones among the victims. And when, on the impulse of grief, someone rushed to his loved ones, he would be killed on the spot. It was under these conditions that we constructed death chambers for our brethren and ourselves.

This went on for five weeks. After the work on the gas chambers had been completed, I was transferred back to Camp No. 1, where I had to set up a barber shop. Before killing the women, the Germans cut off their hair and gathered it all up carefully. I never learned for what purpose the hair was used.

My quarters were still in Camp No. 2 but, because of a shortage of craftsmen, I was taken each day to Camp No. 1, with Unterscharführer Hermann as my escort. He was about 50 years old, tall and kind. He understood us and was sorry for us. The first time he came to Camp No. 2 and saw the piles of gassed corpses, he turned pale and looked at them with horror and pity. He left with me at once in order to get away from the gruesome scene. He treated us workers very well. Often, he surreptitiously brought us some food from the German kitchen. There was so much kindness in his eyes that one might have been tempted to pour one's heart out to him, but he never talked to the inmates. He was afraid of his colleagues. But his every move and action showed his forthright character.

While I was working in Camp No. 1 many transports arrived. Each time a new transport came, the women and children were herded into the barracks at once, while the men were kept in the yard. The men were ordered to undress, while the women, naively anticipating a chance to take a shower, unpacked towels and soap. The brutal guards, however, shouted orders for quiet, and kicked and dealt out blows. The children cried, while the grownups moaned and screamed. This made things even worse; the whipping only became more cruel.

The women and girls were then taken to the "barber shop" to have their hair clipped. By now they felt sure that they would be taken to have a shower. Then they were escorted, through another exit, to Camp No. 2 where, in freezing weather, they had to stand in the nude, waiting their turn to enter the gas chamber, which had not yet been cleared of the last batch of victims.

All through that winter, small children, stark naked and barefooted, had to stand out in the open for hours on end, awaiting their turn in the increasingly busy gas chambers. The soles of their feet froze and stuck to the icy ground. They stood and cried: some of them froze to death. In the meantime, Germans and Ukrainians walked up and down the ranks, beating and kicking the victims.

One of the Germans, a man named Sepp, was a vile and savage beast, who took special delight in torturing children. When he pushed women around and they begged him to stop because they had children with them, he would frequently snatch a child from the woman's arms and either tear the child in half or grab it by the legs, smash its head against a wall and throw the body away. Such incidents were by no means isolated. Tragic scenes of this kind occurred all the time.

The men endured tortures far worse than the women. They had to undress in the yard, make a neat bundle of their clothing, carry the bundle to a designated spot and deposit it on the pile. They then had to go into the barrack where the women had undressed, and carry the latter's clothes out and arrange them properly. Afterwards, they were lined up and the healthiest, strongest and best-built among them were beaten until their blood flowed freely.

Next, all the men, and women, old people and children had to fall into line and proceed from Camp No. 1 to the gas chambers in Camp No. 2. Along the path leading to the chambers there stood a shack in which some official sat and ordered the people to turn in all their valuables. The unfortunate victims, in the delusion that they would remain alive, tried to hide whatever they could. But the German fiends managed to find everything, if not on the living, then later on the dead. Everyone approaching the shack had to lift his arms high and so the entire macabre procession passed in silence, with arms raised high, into the gas chambers.

A Jew had been selected by the Germans to function as a supposed "bath attendant." He stood at the entrance of the building housing the chambers and urged everyone to hurry inside before the water got cold. What irony! Amidst shouts and blows, the people were chased into the chambers.

As I have already indicated, there was not much space in the gas chambers. People were smothered simply by overcrowding. The motor which generated the gas in the new chambers was defective, and so the helpless victims had to suffer for hours on end before they died. Satan himself could not have devised a more fiendish torture. When the chambers were opened again, many of the victims were only half dead and had to be finished off with rifle butts, bullets or powerful kicks.

Often people were kept in the gas chambers overnight with the motor not turned on at all. Overcrowding and lack of air killed many of them in a very painful way. However, many survived the ordeal of such nights; particularly the children showed a remarkable degree of resistance. They were still alive when they were dragged out of the chambers in the morning, but revolvers used by the Germans made short work of them. . . .

The German fiends were particularly pleased when transports of victims from foreign countries arrived. Such deportations probably caused great indignation abroad. Lest suspicion arise about what was in store for the deportees, these victims from abroad were transported in passenger trains and permitted to take along whatever they needed. These people were well dressed and brought considerable amounts of food and wearing apparel with them. During the journey they had service and even a dining car in the trains. But on their arrival in Treblinka they were faced with stark reality. They were dragged from the trains and subjected to the same procedure as that described above. The next day they had vanished from the scene; all that remained of them was their clothing, their food supplies, and the macabre task of burying them.

The number of transports grew daily, and there were periods when as many as 30,000 people were gassed in one day, with all 13 gas chambers in operation. All we heard was shouts, cries and moans. Those who were left alive to do the work around the camps could neither eat nor control their tears on days when these transports arrived. The less resistant among us, especially the more intelligent, suffered nervous breakdowns and hanged themselves when they returned to the barracks at night after having handled the corpses all day, their ears still ringing with the cries and moans of the victims. Such suicides occurred at the rate of 15 to 20 a day. These people were unable to endure the abuse and tortures inflicted upon them by the overseers and the Germans.

One day a transport arrived from Warsaw, from which some men were selected as workers for Camp No. 2. Among them I saw a few people whom I had known from before the war. They were not fit for this kind of work.

That same day one of our own men by the name of Kuszer could not stand the torture and attacked his tormentor, a German Oberscharführer* named Matthes from Camp No. 2, who was a fiend and a killer, and wounded him. The Hauptsturmführer, on arriving at the scene, dismissed all the craftsmen, and other inmates of the camp were massacred on the spot with blunt tools.

I happened to be working in the woods in between the two camps, dressing lumber. The processions of nude children, men and old people passed that spot in a silent caravan of death. The only sounds we could hear were the shouts of the killers; the victims walked in silence. Now and then, a child would whimper but then some killer's fingers would grasp its thin neck in a vise-like grip, cutting off the last plaintive sobs. The victims walked to their doom with raised arms, stark naked and helpless.

Chapter VIII

Between the two camps there were buildings in which the Ukrainian guards had their quarters. The Ukrainians were constantly drunk, and sold whatever they managed to steal in the camps in order to get more money for brandy. The Germans watched them and frequently took the loot away from them.

When they had eaten and drunk their fill, the Ukrainians looked around for other amusements. They frequently selected the bestlooking

*Technical sergeant.

Jewish girls from the transports of nude women passing their quarters, dragged them into their barracks, raped them and then delivered them to the gas chambers. After being outraged by their executioners, the girls died in the gas chambers with all the rest. It was a martyr's death.

On one occasion a girl fell out of line. Nude as she was, she leaped over a barbed wire fence 3 meters high, and tried to escape in our direction. The Ukrainians noticed this and started to pursue her. One of them almost reached her but he was too close to her to shoot, and she wrenched the rifle from his hands. It wasn't easy to open fire since there were guards all around and there was the danger that one of the guards might be hit. But as the girl held the gun, it went off and killed one of the Ukrainians. The Ukrainians were furious. In her fury, the girl struggled with his comrades. She managed to fire another shot, which hit another Ukrainian, whose arm subsequently had to be amputated. At last they seized her. She paid dearly for her courage. She was beaten, bruised, spat upon, kicked and finally killed. She was our nameless heroine.

On another occasion a transport arrived from Germany. The new arrivals were put through the usual routine. When the people were ordered to undress, one of the women stepped forward with her two children, both of them boys. She presented identity papers showing that she was of pure German stock and had boarded this train by mistake. All her documents were found to be in order and her two sons had not been circumcised. She was a good-looking woman, but there was terror in her eyes. She clung to her children and tried to soothe them, saying that their troubles would soon be cleared up and they would return home to their father. She petted and kissed them, but she was crying because she was haunted by a dreadful foreboding.

The Germans ordered her to step forward. Thinking that this meant freedom for herself and her children, she relaxed. But alas, it had been decided that she was to perish together with the Jews, because she had seen too much and would be liable to tell all about what she had seen, which was supposed to be shrouded in secrecy. Whoever crossed the threshold of Treblinka was doomed to die. Therefore this German woman, together with her children, went to her death along with all the others. Her children cried just as the Jewish children did, and their eyes mirrored the same despair, for in death there is no racial distinction; all are equal. Her husband probably will be killed at the front, and she was killed in the camp.

While I was in Camp No. 1, I managed to find out the identity of certain Jews I had seen wearing yellow patches. They turned out to be professional people and craftsmen who had been left over from earlier transports. They were the ones who had built Treblinka. They had hoped to be liberated after the war, but fate decreed otherwise. It had been decided that whoever had crossed the threshold of this inferno had to die. It would not do to leave witnesses who would be able to identify the spot where these fiendish tortures had been perpetrated.

Among these men there were jewelers who appraised the articles of precious metal which the deportees had brought with them. There was quite a lot of this. The sorting and classifying was done in a separate barrack to which no special guard had been assigned, for there was no reason to expect that these men would be able to steal any of the loot. Where would they dispose of their pilferings? Eventually, whatever they might manage to steal would only get back to the Germans again.

The Ukrainians, by contrast, went wild at the sight of gold. They had no idea of its value, but it was enough to give them something that glittered and to tell them that it was gold. When deportations took place, the Ukrainians broke into the homes of the Jews and demanded gold. They did this without the knowledge of the Germans and, of course, they applied methods of terror. They took whatever was given them. Their faces were greedy and savage and inspired fear and loathing in those who had to deal with them. They hid the loot most carefully in order to have something to show their families as spoils of war. Some of the Ukrainians hailed from nearby villages; others had girl friends in the vicinity to whom they wanted to give gifts. A part of their plunder was always traded in for liquor. They were terrible drunkards.

When the Ukrainians noticed that the Jews were handling the gold under practically no control, they began coercing them to steal. The Jews were compelled to deliver diamonds and gold to the Ukrainian guards or else be killed. Day after day, a gang of Ukrainians took valuables from the room where the valuables of the deportees were kept. One of the Germans noticed this and of course it was the Jews who had to pay the penalty. They were searched, and the search disclosed gold and precious stones on their persons. They could not claim that they had stolen these articles under duress; the Germans would not have believed their story. They were tortured and now they were worse off than the camp laborers. Only half of them—there had been 150 of them—were left alive. Those who survived suffered starvation, misery and incredible tortures.

The entire yard was littered with a variety of articles, for all these people left behind millions of items of wearing apparel and so forth. Since they had all assumed that they were merely going to be resettled at an unknown place and not sent to their death, they had taken their best and most essential possessions with them. The camp yard in Treblinka was filled with everything one's heart might desire. There was everything in plenty. As I passed, I saw a profusion of fountain pens and real tea and coffee. The ground was literally strewn with candy. Transports of people from abroad had come well supplied with fats. All the deportees had been fully confident that they were going to survive.

Jews were put to work at sorting out the plunder, arranging things systematically because every item had to serve a definite purpose. Everything the Jews left behind had its value and its place. Only the Jews themselves were regarded as worthless. Jews had to steal what they could and turn the stolen articles over to the Ukrainians. If they failed to do so, the Ukrainians killed them. On the other hand, if the Jews were caught redhanded, they were killed on the spot. Despite the danger, the traffic continued, a new accomplice taking over where the previous one had left off. In that way a chosen few from among millions survived—between the devil and the deep blue sea.

One day a transport of 80 Gypsies from near Warsaw arrived at the camp. These men, women and children were destitute. All they owned was some soiled underwear and tattered clothes. When they came into the yard, they were very happy. They thought they had entered an enchanted castle. But the hangmen were just as happy, because they wiped out all the Gypsies just as they did the Jews. Within a few hours all was quiet and nothing was left but corpses.

I was still working at Camp No. 1 and was free to move about as I pleased. Though I saw many terrible things there, the sight of those gassed at Camp No. 2 was far more horrible. It was practically decided that I should remain permanently in Camp No. 1. Hermann, the architect, and a master cabinet maker from Bohemia, did what they could to this end, because they had no other craftsmen like myself and they therefore needed me. However, in mid-December 1942 an order came for all inmates of Camp No. 2 to be returned. Since this order could not be appealed, we proceeded to Camp No. 2 without even waiting long enough to eat our noontime meal.

The first sight that met my eyes upon my return was that of the corpses of newly gassed victims on whom "dentists" had worked, extracting their false teeth with pliers. Just one look at this ghastly procedure was enough to make me even more disgusted with life than I had been before. The "dentists" sorted the teeth they extracted according to their value. Of course, whatever teeth the Ukrainians managed to lay their hands on remained in their possession.

I worked for a while in Camp No. 2, doing repair work in the kitchen. The commandant of the kitchen had introduced a new system. During that period fewer transports arrived and no new workers became available. At that time, workers in Camp No. 1 were given numbers and triangular leather identification badges. There was a different color patch for each group. The badges were worn on the left side of the chest. Rumors circulated that we workers in Camp No. 2 would also receive numbers but at the time nothing came of it. At any rate, some system had been introduced so that no stranger from an incoming transport could smuggle himself in, as I had done, to prolong his life.

We began to suffer greatly from the cold and they started issuing blankets to us. While I had been away from Camp No. 2, a carpentry shop had been installed there. A baker from Warsaw served as its foreman. His job was to make up stretchers for carrying the corpses from the gas chambers to the mass graves. The stretchers were constructed very primitively; just two poles with pieces of board nailed at intervals.

The Hauptsturmführer and the two commandants ordered me to build a laundry, a laboratory and accommodations for 15 women. All of these structures were to be built from old materials. Jewish-owned buildings in the vicinity were being dismantled at the time. I could tell them by their house numbers. I selected my crew and began to work. I brought in some of the new lumber from the woods myself. Time flew fast on the job.

But there were new events to upset our emotional balance. This was the period when the Germans talked a lot about Katyn,[1] which they used for anti-Soviet propaganda purposes. One day, by accident, we got hold of a newspaper from which we learned about that mass killing. It was probably these reports that made Himmler decide to visit Treblinka personally and to give orders that henceforth all the corpses of inmates should be cremated. There were plenty of corpses to cremate—there was no one who could have been blamed for the Treblinka killings except the Germans who, for the time being, were the masters of the land which they had wrested from us (Poles) by brute force. They did not want any evidence of the mass murders left.

At any rate, the cremations were promptly begun. The corpses of men, women, children and old people were exhumed from the mass graves. Whenever such a grave was opened, a terrible stench rose from them, because the bodies were already in an advanced stage of decomposition. This work brought continued physical and moral suffering to those who were forced to do it. We, the living, felt renewed grief, even more intensively than before. We were ill-fed, because transports had ceased to arrive, so that the hapless purveyors of food had become a thing of the past. We did not like to draw on our reserves. All we ate was moldy bread, which we washed down with water. The malnutrition caused an epidemic of typhus. Those who became ill needed neither medication nor a bed. A bullet in the neck and all was over.

Work was begun to cremate the dead. It turned out that bodies of women burned more easily than those of men. Accordingly, the bodies of women were used for kindling the fires. Since cremation was hard work, rivalry set in between the labor details as to which of them would be able to cremate the largest number of bodies. Bulletin boards were rigged up and daily scores were recorded. Nevertheless, the results were very poor. The corpses were soaked in gasoline. This entailed considerable expense and the results were inadequate; the male corpses simply would not burn. Whenever an airplane was sighted overhead, all work was stopped, the corpses were covered with foliage as camouflage against aerial observation.

It was a terrifying sight, the most gruesome ever beheld by human eyes. When corpses of pregnant women were cremated, their bellies would burst open. The fetus would be exposed and could be seen burning inside the mother's womb.

All this made no impression whatsoever on the German murderers, who stood around watching as if they were checking a machine which was not working properly and whose production was inadequate.

Then, one day, an Oberscharführer wearing an SS badge arrived at the camp and introduced a veritable inferno. He was about 45 years old, of medium height, with a perpetual smile on his face. His favorite word was *tadellos* (perfect) and that is how he got the by-name Tadellos. His face looked kind and did not show the depraved soul behind it. He got pure pleasure watching the corpses burn; the sight of the flames licking at the bodies was precious to him, and he would literally caress the scene with his eyes.

This is the way in which he got the inferno started: He put into operation an excavator which could dig up 3,000 corpses at one time. A fire grate made of railroad tracks was placed on concrete foundations 100 to 150 meters in length. The workers piled the corpses on the grate and set them on fire.

I am no longer a young man and have seen a great deal in my lifetime, but not even Lucifer could possibly have created a hell worse than this. Can you picture a grate of this length piled high with 3,000 corpses of people who had been alive only a short time before? As you look at their faces it seems as if at any moment these bodies might awaken from their deep sleep. But at a given signal a giant torch is lit and it burns with a huge flame. If you stood close enough, you could well imagine hearing moans from the lips of the sleeping bodies, children sitting up and crying for their mothers. You are overwhelmed by horror and pain, but you stand there just the same without saying anything. The gangsters are standing near the ashes, shaking with satanic laughter. Their faces radiate a truly satanic satisfaction. They toasted the scene with brandy and with the choicest liqueurs, ate, caroused and had a great time warming themselves by the fire.

Thus the Jews were of some use to them even after they had died. Though the winter weather was bitter cold, the pyres gave off heat like an oven. This heat came from the burning bodies of Jews. The hangmen stood warming themselves by the fire, drinking, eating and singing. Gradually, the fire began to die down, leaving only ashes which went to fertilize the silent soil. Human blood and human ashes—what food for the soil! There will be a rich harvest. If only the soil could talk! It knows a lot but it keeps quiet.

Day in and day out the workers handled the corpses and collapsed from physical exhaustion and mental anguish. And while they suffered, the hearts of the fiends were filled with pride and pleasure in the hell they had created. It gave light and warmth, and at the same time it obliterated every trace of the victims, while our own hearts bled. The Oberscharführer who had created this inferno sat by the fire, laughing, caressing it with his eyes and saying, "*tadellos* [perfect]!" To him, these flames represented the fulfillment of his perverted dreams and wishes.

The cremation of the corpses proved an unqualified success. Because they were in a hurry, the Germans built additional fire grates and augmented the crews serving them, so that from 10,000 to 12,000 corpses were cremated at one time. The result was one huge inferno, which from the distance looked like a volcano breaking through the earth's crust to belch forth fire and lava. The pyres sizzled and crackled. The smoke and heat made it impossible to remain close by. It lasted a long time because there were more than half a million dead to dispose of.

The new transports were handled in a simplified manner; the cremation followed directly after the gassing. Transports were now arriving from Bulgaria,* comprising well-to-do people who brought with them large supplies of food: white bread, smoked mutton, cheese, etc. They were killed off just like all the others, but we benefitted from the supplies they had brought. As a result, our diet improved considerably. The Bulgarian Jews were strong and husky specimens. Looking at them, it was hard to believe that in 20 minutes they would all be dead in the gas chambers.

These handsome Jews were not permitted an easy death. Only small quantities of gas were let into the chambers, so that their agony lasted through the night. They also had to endure severe tortures before entering the gas chambers. Envy of their well-fed appearance prompted the hangmen to torment them all the more.

After the Bulgarian transports, more transports began to come from Bialystok and Grodno. In the meantime I had finished the construction of the laboratory, the laundry and the rooms for the women.

One day a transport arrived in Treblinka when we were already locked in our barracks for the night. Accordingly, the Germans and the Ukrainians processed the victims without help. Suddenly we heard yells and heavy rifle fire. We stayed put and waited impatiently for morning to come so that we could learn what had happened. The next morning we saw that the yard was littered with corpses. While we were working, the Ukrainian guards told us that the people who had come on that transport had refused to be led into the gas chambers and had put up a fierce fight. They smashed everything they could lay their hands on and broke open the chests with gold that stood in the corridor leading to the chambers. They grabbed sticks and every weapon they could get hold of to defend themselves. The bullets fell thick and fast, and by morning the yard was strewn with dead bodies and with the improvised weapons the Jews had used in their last fight for life. Those killed while fighting, as well as those who died from gas, were all horribly mutilated. Some of them had had limbs torn from their bodies. By dawn it was all over. The rebels were cremated. To us it was just one more warning that we could not hope to escape our fate.

Chapter IX

About that time, the camp discipline became stricter. A guard station was built, the number of guards increased and a telephone was installed in Camp No. 2. We were short of hands for work, and so men were sent from Camp No. 1. But their work was not considered satisfactory, and so

*Since no Bulgarian Jews were deported, Wiernik must be referring to Jews from Yugoslavian Macedonia and Greek Thrace. Both these regions had been ceded to Bulgaria with German approval.

they were finished off a few days later. Since they were such poor labor material, they were not worth the food required to keep them alive.

The Scharführer, a German master carpenter from Bohemia, whom I have already mentioned, came to me for advice about the construction of a four-story observation tower of the type he had seen in Maidanek. He was very happy when I gave him all the required information and he rewarded me with some bread and sausage. I figured out the specifications for the lumber and screws and proceeded with the construction work. Whenever I started on a new job, I knew that my life would be spared for a few weeks longer because as long as they needed me, they would not kill me.

When I had completed the first tower, the Hauptsturmführer came, praised me extravagantly and ordered me to build three additional towers of the same type around Camp No. 2.

The guard at the camp was increased and it became impossible to get from one camp to the other. Seven men joined in a plot to dig a tunnel through which to escape. Four of them were caught and were tortured for an entire day, which in itself was worse than death. In the evening, when all hands had returned from work, all the inmates were ordered to assemble and witness the hanging of the four men. One of them, Mechel, a Jew from Warsaw, shouted before the noose was tightened around his neck: "Down with Hitler! Long live the Jews!"

Among us workers there were some who were very religious, who recited the daily prayers each day. A German by the name of Karol, who was deputy commandant and a cynic, observed the habit of this little group and made jokes about it. He even gave them a prayer shawl and phylacteries for their devotions, and when one of the men died, he gave permission to give him a traditional Jewish funeral, complete with a tombstone. I advised the men not to do this, because our tormentors would exhume the body and cremate it after they had had their fun watching the ceremony. They refused to heed my advice but they soon found out that I had been right.

In April 1943, transports began to come in from Warsaw. We were told that 600 men in Warsaw were working in Camp No. 1; this report turned out to be based on fact. At the time a typhus epidemic was raging in Camp No. 1. Those who got sick were killed. Three women and one man from the Warsaw transport came to us. The man was the husband of one of the three women. The Warsaw people were treated with exceptional brutality, the women even more harshly than the men. Women with children were separated from the others, led up to the fires and, after the murderers had had their fill of watching the terror-stricken women and children, they killed them right by the pyre and threw them into the flames. This happened quite frequently. The women fainted from fear and the brutes dragged them to the fire half dead. Panic-stricken, the children clung to their mothers. The women begged for mercy, with eyes closed so as to shut out the grisly scene, but their tormentors only leered at them and kept their victims in agonizing suspense for minutes on end. While one batch of women and children were being killed, others were left standing around, waiting their turn. Time and time again children were snatched from their mothers' arms and tossed into the flames alive, while their tormentors laughed, urging the mothers to be brave and jump into the fire after their children and mocking the women for being cowards.

A number of men from Camp No. 1 were sent into our camp as workers. They were terrified and afraid to talk to us, for Camp No. 1 was known to have a very stern discipline. After a while, however, these men calmed down and gave us to understand that a revolt was being planned in Camp No. 1. We wanted to establish contact with the inmates of Camp No. 1, but no opportunity presented itself, for there were watchtowers and guards all around. The food in our camp had improved. We got a shower and even clean linens once a week, and a laundry had been set up in which female inmates were working. We decided that by spring we would either make a try for freedom or perish.

About that time I caught a cold, which developed into pneumonia. All the sick were being killed either by shooting or by injections, but it seems that they needed me. Accordingly, they gave me whatever medical attention was available. A Jewish physician attended me, examined me every day, and gave me medicine and comfort. My German superior, Loeffler, brought me food: white bread, butter and cream. Whenever he confiscated any food from smugglers, he shared it with me. The warm spring weather, the urge to live and the medical help I was getting did their bit and despite the incredible hardships under which I lived, I recovered. I went back to work to finish the construction of the observation towers.

One day the Hauptsturmführer, accompanied by the camp commandant and my superior, Loeffler, came to see me. They asked me whether I would undertake to build a blockhouse. It was to be constructed of logs and serve as a guard station in Camp No. 1. When I began to explain to him how the job should be done, he turned to his companions and remarked that I had understood him in a flash.

There was no lumber or building material on hand. We had to cut the wood with saws. I suggested making a shingle roof, and we had to prepare the shingles ourselves. As a result, I was able to make things easier for a good many camp inmates, who were relieved from the work with the corpses in order to assist me. I built the blockhouse in Camp No. 2 in such a way that it could be taken apart and moved to Camp No. 1. Everybody liked it so much that the Hauptsturmführer and Loeffler bragged to their colleagues that they had done the work themselves.

After a while, the time had come to take the structure apart and move it to Camp No. 1, but the architect Hermann and the master carpenter were unable to reassemble the structure themselves. It was evidently easier for them to kill innocent people than to do this kind of work. Once again, they turned to me for assistance.

This suited me to perfection because in that way I was able to gain access to Camp No. 1 and to make contact with our companions in adversity there. I needed assistance in my work and, although four men would have been enough, I asked for eight.

When I entered Camp No. 1, I did not recognize it at all. It was spotlessly clean and the discipline was extremely strict. Everyone was terror-stricken at the mere sight of a German or a Ukrainian. Not only did the inmates of Camp 1 refuse to speak to us, they were even afraid to look at us.

Starved and ill-treated though they were, they had a secret organization which was functioning efficiently. Everything was carefully planned. A Warsaw baker by the name of Leiteisen, who acted as liaison man between the conspirators, was working near the fence in Camp No. 1. It was difficult to make contact with him because there were German and Ukrainian guards all around and the fence was screened by saplings and you never knew who might be lurking behind them.

The workers in Camp No. 1 were continually under the threat of the whip. Compared with them, we enjoyed complete freedom. For instance, we were permitted to smoke while we worked and even received cigarette rations. We took advantage of our relative freedom for our own purposes. Some of us drew our guard into conversation to divert his attention, while others used that opportunity to make contact with inmates of Camp No. 1.

In due time, we became members of a committee of the secret organization, a circumstance which gave some prospects of deliverance or at least of a heroic death. All this involved considerable risk because of the watchfulness of the guards and the strong fortifications at the camp. However, our motto was "freedom or death." In the meantime, I completed the blockhouse. To celebrate the occasion, the Hauptsturmführer treated us to liquor and sausages. While we worked on the blockhouse, we received additional daily rations of 1/2 kilogram of bread apiece.

Chapter X

In contrast to our camp, the reign of terror in Camp No. 1 was getting worse, with Franz and his man-eating hound lording it over the workers. During my first stay in Camp No. 1 I had noticed a few boys, aged 13 and 14, who had been tending a flock of geese and had been doing odd chores. They were the favorites of the camp. The Hauptsturmführer cared for them almost as a father would for his own children, looking after their needs and often spending hours on end with them. He gave them the best food and the best clothes. Because of the good care, the food and the fresh air they were getting, these boys looked the picture of health and I thought that no harm would come to them, but now, when I returned to Camp No. 1 I immediately noticed that they were no longer around. I was told that after the chief had tired of them, he had had them killed.

Having completed our assignment, we returned to Camp No. 2 in high hopes of being free soon. However, we had nothing definite to go on and the contact was broken off again.

The cremation of corpses had been going on in Camp No. 2 while we had been away, but as there were so many of them, the end was not yet in sight. Two more excavators were brought in, additional fire grates were constructed and the work was speeded up. The fire grates took up almost the entire yard. It was midsummer by then, and the fire grates gave off a terrific heat, turning the place into an inferno. We felt as if we ourselves were on fire. We anxiously waited for the moment when we would be able to force open the gates of the camp.

Several new transports arrived, I did not know from where. Two transports of Poles arrived also, but since I never saw them alive I do not know how they were treated when they had to disrobe and enter the death chambers. They were gassed just as the others had been. When we handled these corpses, we noticed that the men had not been circumcised. Also, we heard the Germans remarking that those "damned Poles" would not rebel again.

The younger inmates of our camp were growing impatient and were anxious to start the revolt, but the time was not ripe. We had not yet completed the plans for the attack, and escape. Contact with Camp No. 1 was difficult, but soon we were able to communicate with them again.

One Sunday afternoon Loeffler, my superior, told me that the Hauptsturmführer wanted to build an additional gate for the blockhouse and that the job would be given to me. He told me to draw up a plan, and I added the necessary information for the Hauptsturmführer, who accepted my suggestions. I submitted my specifications for the materials I would need and I started the job. I eagerly seized this opportunity, for I realized that this was the last chance of establishing contact with the conspirators. I visited Camp No. 1 under all sorts of pretexts and discussed our plans with my fellow conspirators, who, however, did not give any definite information. All they told us was not to give up but to wait. Meanwhile, bigger and better fire grates were set up at the camp, as if they would be needed for centuries to come. Seeing this, the young inmates were eager to take action. Our patience was wearing thin.

In Camp No. 2 we began to organize into groups of five, each group being assigned a specific task such as wiping out the German and Ukrainian garrison, setting the buildings on fire, covering the escape of the inmates, etc. All the necessary paraphernalia was being prepared: blunt tools to kill our keepers, lumber for the construction of bridges, gasoline for setting fires, etc.

The date for starting the revolt was set for June 15, but the zero hour was postponed several times and new dates were set, because the time was not yet ripe. The committee on organization used to meet after we had been locked in the barrack for the night. After the rest of our fellow inmates, worn out by the day's toil and abuse, had fallen asleep, we gathered in a corner of our barrack, in one of the upper bunks, and proceeded to make our plans. We had to keep the younger men in check, because they were eager for action and wanted to get things going even though we were not yet properly prepared.

We decided not to do anything without the inmates of Camp No. 1, since to do so would have been tantamount to suicide. We in Camp No. 2 were only a handful, because not all of us were physically fit for combat. As I have mentioned before, we had better food and treatment than the inmates in Camp No. 1, but we were only about 300 as against their 700.

The inmates of Camp No. 1 were practically starved and had to endure beatings and brutal punishment, which assumed fiendish forms if they were caught doing business with the Ukrainians. I saw with my own eyes how one of them on whom a piece of sausage had been found was tied to a post and forced to stand motionless through a blisteringly hot day. As he was physically quite strong, he survived the ordeal and did not betray the Ukrainian with whom he had done business. In this connection I must add that whenever the Germans found out about a Ukrainian dealing with the inmates and smuggling food to them, they would beat up the Ukrainian, too. The Ukrainians, in turn, took it out on the Jews. Living under such conditions, the inmates did not last long. It was then Franz's chance to drag those poor devils to the fire grates, torture them brutally and, after beating them to a pulp, kill them and throw their corpses into the fire. In view of these conditions, we knew that the inmates of Camp No. 1 would revolt but, since we were unable to accomplish anything without them, we completed our own preparations, and waited for a signal from them.

Chapter XI

In the meantime, "life" ran its "normal" course. There was no end to macabre ideas. The German staff suddenly felt the need for diversion and amusement, since they had no other worries. Accordingly, they organized compulsory theatrical performances, concerts, dance recitals, etc. The "performers" were recruited from among the inmates, who were excused from work for several hours to participate in rehearsals. The "performances" took place on Sundays. They were compulsory, with the audiences consisting of Germans and Ukrainians. Women were forced to sing in choirs, while the orchestra consisted of three musicians who were compelled to play each day at roll call after the whippings. The inmates were forced to sing Jewish songs as they marched off to work. Plans had been made for a new performance and new costumes obtained for it, but the show never took place because of our successful revolt and escape.

While the Germans ate their midday meal, between noon and 1 P.M., the Jews had to stand in the yard, in front of the mess hall, and provide music and song. The members of the choir had to work just as hard as the rest of the inmates, but had special hours for singing and performing their music. By and large, our tormentors had quite a bit of fun with the rest of us, dressing up as clowns and assigning functions which, heartsore though we were, actually made us laugh.

One Jewish watchman, especially selected by the Germans, was stationed in front of the door of our barrack. He wore red pants like those of a Circassian, a tight-fitting jacket and wooden cartridges on both sides of his chest. He wore a tall fur calpac on his head and carried a wooden rifle. He was forced to clown and dance to the point of exhaustion. On Sundays he wore a suit of white linen with red stripes on the pants, red facings and a red sash. The Germans often got him drunk and used him for horseplay. No one was permitted to enter the barrack during working hours, and so he stood on guard at the door. His name was Moritz and he came from Czestochowa.

Another such poor wretch was the so-called Scheissmeister (shitmaster). He was dressed like a cantor and even had to grow a goatee. He wore a large alarm clock on a string around his neck. No one was permitted to remain in the latrine longer than three minutes, and it was his duty to time everyone who used it. The name of this poor wretch was Julian. He also came from Czestochowa, where he had been the owner of a metal products factory. Just to look at him was enough to make one burst out laughing.

Moritz meekly accepted whatever the Germans did with him; he did not even realize what a pitiful figure he cut. Julian was a poised and quiet man, but when they began their horseplay with him, he wept bitterly. He wept also while he worked on the fire grates. His garb, his appearance and the task he had to perform provoked the German fiends to abuse him all the more and to amuse themselves at his expense.

For quite some time I had been working in Camp No. 1, returning every evening to Camp No. 2. This gave me a chance to make contact with the insurgents in Camp No. 1. I was watched less than the others and also treated better. Time and again, the Ukrainian guards entrusted some of their possessions to me for safekeeping because they knew I would not be searched. My superior bought me food himself and saw to it that I did not share it with anyone else. I never acted obsequious toward the Germans. I never took off my cap when I talked to Franz. Had it been another inmate, he would have killed him on the spot. But all he did was whisper to me in German, "When you talk to me, remember to take off your cap." Under these circumstances, I had almost complete freedom of movement and an opportunity to make all the necessary arrangements.

No transports had been coming to Treblinka for quite some time. Then, one day, as I was busy working near the gate, I noticed quite a different spirit among the German garrison and the Ukrainian guards. The Stabscharführer,* a man of about 50, short, stocky and with a vicious face, left the camp several times by car. Then the gate flew open and about 1,000 Gypsies were marched in. This was the third transport of Gypsies to arrive at Treblinka. They were followed by several wagons carrying all their possessions: filthy tatters, torn bedclothes and other junk. They arrived almost unescorted except for two Ukrainians wearing German uniforms, who were not fully aware of what it all meant. They were sticklers for formality and even demanded a receipt, but they were not even admitted into the camp and their insistence on a receipt was met with sarcastic smiles. They learned on the sly from our Ukrainians that they had just delivered a batch of new victims to a death camp. They paled visibly and again knocked on the gate demanding admittance, whereupon the Stabscharführer came out and handed them a sealed envelope which they took and departed. The Gypsies, who had come from Bessarabia, were gassed just like all the others and then cremated.

July was drawing to a close and the weather was blistering hot. The hardest work was at the mass graves, and the men who exhumed the corpses for cremation were barely able to stand on their feet because of the sickening odors. By now about 75 per cent of the corpses had been cremated; all that remained to be done was to grade down the soil so that not a trace would be found of the crimes which had been committed on that spot. Ashes don't talk.

It was our job to fill in the empty ditches with the ashes of the cremated victims, mixed with soil in order to obliterate all traces of the mass graves. The parcel of ground thus gained had to be utilized one way or another. It was fenced in with barbed wire, taking in an additional plot from the other camp to form an area for planting. An experiment was conducted with planting some vegetation in this area; the soil proved to be fertile. The gardeners among us planted lupine, which grew very well. And so the area of the mass graves, after 75 per cent of the corpses buried there had been exhumed and cremated, was leveled, seeded and fenced in with barbed wire. Pine trees were also planted there.

The Germans were full of pride over what they had accomplished and thought that they deserved some modest entertainment as a reward for their troubles. They began by celebrating the "retirement" of the excavator which had been exhuming our dead brethren. It was pointed skyward, its shovel high in the air. The Germans fired salvos: then came a regular banquet with much drinking and merrymaking.

We, too, benefitted from this celebration: we gained a few days' respite from work, but we realized only too well that these would be our last days on earth, since only 25 per cent of the graves still remained to be emptied. Once this would be finished, the few of us who were the sole witnesses to the appalling crimes which had been committed would also be killed. However, we controlled ourselves and waited patiently for deliverance.

* Staff sergeant.

At that time I was working steadily at Camp No. 1. A portion of the area of Camp No. 2 had been joined with Camp No. 1 and one of the towers had to be moved to Camp No. 2. I worked on this job with my men. I was, therefore, able to remain in contact with our comrades in Camp No. 1.

Within a few days work was begun to empty the remaining 25 per cent of the graves and the bodies were cremated. As I pointed out before, the weather was extremely hot, and as each grave was opened, it gave off a nauseating stench. Once the Germans threw some burning object into one of the opened graves just to see what would happen. Clouds of black smoke began to pour out at once and the fire thus started glimmered all day long. Some of the graves contained corpses which had been thrown into them directly after being gassed. The bodies had had no chance to cool off. They were so tightly packed that, when the graves were opened on a scorchingly hot day, steam belched forth from them as if from a boiler.

In one instance, when a batch of corpses was placed on the fire grate, an uplifted arm stuck out. Four fingers were clenched into a tight fist, except for the index finger, which had stiffened and pointed rigidly skyward as if calling God's judgment down upon the hangmen. It was only coincidence, but it was enough to unnerve all those who saw it. Even our tormentors paled and could not turn their eyes from that ghastly sight. It was as if some higher power had been at work. That arm remained pointed upward for a long, long time. Long after part of the pyre had turned to ashes, the uplifted arm was still there, calling to the heavens above for retributive justice. This small incident, seemingly meaningless, spoiled the high good humor of the hangmen, at least for a while.

I continued working at Camp No. 1, returning to Camp No. 2 each night. I was constructing a birchwood enclosure, a low fence around the flower garden where domesticated animals and birds were also kept. It was a quiet, pretty spot. Wooden benches had been placed there for the convenience of the Germans and Ukrainians. But alas, that serene spot was the seat of infamous plotting, the only theme of which undoubtedly was how to torture us, the hopeless wretches.

Chapter XII

The Lagerälteste* of Camp No. 1 frequently watched me at work from a distance. It was forbidden to talk to any of us, but he frequently spoke a few words to me on the sly. He was a Jew of about 45, tall and pleasant, by the name of Galewski. An engineer by profession, he hailed from Lodz. He had been appointed to his office in August 1942, when Jewish

*Camp senior.

camp "authorities" had first been set up. He was the mainstay of the organization work. Because he did not prostitute himself as some of the others had done, but always considered himself one of us unfortunates, he was frequently beaten and hounded like the rest of us.

When he came to me for a brief exchange of words, he had just been set free from a three-day confinement in a prison cell. While there, he had been let out only once each day—in the morning—to empty the ordure bucket. Now, when no one was near me, he took the opportunity and categorically stated that the younger element should be patient because the hour of deliverance was approaching. He repeated this several times. I had the feeling that zero hour was approaching and that the end was really in sight.

On my return home from work that evening, I called a meeting to check the state of our preparedness. Everybody was excited and we did not sleep at all that night, seeing ourselves already outside the gates of the inferno.

The heat was becoming increasingly unbearable. It was almost impossible to keep standing on our feet. The terrible stench and the heat radiating from the furnaces was maddening. The Germans therefore decided that we were to work from 4 A.M. till noon, at which time they herded us into the barracks area. Once again, we came close to despair. We were afraid that now we would never be able to get out. However, we managed to find a way. We convinced the Germans that it would be better if the corpses would be cremated as soon as possible and said that there were volunteers among us who, for extra bread rations, would gladly work overtime. The Germans agreed.

We arranged two shifts, from noon till 3 P.M. and from 3 P.M. to 6 P.M. We selected the right men and waited from day to day for the signal. Beyond the area of our barrack there was a well that supplied the kitchen and laundry with water. We made use also of this "gateway," although it was guarded all the time. We made frequent trips to that well, even when we did not need water, in order to get the guards used to seeing us come and go.

At that time no transports at all came in, and so the only executions performed were those of individual Jews. After all, our executioners simply could not remain idle. But in due time the Germans were all in a good mood once more because new victims had arrived: a transport from Warsaw which was supposed to have been sent abroad. All the people in that transport were well-to-do and looked prosperous. They numbered about 1,000 men, women and children. We understood that it was a transport of people who had paid plenty of money to be taken to a place of safety. As I subsequently learned, they had been housed in the Hotel Polski, a first-class establishment on Dluga Street in Warsaw, but then they were taken to Treblinka. We learned who they were when we sorted out their possessions and found their personal papers. These people were killed like all the others.

The same fate befell transports coming in from other countries. These people had been told that they were going to be "resettled" in a place called Treblinka. Whenever they passed a station, the poor wretches would poke their heads out of the train windows and casually ask how much longer it was to Treblinka. Spent as they were, they looked forward to reaching a haven where they would be able to rest from their arduous journey. When they finally got to Treblinka, they were put to rest—forever—before they even had time to feel surprise or terror. At this writing, lupine grows over the spot where their ashes were buried.

Next came a transport from the Treblinka Penal Camp. It consisted of about 500 Jews, all barely alive, worked to the bone and brutally tortured. They looked as if they were begging to die and they were killed like all the others.

However, we were drawing closer to the end of our suffering. The day of our deliverance was approaching. Just then, my superior, Loeffler, who had been treating me so well, was transferred to Maidanek. He was bent on taking me with him to work there, and I was in a terrible predicament. I knew that a cruel death awaited each one of us. In Maidanek, I would be unable to find a quick way to freedom in the new surroundings and it would take me a long time to become acquainted with new people and new conditions. However, the decision did not rest with me: what was more, I had to pretend that I was elated over Loeffler's honoring me with such an offer. Luckily for me, the Hauptsturmführer refused to let me go. He still needed me. I, for my part, was very happy about that.

At about that time, for some reason unknown to us, we were ordered to write letters. Some among us were naive enough to do it. Later on I saw with my own eyes how the letters were burned. I do not know whether it had just been a game, a practical joke, or whatever.

Chapter XIII

The final, irrevocable date for the outbreak of the revolt was set for August 2, and we instinctively felt that this would really be the day. We got busy with our preparations, checking whether everything was in readiness and whether each of our men knew the part he had to play.

It so happened that I did not go to Camp No. 1 for several days because I was busy constructing an octagonal building with a suspended roof, resembling a guard station, that was to house a well. I was also constructing a portable building in Camp No. 2 which could be taken apart and which I subsequently had to move to Camp No. 1, where it was supposed to remain permanently. I was becoming impatient because I was unable to get in touch with Camp No. 1 and zero hour was approaching.

August 2, 1943, was a sizzling hot day. The sun shone brightly through the small, grated windows of our barrack. We had practically no sleep that night; dawn found us wide awake and tense. Each of us realized the importance of the moment and thought only of gaining freedom. We were sick of our miserable existence, and all that mattered was to take revenge on our tormentors and to escape. As for myself, all I hoped for was to be able to crawl into some quiet patch of woodland and get some quiet, restful sleep.

At the same time, we were fully aware of the difficulties we would have to overcome. Observation towers, manned by armed guards, stood all around the camp, and the camp itself was teeming with Germans and Ukrainians armed with rifles, machine guns and revolvers. They would lock us up in our barracks as early as 12 noon. The camp was surrounded by several rows of fences and trenches.

However, we decided to risk it, come what may. We had had enough of the tortures, of the horrible sights. I, for one, was determined to live to present to the world a description of the inferno and a sketch of the layout of that accursed hellhole. This resolve had given me the strength to struggle against the hangmen and the endurance to bear the misery. Somehow I felt that I would survive our break for freedom.

A presentiment of the coming storm was in the air and our nerves were at high tension. The Germans and the Ukrainians noticed nothing unusual. Having wiped out millions of people, they did not feel they had to fear a paltry handful of men such as we. They barked orders which were obeyed as usual. But those of us who belonged to the committee were worried because we had no instructions about the timing of the outbreak. I was fidgety. I kept on working but all the time I worried that we might fail to establish contact which, in turn, would mean that we would perish miserably and in vain.

However, I found a way of communicating with Camp No. 1. My superior, Loeffler, was no longer there; he had been replaced by a new man whose name I did not know. We nicknamed him "Brown Shirt." He was very kind to me. I walked up to him and asked him for some boards. Boards were stored in Camp No. 1 and he, not wanting to interrupt our work, went off with some workers to get them. The boards were brought. I inspected and measured them, and then said they weren't right for the job. I volunteered to go over myself to select the material I needed, but I made a wry face as if I did not like the idea. And so I went to the storage shed with my superior, all the while shaking with excitement. I felt that unless I made the most of this opportunity, all would be lost.

Presently I found myself in Camp No. 1 and nervously looked around, appraising our chances. Three other men were with me. The storage shed was guarded by a Jew about 50 years of age, wearing spectacles. Because he was an inmate of Camp No. 1, I knew nothing about him, but he was a participant in the conspiracy. My three helpers engaged the German superior in a conversation to divert his attention, while I pretended to be selecting boards. I deliberately went away from the others, continuing to select boards. Suddenly, someone whispered in my ear: "Today, at 5:30 P.M." I turned around casually and saw the Jewish guard of the storage shed before me. He repeated these words and added: "There will be a signal."

In feverish haste I collected whatever boards were nearest to me, told my comrades to pick them up and started to work, trembling with fear lest I betray my emotions. Thus time went by until noon, when all hands returned from work. Again our committee met furtively and the word was passed around. I asked everyone to keep cool and remember their individual assignments. The younger ones among us were greatly agitated. As I looked at our group, I began to believe that we would really win.

Volunteers for the afternoon work shift were then selected. We assigned the weaker and less capable men to the first shift because it had no task to perform. The first afternoon shift returned from work at 3 P.M. The men we had picked then went to work, thirty in number. They were the bravest, the pluckiest and the strongest in the lot. Their task was to pave the way for the others to escape. A crew was also picked for fetching water from the well. At around 5 P.M. there suddenly was a great need for water. The gate leading to the well was opened wide and the number of water carriers was considerably augmented.

All those assigned to work with the corpses wore only striped overalls. A penalty of 25 lashes was meted out for wearing any other clothing while doing this particular job. On that day, however, the men wore their clothes under their overalls. Before escaping, they would have to get rid of the overalls, which would have given them away at once.

We remained in our barracks, sitting close together and exchanging glances; every few minutes someone would remark that the time was drawing near. Our emotions at that point defied description. We silently bade farewell to the spot where the ashes of our brethren were buried. Sorrow and suffering had bound us to Treblinka, but we were still alive and wanted to escape from this place where so many innocent victims had perished. The long processions, those ghastly caravans of death, were still before our eyes, crying out for vengeance. We knew what lay hidden beneath the surface of this soil. We were the only ones left alive to tell the story. Silently, we took our leave of the ashes of our fellow Jews and vowed that, out of their blood, an avenger would arise.

Suddenly we heard the signal—a shot fired into the air.

We leaped to our feet. Everyone fell to his prearranged task and performed it with meticulous care. Among the most difficult tasks was to lure the Ukrainians from the watchtowers. Once they began shooting at us from above, we would have no chance of escaping alive. We knew that gold held an immense attraction for them, and they had been doing business with the Jews all the time. So, when the shot rang out, one of the Jews sneaked up to the tower and showed the Ukrainian guard a gold coin. The Ukrainian completely forgot that he was on guard duty. He dropped his machine gun and hastily clambered down to pry the piece of gold from the Jew. They grabbed him, finished him off and took his revolver. The guards in the other towers were also dispatched quickly.

Every German and Ukrainian whom we met on our way out was killed. The attack was so sudden that before the Germans were able to gather their wits, the road to freedom lay wide open before us. Weapons were snatched from the guard station and each one of us grabbed all the arms he could. As soon as the signal shot rang out, the guard at the well had been killed and his weapons taken from him. We all ran out of our barracks and took the stations that had been assigned to us. Within a matter of minutes, fires were raging all around. We had done our duty well.

I grabbed some guns and let fly right and left, but when I saw that the road to escape stood open, I picked up an ax and a saw, and ran. At first we were in control of the situation. However, within a short time pursuit got under way from every direction, from Malkinia, Kosow and from the Treblinka Penal Camp. It seemed that when they saw the fires and heard the shooting, they sent help at once.

Our objective was to reach the woods, but the closest patch was five miles away. We ran across swamps, meadows and ditches, with bullets pursuing us fast and furious. Every second counted. All that mattered was to reach the woods because the Germans would not want to follow us there.

Just as I thought I was safe, running straight ahead as fast as I could, I suddenly heard the command "Halt!" right behind me. By then I was exhausted but I ran faster just the same. The woods were just ahead of me, only a few leaps away. I strained all my will power to keep going. The pursuer was gaining and I could hear him running close behind me.

Then I heard a shot; in the same instant I felt a sharp pain in my left shoulder. I turned around and saw a guard from the Treblinka Penal Camp. He again aimed his pistol at me. I knew something about firearms and I noticed that the weapon had jammed. I took advantage of this and deliberately slowed down. I pulled the ax from my belt. My pursuer—a Ukrainian guard—ran up to me yelling in Ukrainian: "Stop or I'll shoot!" I came up close to him and struck him with my ax across the left side of his chest. Yelling: "Yob tvayu mat" (you motherfucker!) he collapsed at my feet.

I was free and ran into the woods. After penetrating a little deeper into the thicket, I sat down among the bushes. From the distance I heard a lot of shooting. Believe it or not, the bullet had not really hurt me. It had gone through all of my clothing and stopped at my shoulder, leaving only a scratch. I was alone. At last, I was able to rest.

1. In 1943 German forces occupying the village of Katyn announced that they had found in the woods nearby a mass grave of some 10,000 Polish officers. They claimed that these Poles had been captured and murdered by Russians. The Russians later accused the Germans of this wholesale murder. [Subsequently, it was proved that the Russians had murdered them.]

2. Kulmhof (Chelmno) Deathcamp

These brief descriptions of Chelmno by the *agents* rather than the victims of destruction come from a volume ironically titled *"The Good Old Days": The Holocaust as Seen by the Perpetrators and Bystanders,* a translation of "Schöne Zeiten," the heading of a photograph album kept by Kurt Franz, the last commandant of the Treblinka deathcamp. Had the Germans won the war, souvenirs like this would have contributed to a new category of memory, what we might call the nostalgia for atrocity, to enshrine for future generations the Third Reich's "magnificent achievement" in making the world free of Jews.

Fortunately, Hitler lost, so Franz's Treblinka album, together with the material about Chelmno in *"The Good Old Days,"* leaves us a far grimmer legacy. From diaries, journals, letters, pretrial interrogations, trial testimonies, and even reports from Protestant and Catholic chaplains, the editors have gathered a horrifying series of eyewitness accounts and photos, chiefly of the destruction of European Jewry. One would hope (though with slim expectation) that this could put to rest forever the persistent charges levelled by some antisemitic circles that there never were mass murders, that no gas chambers existed, and that the Holocaust did not happen. This evidence probably will not convince the authors of such irrational denials, but it may detach the more hesitant skeptics from the ranks of their followers.

Of course, the real value of this information is not to confirm the reality of the Holocaust; the testimony of surviving victims is enough for that. Anyone with sufficient stamina to read *"The Good Old Days"* to the end will gain a rare and memorable glimpse into the workings of the minds of men who could kill or witness killings without feeling pity and then tell about it, sometimes decades later, without remorse. These responses seem to refute the notion that only through a complex psychological self-manipulation called "doubling," which enables the murderers to separate their "normal" self from their "perpetrator" self, could they engage in their bloody work. The explanation offered by Kurt Möbius, one of the perpetrators whose testimony appears in the following selection, is much simpler, but it so offends our devotion to universal systems of moral behavior that it is difficult to accept:

> Although I am aware that it is the duty of the police to protect the innocent I was however at that time convinced that the Jewish people were not innocent but guilty. I believed in the propaganda that Jews were criminals and subhuman and that they were the cause of Germany's decline after the First World War. *The thought that one should oppose or evade the order to take part in the extermination of the Jews never entered my head either* [italics added]. I followed these orders because they came from the highest leaders of the state and not because I was in any way afraid.

Fearless loyalty such as this negates the need for "doubling." For men like Möbius, it is a waste of time to ask why the "guilt" of the Jewish people must "inevitably" lead to their death in the gas vans of Chelmno.

After hearing this "honest" testimony, the court in 1965 sentenced Möbius to eight years' imprisonment as an accessory to the murder of at least 100,000 people. The following selection challenges us to consider what he and his fellow perpetrators do *not* say, as well as what they do.

'Their Soldierly Conduct Is Exemplary': Kulmhof (Chelmno) Extermination Camp in the Wartheland Reichsgau

Between December 1941 and March 1943 at least 145,000 people—according to criminal proceedings in the Laabs et al. (8 Ks 3/62) trial in Bonn on 23 July 1965—were murdered here in gas-vans. On 7 April 1943 the 'castle' was blown up. Between April 1944 and January 1945 several thousand more people were killed. Polish estimates put the total number of victims as high as 300,000.

Theodor Malzmüller on the 'Plague Boils of Humanity'

When we arrived we had to report to the camp commandant, SS-Hauptsturmführer Bothmann. The SS-Hauptsturmführer addressed us in his living quarters, in the presence of SS-Untersturmführer Albert Plate (Bothmann's deputy). He explained that we had been detailed to the Kulmhof [Chelmno] extermination camp as guards and added that in this camp the plague boils of humanity, the Jews, were exterminated. We were to keep quiet about everything we saw or heard, otherwise we would have to reckon with our families' imprisonment and the death penalty.

We were then allocated our places in the guard unit [Wachkommando], which consisted of about fifty to sixty police officers from 1st Company Litzmannstadt Police Battalion. As I recall, there were also some officers from 2nd Company in it. The officer in charge of the guard unit was Oberleutnant Gustav Hüfing. He was from Wesel. . . .

The guardroom was situated in the village of Kulmhof. The unit members were accommodated in houses in the village. The duties of the guard unit consisted of (1) maintaining the security of the guardroom, (2) guarding the so-called 'castle' yard and (3) guarding the so-called 'camp in the wood'.

The extermination camp was made up of the so-called 'castle' and the camp in the wood. The castle was a fairly large stone building at the edge of the village of Kulmhof. It was here that the Jews who had been transported by lorry or railway were first brought. The Jews were addressed by a member of the Sonderkommando in the castle courtyard. I myself

> During my visit to Kulmhof I also saw the extermination installation, with the lorry which had been set up for killing by means of motor exhaust fumes. The head of the Kommando told me that this method, however, was very unreliable, as the gas build-up was very irregular and was often insufficient for killing.
>
> Rudolf Höss, Commandant of Auschwitz,
> on a visit to Chelmno on 16 September 1942

once heard one of these speeches when I was on guard duty in the castle courtyard for a day in December 1942. . . .

When a lorry had arrived the following members of the SS-Sonderkommando addressed the Jews: (1) camp commandant Bothmann, (2) SS-Untersturmführer Albert Plate from North Germany, (3) Polizei-Meister Willi Lenz from Silesia, (4) Polizei-Meister Alois Häberle from Württemberg. They explained to the Jews that they would first of all be given a bath and deloused in Kulmhof and then sent to Germany to work. The Jews then went inside the castle. There they had to get undressed. After this they were sent through a passageway on to a ramp to the castle yard where the so-called 'gas-van' was parked. The back door of the van would be open. The Jews were made to get inside the van. This job was done by three Poles, who I believe were sentenced to death. The Poles hit the Jews with whips if they did not get into the gas-van fast enough. When all the Jews were inside[,] the door was bolted. The driver then switched on the engine, crawled under the van and connected a pipe from the exhaust to the inside of the van. The exhaust fumes now poured into the inside of the truck so that the people inside were suffocated. After about ten minutes, when there were no further signs of life from the Jews, the van set off towards the camp in the wood where the bodies were then burnt. . . .

During the period that I was in the guard unit most of the time I did sentry duty in the interior of the camp in the wood. The camp was in a clearing in the woods between Kulmhof and Warthbrücken. . . . As a guard just within the camp perimeter I frequently saw mass graves, filled with the bodies of Jews who had been exterminated, being dug up by the Jewish Arbeitskommando. The bodies were then burnt in two incinerators. . . .

At the end of March 1943, shortly before the dismantling of Kulmhof extermination camp in April, Gauleiter Greiser suddenly appeared at the camp together with his staff (consisting of fifteen high-ranking SS officers). All members of the SS-Sonderkommando and the Wachkommando had to assemble in the courtyard of the castle where they were addressed by Greiser. In the presence of his staff he explained that Kulmhof extermination camp would shortly be dismantled and he wanted to thank us on behalf of the Führer for the work we had done in Kulmhof. He went on to say that everybody would be given four weeks' special leave and that we were welcome to spend it free of charge on one of his estates. He then invited all those present to a farewell party at a hotel in Warthbrücken. The farewell party was held in a big room in the hotel. After a short while everyone was drunk and fell asleep at the table. The party ended at about one or two in the morning. . . .

A few days after Greiser's farewell party all members of the SS-Sonderkommando and the police guards received four weeks' special leave. Only a few members of the SS-Sonderkommando stayed behind in Kulmhof. One of these was Polizei-Meister Lenz. Then everybody had to report to SS-Obergruppenführer Kaltenbrunner at state security headquarters in Berlin on a particular day. He addressed us all and we were once again thanked on behalf of the Führer for our work in Kulmhof.

We were then all detailed together to Yugoslavia to SS-Division Prinz Eugen, under the command of Bothmann. Here we were deployed against partisans in Yugoslavia and suffered very heavy losses. As far as I can recall, SS-Untersturmführer Plate committed suicide in Serbia after being severely wounded.

In the middle of 1944 some of those former members of the SS-Sonderkommando who were still alive were withdrawn from the SS-Division and sent back to Kulmhof to start up the extermination camp once again.

Gas-van Driver Walter Burmeister on Whether He Ever Thought About What He Was Doing

As soon as the ramp had been erected in the castle, people started arriving in Kulmhof from Litzmannstadt in lorries. . . . The people were told that they had to take a bath, that their clothes had to be disinfected and that they could hand in any valuable items beforehand to be registered. On the instructions of Kommandoführer Lange [Bothmann's predecessor] I also had to give a similar talk in the castle to the people waiting there—how often exactly I can no longer say today. The purpose of the talk was to keep the people in the dark about what lay before them. When they had undressed they were sent to the cellar of the castle and then along a passageway on to the ramp and from there into the gas-van. In the castle there were signs marked 'To the baths'. The gas-vans were large vans about 4–5 m long, 2.20 m wide and 2 m high. The interior walls were lined with sheet metal. On the floor there was a wooden grille. The floor of the van had an opening which could be connected to the exhaust by means of a removable metal pipe. When the lorries were full of people the double doors at the back were closed and the exhaust connected to the interior of the van. . . .

The Kommando member detailed as driver would start the engine straight away so that the people inside the lorry were suffocated by the exhaust gases. Once this had taken place, the union between the exhaust and the inside of the lorry was disconnected and the van was driven to the camp in the woods where the bodies were unloaded. In the early days they were initially buried in mass graves, later incinerated. . . . I then drove the van back to the castle and parked it there. Here it would be cleaned of the excretions of the people that had died in it. Afterwards it would once again be used for gassings. . . .

I can no longer say today what I thought at the time or whether I thought of anything at all. I can also no longer say today whether I was too influenced by the propaganda of the time to have refused to have carried out the orders I had been given.

Kurt Möbius on the Guilt of the Jews and His Own Lack of Blame

. . . In addition Hauptsturmführer Lange said to us that the orders to exterminate the Jews had been issued by Hitler and Himmler. We had been drilled in such a way that we viewed all orders issued by the head of state as lawful and correct. We police went by the phrase, 'Whatever serves the state is right, whatever harms the state is wrong.' I would also like to say that it never even entered my head that these orders could be wrong. Although I am aware that it is the duty of the police to protect the innocent I was however at that time convinced that the Jewish people were not innocent but guilty. I believed all the propaganda that Jews were criminals and subhuman [*Untermenschen*] and that they were the cause of Germany's decline after the First World War. The thought that one should oppose or evade the order to take part in the extermination of the Jews never entered my head either. I followed these orders because they came from the highest leaders of the state and not because I was in any way afraid.

Interrogation of Adolf Eichmann

E I just know the following, that I only saw the following: a room, if I still recall correctly, perhaps five times as big as this one, or it may have been four times as big. There were Jews inside it, they had to get undressed and then a van, completely sealed, drew up to a ramp in front of the entrance. The naked Jews then had to get inside. Then the lorry was closed and it drove off.

L How many people did the van hold?

E I can't say exactly. I couldn't bring myself to look closely, even once. I didn't look inside the entire time. I couldn't, no, I couldn't take any more. The screaming and, and, I was too upset and so on. I also said that to [SS-Obergruppenführer] Müller when I submitted my report.

He did not get very much from my report. I then followed the van—I must have been with some of the people from there who knew the way. Then I saw the most horrifying thing I have ever seen in my entire life.

The van drove up to a long trench, the doors were opened and bodies thrown out. They still seemed alive, their limbs were so supple. They were thrown in, I can still remember a civilian pulling out teeth with some pliers and then I just got the hell out of there. I got into the car, went off and did not say anything else. . . . I'd had more than I could take. I only know that a doctor there in a white coat said to me that I should look through a peep-hole at them in the lorry. I refused to do that. I could not, I could not say anything, I had to get away.

I went to Berlin, reported to Gruppenführer Müller. I told him exactly what I've just said, there wasn't any more I could tell him. . . . Terrible . . . I'm telling you . . . the inferno, can't, that is, I can't take this, I said to him.

Gauleiter Greiser to Himmler, 19 March 1943

Reichsführer!

A few days ago I visited Lange's former Sonderkommando, which today is under the command of SS-Hauptsturmführer Kriminalkommissar Bothmann and stationed in Kulmhof, Kreis Warthbrücken, until the end of the month. During my visit I was so struck by the conduct of the men of the Sonderkommando that I would not like to fail to bring it to your attention. The men have not only fulfilled the difficult task that has been set for them loyally, bravely and in all respects appropriately, but also their soldierly conduct is exemplary.

For example during a social evening to which I had invited them they gave me a contribution of 15,150 RM in cash which they had that day collected spontaneously. That means that each of these eighty-five men in the Sonderkommando had contributed about 180 RM. I have given instructions for the money to be put in the fund set up for the children of murdered ethnic Germans, unless you, Reichsführer, wish it to be put to another or better use.

The men further expressed the wish that all of them, if possible, be put under the command of their Hauptsturmführer Bothmann when they are transferred to their new assignment. I promised the men that I would communicate this wish to you, Reichsführer.

I should be grateful if you would give me permission to invite some of these men to be my guests on my country estate during their leave and to give them a generous allowance to make their leave more enjoyable.

Heil Hitler
(signed) Greiser

3. František R. Kraus

František R. Kraus was born in Prague in 1905. He was employed as a journalist for several Prague newspapers, for news agencies, and for radio. He also published fiction and volumes of reportage. In 1941 he was deported to Terezín (Theresienstadt), where he worked as a laborer. In the fall of 1944, he was sent to Auschwitz and then to two of its subcamps, Gleiwitz and Blechhammer. After his liberation, he returned to Prague, where he was employed by Czechoslovak Radio and resumed his writing activities.

"But Lidice Is in Europe!" offers a nightmarish eyewitness account of the burial of the massacred male population of the Czech village of Lidice in the aftermath to one of the most gruesome Nazi atrocities of World War II. On May 27, 1942, Czech freedom fighters who had parachuted in from England intercepted, near Prague, the motorcade of Reinhard Heydrich, deputy Reich protector of Bohemia and Moravia (annexed provinces of Czechoslovakia). They threw a bomb at the limousine carrying Heydrich, who, severely wounded, died seven days later. In addition to his civil post, Heydrich was head of the Reich Security Main Office under Heinrich Himmler, in which position he was one of the chief architects of the Final Solution. His assassination roused the fury of Nazi leaders against the Jews, whom they erroneously accused of masterminding the plot. Hundreds of Jewish hostages were executed throughout the Reich. More than 1,000 Czechs were arrested in Prague, of whom 100 were shot. But the severest reprisal was reserved for the village of Lidice, which was located about twenty miles northwest of Prague. The Germans insisted (again mistakenly) that the assassins had been hidden there before the assassination.

The Nazis decided to make Lidice an example of the punishment in store for those aiding the resistance, especially in Czechoslovakia, but also in the rest of occupied Europe. They publicized the doom of its inhabitants. The Germans shot or burned alive more than 170 men and boys from the village; killed 71 women and deported nearly 200 more to the concentration camp at Ravensbrück; and shipped about 100 small children to Germany, to an uncertain fate. They even slaughtered or drove out most of the animals before razing the village to the ground.

It is worth mentioning that though extreme, the viciousness of Nazi Germany's response to the killing of Heydrich is not unique; it illustrates the regime's utterly merciless reaction to any effort, by whatever form of resistance, to thwart its policies.

But Lidice Is in Europe!

Once, under the old Hapsburg monarchy, a barracks used to exist here, quartering at the most 150 men of a Maria Theresia guard gunners' corps. Josephus II. Aug. P.P. Aeterni Huius Operis Fundamenta Iecit VI. ID. Oct. MDCCLXXX. is inscribed on the large greyish-brown marble tablet on the wall of the mighty casemate, the awesome spirit of the Middle Ages emanates from the crumbling red brick walls, ramparts and entrenchments. Heavy black wings beating, a flock of ravens flies thunderingly across the sky and disappears on the fiery-red horizon. In 1918, when the Austro-Hungarian monarchy broke apart, young Czechoslovak artillerists took over the barracks. There were at the most 200 of them, and in 1939, the Prussians came, Hitler's gunners. About a hundred moved into the "Sudeten" barracks of Terezín. But today, about 6,500 male prisoners live in the very same barracks, mostly younger men, still fit for work, Jews from Prague, Bohemia and Moravia. The "Dresden" barracks as well as the "Hannover," "Bodenbach" and "Engineers" barracks are already occupied by Jewish prisoners. All barracks are still locked in Terezín. The workers from the workers' barracks cannot see their wives in the women's barracks. The children from the children's homes cannot get across to their mothers . . . But soon, very soon, after the forced evacuation of the Terezín population, the shoemakers, tailors, landlords, the grocers and greengrocers, there will be only one town of Jews, surrounded by barbed wire entanglements, entrenchments and walls, the "Altersparadies der europäischen Judenheit" (old-age paradise of European Jews) as the Nazis proclaim, as it resounds from the radio, telling of the ghetto of the twentieth century.

Our works commando is marching "home." Swish, swash. The threefold echo of our steps is thrown backwards and forwards phantom-like between the facades of the tumble-down houses, most of them already empty, swish swash, a red rain-pipe, twisting into an obtuse angle at the bottom and rust-eaten, clatters down to the pavement. We go marching on, swish, swash. The yellow stars of David shine through the misty twilight, they arch over our hearts, we march in the middle of the uneven road, over the cobbles full of rubbish and dirt. Stifling stench of latrines and of rat excrements rises from the canals . . . Swish, swash, the large yellow stars shine in the sun: it is June, June 11, 1942. We are marching through the dusty, crumbling streets of the future ghetto. Pot-bellied landlords have closed their pubs, have been forced to evacuate. The rusty shutters are drawn over the shops. Houses and schools stand empty. Soon, very soon, a new life will arise from the ruins, life in the ghetto, life in the Middle Ages, a bold jump backwards, centuries back . . . Our guards, regular Protectorate gendarmes, still with the silver lion of Bohemia on their greenish-grey helmets, are allowed to walk on the pavement. We must not set foot on it. Only right in the middle of the road, swish, swash. The gendarmes smoke, spit, and with bayonets up, accompany the group of young Jewish prisoners, all of whom have sprung from David's golden cradle . . .

And suddenly there appears on the corner of the house that turns blind windows towards the barracks, and that once served soldiers' love—up to this day, if you look carefully, you may decipher on the round red lamp above the entrance the inscription "Salon Kairo"—suddenly there is a huge, red poster. It still glistens damply, and has apparently just been put up: on top the black, repulsive eagle, below on the left in German, on the right in Czech, a long litany. We decipher with difficulty a few sentences printed in italics . . . "SS-Obergruppenführer Reinhard Heydrich, Vice Reichsprotector of Bohemia and Moravia and chief of SD and Sicherheitspolizei . . . on May 27, 1942, murdered by a treacherous bomb . . . one of the greatest defenders of the idea of the German Reich . . . died in defence of the preservation and safeguarding of the Third Reich . . . an irreproachable National Socialist . . . inferior men, Jews, criminals . . ." There is suddenly disquiet in our ranks, passing like waves through our commando. The escort of gendarmes, armed with helmet, chin-strap, rifles, bayonets, revolvers, and 60 shots in their cartridge belts of brightly-yellow, sharp-smelling leather, they too become nervous . . . Heydrich justly punished! Swish, swash . . .

The huge gate of the "Sudeten" barracks gapes widely. According to regulation the ghetto police quickly take off their black-and-yellow stocking caps in front of the fat, plumed-hatted gendarmes and we march into the huge, wind-swept yard, crammed full with waiting prisoners, their tin plate or food tin in their callused hands. They are standing in queues, waiting before the two kitchens, the "Prague" and the "Brno" kitchen. On the left a group of unshaven prisoners in crumpled clothes is being taken to the latrine by some gendarmes. They are waiting for tomorrow's execution in the cells behind the gendarmes' guard-room, they are waiting for the rope behind the casemates of the "Aussig" barracks. The gallows have already been put up, they stand like an index-finger threatening against the sky. Hunch-backed, grinning, already put under alcohol today by the SS, the former servant of the post-mortem department of the Prague clinics on Charles Square, by the name of Fischer, gap-toothed, reeking, is constantly kept under "pressure and alcohol" by

Camp Commander Obersturmführer* Dr. Seidl, a mediocre Viennese actor and Camp Inspector Bergel, who was one of those responsible for the assassination of Bundes Chancellor Dr. Dollfuss† . . . The "moribundi" . . . taken back from the latrine, already bear the pale, yellowish face of Hippocrates. It is just on noon. The Jewish "Labour Office" suddenly looks for a works group, quickly, quickly. Orderly Stampf runs excitedly to the first floor, to the rooms of the "Parille" hundred, the command of the "A.K. ists," young Jews from Prague, who had come to Terezín as the very first transport, the so-called Aufbaukommando‡ transport. "Earthwork, earthwork" shouts Stampf raspingly, excited and nervous. Tall Stampf is looking for Karl Parille, the leader of the hundred. He cannot find him, and so he instructs his second-in-command, "company-leader" Franta Krása, to get a group of thirty workers together.

"Fall in, fall in!" Stampf shouts excitedly. One, two, three, four, five, six, whoever crosses his path is lined up. Soon there are thirty men. I am one of them . . .

We set out, come to the gate where we have to stop. "Wait for the gendarmes!" And there they come, stumbling over one another in their hurry to get out of the guard room, putting on their knights' helmets, shouldering their rifles, fixing their chin-straps under their smoothly shaven chins. Oberwachtmeister Kubiňák curses, spits, sending yellow sputum in a wide arc through the gate. He is in command of the ten-men-strong group of gendarmes. We march through the yawning, windy gate. A green truck driven by always-drunk, toothless driver Řeháček is already drawn up outside. Řeháček is a civilian, owner of a truck that he uses for transport purposes, but in the services of the SS. We quickly jump in, Oberwachtmeister Kubiňák and two very young gendarmes sit in front with the driver, the other seven gendarmes jump in with us . . . We drive off. We stand on top close together, we must not sit down, only the gentlemen of the gendarmerie are sitting, propped against the boards, smoking, spitting, taking a pull, now and then, from their flat field-grey bottles filled with cheap rum, again leaning against the boards, laughing, dozing . . . The engine hums, the truck clatters and clanks over the dirty ghetto pavement, we hang on to each other . . .

We stop in front of the "Construction Yard." "Eight men down and load up, quickly, quickly!" these are gendarme orders. We get hold of shovels, pick-axes, cramp-irons, two barrels filled with caustic lime . . . Lime, Lime! Yes, we guess already, yes, we know already: "Lime, lime!" I say quietly to Langendorf, the young composer from Prague and best of comrades. "Yes, Franta, lime!"

The journey goes on. Not towards Litoměřice, as usual, when we march or drive to work, no, this time in the opposite direction. Yellow name plates with black lettering . . . Hrobce, Slaný, . . . Villages lie qui-

*SS rank equal to first lieutenant.

†Engelbert Dollfuss, chancellor of Austria, was murdered by Austrian Nazis in 1934.

‡Construction squad.

etly in the noon-day heat, cackling hens run out of our way, dogs bark and chase us to the end of the village, then they stop and look after us, their tails between their legs, fair-haired children play on green, flower-dotted meadows, white butterflies whir over sun-kissed, fresh young corn, yellow ducklings swim nimbly over dainty little village ponds, bumblebees buzz by, the tender blades of grass bend in the warm wind . . . highways, villagers look up to us, mouths gaping, old people sit in the warm sun in front of their small, red-roofed cottages. A green truck with helmeted lansquenets, blinking bayonets on their rifles, in their midst a grey heap of prisoners supporting one another, on their breasts the yellow glistening star of David, turning towards the eternal sun . . .

Suddenly I am hungry. Each of us got one slice of black bread and a quarter of margarine for the hot journey. I start to eat, standing up, the truck jogs along, the artificial butter has melted in the paper, like a hungry dog I lick it out of the paper. "Direction Kladno!" says the inscription at the crossroads. And there, suddenly, smoke and flames, far, faraway, but already visible, on the right a village must be burning . . . And here the yellow signpost that bears the inscription "Lidice." I have never heard or read the name before in my life. The brakes whine, the truck stops. The air is thick with fog and smoke, our faces change, suddenly a stream of fire shoots up to the sky, one can taste bitter powder on one's dry tongue, our nerves are tight, our bodies tense . . . What is going to happen to us?

Our gendarmes jump off the truck and on orders from an unknown SS-man lie down in the burned grass under a blackened pear tree. Drunken driver Řeháček, whose flat feet are lost in huge SS-boots, is relieved at the wheel by a helmeted policeman. We drive on, slowly, at a walk; we drive into a brightly burning town. We hear the beams bursting and cracking, our truck shakes, beside us, before us, a spray of destruction, a terrifying struggle of horror, we are driving straight into Dante's Inferno, the picture of our time is reflected in the shimmer of the horizon, we walk, but not guided by Virgil, through the Inferno, St. Bernhard is not our guide today through purgatory . . . Will Beatrice, the beloved, take us to Paradise? Splinters are flying through the air, clods fly sky-high and cover our face with mud . . .

"Get out!" Everywhere black and grey-green SS, Schupo, Sipo (Security police), Kripo (Criminal police) and Gestapo, many of them dead drunk, reeling through the ruins. We crawl along, through the blackened garden of a country-homestead . . . here . . . here it is, the Steyrcabriolet of our Camp Commander, party member Obersturmführer Dr. Seidl, yes, indeed, and beside his car, Seidl himself on tottering spider legs, his well-polished Prussian riding boots glisten through the blackish-grey smoke, he seems drunk, next to him his driver, Hauptscharführer Wostrell of Linz, further Obersturmführer Bergel, the Camp Inspector, and SS-men Polljak, Habenich, Schterba.

We walk on as in a dream . . . and here . . . here the dead lie on the ground! Shot! A colourful carpet bleeding from a thousand wounds . . .

they are lying on their backs, on their chests, on the side, caps and hats have fallen off, hands under their bodies or arms out stretched, their eyes glassy, the white teeth in their red faces showing behind their drawn-up lips . . . the terrible white of their eyes, it glistens, shines, through the smoke and flames, the martyrs of Lidice, heroes of Bohemia, heroically dying miners and small householders, peasants and pensioners, the old parish priest, a boy, a dog . . . shot up and burnt trees, mattresses and pallets piled up in the background, formed the execution wall . . . The camp commander of Terezín makes a sign on the steaming ground with his riding whip with the silver bulldog handle. This is the spot where we shall dig. Franta Krása, our company leader, takes the order: "Twelve metres long, nine metres wide and four metres deep! You understand, you sons of bitches from Jericho, you pig-eaters, you world criminals and arse-holes?"

We start to dig. Mother Earth, dear, kind Mother of all Mothers, how we defame your holy name . . . again we tear wounds into your holy body, like that time, when the first of our brethren were hanged behind the "Aussig" barracks in the garrison town of her Majesty, the German Empress, Queen of Bohemia and Hungary, the Austrian Arch-Duchess, and we had to dig them in, to the cursings of the "Guard, whom the Führer loves," in the light of the gendarmes' torches, that time when the fat Bergel with the leather whip beat our bare backs, and his order in the language of Goethe resounded over the red casemates . . . round about us there is a clattering and groaning, thundering and cracking, walls burst, the end of the world seems near . . . burning beams hurtle rattling to the ground . . . we sink deeper and ever deeper into the darkness of black earth . . . Above us flames and smoke, scared pigeons flutter from one burning point to the other, can find no rest, for everything glows, growls, shouts, vomits, groans, weeps and roars, it is the misery of the world, a wild and gruesome pain rides a shy, emaciated nag through the world, and this pain has taken on the shape of a yellow skeleton in fluttering, white, blood-stained overall, a snow-white pigeon is quite alone, fluttering to and fro, the drunk monster in the Führer's uniform, Parteigenosse Obersturmführer Dr. Seidl draws his small revolver . . . puff . . . the pigeon in white, in the tender bride's veil of peace, drops to his feet, a gentle shudder passes through the white head, the pink eyes turn glassy, Seidl kicks the tiny bird down to us into the yawning mass-grave, kicks the small, pure, dead bird down with his Prussian, highly polished, elegant riding boot into the war grave of humanity . . . Suddenly thunder fills the air above us, the flames howl, invisible doors spring open, air pressure tears them out of invisible hinges, all the sins of humanity pass swirling down, a witches' dance has started with howling, whistling, thumping, roaring, ringing and groaning, unknown voices shout, martyred mankind revolts against the terror of this century! We sink deeper and deeper. It is hot so near the glow. We sweat and starve. We work half-naked. We dig throughout the night. In the morning a fat Schutzpoliceman in the helmet of the Teutenburger Forest throws each of us a piece of black bread, it soars through the air and we catch it like wild cats behind iron bars, eat it avidly within a few minutes and go on digging . . . Fog and smoke hang over us, a red glow stands above the garden over the farm, dawn breaks slowly . . . The yellow wax-figures of Seidl and Wostrell appear above the abyss. Both are smoking, bead drunk. Schutzpolizei are making aromatic soup, from stolen poultry, in the huge pots of the Lidice housewives . . . Suddenly the church breaks apart: a new metallic thundering breaks up the walls, the ringing of the bells resounds clearly, there is a thumping in the tower, flames roar up again, then suddenly the ringing stops, torn away from the roof the bell hurtles down, breaks through the wooden floor and ends with huge clattering on the stone floor, white smoke rolls out of the fallen nave . . . Next to me stands Karl Langendorf, young, beautiful, the composer, he stands there like a marble statue, his mouth wide open, he raises and lowers his fists . . . Then low singing sounds from his lips, it is Antonín Dvořák's Requiem* . . . Requiem aeternam dona eis domine et lux perpetua luceat eis† . . . A cloud of decay, dust and powder stands over his head, beside him the red poppies fade and marguerites lower tired, innocent heads, to lie down and die . . . Te decet hymnus, Deus in Sion, et Tibi reddetur‡ . . . Terrible is the roaring sky over Karl Langendorf, but he goes on singing, he drowns out the terror of this time . . . Dies irae, dies illa solvet saeclum in favilla, teste David et Sibylla§ . . . Ploughed up and torn up, rent asunder and treeless, is this spot of Bohemian land in the heart of Europe . . . A new fireworks: on pale stems bright spheres shoot up from the ground, blossom forth and are blown away, deadly seed over blood-soaked fields . . . a windy morning rises from the blood-drenched east and Karl Langendorf sings Sanctus, Sanctus, Sanctus . . . Dominus Deus Sabaoth‖ . . . the earth shakes and shivers and cries from its depth and revolts and gives forth tongues of flames from glowing volcanoes and jumps sky-high from fountains and roars up and breaks into myriad cliffs, raises stones and smoke and poison and earth rearing into the air . . . and the air rages like a wounded cyclop and hurtles the deadly rocks down to us again . . . bricks drop onto the empty church benches, jump high again and dance to and fro as if it were a festive church holiday, then the beams clatter down and break the roof, walls and vaultings shake, pictures of saints in gold frames fall from the old walls, and thunder to the ground . . . the priestless chancel burns . . . mass is being celebrated for the last time here . . . Hosana in

*Kraus interposes fragments from Dvořák's *Requiem* to commemorate the burning of the village of Lidice and the burial of the bodies of its murdered inhabitants.

†Eternal rest grant them, O Lord.

‡A hymn, O God, becometh Thee in Zion.

§The day of wrath, that day, / will dissolve the world in ashes, / as David prophesied with the Sibyl.

‖Holy, Holy, Holy, Lord God of the Sabbath.

excelsis . . . Benedictus, qui venit in nomine Domini. Hosana in excelsis* . . .

We are down below. Finished . . . The work has been done . . . Now we shall stand at the edge to be shot. We shall tumble head first into the pit, then lime, then the others over us, the martyrs of Lidice, then lime and lime again, then SS and Schupo to dig us in . . . Lidice was the name of this little town in Bohemia, Lidice, it was in Europe!

Seidl starts to scream in his shrill, effeminate voice: "Out of there, get moving, you Hebrews!" We lift each other up, crawl over our comrades' backs, then they drag us out from above, we must take the dead's blood-stained shoes off, quickly there is a heap of these shoes that will soon set out on their march, their march through the world . . . Then Seidl screams: "Take their identity cards." We set to. The red citizens' cards must be taken out of the blood-soaked clothes . . . Paper money must be handed to Seidl, he stuffs it in all the pockets of his uniform . . . pushes it down his elegant, high, Prussian riding boots, then Wostrell stuffs his pockets, pushes the money under the "Ostmark" cap with the pointed rim. "Dig them in!" screams Seidl. We dig and scrape, we work thirty-six hours. Without break. Seidl has picked on Langendorf: "You hairy orangoutang, you! You swindler from Jericho!" His sharp, leather riding whip has split Langendorf's scalp. The blood runs down over his back, he stands upright, looking at the sky, his lips move . . . Agnus Dei qui tollis peccata mundi dona eis requiem sempiternam† . . .

"Fall out!" Schutzpolizei takes us out of the smoking ruins. The green truck, Řeháček on reeling-drunk, bent legs, toothless, laughing a pagoda-like, poisonous laugh, gendarmes jump up from the blackened grass where they have been lying, click their heels before the Nazis, we crawl up and the journey starts into the falling night, there is a roaring in our ears, as if sea water were entering a drowning man's brain, up in the truck we tumble down over each other, sleep a dozing bloodsoaked leaden sleep, but a kind of dozing only, nevertheless, we drive through the greenish-blue night to the ghetto, past the stinking canals, rats rush away, the huge barrack gates open up like the jaws of a phantastic monster, the truck stops. Gendarmes quickly jump down, spit, smoke, curse, "Grave diggers, queers!" We totter through the old, wind-swept yard to the stairs on the right, our steps drag over the corridor, echo from the mosaic-covered stones, room 61, comrades are waiting, those that have remained at home. At the head of our bunks there already stand burning candles, the tiny yellow flames lick and flutter in the draught of the opening door, just as that time, when the first of our comrades were hanged. Comrades are singing the monotonous melody of the Kadish, stop, smother their joy, the tiny yellow flames are blown out, rest, worn-out

*Hosanna in the highest. Blessed is He that cometh in the name of the Lord. Hosanna in the highest.

†O Lamb of God that takest away the sins of the world grant them eternal rest.

rest and embraces under the dark cover of the night, we fall on our bunks . . .

Rest, worn-out rest, and then a voice is raised in the dark. It is the physician Dr. Grünfeld who speaks: "Comrades, you smell of blood, of martyrs' blood . . . mix it with your heart's blood and let it run through the veins and hearts of all humanity . . ." I sink back. My eyes pass over the barred windows. Outside the night is of deepest black. And beneath me, on the lower bunk, Karl Langendorf sings quietly: "Requiem aeternam dona eis, Domine, et lux perpetua luceat eis* . . ." Then he adds in a low voice:

"But Lidice is in Europe!"

A spray of tiny stars glitters outside beyond the bars of the barrack windows above the steep gables of the black little ghetto houses . . .

*Rest eternal grant them, O Lord, and may light perpetual shine upon them.

4. Jacques Furmanski

Although we have many accounts of the selection process for the gas chamber by those who survived, Jacques Furmanski's is unusual for a number of reasons. He is an agent for the dead as well as for the living, giving a vivid account of the thin line separating the two and providing a voice for his doomed friend, who would otherwise remain silent and anonymous forever. He also evokes the dreadful serenity of despair that fell like a pall over the condemned as they faced the illogic of their fate with a helpless inertia that even the narrator can only convey without entirely understanding it.

In "Conversation with a Dead Man," from a collection of eyewitness reports on the persecution of Jews in the Third Reich, Furmanski's description of the paralysis that seized the victims who awaited such abnormal death brings us to the vestibule of insight: "They've lost their vitality; their existence crushes them, and the daily drama of death has destroyed in them every response." For the inmates of the deathcamps, such an end was never averted, only postponed; someone faced it nearly every day, so it was only a matter of time before your own turn came. In this relentless milieu, Camus's Sisyphean gesture of defiance would seem both futile and absurd. Deprived of a sense of individuality that in a "normal" universe, however indifferent to man, might lend dignity to such a gesture, the victim facing the gas chamber was left without a meaningful antagonist to defy. This terse narrative gives us a sense of the psychological effects of this situation and evokes a feeling of speechless melancholy that reaches us from the very edge of mortality.

Furmanski also offers a glimpse into a kind of daring that, because of the circumstances in which it unfolds, cannot be called heroic without falsifying the context of such action. Although it was possible to save a friend by disguising him as an Aryan or persuading the recorder to substitute a fake number for a genuine one, all this usually meant was a brief reprieve rather than a final rescue. Moreover, when the Germans discovered that the count was off, they would simply substitute another candidate for death to replace the one who had been salvaged. A cause to celebrate was thus always balanced by a reason to grieve. What consolation could there be, when pleading for the few who had connections implied ignoring all the others who had none?

Although it may not help us to *understand* what the final encounter between these two friends must have been like, testimony such as this—one is tempted to say *only* testimony such as this—compels us to try to *imagine* sympathetically that unimaginable moment. Any text that invites us to the brink of such a fearful endeavor without lapsing into maudlin and thus escapist language justifies its existence and merits our serious regard.

Conversation with a Dead Man

They're beginning this time in Barrack 10. It will be rough, they're taking a lot. . . .

I look around me. Faces are pale, appearances feverish: We can't speak any more. Even the bravest and most talkative have grown quiet. We wait. Who can describe and report on the atmosphere of this waiting, what's going on inside us? There they are, they're coming. We are all naked and lined up in ranks. The barrack elder counts, counts again, and when these bandits appear, we're nothing but "living dead men": They hold our lives in their hands. We look around us, and with this look we say to ourselves: "Whose turn is it? Which of us will be condemned to death?"

An order: "Attention! Stand at attention!" and the pernicious words resound in my ears: "How many Jews? All there?"

Like cattle for slaughter we're sorted out. They examine us from all sides. We have to turn around, and in the course of half a minute a man's fate is decided.

"Your number? Are you ill? How long in the camp?" The recorder notes down the number of those condemned. . . .

I am one of the first to get through. In order not to mix up those already "seen" with the others, they crowd us near the entrance. My heart beats rapidly, I am anxious about the fate of my friends.

One of them is so weak and thin from typhus that he doesn't dare to let himself be seen; otherwise he would certainly be selected for death. At the last moment I convince him to put on an "Aryan" jacket and to mix himself in with the twelve "Aryans" [non-Jews] who fortunately have remained in our barrack. The barrack elder helps us, and counts so well that the SS don't notice his absence. The minutes seem like years to me. I tremble for fear that the deception will be discovered. We all know the punishment: They will seize all of us, without distinction. The responsibility is terrible. At the moment I hadn't made clear to myself my reasons for such daring and such risk. A communal feeling, and the firm desire to save my friend had driven me to behave in that way.

The selection is over. We join each other again as if nothing had happened. We're called: The soup is ready. Those who have been condemned to death and we who have been saved are right next to each other, and surprising as it sounds, there is no revolt, no protest, no scene of despair. It's so inconceivable that I still don't understand it today, in spite of all the arguments that I try to find. A deathly silence surrounds me; those condemned to the gas chamber seek out their friends. I am depressed and lie down on my bed. After a few moments I too run to my friends and acquaintances to know whether I can do anything. I see the crowd by the office where the papers with the names of those selected are and where they are beginning to draw up the list of those to be killed. Everyone is pressing forward, to intervene, to plead. I see in front of me the head of my friend K., who risks his life and achieves the impossible: He crosses out some numbers from the list and enters false ones, to save a few more.

Whoever was not present at this process, whoever has not seen the distressed eyes of the condemned who await the results of a friend's intervention, knows nothing of hopelessness and nothing of genuine despair.

I stay in the barrack. They say that they will separate the "condemned" from us toward evening, and take them to the gas chamber tomorrow.

One of my friends, one of those "selected" to die, comes up to me and says: "It's all over for me. I don't have to suffer any more. Poor fellow, you have to go on suffering, and the result will be the same." I understood: He's consoling himself. I don't dare to say anything. He sits down next to me and remains quiet. "Give me something to eat." I give him something, and he says nothing more. A quarter of an hour passed. "I'm still hungry," he says; I look for some bread, and he goes on eating. I don't dare to say anything, although I would like to say so much, because this fellow has often repeated to me: "They'll never get me. They'll have to pay dearly for my skin—I'll know how to die." I wait for him to share his thoughts or plans with me, what he will do, something unusual that no one in the camp has ever tried before.

Nothing—not a single word. He understands, and senses what's going on inside me. He says to me: "You're surprised that I'm eating. I eat and say nothing." "Yes," I say, and wait nervously for him to continue.

"I'm thinking of death," he says, "but not as you think. I know that even criminals and bandits legally condemned to death are asked if they have a last wish. Usually, alcohol or cigarettes. But for us, nothing, nothing at all. I too would like to have one last pleasure before I die. The only thing left to me is to eat! To eat! To die full! I've suffered so much from hunger!"

Something terrible happens inside me. I had another view of death. I thought he would speak of his family, of vengeance, of greetings he wanted me to convey. Nothing . . . he is completely drained. All the words that he had been fond of using earlier had only been attempts to cheer himself up.

At the crucial moment nothing remained, and suddenly I understand why so many thousands before him marched to their death without objecting.

They've lost their vitality; their existence crushes them, and the daily drama of death has destroyed in them every response. What shall one say later to the wife or mother of a friend whose death one has to announce and above all how is one to answer that question which is asked so often: "Did he send any messages for me? Did he think of me?"

My friend stays at my side for a long time. We don't speak. At last he presses my hand, looks in my eyes and says: "Be good, old man, be brave."

I am silent, numb, faced with the most terrible dilemma of my life. I don't know what to answer. I feel an urgent need to tell him: "Defend yourself, show at least something, we'll work together!" But as he stands in front of me, I feel that he's already far away from us, that he's already gone and is not thinking of anything any more.

Silence loomed between us. I hugged him, and he left.

5. Charlotte Delbo

Charlotte Delbo was born on the outskirts of Paris in 1913. As a young woman, she worked as assistant to theater impresario Louis Jouvet, and was on tour in South America with his theatrical company when the Germans occupied her country in 1940. After learning that the Gestapo had executed a friend, she decided to return, and in November 1941, despite Jouvet's strong opposition, she made her way back to Paris (via Portugal and Spain and the unoccupied zone of France) to rejoin her husband, Georges Dudach, who was working with the Resistance.

In March 1942, French police arrested them in their apartment and turned them over to the Germans. Georges was executed by firing squad in May, after his wife was allowed a brief visit (an encounter she has written about movingly in several works). Delbo remained in prisons in France until January 1943, when she was sent with a convoy of 230 Frenchwomen to Auschwitz. Only 49 returned. She offers an account of this journey in *Le Convoi du 24 janvier* (1965), together with brief biographies of all but one of her fellow deportees, and some interesting statistical appendices on the relationship between survival and age, profession, education, and political affiliation.

Delbo remained in Auschwitz and a satellite camp called Raisko until January 1944, when she was sent with a small contingent of her compatriots to Ravensbrück. Near the end of the war, she was released to the Red Cross, which moved her to Sweden to recuperate. Although she has also written plays and essays, Delbo's masterwork is the trilogy *Auschwitz et après (Auschwitz and Afterward),* of which only the first volume, *None of Us Will Return* (written in 1946, but not published until 1965), has appeared in English.*

The stylistic originality and visual intensity of Delbo's trilogy are well illustrated by the following selection from her last work, *La Mèmoire et les jours (Days and Memory),* which was published in 1985, the year of Delbo's death. Also notable is her subtle assembly of a variety of voices, intricately linked and differentiated as she shifts from her own point of

*Rosette Lamont is translating all three volumes for Yale University Press.

view to narrative stances depicting other genders, ages, religions, and family ties. The result is a medley of memories designed to wreck our penchant for universalizing the survivor and to force us instead to see the Auschwitz experience as a miscellany of personal ordeals that cannot be reduced to a common nucleus.

Voices

I

Explaining the inexplicable. There comes to mind the image of a snake shedding its old skin, emerging from beneath it in a fresh, glistening one. In Auschwitz I took leave of my skin—it had a bad smell, that skin—worn from all the blows it had received, and found myself in another, beautiful and clean, although with me the molting was not as rapid as the snake's. Along with the old skin went the visible traces of Auschwitz: the leaden stare out of sunken eyes, the tottering gait, the frightened gestures. With the new skin returned the gestures belonging to an earlier life: the using of a toothbrush, of toilet paper, of a handkerchief, of a knife and fork, eating food calmly, saying hello to people upon entering a room, closing the door, standing up straight, speaking, later on smiling with my lips and, still later, smiling both at once with my lips and my eyes. Rediscovering odors, flavors, the smell of rain. In Birkenau, rain heightened the odor of diarrhea. It is the most fetid odor I know. In Birkenau, the rain came down upon the camp, upon us, laden with soot from the crematoriums, and with the odor of burning flesh. We were steeped in it.

It took a few years for the new skin to fully form, to consolidate. Rid of its old skin, it's still the same snake. I'm the same too, apparently. However . . .

How does one rid oneself of something buried far within: memory and the skin of memory. It clings to me yet. Memory's skin has hardened, it allows nothing to filter out of what it retains, and I have no control over it. I don't feel it anymore.

In the camp one could never pretend, never take refuge in the imagination. I remember Yvonne Picart, a morning when we were carrying bricks from a wrecker's depot. We carried two bricks at a time, from one pile to another pile. We were walking side by side, our bricks hugged to our chests, bricks we had pried from a pile covered with ice, scraping our hands. Those bricks were heavy, and got heavier as the day wore on. Our hands were blue from cold, our lips cracked. Yvonne said to me: "Why can't I imagine I'm on the Boulevard Saint-Michel, walking to class with an armful of books?" and she propped the two bricks inside her forearm, holding them as students do books. "It's impossible. One can't imagine either being somebody else or being somewhere else."

I too, I often tried to imagine I was somewhere else. I tried to visualize myself as someone else, as when in a theatrical role you become another person. It didn't work.

In Auschwitz reality was so overwhelming, the suffering, the fatigue, the cold so extreme, that we had no energy left for this type of pretending. When I would recite a poem, when I would tell the comrades beside me what a novel or a play was about while we went on digging in the muck of the swamp, it was to keep myself alive, to preserve my memory, to remain me, to make sure of it. Never did that succeed in nullifying the moment I was living through, not for an instant. To think, to remember was a great victory over the horror, but it never lessened it. Reality was right there, killing. There was no possible getting away from it.

How did I manage to extricate myself from it when I returned? What did I do so as to be alive today? People often ask me that question, to which I continue to look for an answer, and still find none.

Auschwitz is so deeply etched in my memory that I cannot forget one moment of it. —So you are living with Auschwitz? —No, I live next to it. Auschwitz is there, unalterable, precise, but enveloped in the skin of memory, an impermeable skin that isolates it from my present self. Unlike the snake's skin, the skin of memory does not renew itself. Oh, it may harden further . . . Alas, I often fear lest it grow thin, crack, and the camp get hold of me again. Thinking about it makes me tremble with apprehension. They claim the dying see their whole life pass before their eyes . . .

In this underlying memory sensations remain intact. No doubt, I am very fortunate in not recognizing myself in the self that was in Auschwitz. To return from there was so improbable that it seems to me I was never there at all. Unlike those whose life came to a halt as they crossed the threshold of return, who since that time survive as ghosts, I feel that the one who was in the camp is not me, is not the person who is here, facing you. No, it is all too incredible. And everything that happened to that other, the Auschwitz one, now has no bearing upon me, does not concern me, so separate from one another are this deep-lying memory and ordinary memory. I live within a twofold being. The Auschwitz double doesn't bother me, doesn't interfere with my life. As though it weren't I at all. Without this split I would not have been able to revive.

The skin enfolding the memory of Auschwitz is tough. Even so it gives way at times, revealing all it contains. Over dreams the conscious will has no power. And in those dreams I see myself, yes, my own self such as I know I was: hardly able to stand on my feet, my throat tight, my heart beating wildly, frozen to the marrow, filthy, skin and bones; the suffering I feel is so unbearable, so identical to the pain endured there, that I feel it physically, I feel it throughout my whole body which becomes a mass of suffering; and I feel death fasten on me, I feel that I am dying. Luckily, in my agony I cry out. My cry wakes me and I emerge from the nightmare, drained. It takes days for everything to get back to normal, for everything to get shoved back inside memory, and for the skin of memory to mend again. I become myself again, the person you know, who can talk to you about Auschwitz without exhibiting or registering any anxiety or emotion.

Because when I talk to you about Auschwitz, it is not from deep memory my words issue. They come from external memory, if I may put it that way, from intellectual memory, the memory connected with thinking processes. Deep memory preserves sensations, physical imprints. It is the memory of the senses. For it isn't words that are swollen with emotional charge. Otherwise, someone who has been tortured by thirst for weeks on end could never again say "I'm thirsty. How about a cup of tea." This word has also split in two. *Thirst* has turned back into a word for commonplace use. But if I dream of the thirst I suffered in Birkenau, I once again see the person I was, haggard, halfway crazed, near to collapse; I physically feel that real thirst and it is an atrocious nightmare. If, however, you'd like me to talk to you about it . . .

This is why I say today that while knowing perfectly well that it corresponds to the facts, I no longer know if it is real.

II

She says: "One doesn't die from grief." In her smooth, colorless voice, smooth as her evenly aged face is smooth, as colorless as her eyes where you occasionally find a glint of their former blue.

"No, it simply isn't so. One doesn't die from grief." In a lower voice she repeats: "You go on living." You go on living, yes. It's worse. She lives in her grief, lives with her sorrow, that unaltering double of herself. She bears her grief ever since she bore in her arms that sister of hers, who died in the night. The night of all nights whence those who have returned have not issued forth. She held her dying sister in her arms, hugged her to herself in order to keep her back, prevent her from slipping out of life. Softly she blew her breath upon her sister's face to warm the lips that were turning blue, to impart her own breath to them, and when her sister's heart stopped, she was filled with anger at her own which continued to beat. Which yet beats today, after all these years she has spent on the borders of life. And when she says that you don't die from grief, she's apologizing for being alive. Barely alive.

Ever since she carried her sister out into the snow for the corpse collection squad to pick up and dump on the pile, one more for the night's pile of bodies, ever since she had loosened her grip and let go her sister's still faintly warm body that she had clasped, holding it as tight as she could, sapped as she was from several weeks of camp, the desperate struggle lasting all night—; ever since that morning she knows that you don't die from grief.

Would she be dead had the others not held her upright and got her to roll-call, kept her from falling into the ditch between the barracks and the mustering yard, kept her from fainting, held her on her feet till the end of the roll-call, helped her walk upon the icy, road and reach the marsh, would she be dead? No. That night she had drawn in her sister's last breath, inside her she bore her sister, alive from now on through her alone.

Like all survivors, she wonders how she returned, and why. The others feel they fought with superhuman determination in order to come back; she doesn't know. She just came back, that's all. Within her heart the whole weight of that night, her inability to share her living breath with her sister, and then the weight of the girl she carried out of the barracks and laid upon the snow, delicately, maternally, a kind of burial, a . . . sacrament of tenderness, before that body became an object to be burned that they shovel onto the pile of last night's dead who will be burned in the course of the day or who, if today there are too many of them, will wait until tomorrow, exposed to the rats.

She came back home, but not back to life. Life flowed over her the way a stream's water flows over the stones it polishes, it wore her away, day by day. Her gaze faded, her voice lost its color, her hair grew gray. How many years now? She counted them, but it's not the right count. Auschwitz was yesterday. That night was last night.

For all these years she has done little things, gone through the little motions of everyday life, she listens to the sounds of the life moving by around her. She hears only the wind blowing across the icy plain, the shouts of the female guards overseeing the prisoners in the frozen marshes, the barking of the dogs. She smells only the smell of the crematorium. She hears the voices of her friends who tore her away from her dead sister's body: "Come on! Come on! We've got to get to roll-call" and who dragged her off, propped her up in the ranks, told her to cry, but she hadn't been able to cry, neither that morning nor since. For lack of tears her gaze has dulled.

III

What is she holding in her arms
hugged to her breast
that one
in the front row
there, in the row facing ours
yes that one in the front row.

The ranks facing ours
are still Gypsies.
Yes, the Gypsies.
How do you know it's a Gypsy when all that's left of it is a
skeleton?
Since the middle of the night they have been standing
over there in the snow that thousands of feet have trampled
into hardened slippery sheets
Since the middle of the night
we've been here standing in the snow,
standing in the night
the night broken by the spaced floodlights
on the barbed wire fences

IV

The projectors light the barbed wire strung between high white poles. Encircled by light, the camp lies in darkness and in this black abyss nothing can be distinguished

nothing except darker shapes swaying
ghostlike upon the ice.

The roll-call siren has emptied the barracks. By swaying clusters, the women have all stumbled out, clinging to each other so as not to fall

And when one does fall, the whole cluster reels and falls and gets back up, falls again and rises, and in spite of it all moves on.

Without a word.

There is only the screaming of the furies who want the barracks to empty faster, want the reeling shades to move faster from the barracks to the space where the roll is called.

In the darkness, for the beams of the projectors do not reach the spaces between the barracks. They light only the gate and the barbed wire enclosure so that the sentinels up in the watchtowers may spot those trying to escape and shoot

as if one could escape
as if one could cut through the fence of high-tension live barbed wire
as if . . .

In the dark you cannot see where you step, you fall into holes, stumble into drifts of snow.

Clutching one another, guided by shouts and blows of clubs, the shades of the night take the places where they must be to await the break of day.

Panic sometimes. Where are you? I'm right here.

Hold on to me, or a voice full of despair: My galoshes. I've lost my galoshes. It's from a woman who slipped, got up, but without her galoshes, flown off who knows in what direction, and the whole group stops, stoops over the snow, gropes unseeingly. The galoshes must be found.

Barefoot at roll-call is certain death. Barefoot in the snow for hours—death.

They're all hunting for the galoshes and others behind them grow impatient and shove because the fury, the barracks leader, in her boots and steady on her feet, comes down on them with her club, screaming as she swings at everyone within reach. In the dark she can't see where she's hitting, the club always strikes someone, lands on those shapes that are squatting and groping and straightening up and, beneath the blows, there fall again those who had succeeded in rising to their feet.

Here, I've got one of them. I've got the other. The galoshes pass from hand to hand until they reach the one they belong to, who murmurs her thanks but is not heard, so exhausted is she by fear, by falls, clubbing, and the hallucinations of the night.

The group sets off again, a chaotic procession. Tortuous wending of its way between obstacles: piles of brick covered with frozen snow, piles of snow, holes full of water turned into ice

a ditch to step over

and the frozen earth, bristling and jagged, like a plowed field petrified into clods of ice.

At last they're all in place, lined up for the roll-call, on each side of the road running down the middle of the camp.

They shall have to wait hours and hours before daybreak, before the counting. The SS do not arrive until it is light.

First you stamp your feet. The cold pierces to the marrow. You no longer feel your body, you no longer feel anything of your self. First you stamp your feet. But it's tiring to stamp your feet. Then you huddle over, arms crossed over your chest, shoulders hunched, and all squeeze close to one another

but keeping in rank because the club-wielding furies are there and watching.

Even the strength to raise one's eyes, to look to see if there are stars in the sky

there's a chilling effect to stars

to cast a glance about

even the strength that takes must be saved and no one looks up.

And to see what in that darkness?

The women from the other barracks, also reeling and falling, trying to form their ranks?

all these ranks stretching from one end of the camp to the other, on each side of the road,

that makes how many women, how many thousands of women, all these ranks?

these ranks bobbing up and down because the women are stamping their feet and then they halt because to stamp one's feet is exhausting

You see nothing, each one is enclosed in the shroud of her own skin,

you feel nothing, neither the person next to you, huddling against you, nor that other who has fallen and is being helped up.

You don't speak because the cold would freeze your saliva.

Each feels she is dying, crumbling into confused images, dead to herself already, without a past, any reality, without anything,

the sky must have grown light without anyone noticing.

And now, in the pallid light of the night drawing to a close,

the ranks across the way suddenly emerge, the ranks of the Gypsy women

like ourselves all blue from cold.

How would you know a Gypsy if not by her tattered dress? The Jewish women do not have striped uniforms either, they have grotesque clothing, coats too long or too tight, mudspattered, torn, with a huge red cross painted on the back.

The Gypsy women have tatters, what's left of their full skirts and their scarves.

And suddenly there she is, you can make her out, the one in the front row, holding, clutched to her breast, a bundle of rags.

In her gaunt face, eyes gleaming so bright that you must look away not to be pierced by them

her eyes gleaming with fever, with hatred, a burning, unbearable hatred.

And what else but hatred is holding together these rags this spectre of a woman is made of, with her bundle pressed against her chest by hands purple from cold?

She holds the bundle of rags to her, in the crook of her arm, the way a baby is held, the baby's head against its mother's breast.

Daylight.

The Gypsy stands straight, so tense that it is visible through her tatters, her left hand placed upon the baby's face. It is an infant, that bundle of rags she is clutching. It became obvious when she shifted the upper part of the bundle, turning it outward a little, to help it breathe perhaps, now that daylight has come.

Quickly she shelters the baby's face again and hugs it tighter then she shifts the bundle of rags to her other arm, and we see the infant's head lolling, bluish, almost black.

With a gentle movement she raises the baby's head, props it in the hollow between her arm and her breast,

and again she lifts her eyes, and again the impression she gives is of tension and fierceness, with her unbearable stare.

The SS arrive. All the women stiffen as they move down the ranks, counting. That lasts a long time. A long time. Finally, one side is done. You can put your hands back in the sleeves of your jacket, you can hunch up your shoulders, as if it were possible to make yourself a smaller target for the cold.

I look at the Gypsy holding her baby pressed against her. It's dead, isn't it?

Yes, it's dead. Its purplish head, almost black, falls back when not supported by the Gypsy's hand.

For how long has it been dead, cradled in its mother's arms, this rag-swaddled infant? For hours, perhaps for days.

The SS move past, counting the ranks of the Gypsies. They do not see the woman with the dead baby and the frightening eyes.

A whistle blows. The roll-call is over. We break formation. Again we slide and fall on the sheet of ice, now spotted here and there with diarrhea.

The Gypsies' formation breaks up too. The woman with the baby runs off. Where is she heading for shelter?

The Gypsies are not marched out of the camp for work. Men, women, children are mixed together in a separate enclosed area. The camp for families. And why are there Gypsy women over here, in our camp? Nobody knows.

When the roll was called that evening she was there, with her dead baby in her arms. Standing in the front row. Standing straight.

The following morning at roll-call she was there, hugging her bundle of rags, her eyes still brighter, still wilder.

Then she stopped coming to roll-call.

Someone saw the bundle of rags, the dead baby, on the garbage heap by the kitchen.

The Gypsy had been clubbed to death by a policewoman who'd tried to pull the dead baby away from her.

This woman, hugging her baby to her, had fought, butting her head, kicking, protecting herself and then striking with her free hand . . . a struggle in which she had been crushed despite the hate that gave her the strength of a lioness defending her brood.

The Gypsy had fallen dead in the snow. The corpse collection squad had picked up her body and carried it to where the corpses are stacked before being loaded on the truck which dumps them at the crematorium.

The mother killed, the policewoman had torn the baby from her arms and tossed it on the garbage heap in front of which the struggle had taken place. The Gypsy woman had raced to the edge of the camp, tightly cradling the baby in her crossed arms, had run till she was out of breath and it was when she was blocked by the garbage pile that she turned to face the fury and her club.

The corpse collection squad picked up the mother. The baby, in its rags, remained on the garbage heap, mixed with the refuse.

All the Gypsies disappeared very fast. All gassed. Thousands of them. The family camp was emptied out, that made room for the next arrivals. Not Gypsies. We saw nothing more of Gypsies at Birkenau. Gypsies are less numerous than Jews, it didn't take much time to dispose of them.

V

My mother, the stars

The whole while that you were there
that I didn't know where you were
I never closed my bedroom shutters
at night
I never drew the curtains.
From my bed
I would look at a star,
always the same star.
The minute it appeared
I'd recognize it.
I kept thinking
Charlotte too is looking at the sky
She too sees that star.
Wherever she is, she sees it.
She knows that I think of her
that I'm thinking of her every minute
every second
I didn't want to go to sleep
for fear my thoughts toward you would slumber.
I'd fall asleep in the morning
when daylight erased my star.
On the nights the sky was covered
I'd follow the shifting clouds
so as not to miss my star
when it emerged
from a break in the clouds.
To see it
just for a minute
to see it.
It told me you were alive.
I didn't like starless nights
nor full moon nights
when glowing moonlight devoured the stars.
It seemed to me I was losing sight of you.

Now that you are back
I'll close the shutters.
They must be very rusty.

During all that while, that long while

My mother never again talked to me about the camp, never
asked me anything about Auschwitz.

VI

What killed me was to find out my mother had not been gassed.

For everybody, gassing was the worst—a terrifying image, scenes of horror. Not for me. I don't know how to put it to you. I mean I feel it differently.

When I was arrested I didn't feel afraid. My mother had been deported six months before. We knew she had been sent off. She had tossed a note from the train, we got it. I hadn't had any further news from her, of course. I spent only three days in Drancy.* Then right away the cattle-car. Destination unknown. Who had even heard the name of Auschwitz? With me, in the packed cattle-car, women and children. The children yelled or moaned. The women cried silently. Three nights, three days without a drop of water to wash the babies, nothing for the bottles, dry bread for the bigger ones. The women cried quietly, without stirring, keeping a tender, reassuring hand on their children. They had no idea what would become of them, of the little ones. They were frightened. I wasn't. I took it easy, I was almost happy: I was going to see Mama again. It didn't occur to me that the convoys leaving Drancy might be sent to several places. I assumed they were putting all the Jews in the same place. Mama was there already, I was going to join her. It was as simple as that to my girlish mind. The funny part was that I was right. From France, the Jews were dispatched to Auschwitz. They all got on my nerves, those women who moaned or else sat in a daze. The men were in another car. At every stop everyone shouted family or first names to find out how the others were doing. I thought to myself that judging from the way they treated us from the start, we obviously couldn't expect anything good to happen. Well, so what? We'll see by and by. The vitality you have when you're fifteen!

I held on to the food I had been given for the trip. Good stuff from the country: honey, paté, packed by my nurse. I kept it all for Mama.

Getting down out of the train, on the ballast, shouts, dogs, boots, guns. You remember that too, but when I got there it wasn't the middle of winter, it wasn't so awful. Besides, my heart was so full of joy. To see Mama again. Together we'd overcome any hardship.

The journey hadn't tired me that much. I had just turned fifteen, strong as a little ox. I had the apple cheeks of the country girl I had become during the two years I'd been with the farm woman my parents had entrusted me to for safekeeping. Alert, eyes open, I observed, I watched—as if I'd had eyes on all sides of my head. Nothing escaped me and I ducked the blows almost automatically. At the sight of the dogs the others backed off right into the clubs; as for me, I slipped off to the side. I must say that I wasn't weighted down by anything. The others had their kids. All I had was my small bundle.

When they ordered the women with children and the old people to

*Transit camp near Paris from which deportees were sent to Auschwitz.

line up on one side—I understood because German is a lot like Yiddish—I said to myself—and it came to me like that, quick as lightning—"Oh no, I'm not going to line up with these whining women and their kids. Not with the old folks either. And end up in some kind of nursing home, emptying chamberpots—not me!" When they called the bigger children I didn't budge either. I stuck out my chest, already round, and stood up tall. If I wanted to be with Mama I had to line up with the women. That's how I made it to Birkenau. There, of course, it was really something. The corpses, the stench, the rot. Still, there were women in striped dresses going this way and that. So, those who had entered the camp had not all died. My mother would be among them. I refused to see any of the horror. I kept my eyes wide open, I stared at every woman prisoner who went by. Even dressed that way, even with her head shaved, even limping, I'd recognize Mama if I saw her.

They put us into an already overcrowded barracks. Except for our small group there was no one French. Slovaks, Hungarians. I questioned everyone I could. Nobody had seen Mama. In six months the camp population had turned over completely, as you know. But Mama was strong. With all these barracks, how was I to find her? I'd have to go everywhere, and so as soon as I could, that evening after the roll-call, I ran about from one barracks to the next. They didn't even answer me. A tall blond woman, with beautiful gray eyes. As if there were still blonds and gray eyes in any such place as that.

It wasn't until at least two weeks after we got there that I heard talk about gas chambers. I didn't believe it, of course. Famished, worn out, in bad shape as they all were, these women weren't in their right minds. Who can believe such a horror story? At the same time, I wondered where all those who had lined up along the tracks had gone to. I tried reasonable answers to my questions: the aged were in a nursing home, the babes in arms and their mothers in a . . . in a what? And the bigger children, the ones under fourteen or fifteen who had been called to one side? To go on from there to supposing they'd been asphyxiated in a gas chamber . . . Hundreds of them? Impossible. Finally I had to face the truth. A couple of hundred people were plainly dying in the camp every day, but the smoke constantly coming from those enormous smokestacks, the odor of burned flesh . . . But these trains we saw arrive in the morning as we were starting out to work; and yet on certain days not a single new face in the camp. The work column would halt and mark time while they formed into ranks. These thousands of people . . . who arrived every day, without the camp population getting any larger. You could check it by the tattooed numbers. And those lines waiting in front of a little door, that you never saw come out again . . . One day we'd been working close. They'd gone in all day long. Women patting the little ones to keep them quiet, the bigger ones well behaved, with serious faces like children who don't want to be treated like babies. Old people, barely able to keep on their feet, and others too, not all that old, as I think of them today. The door opened only when the line was long enough to make a complete batch. Just the way you wait to have enough dirty clothes to make a load for the machine. Showers? You don't take a shower packed together as in the subway at rush hour. So when a Czech woman who worked with the clothes—you know, the team that sorted out and stacked what the people took off in order to go naked into the gas chamber—, when that Czech woman explained the system to me, I told myself that my mother, maybe because she was carrying a baby for someone who had several of them, or perhaps because she was helping some old lady having trouble walking, I told myself that my mother had got herself into the wrong line and that she'd been gassed.

You won't believe it, but I felt relieved. By then I had become acquainted with life and death in the camp. I was filthy, in rags, covered with lice, a stinking scarecrow. I was at the end of my rope. At least Mama wouldn't have had to endure that. She'd been spared this degradation.

Following my return I adjusted myself to that idea, a comforting one all told. I see you don't understand. For you the worst of all is the gas chamber, isn't it?

Well, recently . . . You know they've found the lists of those in all the shipments that left from Drancy, and they even know what happened to those shipments. Well, I recently learned that for the convoy my mother was in—her name is there on the list—there was no selection upon arrival, no special line for the young and the old. An exceptional case, you can call it that. The whole shipment went into the camp. Perhaps their stock of gas had run out.

She had tried to make a joke. There was a pause in her voice as she repressed a sob.

"Your mother died almost forty years ago. Why do you torture yourself? Gassed or not gassed—what difference does that make?"

I had spoken too quickly. I recall that in my prison cell, where we were several women with husbands who were or who would be condemned to death, one of us said, "Me, I would prefer the guillotine. It's cleaner, quicker. I'm sure you suffer less." And I had a picture of the severed head, its eyes rolled back. I shivered with horror and my heart climbed up into my mouth. Beheaded, shot—these young men not yet in their thirties. And even so, I thought, the guillotine is worse than bullets in the chest. My husband was shot.

She went on. You don't understand, she said, you don't understand that that changes everything. When I was positive that my mother had been gassed without entering the camp, I was less unhappy. I could put up with everything, telling myself that Mama had not witnessed this, had not undergone this. She died without realizing what this place was. She'd been told to undress, and five minutes later it was over with. She entered the gas chamber the way we went into the showers when we arrived. You remember. When they told us to get undressed we got undressed without suspecting anything wrong. All I was worried about was not losing my little package of food. I had to turn it in along with my clothing. Oh, how that hurt! When they ordered us into the showers we went in there without seeing anything odd about it. It was a real shower. Fine. So, for Mama . . . for Mama it was gas; but she didn't know it. That's what I said to myself and I felt almost reassured.

But no, she wasn't that lucky. Today, the full horror of Birkenau comes back to me through my mother. I'd forgotten. No, that's not the right word. You can't forget anything. I mean I had just stopped thinking about it. No, I was able to think about it without it hurting. Sometimes a nightmare, otherwise nothing painful.

When I returned I was seventeen. My mother had died without having had time to suffer, believing I was safe. I wanted to live. Life is strong in you at seventeen. And now, the truth.

I wonder now how she died. And when. How long she suffered. I've tried without success to locate women in her convoy. Not one survivor. I would have seen them in the camp had they held on until my arrival. If you knew how I searched for her there! Everything pointed to my mother having been gassed with her whole convoy. Since that isn't so, I wonder how long Mama held on. Perhaps she died on the eve of my arrival. How long did she suffer? She wasn't forty—but for Birkenau that's old. But she was so brave. And strengthened by the longing to see her children again, above all me, whom she had had to leave so young. I am the last-born. I am sure she fought to the last drop of her strength. And you know how far that last drop could take those who wanted to get back. Today, it all comes back to me in visions of Mama in its midst. Is it my mother, that woman a dog drags for thirty feet over a stony road and then leaves there, her throat torn open, moaning, then ceasing to moan? Or this other one, unable to get up under the beating she is receiving? The SS go on bludgeoning an already lifeless rag. Or is she one of these skeleton-like typhus cases in a group being removed from the infirmary to Block 25? Or one of the corpses piling up in the mud? And if she died in the infirmary, stark naked, covered with lice, her skin scraped by the rough boards of the bunks, with the rats busy all around? Mama . . . She dragged herself to the swamp, exhausted by dysentery, and nasty, stinking, she who used to be so fastidious. There is no end to all those atrocious images. I don't need to remind you of all the ways there were to die in Birkenau, every one of them atrocious. The one thing I am sure of is that she didn't take her own life, didn't end it by throwing herself onto the electrified barbed wire. She was far too tenacious for that.

Now I'll never know. Indeed, even if I came across someone who was with her, I wouldn't be told the truth. Have you told the truth to the kin of those who died back there? Without passing the word along, all the survivors have said the same thing: he or she died of typhus. In a coma, without suffering. So as not to cause pain to the members of the family, at least so as not to worsen it.

No, I won't ever know.

"Come, you mustn't go on torturing yourself. What does it matter today? Time has healed the wounds."

And it's you who say such a thing? You? You don't understand. I lived with the idea my mother had died without suffering. Sure, I grieved for her. It's always terrible to lose one's mother. That pain, yes, time had lessened it. But today the wound opens again and it hurts all the more because the scar had hardened.

VII

No, I'm not sorry I was sent there. I'm even rather glad I was. That seems odd to you? You're going to understand. No, I'm not the guy who comes back proud of having lived through something unusual. There's none of that stuff between you and me.

You know that I was arrested a first time along with my mother. I haven't told you about that? It was one of the first big roundups, in '42. One day there's a rumor going around: a roundup of Jews, for tomorrow or the day after. Where did the rumor originate? Perhaps from central headquarters of the Paris police. I'm just trying to save its honor . . . In our house we never altogether believed it. We weren't living in the Jewish neighborhood: rue des Rosiers, rue des Ecouffes, the Temple. We lived near the Bastille. My mother had persuaded my father to leave the house and hide out somewhere at least long enough for things to blow over. He left without saying where he was going. For at the beginning it was thought they only took the men. Women and children weren't supposed to be in any danger. I stayed with my mother, listening, on the look out. Nothing. And the following day the cops came to pick us up. Who had tipped them off? Even today I've no idea. My mother got some things together, pretty much haphazardly, and put them in a big shopping bag. She wanted to empty the larder. "That'll do it, let's get going!" A bus, an ordinary green bus, public transportation, with its open back platform was waiting down there, already crowded. We found our way to some vacant seats inside. Nobody looked at anybody else, each pretending he was there through some mistake. There were old people, young people, women, children. All aboard! We drive through Paris and wind up in front of the Vel d'Hiv.* Other buses crammed with people were arriving at the same time. You had to wait in line to get in. We got in line, my mother and I. I didn't fuss—I was fourteen and acting like a little man. My mother was calm, didn't say a word. I didn't notice that she was watching everything going on around us. Suddenly, without turning my way, she whispered: "Get out of here. Take this"—a light wool vest she had over her arm, something she'd grabbed at the last minute: you know, in the middle of the summer, she'd thought of something for the cool evenings—"hold it in front of you to hide your star. Go along quietly. Don't look back, don't run. Goodbye. Now go!" I slipped out of the line,

*Velodrome in Paris where Jews were assembled before being deported.

just where my mother had noticed a gap between the cops. I tried to look natural. After a couple of blocks I couldn't prevent myself from running. Then I asked myself where I was running to. I slowed down. My mother had told me to go along quietly. Go where? Not home, obviously. Not to the homes of relatives or Jewish friends. It seemed to me all of a sudden that I didn't know anyone. I walked through the streets. As evening was falling I knocked on the door of a school friend—it was the only place that finally occurred to me. His parents—they weren't Jews—invited me right in. I told them what had happened. They fed me. I was dying of hunger. They fixed a cot for me in my friend's room.

The next day I tore off my star. I didn't go back to school. I looked for my father. Meanwhile he had been looking for us, but it took over a week before we found one another. Thank heavens. I was beginning to feel embarrassed about taking advantage of my friend's hospitality. At fourteen I had done enough standing in line in front of empty shops to know what it meant to have an extra mouth to feed. My father, who had thought things over since leaving us to go into hiding, had a clear view of the situation. "What we're not going to do," he said to me, "is hide like scared rabbits waiting for all of us to be caught before doing something. If we've got to hide, very well, let it be for a purpose."

I went by my friend's house to thank his parents, especially his mother who'd been so warm to me. With my friend I took on a knowing and mysterious air, hinting that I was joining an organization.

I hid out with my father. It wasn't long, though, before he made the contacts that would get us across the demarcation line, and we joined up with a resistance group that was then forming. There it was easier. The guys had everything set up for ration cards, forged papers, hideouts. At first it was propagandizing in town. Then a *maquis* took shape: sabotage, attacks on German soldiers, radio liaison with London. On the whole it worked well. And I had the impression we were going to pull through, that the end of the war was approaching—and then I was captured with several others in my group. That was in the Spring of '44. The Gestapo identified me in no time. It was the classic itinerary: Drancy, Auschwitz. No, I was forgetting: the guys who were captured with me, being adults, were shot. I was sixteen years old.

At Auschwitz the system had reached perfection. The gas chambers were running round the clock. Naturally, we—the passengers in my cattle-car—knew nothing of their existence and the selection upon our arrival didn't frighten us. Moreover, nothing frightened me anymore. In the Spring of '44 Hitler had lost the war. It was only a question of a few months.

Big for my age, I was put into the column of able-bodied men which marched into the camp. The others were marched to the showers—as I found out a few weeks afterwards. You're not someone who needs to be told how it was, what I endured—the standard experience.

In the men's camp it was difficult if not impossible to communicate with Birkenau, the women's camp. My mother must have been there at some time. When we passed the women's columns, on the way to the marshes, the idea that my mother could have held out for two years simply struck me as ridiculous. She was surely already dead. I never managed to find out anything at all about her.

"And later, when you got home?"

"I didn't expect my mother to come back. No, I had no hope, not even a subconscious one, of her return. To hold out in Birkenau from July of '42 until '45—it would have been a miracle. We were terrible realists in my family, you remember. I don't know how my mother died. I left it at that."

"And your father?"

"He came through the whole business. Talk about luck! Incredible."

"He didn't try to find out, like just about everyone else?"

"No. I told him what the situation had been, as briefly as possible. We didn't much enjoy talking when we got back. I think he understood right away. He hasn't inquired."

"Even now with what they know about the destinations of the convoys?"

"When I said that I was rather glad to have been there, it's—how should I put it? Anyhow, you surely know what I mean, you of all people. For me, my mother didn't just vanish into some black hole, she wasn't just gobbled up into nothingness, some unimaginable place you can piece together from the accounts of survivors. I know what my mother went through, what she saw, what she suffered, and it's something I feel I shared with her."

6. Christopher R. Browning

Christopher R. Browning is a professor of history at Pacific Lutheran University in Tacoma, Washington. His most recent work is *The Path to Genocide: Essays on Launching the Final Solution* (1993). He is also the author of *The Final Solution and the German Foreign Office* (1978), *Fateful Months: Essays on the Emergence of the Final Solution* (1985), and *Ordinary Men: Reserve Police Battalion 101 and the Final Solution in Poland* (1992), a much-expanded account of the work of the German police battalion whose exploits in a Polish village in the summer of 1942 is described in the following essay, "One Day in Józefów: Initiation to Mass Murder."

Browning helps put to rest the myth that the killers were driven to their deeds by fear of punishment, though after reading his analysis we may feel no closer to understanding how ordinary men, many of them husbands and fathers themselves, were able to engage in the routine and often daily slaughter in cold blood of thousands of helpless and innocent men, women, and children. Instead of finding their grisly work more and more oppressive, they seem merely to have grown used to it as part of their "job." Since most of them were not members of the SS or even the Nazi party, ideology did not motivate their behavior. Readers interested in pursuing this question might examine the testimony to be found in *"The Good Old Days": The Holocaust as Seen by Its Perpetrators and Bystanders.* (1991).

Some reviewers have criticized Browning for relying solely on German documents in his research and ignoring the testimony of non-Jews who lived in the vicinity of Józefów, and of the survivors among the young male Jews from the village who were sent to a labor camp while members of their families were taken to the woods and shot. Historians differ about the reliability and value of such testimony, since it is based on memory; but much of the documentary evidence that Browning uses from trials held decades after the war must also have been based on the witnesses' memories and hence is vulnerable to the same charges.

As Charlotte Delbo insisted, how we remember what we remember about the Holocaust involves not only the truth of the event, but also its impact on the subsequent life and imagination of the witness, whether victim or perpetrator. The variety of sources represented in this anthology should help the reader to make an independent assessment of this issue. As for understanding the behavior of the murderers—Browning's study reminds us how elusive and complex that question remains, despite decades of research and commentary on the dilemma.

One Day in Józefów: Initiation to Mass Murder

In mid-March of 1942, some 75 to 80 percent of all victims of the Holocaust were still alive, while some 20 to 25 percent had already perished. A mere eleven months later, in mid-February 1943, the situation was exactly the reverse. Some 75 to 80 percent of all Holocaust victims were already dead, and a mere 20 to 25 percent still clung to a precarious existence. At the core of the Holocaust was an intense eleven-month wave of mass murder. The center of gravity of this mass murder was Poland, where in March 1942, despite two and a half years of terrible hardship, deprivation, and persecution, every major Jewish community was still intact; eleven months later, only remnants of Polish Jewry survived in a few rump ghettos and labor camps. In short, the German attack on the Polish ghettos was not a gradual or incremental program stretched over a long period of time, but a veritable blitzkrieg, a massive offensive requiring the mobilization of large numbers of shock troops at the very period when the German war effort in Russia hung in the balance.

The first question I would like to pose, therefore, is what were the manpower sources the Germans tapped for their assault on Polish Jewry? Since the personnel of the death camps was quite minimal, the real question quite simply is who were the ghetto-clearers? On close examination one discovers that the Nazi regime diverted almost nothing in terms of real military resources for this offensive against the ghettos. The local German authorities in Poland, above all SS and Police Leader (SSPF) Odilo Globocnik, were given the task but not the men to carry it out. They had to improvise by creating ad hoc "private armies." Coordination and guidance of the ghetto-clearing was provided by the staffs of the SSPF and commander of the security police in each district in Poland. Security police and gendarmerie in the branch offices in each district provided local expertise. But the bulk of the manpower had to be recruited from two sources. The first source was the Ukrainians, Lithuanians, and Latvians recruited out of the prisoner of war camps and trained at the SS camp in Trawniki. A few hundred of these men, among them Ivan Demjanjuk,*

*John Demjanjuk was extradited to Israel from the United States and charged with having been the Ivan the Terrible who had run the gas chambers at Treblinka. He was convicted and sentenced to death. On the basis of new evidence, the Israeli Supreme Court reversed the conviction.

were then sent to the death camps of Operation Reinhard, where they outnumbered the German staff roughly 4 to 1. The majority, however, were organized into mobile units and became itinerant ghetto-clearers, traveling out from Trawniki to one ghetto after another and returning to their base camp between operations.

The second major source of manpower for the ghetto-clearing operations was the numerous battalions of Order Police (*Ordnungspolizei*) stationed in the General Government. In 1936, when Himmler gained centralized control over all German police, the Secret State Police (Gestapo) and Criminal Police (Kripo) were consolidated under the Security Police Main Office of Reinhard Heydrich. The German equivalent of the city police (*Schutzpolizei*) and county sheriffs (*Gendarmerie*) were consolidated under the Order Police Main Office of Kurt Daluege. The Order Police were far more numerous than the more notorious Security Police and encompassed not only the regular policemen distributed among various urban and rural police stations in Germany, but also large battalion-size units, which were stationed in barracks and were given some military training. As with National Guard units in the United States, these battalions were organized regionally. As war approached in 1938–39, many young Germans volunteered for the Order Police in order to avoid being drafted into the regular army.

Beginning in September 1939, the Order Police battalions, each of approximately five hundred men, were rotated out from their home cities on tours of duty in the occupied territories. As the German empire expanded and the demand for occupation forces increased, the Order Police was vastly expanded by creating new reserve police battalions. The career police and prewar volunteers of the old battalions were distributed to become the noncommissioned officer cadres of these new reserve units, whose rank and file were now composed of civilian draftees considered too old by the Wehrmacht for frontline military service.

One such unit, Reserve Police Battalion 101 from Hamburg, was one of three police battalions stationed in the district of Lublin during the onslaught against the Polish ghettos. Because no fewer than 210 former members of this battalion were interrogated during more than a decade of judicial investigation and trials in the 1960s and early 1970s, we know a great deal about its composition. First let us examine the officer and noncommissioned officer (NCO) cadres.

The battalion was commanded by Major Wilhelm Trapp, a fifty-three-year-old career policeman who had risen through the ranks and was affectionately referred to by his men as "Papa Trapp." Though he had joined the Nazi Party in December 1932, he had never been taken into the SS or even given an SS-equivalent rank. He was clearly not considered SS material. His two captains, in contrast, were young men in their late twenties, both party members and SS officers. Even in their testimony twenty-five years later they made no attempt to conceal their contempt for their commander as both weak and unmilitary. Little is known about the first lieutenant who was Trapp's adjutant, for he died in the spring of

1943. In addition, however, the battalion had seven reserve lieutenants, that is men who were not career policemen but who, after they were drafted into the Order Police, had been selected to receive officer training because of their middle-class status, education, and success in civilian life. Their ages ranged from thirty-three to forty-eight; five were party members, but none belonged to the SS. Of the thirty-two NCOs on whom we have information, twenty-two were party members but only seven were in the SS. They ranged in age from twenty-seven to forty years old; their average was thirty-three and a half.

The vast majority of the rank and file had been born and reared in Hamburg and its environs. The Hamburg element was so dominant and the ethos of the battalion so provincial that contingents from nearby Wilhelmshaven and Schleswig-Holstein were considered outsiders. Over 60 percent were of working-class background, but few of them were skilled laborers. The majority of them held typical Hamburg working-class jobs: dock workers and truck drivers were most numerous, but there were also many warehouse and construction workers, machine operators, seamen and waiters. About 35 percent were lower-middle class, virtually all of whom were white-collar workers. Three-quarters of them were in sales of some sort; the other one-quarter performed various office jobs, both in the government and private sectors. The number of independent artisans, such as tailors and watch makers, was small; and there were only three middle-class professionals—two druggists and one teacher. The average age of the men was thirty-nine; over half were between thirty-seven and forty-two, the *Jahrgänge** most intensively drafted for police duty after September 1939.

The men of Reserve Police Battalion 101 were from the lower orders of German society. They had experienced neither social nor geographic mobility. Very few were economically independent. Except for apprenticeship or vocational training, virtually none had any education after leaving the Volksschule at age fourteen or fifteen. About 25 percent were Nazi Party members in 1942, most having joined in 1937 or later. Though not questioned about their pre-1933 political affiliation during their interrogations, presumably many had been Communists, Socialists, and labor union members before 1933. By virtue of their age, of course, all went through their formative period in the pre-Nazi era. These were men who had known political standards and moral norms other than those of the Nazis. Most came from Hamburg, one of the least Nazified cities in Germany, and the majority came from a social class that in its political culture had been anti-Nazi.

These men would not seem to have been a very promising group from which to recruit mass murderers of the Holocaust. Yet this unit was to be extraordinarily active both in clearing ghettos and in massacring Jews outright during the blitzkrieg against Polish Jewry. If these middle-aged reserve policemen became one major component of the murderers, the

*Referring to men in their late thirties or early forties.

second question posed is how? Specifically, what happened when they were first assigned to kill Jews? What choices did they have, and how did they react?

Reserve Police Battalion 101 departed from Hamburg on June 20, 1942, and was initially stationed in the town of Bilgoraj, fifty miles south of Lublin. Around July 11 it received orders for its first major action, aimed against the approximately 1,800 Jews living in the village of Jozefow, about twenty miles slightly south and to the east of Bilgoraj. In the General Government a seventeen-day stoppage of Jewish transports due to a shortage of rolling stock had just ended, but the only such trains that had been resumed were several per week from the district of Krakau to Belzec. The railway line to Sobibor was down, and that camp had become practically inaccessible. In short the Final Solution in the Lublin district had been paralyzed, and Globocnik was obviously anxious to resume the killing. But Jozefow could not be a deportation action. Therefore the battalion was to select out the young male Jews in Jozefow and send them to a work camp in Lublin. The remaining Jews—about 1,500 women, children, and elderly—were simply to be shot on the spot.

On July 12 Major Trapp summoned his officers and explained the next day's assignment. One officer, a reserve lieutenant in 1st company and owner of a family lumber business in Hamburg, approached the major's adjutant, indicated his inability to take part in such an action in which unarmed women and children were to be shot, and asked for a different assignment. He was given the task of accompanying the work Jews to Lublin. The men were not as yet informed of their imminent assignment, though the 1st company captain at least confided to some of his men that the battalion had an "extremely interesting task" (*hochinteressante Aufgabe*) the next day.

Around 2 A.M. the men climbed aboard waiting trucks, and the battalion drove for about an hour and a half over an unpaved road to Jozefow. Just as daylight was breaking, the men arrived at the village and assembled in a half-circle around Major Trapp, who proceeded to give a short speech. With choking voice and tears in his eyes, he visibly fought to control himself as he informed his men that they had received orders to perform a very unpleasant task. These orders were not to his liking, either, but they came from above. It might perhaps make their task easier, he told the men, if they remembered that in Germany bombs were falling on the women and children. Two witnesses claimed that Trapp also mentioned that the Jews of this village had supported the partisans. Another witness recalled Trapp's mentioning that the Jews had instigated the boycott against Germany. Trapp then explained to the men that the Jews in the village of Jozefow would have to be rounded up, whereupon the young males were to be selected out for labor and the others shot.

Trapp then made an extraordinary offer to his battalion: if any of the older men among them did not feel up to the task that lay before him, he could step out. Trapp paused, and after some moments, one man stepped forward. The captain of 3rd company, enraged that one of his men had broken ranks, began to berate the man. The major told the captain to hold his tongue. Then ten or twelve other men stepped forward as well. They turned in their rifles and were told to await a further assignment from the major.

Trapp then summoned the company commanders and gave them their respective assignments. Two platoons of 3rd company were to surround the village; the men were explicitly ordered to shoot anyone trying to escape. The remaining men were to round up the Jews and take them to the market place. Those too sick or frail to walk to the market place, as well as infants and anyone offering resistance or attempting to hide, were to be shot on the spot. Thereafter, a few men of 1st company were to accompany the work Jews selected at the market place, while the rest were to proceed to the forest to form the firing squads. The Jews were to be loaded onto battalion trucks by 2nd company and shuttled from the market place to the forest.

Having given the company commanders their respective assignments, Trapp spent the rest of the day in town, mostly in a school room converted into his headquarters but also at the homes of the Polish mayor and the local priest. Witnesses who saw him at various times during the day described him as bitterly complaining about the orders he had been given and "weeping like a child." He nevertheless affirmed that "orders were orders" and had to be carried out. Not a single witness recalled seeing him at the shooting site, a fact that was not lost upon the men, who felt some anger about it. Trapp's driver remembers him saying later, "If this Jewish business is ever avenged on earth, then have mercy on us Germans" (Wenn sich diese Judensache einmal auf Erden rächt, dann gnade uns Deutschen).

After the company commanders had relayed orders to the men, those assigned to the village broke up into small groups and began to comb the Jewish quarter. The air was soon filled with cries, and shots rang out. The market place filled rapidly with Jews, including mothers with infants. While the men of Reserve Police Battalion 101 were apparently willing to shoot those Jews too weak or sick to move, they still shied for the most part from shooting infants, despite their orders. No officer intervened, though subsequently one officer warned his men that in the future they would have to be more energetic.

As the roundup neared completion, the men of 1st company were withdrawn from the search and given a quick lesson in the gruesome task that awaited them by the battalion doctor and the company's first sergeant. The doctor traced the outline of a human figure on the ground and showed the men how to use a fixed bayonet placed between and just above the shoulder blades as a guide for aiming their carbines. Several men now approached the 1st company captain and asked to be given different assignment; he curtly refused. Several others who approached the first sergeant rather than the captain fared better. They were given guard duty along the route from the village to the forest.

The first sergeant organized his men into two groups of about thirty-five men, which was roughly equivalent to the number of Jews who could be loaded into each truck. In turn each squad met an arriving truck at the unloading point on the edge of the forest. The individual squad members paired off *face-to-face* with the individual Jews they were to shoot, and marched their victims into the forest. The first sergeant remained in the forest to supervise the shooting. The Jews were forced to lie face down in a row. The policemen stepped up behind them, and on a signal from the first sergeant fired their carbines at point-blank range into the necks of their victims. The first sergeant then moved a few yards deeper into the forest to supervise the next execution. So-called mercy shots were given by a noncommissioned officer, as many of the men, some out of excitement and some intentionally, shot past their victims. By mid-day alcohol had appeared from somewhere to "refresh" the shooters. Also around mid-day the first sergeant relieved the older men, after several had come to him and asked to be let out. The other men of 1st company, however, continued shooting throughout the day.

Meanwhile the Jews in the market place were being guarded by the men of 2nd company, who loaded the victims onto the trucks. When the first salvo was heard from the woods, a terrible cry swept the market place, as the collected Jews now knew their fate. Thereafter, however, a quiet—indeed "unbelievable"—composure settled over the Jews, which the German policemen found equally unnerving. By mid-morning the officers in the market place became increasingly agitated. At the present rate, the executions would never be completed by nightfall. The 3rd company was called in from its outposts around the village to take over close guard of the market place. The men of 2nd company were informed that they too must now go to the woods to join the shooters. At least one sergeant once again offered his men the opportunity to report if they did not feel up to it. No one took up his offer. In another unit, one policeman confessed to his lieutenant that he was "very weak" and could not shoot. He was released.

In the forest the 2nd company was divided into small groups of six to eight men rather than the larger squads of thirty-five as in 1st company. In the confusion of the small groups coming and going from the unloading point, several men managed to stay around the trucks looking busy and thus avoided shooting. One was noticed by his comrades, who swore at him for shirking, but he ignored them. Among those who began shooting, some could not last long. One man shot an old woman on his first round, after which his nerves were finished and he could not continue. Another discovered to his dismay that his second victim was a German Jew—a mother from Kassel with her daughter. He too then asked out. This encounter with a German Jew was not exceptional. Several other men also remembered Hamburg and Bremen Jews in Jozefow. It was a grotesque irony that some of the men of Reserve Police Battalion 101 had guarded the collection center in Hamburg, the confiscated freemason lodge house on the Moorweide next to the university library, from which the Hamburg Jews had been deported the previous fall. A few had even guarded the deportation transports to Lodz, Riga, and Minsk. These Hamburg policemen had now followed other Jews deported from northern Germany, in order to shoot them in southern Poland.

A third policeman was in such an agitated state that on his first shot he aimed too high. He shot off the top of the head of his victim, splattering brains into the face of his sergeant. His request to be relieved was granted. One policeman made it to the fourth round, when his nerves gave way. He shot past his victim, then turned and ran deep into the forest and vomited. After several hours he returned to the trucks and rode back to the market place.

As had happened with 1st company, bottles of vodka appeared at the unloading point and were passed around. There was much demand, for among the 2nd company, shooting instructions had been less explicit and initially bayonets had not been fixed as an aiming guide. The result was that many of the men did not give neck shots but fired directly into the heads of their victims at pointblank range. The victims' heads exploded, and in no time the policemen's uniforms were saturated with blood and splattered with brains and splinters of bone. When several officers noted that some of their men could no longer continue or had begun intentionally to fire past their victims, they excused them from the firing squads.

Though a fairly significant number of men in Reserve Police Battalion 101 either did not shoot at all or started but could not continue shooting, most persevered to the end and lost all count of how many Jews they had killed that day. The forest was so filled with bodies that it became difficult to find places to make the Jews lie down. When the action was finally over at dusk, and some 1,500 Jews lay dead, the men climbed into their trucks and returned to Bilgoraj. Extra rations of alcohol were provided, and the men talked little, ate almost nothing, but drank a great deal. That night one of them awoke from a nightmare firing his gun into the ceiling of the barracks.

Following the massacre at Jozefow, Reserve Police Battalion 101 was transferred to the northern part of the Lublin district. The various platoons of the battalion were stationed in different towns but brought together for company-size actions. Each company was engaged in at least one more shooting action, but more often the Jews were driven from the ghettos onto trains bound for the extermination camp of Treblinka. Usually one police company worked in conjunction with a Trawniki unit for each action. The "dirty work"—driving the Jews out of their dwellings with whips, clubs, and guns; shooting on the spot the frail, sick, elderly, and infants who could not march to the train station; and packing the train cars to the bursting point so that only with the greatest of effort could the doors even be closed—was usually left to the so-called Hiwis (*Hilfswilligen* or "volunteers") from Trawniki.

Once a ghetto had been entirely cleared, it was the responsibility of the men of Reserve Police Battalion 101 to keep the surrounding region *"judenfrei."* Through a network of Polish informers and frequent search patrols—casually referred to as *Judenjagden* or "Jew hunts"—the policemen remorselessly tracked down those Jews who had evaded the roundups and fled to the forests. Any Jew found in these circumstances was simply shot on the spot. By the end of the year there was scarcely a Jew alive in the northern Lublin district, and Reserve Police Battalion 101 increasingly turned its attention from murdering Jews to combatting partisans.

In looking at the half-year after Jozefow, one sees that this massacre drew an important dividing line. Those men who stayed with the assignment and shot all day found the subsequent actions much easier to perform. Most of the men were bitter about what they had been asked to do at Jozefow, and it became taboo even to speak of it. Even thirty years later they could not hide the horror of endlessly shooting Jews at point-blank range. In contrast, however, they spoke of surrounding ghettos and watching the Hiwis brutally drive the Jews onto the death trains with considerable detachment and a near-total absence of any sense of participation or responsibility. Such actions they routinely dismissed with a standard refrain: "I was *only* in the police cordon there." The shock treatment of Jozefow had created an effective and desensitized unit of ghetto-clearers and, when the occasion required, outright murderers. After Jozefow nothing else seemed so terrible. Heavy drinking also contributed to numbing the men's sensibilities. One nondrinking policemen noted that "most of the other men drank so much solely because of the many shootings of Jews, for such a life was quite intolerable sober" (die meisten der anderen Kameraden lediglich auf Grund der vielen Judenerschiessungen soviel getrunken haben, da ein derartiges Leben nüchtern gar nicht zu ertragen war).

Among those who either chose not to shoot at Jozefow or proved "too weak" to carry on and made no subsequent attempt to rectify this image of "weakness," a different trend developed. If they wished they were for the most part left alone and excluded from further killing actions, especially the frequent "Jew hunts." The consequences of their holding aloof from the mass murder were not grave. The reserve lieutenant of 1st company who had protested against being involved in the Jozefow shooting and been allowed to accompany the work Jews to Lublin subsequently went to Major Trapp and declared that in the future he would not take part in any *Aktion* unless explicitly ordered. He made no attempt to hide his aversion to what the battalion was doing, and his attitude was known to almost everyone in the company. He also wrote to Hamburg and requested that he be recalled from the General Government because he did not agree with the "nonpolice" functions being performed by the battalion there. Major Trapp not only avoided any confrontation but protected him. Orders involving actions against the Jews were simply passed from battalion or company headquarters to his deputy. He was, in current terminology, "left out of the loop." In November 1942 he was recalled to Hamburg, made adjutant to the Police President of that city, and subsequently promoted!

The man who had first stepped out at Jozefow was sent on almost every partisan action but not on the "Jew hunts." He suspected that this pattern resulted from his earlier behavior in Jozefow. Another man who had not joined the shooters at Jozefow was given excessive tours of guard duty and other unpleasant assignments and was not promoted. But he was not assigned to the "Jew hunts" and firing squads, because the officers wanted only "men" with them and in their eyes he was "no man." Others who felt as he did received the same treatment, he said. Such men could not, however, always protect themselves against officers out to get them. One man was assigned to a firing squad by a vengeful officer precisely because he had not yet been involved in a shooting.

The experience of Reserve Police Battalion 101 poses disturbing questions to those concerned with the lessons and legacies of the Holocaust. Previous explanations for the behavior of the perpetrators, especially those at the lowest level who came face-to-face with the Jews they killed, seem inadequate. Above all the perpetrators themselves have constantly cited inescapable orders to account for their behavior. In Jozefow, however, the men had the opportunity both before and during the shooting to withdraw. The battalion in general was under orders to kill the Jews of Jozefow, but each individual man was not.

Special selection, indoctrination, and ideological motivation are equally unsatisfying as explanations. The men of Reserve Police Battalion 101 were certainly not a group carefully selected for their suitability as mass murderers, nor were they given special training and indoctrination for the task that awaited them. They were mainly apolitical, and even the officers were only partly hard-core Nazi. Major Trapp in particular made no secret of his disagreement with the battalion's orders, and by Nazi standards he displayed shameful weakness in the way he carried them out. Among the men who did the killing there was much bitterness about what they had been asked to do and sufficient discomfort that no one wished to talk about it thereafter. They certainly did not take pride in achieving some historic mission.

While many murderous contributions to the Final Solution—especially those of the desk murderers—can be explained as routinized, depersonalized, segmented, and incremental, thus vitiating any sense of personal responsibility, that was clearly not the case in Jozefow, where the killers confronted the reality of their actions in the starkest way. Finally, the men of Reserve Police Battalion 101 were not from a generation that had been reared and educated solely under the Nazi regime and thus had no other political norms or standards by which to measure their behavior. They were older; many were married family men; and many came from a social and political background that would have exposed them to anti-Nazi sentiments before 1933.

What lessons, then, can one draw from the testimony given by the perpetrators of the massacre of the Jews in Jozefow?[1] Nothing is more elusive in this testimony than the consciousness of the men that morning of July 13, 1942, and above all their attitude toward Jews at the time. Most simply denied that they had had any choice. Faced with the testimony of others, they did not contest that Trapp had made the offer but repeatedly claimed that they had not heard that part of his speech or could not remember it. A few who admitted that they had been given the choice and yet failed to opt out were quite blunt. One said that he had not wanted to be considered a coward by his comrades. Another—more aware of what truly required courage—said quite simply: "I was cowardly." A few others also made the attempt to confront the question of choice but failed to find the words. It was a different time and place, as if they had been on another political planet, and the political vocabulary and values of the 1960s were helpless to explain the situation in which they had found themselves in 1942. As one man admitted, it was not until years later that he began to consider that what he had done had not been right. He had not given it a thought at the time.

Several men who chose not to take part were more specific about their motives. One said that he accepted the possible disadvantages of his course of action "because I was not a career policeman and also did not want to become one, but rather an independent skilled craftsman, and I had my business back home. . . . thus it was of no consequence that my police career would not prosper" (denn ich war kein aktiven Polizist und wollte auch keiner werden, sondern selbstständiger Handwerksmeister und ich hatte zu Hause meinen Betrieb. . . . deshalb macht es mir nichts aus, dass meine Karriere keinen Aufstieg haben würde). The reserve lieutenant of 1st company placed a similar emphasis on the importance of economic independence when explaining why his situation was not analogous to that of the two SS captains on trial. "I was somewhat older then and moreover a reserve officer, so it was not particularly important to me to be promoted or otherwise to advance, because I had my prosperous business back home. The company chiefs . . . on the other hand were young men and career policemen, who wanted to become something. Through my business experience, especially because it extended abroad, I had gained a better overview of things." He alone then broached the most taboo subject of all: "Moreover through my earlier business activities I already knew many Jews" (Ich war damals etwas älter und ausserdem Reserveoffizier, mir kam es insbesondere nicht darauf an, befördert zu werden oder sonstwie weiterzukommen, denn ich hatte ja zuhause mein gutgehendes Geschäft. Die Kompaniechefs . . . dagegen waren junge Leute vom aktiven Dienst, die noch etwas werden wollten. Ich hatte durch meine kaufmännishce Tätigkeit, die sich insbesondere auch auf das Ausland erstreckte, einen besseren Überblick über die Dinge. Ausserdem kannte ich schon durch meine geschäftliche Tätigkeit von frühen viele Juden).

Crushing conformity and blind, unthinking acceptance of the political norms of the time on the one hand, careerism on the other—these emerge as the factors that at least some of the men of Reserve Police Battalion 101 were able to discuss twenty-five years later. What remained virtually unexamined by the interrogators and unmentioned by the policemen was the role of antisemitism. Did they not speak of it because antisemitism had not been a motivating factor? Or were they unwilling and unable to confront this issue even after three decades, because it had been all too important, all too pervasive? One is tempted to wonder if the silence speaks louder than words, but in the end—as Claudia Koonz reminds us—the silence is still silence, and the question remains unanswered.

Was the incident at Jozefow typical? Certainly not. I know of no other case in which a commander so openly invited and sanctioned the nonparticipation of his men in a killing action. But in the end the most important fact is not that the experience of Reserve Police Battalion 101 was untypical, but rather that Trapp's extraordinary offer did not matter. Like any other unit, Reserve Police Battalion 101 killed the Jews they had been told to kill.

1. This study is based entirely on the judicial records in the Staatsanwaltschaft Hamburg that resulted from two investigations of Reserve Police Battalion 101: 141 Js 1957/62 and 141 Js 128/65. German laws and regulations for the protection of privacy prohibit the revealing of names from such court records. Thus, with the exception of Major Trapp, who was tried, convicted, and executed in Poland after the war, I have chosen simply to refer to individuals generically by rank and unit rather than by pseudonyms.

7. Primo Levi

Primo Levi was born in Turin, in 1919. He studied chemistry at the University of Turin, but his involvement with an anti-Fascist group led to his arrest by the Germans in December 1943 (Italy had signed a separate armistice with the Allies three months earlier). In February 1944, he was deported to Auschwitz with other Italian Jews. Virtually ignored when it was published in Italy in 1947 as *If This Be a Man, Survival in Auschwitz,* Levi's first book, has since become a classic. It was reprinted there to general acclaim in 1958. Levi's account of his ordeal in the camp describes his work in Buna/Monowitz, first outdoors as a laborer and then indoors in a chemical laboratory, a piece of good luck that, he says, contributed greatly to his survival. In the "Vanadium" chapter of *The Periodic Table* (1975), Levi offers a fascinating report on an unusual postwar encounter with one of his German supervisors in this laboratory.

Levi says that he wrote *Survival in Auschwitz* soon after the war as an "interior liberation." How little he succeeded in freeing himself of the burden of Auschwitz is confirmed by his last and perhaps finest work, *The Drowned and the Saved,* published posthumously, from which the following selection, "Shame," is taken. Here he continued to assess the meaning and nature of the Auschwitz experience for our time, though this is also a persistent theme in *The Reawakening* (1963), the sequel to *Survival in Auschwitz,* describing his circuitous return to Italy from the camp, and in *Moments of Reprieve* (1986).

Levi retired from his job as manager of a chemical factory in 1977, to devote full time to writing. Chiefly in his collections of essays, he sketched a sober, unromantic portrait of the psychological and moral impact of the Holocaust on his own view of civilization and, by extension, on the modern sensibility. Like Jean Améry, about whose work on Auschwitz and later suicide he wrote with critical vigor in *The Drowned and the Saved,* Levi seems to have felt as he grew older an increasing disenchantment with his earlier hope that his voice might awaken the world to the private and political perils unleashed in a place like Auschwitz. His death in 1987, apparently by suicide, remains a controversial issue among those who knew him well.

Levi wrote *The Drowned and the Saved* decades after the war, but it is clear that the inner anguish kindled by his stay in Auschwitz was never stifled—indeed, it may have ignited anew and with even greater fervor during the intervening years. One of the dismal ironies of the event we call the Holocaust is that it left a far greater legacy of guilt among the surviving victims than it did among those responsible for their ordeal. Levi wrestles in these pages with a feeling of shame that has no rational source but lingers in his life nonetheless. Long after the individual agents of mass murder have dismissed or forgotten their role in the destruction of European Jewry, Levi still feels haunted by the misery he witnessed in others and could do nothing to allay. Our knowledge of his end adds a particular pathos to the painful self-examination from which he never relented.

Shame

A certain fixed image has been proposed innumerable times, consecrated by literature and poetry, and picked up by the cinema: "the quiet after the storm," when all hearts rejoice. "To be freed from pain / is delightful for us." The disease runs its course and health returns. To deliver us from imprisonment "our boys," the liberators, arrive just in time, with waving flags; the soldier returns and again finds his family and peace.

Judging by the stories told by many who came back and from my own memories, Leopardi the pessimist stretched the truth in this representation; despite himself, he showed himself to be an optimist. In the majority of cases, the hour of liberation was neither joyful nor lighthearted. For most it occurred against a tragic background of destruction, slaughter, and suffering. Just as they felt they were again becoming men, that is, responsible, the sorrows of men returned: the sorrow of the dispersed or lost family; the universal suffering all around; their own exhaustion, which seemed definitive, past cure; the problems of a life to begin all over again amid the rubble, often alone. Not "pleasure the son of misery," but misery the son of misery. Leaving pain behind was a delight for only a few fortunate beings, or only for a few instants, or for very simple souls; almost always it coincided with a phase of anguish.

Anguish is known to everyone, even children, and everyone knows that it is often blank, undifferentiated. Rarely does it carry a clearly written label that also contains its motivation; any label it does have is often mendacious. One can believe or declare oneself to be anguished for one reason and be so due to something totally different. One can think that one is suffering at facing the future and instead be suffering because of one's past; one can think that one is suffering for others, out of pity, out of compassion, and instead be suffering for one's own reasons, more or less profound, more or less avowable and avowed, sometimes so deep that only the specialist, the analyst of souls, knows how to exhume them.

Naturally, I dare not maintain that the movie script I referred to before is false in every case. Many liberations were experienced with full, authentic joy—above all by combatants, both military and political, who at that moment saw the aspirations of their militancy and their lives realized, and also on the part of those who had suffered less and for less time, or only in their own person and not because of their family, friends, or loved ones. And besides, luckily, human beings are not all the same: there are among us those who have the virtue and the privilege of extracting, isolating those instants of happiness, of enjoying them fully, as though they were extracting pure gold from dross. And finally, among the testimonies, written or spoken, some are unconsciously stylized, in which convention prevails over genuine memory: "Whoever is freed from slavery rejoices. I too was liberated, hence I too rejoice over it. In all films, all novels, just as in *Fidelio,* the shattering of the chains is a moment of solemn or fervid jubilation, and so was mine." This is a specific case of that drifting of memory I mentioned in the first chapter, and which is accentuated with the passing of years and the piling up of the experiences of others, true or presumed, on one's own. But anyone who, purportedly or by temperament, shuns rhetoric, usually speaks in a different voice. This, for example, is how, on the last page of his memoir, *Eyewitness Auschwitz: Three Years in the Gas Chambers,* Filip Müller, whose experience was much more terrible than mine, describes his liberation:

> Although it may seem incredible, I had a complete letdown or depression. That moment, on which for three years all my thoughts and secret desires were concentrated, did not awaken happiness or any other feeling in me. I let myself fall from my pallet and crawled to the door. Once outside I tried vainly to go further, then I simply lay down on the ground in the woods and fell asleep.

I now reread the passage from my own book, *The Reawakening,* which was published in Italy only in 1963, although I had written these words as early as 1947. In it is a description of the first Russian soldiers facing our Lager packed with corpses and dying prisoners.

> They did not greet us, nor smile; they seemed oppressed, not only by pity but also by a confused restraint which sealed their mouths, and kept their eyes fastened on the funereal scene. It was the same shame which we knew so well, which submerged us after the selections, and every time we had to witness or undergo an outrage: the shame that the Germans never knew, the shame which the just man experiences when confronted by a crime committed by another, and he feels remorse because of its existence, because of its having been irrevocably introduced into the world of existing things, and because his will has proven nonexistent or feeble and was incapable of putting up a good defense.

I do not think that there is anything I need erase or correct, but there is something I must add. That many (including me) experienced "shame," that is, a feeling of guilt during the imprisonment and afterward, is an ascertained fact confirmed by numerous testimonies. It may seem absurd, but it is a fact. I will try to interpret it myself and to comment on the interpretations of others.

As I mentioned at the start, the vague discomfort which accompanied liberation was not precisely shame, but it was perceived as such. Why? There are various possible explanations.

I will exclude certain exceptional cases: the prisoners who, almost all of them political, had the strength and opportunity to act within the Lager in defense of and to the advantage of their companions. We, the almost total majority of common prisoners, did not know about them and did not even suspect their existence, and logically so: due to obvious political and police necessity (the Political Section of Auschwitz was simply a branch of the Gestapo) they were forced to operate secretly, not only where the Germans were concerned but in regard to everyone. In Auschwitz, the concentrationary empire which in my time was constituted by 95 percent Jews, this political network was embryonic; I witnessed only one episode that should have led me to sense something had I not been crushed by the everyday travail.

Around May 1944 our almost innocuous Kapo was replaced, and the newcomer proved to be a fearsome individual. All Kapos gave beatings: this was an obvious part of their duties, their more or less accepted language. After all, it was the only language that everyone in that perpetual Babel could truly understand. In its various nuances it was understood as an incitement to work, a warning or punishment, and in the hierarchy of suffering it had a low rank. Now, the new Kapo gave his beatings in a different way, in a convulsive, malicious, perverse way: on the nose, the shin, the genitals. He beat to hurt, to cause suffering and humiliation. Not even, as with many others, out of blind racial hatred, but with the obvious intention of inflicting pain, indiscriminately, and without pretext, on all his subjects. Probably he was a mental case, but clearly under those conditions the indulgence that we today consider obligatory toward such sick people would have been out of place. I spoke about it with a colleague, a Jewish Croatian Communist: What should we do? How to protect ourselves? How to act collectively? He gave me a strange smile and simply said: "You'll see, he won't last long." In fact, the beater vanished within a week. But years later, during a meeting of survivors, I found out that some political prisoners attached to the Work Office inside the camp had the terrifying power of switching the registration numbers on the lists of prisoners destined to be gassed. Anyone who had the ability and will to act in this way, to oppose in this or other ways the machine of the Lager, was beyond the reach of "shame"—or at least the shame of which I am speaking, because perhaps he experiences something else.

Equally protected must have been Sivadjan, a silent and tranquil man whom I mentioned in passing in *Survival in Auschwitz,* in the chapter "The Canto of Ulysses," and about whom I discovered on that same occasion that he had brought explosives into the camp to foment a possible insurrection.

In my opinion, the feeling of shame or guilt that coincided with reacquired freedom was extremely composite: it contained diverse elements, and in diverse proportions for each individual. It must be remembered that each of us, both objectively and subjectively, lived the Lager in his own way.

Coming out of the darkness, one suffered because of the reacquired consciousness of having been diminished. Not by our will, cowardice, or fault, yet nevertheless we had lived for months and years at an animal level: our days had been encumbered from dawn to dusk by hunger, fatigue, cold, and fear, and any space for reflection, reasoning, experiencing emotions was wiped out. We endured filth, promiscuity, and destitution, suffering much less than we would have suffered from such things in normal life, because our moral yardstick had changed. Furthermore, all of us had stolen: in the kitchen, the factory, the camp, in short, "from the others," from the opposing side, but it was theft nevertheless. Some (few) had fallen so low as to steal bread from their own companions. We had not only forgotten our country and our culture, but also our family, our past, the future we had imagined for ourselves, because, like animals, we were confined to the present moment. Only at rare intervals did we come out of this condition of leveling, during the very few Sundays of rest, the fleeting minutes before falling asleep, or the fury of the air raids, but these were painful moments precisely because they gave us the opportunity to measure our diminishment from the outside.

I believe that it was precisely this turning to look back at the "perilous water" that gave rise to so many suicides after (sometimes immediately after) Liberation. It was in any case a critical moment which coincided with a flood of rethinking and depression. By contrast, all historians of the Lager—and also of the Soviet camps—agree in pointing out that cases of suicide *during* imprisonment were rare. Several explanations of this fact have been put forward; for my part I offer three, which are not mutually exclusive.

First of all, suicide is an act of man and not of the animal. It is a meditated act, a noninstinctive, unnatural choice, and in the Lager there were few opportunities to choose: people lived precisely like enslaved animals that sometimes let themselves die but do not kill themselves.

Secondly, "there were other things to think about," as the saying goes. The day was dense: one had to think about satisfying hunger, in some way elude fatigue and cold, avoid the blows. Precisely because of the constant imminence of death there was no time to concentrate on the idea of death. Svevo's remark in *Confessions of Zeno,* when he ruthlessly describes his father's agony, has the rawness of truth: "When one is dying, one is much too busy to think about death. All one's organism is devoted to breathing."

Thirdly, in the majority of cases, suicide is born from a feeling of guilt that no punishment has attenuated; now, the harshness of imprisonment was perceived as punishment, and the feeling of guilt (if there is punishment, there must have been guilt) was relegated to the background, only to re-emerge after the Liberation. In other words, there was no need to punish oneself by suicide because of a (true or presumed) guilt: one was already expiating it by one's daily suffering.

What guilt? When all was over, the awareness emerged that we had not done anything, or not enough, against the system into which we had been absorbed. About the failed resistance in the Lagers, or, more accurately, in some Lagers, too much has been said, too superficially, above all by people who had altogether different crimes to account for. Anyone who made the attempt knows that there existed situations, collective or personal, in which active resistance was possible, and others, much more frequent, in which it was not. It is known that, especially in 1941, millions of Soviet military prisoners fell into German hands. They were young, generally well nourished and robust; they had military and political training, and often they formed organic units with soldiers with the rank of corporal and up, noncommissioned officers, and officers. They hated the Germans who had invaded their country, and yet they rarely resisted. Malnutrition, despoilment, and other physical discomforts, which it is so easy and economically advantageous to provoke and at which the Nazis were masters, are rapidly destructive and paralyze before destroying, all the more so when they are preceded by years of segregation, humiliation, maltreatment, forced migration, laceration of family ties, rupture of contact with the rest of the world—that is to say, the situation of the bulk of the prisoners who had landed in Auschwitz after the introductory hell of the ghettos or the collection camps.

Therefore, on a rational plane, there should not have been much to be ashamed of, but shame persisted nevertheless, especially for the few bright examples of those who had the strength and possibility to resist. I spoke about this in the chapter "The Last" in *Survival in Auschwitz,* where I described the public hanging of a resistor before a terrified and apathetic crowd of prisoners. This is a thought that then just barely grazed us, but that returned "afterward": you too could have, you certainly should have. And this is a judgment that the survivor believes he sees in the eyes of those (especially the young) who listen to his stories and judge with facile hindsight, or who perhaps feel cruelly repelled. Consciously or not, he feels accused and judged, compelled to justify and defend himself.

More realistic is self-accusation, or the accusation of having failed in terms of human solidarity. Few survivors feel guilty about having deliberately damaged, robbed, or beaten a companion. Those who did so (the Kapos, but not only they) block out the memory. By contrast, however, almost everybody feels guilty of having omitted to offer help. The presence at your side of a weaker—or less cunning, or older, or too young—companion, hounding you with his demands for help or with his simple presence, in itself an entreaty, is a constant in the life of the Lager. The demand for solidarity, for a human word, advice, even just a listening ear, was permanent and universal but rarely satisfied. There was no time, space, privacy, patience, strength; most often, the person to whom the request was addressed found himself in his turn in a state of need, entitled to comfort.

I remember with a certain relief that I once tried to give courage (at a moment when I felt I had some) to an eighteen-year-old Italian who had just arrived, who was floundering in the bottomless despair of his first days in camp. I forget what I told him, certainly words of hope, perhaps a few lies, acceptable to a "new arrival," expressed with the authority of my twenty-five years and my three months of seniority; at any rate, I made him the gift of a momentary attention. But I also remember, with disquiet, that much more often I shrugged my shoulders impatiently at other requests, and this precisely when I had been in camp for almost a year and so had accumulated a good store of experience: but I had also deeply assimilated the principal rule of the place, which made it mandatory that you take care of yourself first of all. I never found this rule expressed with as much frankness as in *Prisoners of Fear* by Ella Lingens-Reiner (where, however, the woman doctor, regardless of her own statement, proved to be generous and brave and saved many lives): "How was I able to survive in Auschwitz? My principle is: I come first, second, and third. Then nothing, then again I; and then all the others."

In August of 1944 it was very hot in Auschwitz. A torrid, tropical wind lifted clouds of dust from the buildings wrecked by the air raids, dried the sweat on our skin, and thickened the blood in our veins. My squad had been sent into a cellar to clear out the plaster rubble, and we all suffered from thirst: a new suffering, which was added to, indeed, multiplied by the old one of hunger. There was no drinkable water in the camp or often on the work site; in those days there was often no water in the wash trough either, undrinkable but good enough to freshen up and clean off the dust. As a rule, the evening soup and the ersatz coffee distributed around ten o'clock were abundantly sufficient to quench our thirst, but now they were no longer enough and thirst tormented us. Thirst is more imperative than hunger: hunger obeys the nerves, grants remission, can be temporarily obliterated by an emotion, a pain, a fear (we had realized this during our journey by train from Italy); not so with thirst, which does not give respite. Hunger exhausts, thirst enrages; in those days it accompanied us day and night: by day, on the work site, whose order (our enemy, but nevertheless order, a place of logic and certainty) was transformed into a chaos of shattered constructions; by night, in the hut without ventilation, as we gasped the air breathed a hundred times before.

The corner of the cellar that had been assigned to me by the Kapo and where I was to remove the rubble was next to a large room filled with chemical equipment in the process of being installed but already damaged by the bombs. Along the vertical wall ran a two-inch pipe, which ended in a spigot just above the floor. A water pipe? I took a chance and tried to open it. I was alone, nobody saw me. It was blocked, but using a stone for a hammer I managed to shift it a few millimeters. A few drops came out, they had no odor, I caught them on my fingers: it really seemed water. I had no receptacle, and the drops came out slowly, without pressure: the pipe must be only half full, perhaps less. I stretched out on the floor with my mouth under the spigot, not trying to open it further: it was water made tepid by the sun, insipid, perhaps distilled or the result of condensation; at any rate, a delight.

How much water can a two-inch pipe one or two meters high contain? A liter, perhaps not even that. I could have drunk all of it immediately;

that would have been the safest way. Or save a bit for the next day. Or share half of it with Alberto. Or reveal the secret to the whole squad. I chose the third path, that of selfishness extended to the person closest to you, which in distant times a friend of mine appropriately called us-ism. We drank all the water, in small, avaricious gulps, changing places under the spigot, just the two of us. On the sly. But on the march back to camp at my side I found Daniele, all gray with cement dust, his lips cracked and his eyes feverish, and I felt guilty. I exchanged a look with Alberto; we understood each other immediately and hoped nobody had seen us. But Daniele had caught a glimpse of us in that strange position, supine near the wall among the rubble, and had suspected something, and then had guessed. He curtly told me so many months later, in Byelorussia, after the Liberation: Why the two of you and not I? It was the "civilian" moral code surfacing again. The same according to which I the free man of today perceive as horrifying the death sentence of the sadistic Kapo, decided upon and executed without appeal, silently, with the stroke of an eraser. Is this belated shame justified or not? I was not able to decide then and I am not able to decide even now, but shame there was and is, concrete, heavy, perennial. Daniele is dead now, but in our meetings as survivors, fraternal, affectionate, the veil of that act of omission, that unshared glass of water, stood between us, transparent, not expressed, but perceptible and "costly."

Changing moral codes is always costly: all heretics, apostates, and dissidents know this. We cannot judge our behavior or that of others, driven at that time by the code of that time, on the basis of today's code; but the anger that pervades us when one of the "others" feels entitled to consider us "apostates," or, more precisely, reconverted, seems right to me.

Are you ashamed because you are alive in place of another? And in particular, of a man more generous, more sensitive, more useful, wiser, worthier of living than you? You cannot block out such feelings: you examine yourself, you review your memories, hoping to find them all, and that none of them are masked or disguised. No, you find no obvious transgressions, you did not usurp anyone's place, you did not beat anyone (but would you have had the strength to do so?), you did not accept positions (but none were offered to you . . .), you did not steal anyone's bread; nevertheless you cannot exclude it. It is no more than a supposition, indeed the shadow of a suspicion: that each man is his brother's Cain, that each one of us (but this time I say "us" in a much vaster, indeed, universal sense) has usurped his neighbor's place and lived in his stead. It is a supposition, but it gnaws at us; it has nestled deeply like a woodworm; although unseen from the outside, it gnaws and rasps.

After my return from imprisonment I was visited by a friend older than myself, mild and intransigent, the cultivator of a personal religion, which, however, always seemed to me severe and serious. He was glad to find me alive and basically unhurt, perhaps matured and fortified, certainly enriched. He told me that my having survived could not be the work of chance, of an accumulation of fortunate circumstances (as I did then and still do maintain) but rather of Providence. I bore the mark, I was an elect: I, the nonbeliever, and even less of a believer after the season of Auschwitz, was a person touched by Grace, a saved man. And why me? It is impossible to know, he answered. Perhaps because I had to write, and by writing bear witness: Wasn't I in fact then, in 1946, writing a book about my imprisonment?

Such an opinion seemed monstrous to me. It pained me as when one touches an exposed nerve, and kindled the doubt I spoke of before: I might be alive in the place of another, at the expense of another; I might have usurped, that is, in fact, killed. The "saved" of the Lager were not the best, those predestined to do good, the bearers of a message: what I had seen and lived through proved the exact contrary. Preferably the worst survived, the selfish, the violent, the insensitive, the collaborators of the "gray zone," the spies. It was not a certain rule (there were none, nor are there certain rules in human matters), but it was nevertheless a rule. I felt innocent, yes, but enrolled among the saved and therefore in permanent search of a justification in my own eyes and those of others. The worst survived, that is, the fittest; the best all died.

Chaim died, a watchmaker from Krakow, a pious Jew who despite the language difficulties made an effort to understand and be understood, and explained to me, the foreigner, the essential rules for survival during the first crucial days of captivity; Szabo died, the taciturn Hungarian peasant who was almost two meters tall and so was the hungriest of all, and yet, as long as he had the strength, did not hesitate to help his weaker companions to pull and push; and Robert, a professor at the Sorbonne who spread courage and trust all around him, spoke five languages, wore himself out recording everything in his prodigious memory, and had he lived would have answered the questions which I do not know how to answer; and Baruch died, a longshoreman from Livorno, immediately, on the first day, because he had answered the first punch he had received with punches and was massacred by three Kapos in coalition. These, and innumerable others, died not despite their valor but because of it.

My religious friend had told me that I survived so that I could bear witness. I have done so, as best I could, and I also could not have done so; and I am still doing so, whenever the opportunity presents itself; but the thought that this testifying of mine could by itself gain for me the privilege of surviving and living for many years without serious problems troubles me because I cannot see any proportion between the privilege and its outcome.

I must repeat: we, the survivors, are not the true witnesses. This is an uncomfortable notion of which I have become conscious little by little, reading the memoirs of others and reading mine at a distance of years. We survivors are not only an exiguous but also an anomalous minority: we are those who by their prevarications or abilities or good luck did not touch bottom. Those who did so, those who saw the Gorgon, have not returned to tell about it or have returned mute, but they are the "Muslims," the submerged, the complete witnesses, the ones whose deposition would have a general significance. They are the rule, we are the exception. Under another sky, and returned from a similar and diverse slavery, Solzhenitsyn also noted: "Almost all those who served a long sentence and whom you congratulate because they are survivors are unquestionably *pridurki* or were such during the greater part of their imprisonment. Because Lagers are meant for extermination, this should not be forgotten."

In the language of that other concentrationary universe, the *pridurki* are the prisoners who, in one way or another, won a position of privilege, those we called the Prominent.

We who were favored by fate tried, with more or less wisdom, to recount not only our fate but also that of the others, indeed of the drowned; but this was a discourse "on behalf of third parties," the story of things seen at close hand, not experienced personally. The destruction brought to an end, the job completed, was not told by anyone, just as no one ever returned to describe his own death. Even if they had paper and pen, the drowned would not have testified because their death had begun before that of their body. Weeks and months before being snuffed out, they had already lost the ability to observe, to remember, to compare and express themselves. We speak in their stead, by proxy.

I could not say whether we did or do so out of a kind of moral obligation toward those who were silenced or in order to free ourselves of their memory; certainly we do it because of a strong and durable impulse. I do not believe that psychoanalysts (who have pounced upon our tangles with professional avidity) are competent to explain this impulse. Their knowledge has been built up and tested "outside," in the world that, for the sake of simplicity, we call civilian: psychoanalysis traces its phenomenology and tries to explain it; studies its deviations and tries to heal them. Their interpretations, even those of someone like Bruno Bettelheim,* who went through the trials of the Lager, seem to me approximate and simplified, as if someone wished to apply the theorems of plane geometry to the solution of spheric triangles. The mental mechanisms of the *Häftlinge*† were different from ours; curiously, and in parallel, different also were their physiology and pathology. In the Lager colds and influenza were unknown, but one died, at times suddenly, from illnesses that the doctors never had an opportunity to study. Gastric ulcers and mental illnesses were healed (or became asymptomatic), but everyone suffered from an unceasing discomfort that polluted sleep and was nameless. To define this as a "neurosis" is reductive and ridiculous. Perhaps it would be more correct to see in it an atavistic anguish whose echo one hears in the second verse of Genesis: the anguish inscribed in everyone of the

*Bettelheim, a child psychologist, spent a year in Dachau and Buchenwald (1938–1939). He was released and made his way to the United States. Unlike Levi, he had no experience of a concentration or extermination camp during World War II.

†Prisoners.

"tohu-bohu"* of a deserted and empty universe crushed under the spirit of God but from which the spirit of man is absent; not yet born or already extinguished.

And there is another, vaster shame, the shame of the world. It has been memorably pronounced by John Donne, and quoted innumerable times, pertinently or not, that "no man is an island," and that every bell tolls for everyone. And yet there are those who, faced by the crime of others or their own, turn their backs so as not to see it and not feel touched by it. This is what the majority of Germans did during the twelve Hitlerian years, deluding themselves that not seeing was a way of not knowing, and that not knowing relieved them of their share of complicity or connivance. But we were denied the screen of willed ignorance, T. S. Eliot's "partial shelter": we were not able not to see. The ocean of pain, past and present, surrounded us, and its level rose from year to year until it almost submerged us. It was useless to close one's eyes or turn one's back to it because it was all around, in every direction, all the way to the horizon. It was not possible for us nor did we want to become islands; the just among us, neither more nor less numerous than in any other human group, felt remorse, shame, and pain for the misdeeds that others and not they had committed, and in which they felt involved, because they sensed that what had happened around them and in their presence, and in them, was irrevocable. Never again could it be cleansed; it would prove that man, the human species—we, in short—had the potential to construct an infinite enormity of pain, and that pain is the only force created from nothing, without cost and without effort. It is enough not to see, not to listen, not to act.

We are often asked, as if our past conferred a prophetic ability upon us, whether Auschwitz will return: whether, that is, other slaughters will take place, unilateral, systematic, mechanized, willed, at a governmental level, perpetrated upon innocent and defenseless populations and legitimized by the doctrine of contempt. Prophets, to our good fortune, we are not, but something can be said. That a similar tragedy, almost ignored in the West, did take place, in Cambodia, in about 1975. That the German slaughter could be set off—and after that feed on itself—out of a desire for servitude and smallness of soul, thanks to the concurrence of a number of factors (the state of war, German technological and organizational perfectionism, Hitler's will and inverted charisma, the lack in Germany of solid democratic roots), not very numerous, each of them indispensable but insufficient if taken singly. These factors can occur again and are already recurring in various parts of the world. The convergence again of all of them within ten or twenty years (there is no sense in speaking of a more remote future) is not very likely but also not impossible. In my opinion, a mass slaughter is particularly unlikely in the Western world, Japan, and also the Soviet Union: the Lagers of World War II are still part of the

*Without form and void (Genesis 1:2).

memory of many, on both the popular and governmental levels, and a sort of immunizational defense is at work which amply coincides with the shame of which I have spoken.

As to what might happen in other parts of the world, or later on, it is prudent to suspend judgment. And the nuclear apocalypse, certainly bilateral, probably instantaneous and definitive, is a greater and different horror, strange, new, which stands outside the theme I have chosen.

8. Jean Améry

Jean Améry, an acronym for Hans Maier, was born in Vienna in 1912. His father, who was killed in World War I, was Jewish; his mother was Catholic. He studied literature and philosophy at the University of Vienna, but economic need prevented him from completing work for a degree. As Nazi antisemitic laws in Germany grew more stringent and began to influence life in Austria, Améry moved from assimilation to accepting his Jewish identity. In 1937 he married a Jewish woman, and in 1938, after the German annexation of Austria, he fled with her to Belgium. His mother remained in Vienna and died in 1939.

Ironically, the Belgians deported Améry as a German alien in 1940 to southern France, where he was imprisoned in Gurs and other camps. He escaped in 1941, returned to Belgium, and for the next two years worked with the Belgian resistance. He was eventually captured and sent to the notorious Gestapo prison at Fort Breendonk, located between Antwerp and Brussels, where he was brutally tortured. He describes this experience in the following selection, "Torture," taken from *At the Mind's Limits.* Originally published in 1966 as *Jenseits von Schuld und Sühne (Beyond Guilt and Atonement),* the title is an allusion not only to Nietzsche's *Beyond Good and Evil,* but also to Dostoevsky's *Crime and Punishment,* which in some early German translations was called *Schuld und Sühne.*

One of the facts impressed on Améry by his torture was how far the Nazi world had drifted from the universe of Nietzsche and Dostoevsky and their intellectual heirs and ancestors. In Auschwitz, where he was sent from Fort Breendonk, Améry's unsettling discovery that "no bridge led from death in Auschwitz to *Death in Venice* [Thomas Mann's famous novella]" forced him to reevaluate the vaunted power of mind in a milieu where "the intellectual was alone with his intellect," virtually deprived of the chance of believing in "the reality of the world of the mind."

Perhaps one motive for Améry's decision to choose journalism after the British "liberated" him from Bergen Belsen in 1945 and he returned to Brussels was the poverty of the existence he had endured in the camps when he had been unable to give social expression to his thoughts. For twenty years, his journalism was directed toward Swiss audiences, as he refused to write for Germans or visit Germany. All this changed when, at the age of fifty-four, he published *At the Mind's Limits,* which established his reputation in Germany as a leading author/essayist on the implications of the camp experience for modern European life and thought. Here and in his subsequent books, he wrestles with the question of how such a great literary and intellectual tradition could have resulted in the moral and physical disaster of the Third Reich.

Two years before his own suicide in 1978, Améry had completed a book on that subject. The reasons for his self-inflicted death remain a mystery, though we know that, like that of many former victims of the German concentration and death camps, his health was impaired by the harsh physical conditions of those years. In addition, one cannot help wondering whether a political humanist like Améry might have been dejected by the world's continued indifference to the persecuted of all nations, whose spokesman he became. The constant outbursts of violence and terrorism on the international scene may have convinced this survivor of Gurs, Fort Breendonk, Auschwitz, and Bergen Belsen that taking his own life was a logical gesture of protest, a final affirmation of what he called "the last path toward freedom." His silenced voice abides in the unanswered challenge.

But there is a negative side to Améry's gesture that we may never be able to understand. No one has successfully measured the long-term effects of the Holocaust experience on the spirit of those who, like Améry, Primo Levi, Paul Celan, and other less well-known surviving victims, took their own lives many years after the ordeal seemingly had ended. In the selection included here, Améry writes: "Whoever was tortured stays tortured. Torture is ineradicably burned into him, even when no clinically objective traces can be detected." The lingering sensation of having been reduced to nothing more than a creature of helpless flesh—an experience shared by countless inmates of the deathcamps—remains a source of permanent humiliation to many who endured it. Rereading the conclusion to "Torture" in light of Améry's subsequent suicide, do we detect a clue to the mood that may have driven him to surrender his life:

> Whoever has succumbed to torture can no longer feel at home in the world. The shame of destruction cannot be erased. Trust in the world, which already collapsed in part at the first blow, but in the end, under torture, fully, will not be regained. That one's fellow man [his torturers] was experienced as the antiman remains in the tortured person as an accumulated horror. It blocks the view into a world in which the principle of hope rules.

During torture, Améry insists (and would it have been different in Auschwitz?), man learns that he is nothing more than the prey of death. Lacking a meaningful form of revenge, keeping vigil over an unvented resentment, did Améry finally find that the memory of "accumulated horror" without remedy was more than he could bear, choosing to embrace his earlier doom instead? We are left trying to decide whether his suicide was a victory or a defeat.

Torture

Whoever visits Belgium as a tourist may perhaps chance upon Fort Breendonk, which lies halfway between Brussels and Antwerp. The compound is a fortress from the First World War, and what its fate was at that time I don't know. In the Second World War, during the short eighteen days of resistance by the Belgian army in May 1940, Breendonk was the last headquarters of King Leopold. Then, under German occupation, it became a kind of small concentration camp, a "reception camp," as it was called in the cant of the Third Reich. Today it is a Belgian National Museum.

At first glance, the fortress Breendonk makes a very old, almost historic impression. As it lies there under the eternally rain-gray sky of Flanders, with its grass-covered domes and black-gray walls, it gives the feeling of a melancholy engraving from the 1870s war. One thinks of Gravelotte and Sedan and is convinced that the defeated Emperor Napoleon III, with kepi in hand, will immediately appear in one of the massive, low gates. One must step closer, in order that the fleeting picture from past times be replaced by another, which is more familiar to us. Watchtowers arise along the moat that rings the castle. Barbed-wire fences wrap around them. The copperplate of 1870 is abruptly obscured by horror photos from the world that David Rousset has called "l'Univers Concentrationnaire." The creators of the National Museum have left everything the way it was between 1940 and 1944. Yellowed wall cards: "Whoever goes beyond this point will be shot." The pathetic monument to the resistance movement that was erected in front of the fortress shows a man forced to his knees, but defiantly raising his head with its oddly Slavic lines. This monument would not at all have been necessary to make clear to the visitor *where* he is and *what* is recollected there.

One steps through the main gate and soon finds oneself in a room that in those days was mysteriously called the "business room." A picture of Heinrich Himmler on the wall, a swastika flag spread as a cloth over a long table, a few bare chairs. The business room. Everyone went about his business, and theirs was murder. Then the damp, cellarlike corridors, dimly lit by the same thin and reddishly glowing bulbs as the ones that used to hang there. Prison cells, sealed by inch-thick wooden doors. Again and again one must pass through heavy barred gates before one finally stands in a windowless vault in which various iron implements lie about. From there no scream penetrated to the outside. There I experienced it: torture.

If one speaks about torture, one must take care not to exaggerate. What was inflicted on me in the unspeakable vault in Breendonk was by far not the worst form of torture. No red-hot needles were shoved under my fingernails, nor were any lit cigars extinguished on my bare chest. What did happen to me there I will have to tell about later; it was relatively harmless and it left no conspicuous scars on my body. And yet, twenty-two years after it occurred, on the basis of an experience that in no way probed the entire range of possibilities, I dare to assert that torture is the most horrible event a human being can retain within himself.

But very many people have preserved such things, and the horrible can make no claim to singularity. In most Western countries torture was eliminated as an institution and method at the end of the eighteenth century. And yet, today, two hundred years later, there are still men and women—no one knows how many—who can tell of the torture they underwent. As I am preparing this article, I come across a newspaper page with photos that show members of the South Vietnamese army torturing captured Vietcong rebels. The English novelist Graham Greene wrote a letter about it to the London *Daily Telegraph,* saying:

> The strange new feature about the photographs of torture now appearing in the British and American press is that they have been taken with the approval of the torturers and are published over captions that contain no hint of condemnation. They might have come out of a book on insect life. . . . Does this mean that the American authorities sanction torture as a means of interrogation? The photographs certainly are a mark of honesty, a sign that the authorities do not shut their eyes to what is going on, but I wonder if this kind of honesty without conscience is really to be preferred to the old hypocrisy.

Every one of us will ask himself Graham Greene's question. The admission of torture, the boldness—but is it still that?—of coming forward with such photos is explicable only if it is assumed that a revolt of public conscience is no longer to be feared. One could think that this conscience has accustomed itself to the practice of torture. After all, torture was, and is, by no means being practiced only in Vietnam during these decades. I would not like to know what goes on in South African, Angolese, and Congolese prisons. But I do know, and the reader probably has also heard, what went on between 1956 and 1963 in the jails of French Algeria. There is a frighteningly exact and sober book on it, *La Question* by Henri Alleg, a work whose circulation was prohibited, the report of an eyewitness who was also personally tortured and who gave evidence of the horror, sparingly and without making a fuss about himself. Around 1960 numerous other books and pamphlets on the subject appeared: the learned criminological treatise by the famous lawyer Alec Mellor, the protest of the publicist Pierre-Henri Simon, the ethical-philosophic investigation of a theologian named Vialatoux. Half the French nation rose up against the torture in Algeria. One cannot say often and emphatically enough that by this the French did honor to themselves. Leftist intellectuals protested. Catholic trade unionists and other Christian laymen warned against the torture, and at the risk of their safety and lives took action against it. Prelates raised their voices, although to our feeling much too gently.

But that was the great and freedom-loving France, which even in those dark days was not entirely robbed of its liberty. From other places the screams penetrated as little into the world as did once my own strange and uncanny howls from the vault of Breendonk. In Hungary there presides a Party First Secretary, of whom it is said that under the régime of one of his predecessors torturers ripped out his fingernails. And where and who are all the others about whom one learned nothing at all, and of whom one will probably never hear anything? Peoples, governments, authorities, names that are known, but which no one says aloud. Somewhere, someone is crying out under torture. Perhaps in this hour, this second.

And how do I come to speak of torture solely in connection with the Third Reich? Because I myself suffered it under the outspread wings of this very bird of prey, of course. But *not only* for that reason; rather, I am convinced, beyond all personal experiences, that torture was not an accidental quality of this Third Reich, but its essence. Now I hear violent objection being raised, and I know that this assertion puts me on dangerous ground. I will try to substantiate it later. First, however, I suppose I must tell what the content of my experiences actually was and what happened in the cellar-damp air of the fortress Breendonk.

In July 1943 I was arrested by the Gestapo. It was a matter of fliers. The group to which I belonged, a small German-speaking organization within the Belgian resistance movement, was spreading anti-Nazi propaganda among the members of the German occupation forces. We produced rather primitive agitation material, with which we imagined we could convince the German soldiers of the terrible madness of Hitler and his war. Today I know, or at least believe to know, that we were aiming our feeble message at deaf ears. I have much reason to assume that the soldiers in field-gray uniform who found our mimeographed papers in front of their barracks clicked their heels and passed them straight on to their superiors, who then, with the same official readiness, in turn notified the security agency. And so the latter rather quickly got onto our trail and raided us. One of the fliers that I was carrying at the time of my arrest bore the message, which was just as succinct as it was propagandistically ineffectual, "Death to the SS bandits and Gestapo hangmen!" Whoever was stopped with such material by the men in leather coats and with drawn pistols could have no illusions of any kind. I also did not allow myself any for a single moment. For, God knows, I regarded myself—wrongly, as I see today—as an old, hardened expert on the system, its men, and its methods. A reader of the *Neue Weltbühne* and the *Neues*

Tagebuch in times past, well up on the KZ* literature of the German emigration from 1933 on, I believed to anticipate what was in store for me. Already in the first days of the Third Reich I had heard of the cellars of the SA barracks on Berlin's General Pape Street. Soon thereafter I had read what to my knowledge was the first German KZ document, the little book *Oranienburg* by Gerhart Segers. Since that time so many reports by former Gestapo prisoners had reached my ears that I thought there could be nothing new for me in this area. What would take place would then have to be incorporated into the relevant literature, as it were. Prison, interrogation, blows, torture; in the end, most probably death. Thus it was written and thus it would happen. When, after my arrest, a Gestapo man ordered me to step away from the window—for he knew the trick, he said, with your chained hands you tear open the window and leap onto a nearby ledge—I was certainly flattered that he credited me with so much determination and dexterity, but, obeying the order, I politely gestured that it did not come into question. I gave him to understand that I had neither the physical prerequisites nor at all the intention to escape my fate in such an adventurous way. I knew what was coming and they could count on my consent to it. But does one really know? Only in part. "Rien n'arrive ni comme on l'espère, ni comme on le craint," Proust writes somewhere. Nothing really happens as we hope it will, nor as we fear it will. But not because the occurrence, as one says, perhaps "goes beyond the imagination" (it is not a quantitative question), but because it is reality and not phantasy. One can devote an entire life to comparing the imagined and the real, and still never accomplish anything by it. Many things do indeed happen approximately the way they were anticipated in the imagination: Gestapo men in leather coats, pistol pointed at their victim—that is correct, all right. But then, almost amazingly, it dawns on one that the fellows not only have leather coats and pistols, but also faces: not "Gestapo faces" with twisted noses, hypertrophied chins, pockmarks, and knife scars, as might appear in a book, but rather faces like anyone else's. Plain, ordinary faces. And the enormous perception at a later stage, one that destroys all abstractive imagination, makes clear to us how the plain, ordinary faces finally become Gestapo faces after all, and how evil overlays and exceeds banality. For there is no "banality of evil," and Hannah Arendt, who wrote about it in her Eichmann book, knew the enemy of mankind only from hearsay, saw him only through the glass cage.

When an event places the most extreme demands on us, one ought not to speak of banality. For at this point there is no longer any abstraction and never an imaginative power that could even approach its reality. That someone is carried away shackled in an auto is "self-evident" only when you read about it in the newspaper and you rationally tell yourself, just at the moment when you are packing fliers: well of course, and what more? It can and it will happen like that to me someday, too. But the

* *Konzentrationslager*, or concentration camp.

auto is different, and the pressure of the shackles was not felt in advance, and the streets are strange, and although you may previously have walked by the gate of the Gestapo headquarters countless times, it has other perspectives, other ornaments, other ashlars when you cross its threshold as a prisoner. Everything is self-evident, and nothing is self-evident as soon as we are thrust into a reality whose light blinds us and burns us to the bone. What one tends to call "normal life" may coincide with anticipatory imagination and trivial statement. I buy a newspaper and am "a man who buys a newspaper." The act does not differ from the image through which I anticipated it, and I hardly differentiate myself personally from the millions who performed it before me. Because my imagination did not suffice to entirely capture such an event? No, rather because even in direct experience everyday reality is nothing but codified abstraction. Only in rare moments of life do we truly stand face to face with the event and, with it, reality.

It does not have to be something as extreme as torture. Arrest is enough and, if need be, the first blow one receives. "If you talk," the men with the plain, ordinary faces said to me, "then you will be put in the military police prison. If you don't confess, then it's off to Breendonk, and you know what that means." I knew, and I didn't know. In any case, I acted roughly like the man who buys a newspaper, and spoke as planned. I would be most pleased to avoid Breendonk, with which I was quite familiar, and give the evidence desired of me. Except that I unfortunately knew nothing, or almost nothing. Accomplices? I could name only their aliases. Hiding places? But one was led to them only at night, and the exact addresses were never entrusted to us. For these men, however, that was far too familiar twaddle, and it didn't pay them to go into it. They laughed contemptuously. And suddenly I felt—*the first blow.*

In an interrogation, blows have only scant criminological significance. They are tacitly practiced and accepted, a normal measure employed against recalcitrant prisoners who are unwilling to confess. If we are to believe the above-cited lawyer, Alec Mellor, and his book *La Torture,* then blows are applied in more or less heavy doses by almost all police authorities, including those of the Western-democratic countries, with the exception of England and Belgium. In America one speaks of the "third degree" of a police investigation, which supposedly entails something worse than a few punches. France has even found an argot word that nicely plays down a beating by the police. One speaks of the prisoner's "passage à tabac." After the Second World War a high French criminal investigator, in a book intended for his subordinates, still explained in extravagant detail that it would not be possible to forgo physical compulsion at interrogations, "within the bounds of legality."

Mostly, the public does not prove to be finicky when such occurrences in police stations are revealed now and then in the press. At best, there may be an interpellation in Parliament by some leftist-oriented deputy. But then the stories fizzle out; I have never yet heard of a police official who had beaten a prisoner and was not energetically covered by his superior officers. Simple blows, which really are entirely incommensurable with actual torture, may almost never create a far-reaching echo among the public, but for the person who suffers them they are still experiences that leave deep marks—if one wishes not to use up the high-sounding words already and clearly say: enormities. The first blow brings home to the prisoner that he is *helpless,* and thus it already contains in the bud everything that is to come. One may have known about torture and death in the cell, without such knowledge having possessed the hue of life; but upon the first blow they are anticipated as real possibilities, yes, as certainties. They are permitted to punch me in the face, the victim feels in numb surprise and concludes in just as numb certainty: they will do with me what they want. Whoever would rush to the prisoner's aid—a wife, a mother, a brother, or friend—he won't get this far.

Not much is said when someone who has never been beaten makes the ethical and pathetic statement that upon the first blow the prisoner loses his human dignity. I must confess that I don't know exactly what that is: human dignity. One person thinks he loses it when he finds himself in circumstances that make it impossible for him to take a daily bath. Another believes he loses it when he must speak to an official in something other than his native language. In one instance human dignity is bound to a certain physical convenience, in the other to the right of free speech, in still another perhaps to the availability of erotic partners of the same sex. I don't know if the person who is beaten by the police loses human dignity. Yet I am certain that with the very first blow that descends on him he loses something we will perhaps temporarily call "trust in the world." Trust in the world includes all sorts of things: the irrational and logically unjustifiable belief in absolute causality perhaps, or the likewise blind belief in the validity of the inductive inference. But more important as an element of trust in the world, and in our context what is solely relevant, is the certainty that by reason of written or unwritten social contracts the other person will spare me—more precisely stated, that he will respect my physical, and with it also my metaphysical, being. The boundaries of my body are also the boundaries of my self. My skin surface shields me against the external world. If I am to have trust, I must feel on it only what I *want* to feel.

At the first blow, however, this trust in the world breaks down. The other person, *opposite* whom I exist physically in the world and *with* whom I can exist only as long as he does not touch my skin surface as border, forces his own corporeality on me with the first blow. He is on me and thereby destroys me. It is like a rape, a sexual act without the consent of one of the two partners. Certainly, if there is even a minimal prospect of successful resistance, a mechanism is set in motion that enables me to rectify the border violation by the other person. For my part, I can expand in urgent self-defense, objectify my own corporeality, restore the trust in my continued existence. The social contract then has another text and other clauses: an eye for an eye and a tooth for a tooth. You can also regulate your life according to that. You *cannot* do it when it is the other one who knocks out the tooth, sinks the eye into a swollen mass, and you yourself suffer on your body the counter-man that your fellow man became. If no help can be expected, this physical overwhelming by the other then becomes an existential consummation of destruction altogether.

The expectation of help, the certainty of help, is indeed one of the fundamental experiences of human beings, and probably also of animals. This was quite convincingly presented decades ago by old Kropotkin, who spoke of "mutual aid in nature," and by the modern animal behaviorist Lorenz. The expectation of help is as much a constitutional psychic element as is the struggle for existence. Just a moment, the mother says to her child who is moaning from pain, a hot-water bottle, a cup of tea is coming right away, we won't let you suffer so! I'll prescribe you a medicine, the doctor assures, it will help you. Even on the battlefield, the Red Cross ambulances find their way to the wounded man. In almost all situations in life where there is bodily injury there is also the expectation of help; the former is compensated by the latter. But with the first blow from a policeman's fist, against which there can be no defense and which no helping hand will ward off, a part of our life ends and it can never again be revived.

Here it must be added, of course, that the reality of the police blows must first of all be accepted, because the existential fright from the first blow quickly fades and there is still room in the psyche for a number of practical considerations. Even a sudden joyful surprise is felt; for the physical pain is not at all unbearable. The blows that descend on us have above all a subjective spatial and acoustical quality: spatial, insofar as the prisoner who is being struck in the face and on the head has the impression that the room and all the visible objects in it are shifting position by jolts; acoustical, because he believes to hear a dull thundering, which finally submerges in a general roaring. The blow acts as its own anesthetic. A feeling of pain that would be comparable to a violent toothache or the pulsating burning of a festering wound does not emerge. For that reason, the beaten person thinks roughly this: well now, that can be put up with; hit me as much as you want, it will get you nowhere.

It got them nowhere, and they became tired of hitting me. I kept repeating only that I knew nothing, and therefore, as they had threatened, I was presently off, not to the army-administered Brussels prison, but to the "Reception Camp Breendonk," which was controlled by the SS. It would be tempting to pause here and to tell of the auto ride from Brussels to Breendonk through twenty-five kilometers of Flemish countryside, of the wind-bent poplars, which one saw with pleasure, even if the shackles hurt one's wrists. But that would sidetrack us, and we must quickly come to the point. Let me mention only the ceremony of driving through the first gate over the drawbridge. There even the Gestapo men had to present their identification papers to the SS guards, and if, despite all, the prisoner had doubted the seriousness of the situation, here, below the watchtowers and at the sight of the submachine guns, in view of the entrance ritual, which did not lack a certain dark solemnity, he had to recognize that he had arrived at the end of the world.

Very quickly one was taken into the "business room," of which I have already spoken. The business that was conducted here obviously was a flourishing one. Under the picture of Himmler, with his cold eyes behind the pince-nez, men who wore the woven initials SD on the black lapels of their uniforms went in and out, slamming doors and making a racket with their boots. They did not condescend to speak with the arrivals, either the Gestapo men or the prisoners. Very efficiently they merely recorded the information contained on my false identity card and speedily relieved me of my rather inconsiderable possessions. A wallet, cuff links, and my tie were confiscated. A thin gold bracelet aroused derisive attention, and a Flemish SS man, who wanted to appear important, explained to his German comrades that this was the sign of the partisans. Everything was recorded in writing, with the precision befitting the occurrences in a business room. Father Himmler gazed down contentedly onto the flag that covered the rough wooden table, and onto his people. They were dependable.

The moment has come to make good a promise I gave. I must substantiate why, according to my firm conviction, torture was the essence of National Socialism—more accurately stated, why it was precisely in torture that the Third Reich materialized in all the density of its being. That torture was, and is, practiced elsewhere has already been dealt with. Certainly. In Vietnam since 1964. Algeria 1957. Russia probably between 1919 and 1953. In Hungary in 1919 the Whites and the Reds tortured. There was torture in Spanish prisons by the Falangists as well as the Republicans. Torturers were at work in the semifascist Eastern European states of the period between the two World Wars, in Poland, Romania, Yugoslavia. Torture was no invention of National Socialism. But it was its apotheosis. The Hitler vassal did not yet achieve his full identity if he was merely as quick as a weasel, tough as leather, hard as Krupp steel. No Golden Party Badge made of him a fully valid representative of the Führer and his ideology, nor did any Blood Order or Iron Cross. He had to *torture,* destroy, in order to be great in bearing the suffering of others. He had to be capable of handling torture instruments, so that Himmler would assure him his Certificate of Maturity in History; later generations would admire him for having obliterated his feelings of mercy.

Again I hear indignant objection being raised, hear it said that not Hitler embodied torture, but rather something unclear, "totalitarianism." I hear especially the example of Communism being shouted at me. And didn't I myself just say that in the Soviet Union torture was practiced for thirty-four years? And did not already Arthur Koestler* . . . ? Oh yes, I know, I know. It is impossible to discuss here in detail the political "Operation Bewilderment" of the postwar period, which defined Com-

*Koestler was the author of *Darkness at Noon,* a novel about the victims of the purge trials in the Soviet Union under Stalin during the 1930s.

munism and National Socialism for us as two not even very different manifestations of one and the same thing. Until it came out of our ears, Hitler and Stalin, Auschwitz, Siberia, the Warsaw Ghetto Wall and the Berlin Ulbricht-Wall were named together, like Goethe and Schiller, Klopstock and Wieland. As a hint, allow me to repeat here in my own name and at the risk of being denounced what Thomas Mann once said in a much attacked interview: namely, that no matter how terrible Communism may at times appear, it still symbolizes an idea of man, whereas Hitler-Fascism was not an idea at all, but depravity. Finally, it is undeniable that Communism could de-Stalinize itself and that today in the Soviet sphere of influence, if we can place trust in concurring reports, torture is no longer practiced. In Hungary a Party First Secretary can preside who was himself once the victim of Stalinist torture. But who is really able to imagine a de-Hitlerized National Socialism and, as a leading politician of a newly ordered Europe, a Röhm* follower who in those days had been dragged through torture? No one can imagine it. It would have been impossible. For National Socialism—which, to be sure, could not claim a single idea, but did possess a whole arsenal of confused, crackbrained notions—was the only political system of this century that up to this point had not only practiced the rule of the antiman, as had other Red and White terror regimes also, but had expressly established it as a principle. It hated the word "humanity" like the pious man hates sin, and that is why it spoke of "sentimental humanitarianism." It exterminated and enslaved. This is evidenced not only by the corpora delicti, but also by a sufficient number of theoretical confirmations. The Nazis tortured, as did others, because by means of torture they wanted to obtain information important for national policy. But in addition they tortured with the good conscience of depravity. They martyred their prisoners for definite purposes, which in each instance were exactly specified. Above all, however, they tortured because they were torturers. They placed torture in their service. But even more fervently they were its servants.

When I recall those past events, I still see before me the man who suddenly stepped into the business room and who seemed to count in Breendonk. On his field-gray uniform he wore the black lapels of the SS, but he was addressed as "Herr Leutnant." He was small, of stocky figure, and had that fleshy, sanguine face that in terms of popular physiognomy would be called "gruffly good-natured." His voice crackled hoarsely, the accent was colored by Berlin dialect. From his wrist there hung in a leather loop a horsewhip of about a meter in length. But why, really, should I withhold his name, which later became so familiar to me? Perhaps at this very hour he is faring well and feels content with his healthily sunburned self as he drives home from his Sunday excursion. I have no reason not to name him. The Herr Leutnant, who played the role of a torture specialist here, was named Praust. P – R – A – U – S – T. "Now it's coming,"

*Ernst Röhm, head of the Nazi storm troopers, was killed together with many of his followers in a purge ordered by Hitler in 1934.

he said to me in a rattling and easygoing way. And then he led me through the corridors, which were dimly lit by reddish bulbs and in which barred gates kept opening and slamming shut, to the previously described vault, the bunker. With us were the Gestapo men who had arrested me.

If I finally want to get to the analysis of torture, then unfortunately I cannot spare the reader the objective description of what now took place; I can only try to make it brief. In the bunker there hung from the vaulted ceiling a chain that above ran into a roll. At its bottom end it bore a heavy, broadly curved iron hook. I was led to the instrument. The hook gripped into the shackle that held my hands together behind my back. Then I was raised with the chain until I hung about a meter over the floor. In such a position, or rather, when hanging this way, with your hands behind your back, for a short time you can hold at a half-oblique through muscular force. During these few minutes, when you are already expending your utmost strength, when sweat has already appeared on your forehead and lips, and you are breathing in gasps, you will not answer any questions. Accomplices? Addresses? Meeting places? You hardly hear it. All your life is gathered in a single, limited area of the body, the shoulder joints, and it does not react; for it exhausts itself completely in the expenditure of energy. But this cannot last long, even with people who have a strong physical constitution. As for me, I had to give up rather quickly. And now there was a crackling and splintering in my shoulders that my body has not forgotten until this hour. The balls sprang from their sockets. My own body weight caused luxation; I fell into a void and now hung by my dislocated arms, which had been torn high from behind and were now twisted over my head. Torture, from Latin *torquere,* to twist. What visual instruction in etymology! At the same time, the blows from the horsewhip showered down on my body, and some of them sliced cleanly through the light summer trousers that I was wearing on this twenty-third of July 1943.

It would be totally senseless to try and describe here the pain that was inflicted on me. Was it "like a red-hot iron in my shoulders," and was another "like a dull wooden stake that had been driven into the back of my head"? One comparison would only stand for the other, and in the end we would be hoaxed by turn on the hopeless merry-go-round of figurative speech. The pain was what it was. Beyond that there is nothing to say. Qualities of feeling are as incomparable as they are indescribable. They mark the limit of the capacity of language to communicate. If someone wanted to impart his physical pain, he would be forced to inflict it and thereby become a torturer himself.

Since the *how* of pain defies communication through language, perhaps I can at least approximately state *what* it was. It contained everything that we already ascertained earlier in regard to a beating by the police: the border violation of my self by the other, which can be neither neutralized by the expectation of help nor rectified through resistance. Torture is all that, but in addition very much more. Whoever is overcome by pain through torture experiences his body as never before. In self-negation, his flesh becomes a total reality. Partially, torture is one of those life experiences that in a milder form present themselves also to the consciousness of the patient who is awaiting help, and the popular saying according to which we feel well as long as we do not feel our body does indeed express an undeniable truth. But only in torture does the transformation of the person into flesh become complete. Frail in the face of violence, yelling out in pain, awaiting no help, capable of no resistance, the tortured person is only a body, and nothing else beside that. If what Thomas Mann described years ago in *The Magic Mountain* is true, namely, that the more hopelessly man's body is subjected to suffering, the more physical he is, then of all physical celebrations torture is the most terrible. In the case of Mann's consumptives, they still took place in a state of euphoria; for the martyred they are death rituals.

It is tempting to speculate further. Pain, we said, is the most extreme intensification imaginable of our bodily being. But maybe it is even more, that is: death. No road that can be travelled by logic leads us to death, but perhaps the thought is permissible that through pain a path of feeling and premonition can be paved to it for us. In the end, we would be faced with the equation: Body = Pain = Death, and in our case this could be reduced to the hypothesis that torture, through which we are turned into body by the other, blots out the contradiction of death and allows us to experience it personally. But this is an evasion of the question. We have for it only the excuse of our own experience and must add in explanation that torture has an indelible character. Whoever was tortured, stays tortured. Torture is ineradicably burned into him, even when no clinically objective traces can be detected. The permanence of torture gives the one who underwent it the right to speculative flights, which need not be lofty ones and still may claim a certain validity.

I speak of the martyred. But it is time to say something about the tormentors also. No bridge leads from the former to the latter. Modern police torture is without the theological complicity that, no doubt, in the Inquisition joined both sides; faith united them even in the delight of tormenting and the pain of being tormented. The torturer believed he was exercising God's justice, since he was, after all, purifying the offender's soul; the tortured heretic or witch did not at all deny him this right. There was a horrible and perverted togetherness. In present-day torture not a bit of this remains. For the tortured, the torturer is solely the other, and here he will be regarded as such.

Who were the others, who pulled me up by my dislocated arms and punished my dangling body with the horsewhip? As a start, one can take the view that they were merely brutalized petty bourgeois and subordinate bureaucrats of torture. But it is necessary to abandon this point of view immediately if one wishes to arrive at an insight into evil that is more than just banal. Were they sadists, then? According to my well-founded conviction, they were not sadists in the narrow sexual-pathologic sense. In general, I don't believe that I encountered a single genuine sadist of this sort during my two years of imprisonment by the Gestapo and in concentration camps. But probably they *were* sadists if we leave sexual pathology aside and attempt to judge the torturers according to the categories of, well, the *philosophy* of the Marquis de Sade. Sadism as the dis-ordered view of the world is something other than the sadism of the usual psychology handbooks, also other than the sadism interpretation of Freudian analysis. For this reason, the French anthropologist Georges Bataille will be cited here, who has reflected very thoroughly on the odd Marquis. We will then perhaps see not only that my tormentors lived on the border of a sadistic philosophy but that National Socialism in its totality was stamped less with the seal of a hardly definable "totalitarianism" than with that of *sadism.*

For Georges Bataille, sadism is to be understood not in the light of sexual pathology but rather in that of existential psychology, in which it appears as the radical negation of the other, as the denial of the social principle as well as the reality principle. A world in which torture, destruction, and death triumph obviously cannot exist. But the sadist does not care about the continued existence of the world. On the contrary: he wants to nullify this world, and by negating his fellow man, who also in an entirely specific sense is "hell" for him, he wants to realize his own total sovereignty. The fellow man is transformed into flesh, and in this transformation he is already brought to the edge of death; if worst comes to worst, he is driven beyond the border of death into Nothingness. With that the torturer and murderer realizes his own destructive being, without having to lose himself in it entirely, like his martyred victim. He can, after all, cease the torture when it suits him. He has control of the other's scream of pain and death; he is master over flesh and spirit, life and death. In this way, torture becomes the total inversion of the social world, in which we can live only if we grant our fellow man life, ease his suffering, bridle the desire of our ego to expand. But in the world of torture man exists only by ruining the other person who stands before him. A slight pressure by the tool-wielding hand is enough to turn the other—along with his head, in which are perhaps stored Kant and Hegel, and all nine symphonies, and the World as Will and Representation—into a shrilly squealing piglet at slaughter. When it has happened and the torturer has expanded into the body of his fellow man and extinguished what was his spirit, he himself can then smoke a cigarette or sit down to breakfast or, if he has the desire, have a look in at the World as Will and Representation.

My boys at Breendonk contented themselves with the cigarette and, as soon as they were tired of torturing, doubtlessly let old Schopenhauer be. But this still does not mean that the evil they inflicted on me was banal. If one insists on it, they were bureaucrats of torture. And yet, they were also much more. I saw it in their serious, tense faces, which were not swelling, let us say, with sexual-sadistic delight, but concentrated in murderous self-realization. With heart and soul they went about their business, and the name of it was power, dominion over spirit and flesh, orgy of unchecked self-expansion. I also have not forgotten that there were moments when I felt a kind of wretched admiration for the agonizing sovereignty they exercised over me. For is not the one who can reduce a person so entirely to a body and a whimpering prey of death a god or, at least, a demigod?

But the concentrated effort of torture naturally did not make these people forget their profession. They were "cops," that was métier and routine. And so they continued asking me questions, constantly the same ones: accomplices, addresses, meeting places. To come right out with it: I had nothing but luck, because especially in regard to the extorting of information our group was rather well organized. What they wanted to hear from me in Breendonk, I simply did not know myself. If instead of the aliases I had been able to name the real names, perhaps, or probably, a calamity would have occurred, and I would be standing here now as the weakling I most likely am, and as the traitor I potentially already was. Yet it was not at all that I opposed them with the heroically maintained silence that befits a real man in such a situation and about which one may read (almost always, incidentally, in reports by people who were not there themselves). I talked. I accused myself of invented absurd political crimes, and even now I don't know at all how they could have occurred to me, dangling bundle that I was. Apparently I had the hope that, after such incriminating disclosures, a well-aimed blow to the head would put an end to my misery and quickly bring on my death, or at least unconsciousness. Finally, I actually did become unconscious, and with that it was over for a while—for the "cops" abstained from awakening their battered victim, since the nonsense I had foisted on them was busying their stupid heads.

It was over for a while. It still is not over. Twenty-two years later I am still dangling over the ground by dislocated arms, panting, and accusing myself. In such an instance there is no "repression." Does one repress an unsightly birthmark? One can have it removed by a plastic surgeon, but the skin that is transplanted in its place is not the skin with which one feels naturally at ease.

One can shake off torture as little as the question of the possibilities and limits of the power to resist it. I have spoken with many comrades about this and have attempted to relive all kinds of experiences. Does the brave man resist? I am not sure. There was, for example, that young Belgian aristocrat who converted to Communism and was something like a hero, namely in the Spanish civil war, where he had fought on the Republican side. But when they subjected him to torture in Breendonk, he "coughed up," as it is put in the jargon of common criminals, and since he knew a lot, he betrayed an entire organization. The brave man went very far in his readiness to cooperate. He drove with the Gestapo men to the homes of his comrades and in extreme zeal encouraged them to confess just everything, but absolutely everything, that was their only hope, and it was, he said, a question of paying any price in order to escape torture. And I knew another, a Bulgarian professional revolutionary, who had been subjected to torture compared to which mine was only a somewhat strenuous sport, and who had remained silent, simply and steadfastly silent. Also the unforgettable Jean Moulin, who is buried in the Pantheon in Paris, shall be remembered here. He was arrested as the first chairman of the French Resistance Movement. If he had talked, the entire Résistance would have been destroyed. But he bore his martyrdom beyond the limits of death and did not betray one single name.

Where does the strength, where does the weakness come from? I don't know. *One* does not know. No one has yet been able to draw distinct borders between the "moral" power of resistance to physical pain and "bodily" resistance (which likewise must be placed in quotation marks). There are more than a few specialists who reduce the entire problem of bearing pain to a purely physiological basis. Here only the French professor of surgery and member of the Collège de France, René Leriche, will be cited, who ventured the following judgment:

"We are not equal before the phenomenon of pain," the professor says.

> One person already suffers where the other apparently still perceives hardly anything. This has to do with the individual quality of our sympathetic nerve, with the hormone of the parathyroid gland, and with the vasoconstrictive substances of the adrenal glands. Also in the physiological observation of pain we cannot escape the concept of individuality. History shows us that we people of today are more sensitive to pain than our ancestors were, and this from a purely physiological standpoint. I am not speaking here of any hypothetical moral power of resistance, but am staying within the realm of physiology. Pain remedies and narcosis have contributed more to our greater sensitivity than moral factors. Also the reactions to pain by various people are absolutely not the same. Two wars have given us the opportunity to see how the physical sensitivities of the Germans, French, and English differ. Above all, there is a great separation in this regard between the Europeans on the one hand and the Asians and Africans on the other. The latter bear physical pain incomparably better than the former . . .

Thus the judgment of a surgical authority. It will hardly be disputed by the simple experiences of a nonprofessional, who saw many individuals and members of numerous ethnic groups suffering pain and deprivation. In this connection, it occurs to me that, as I was able to observe later in the concentration camp, the Slavs, and especially the Russians, bore physical injustice easier and more stoically than did, for example, Italians, Frenchmen, Hollanders, or Scandinavians. As body, we actually are not equal when faced with pain and torture. But that does not solve our problem of the power of resistance, and it gives us no conclusive answer to the question of what share moral and physical factors have in it. If we agree to a reduction to the purely physiological, then we run the risk of finally pardoning every kind of whiny reaction and physical cowardice. But if we exclusively stress the so-called moral resistance, then we would have to measure a weakly seventeen-year-old gymnasium pupil who fails to withstand torture by the same standards as an athletically built thirty-year-old laborer who is accustomed to manual work and hardships. Thus we had better let the question rest, just as at that time I myself did not further analyze my power to resist, when, battered and with my hands still shackled, I lay in the cell and ruminated.

For the person who has survived torture and whose pains are starting to subside (before they flare up again) experiences an ephemeral peace that is conducive to thinking. In one respect, the tortured person is content that he was body only and because of that, so he thinks, free of all political concern. You are on the outside, he tells himself more or less, and I am here in the cell, and that gives me a great superiority over you. I have experienced the ineffable, I am filled with it entirely, and now see, if you can, how you are going to live with yourselves, the world, and my disappearance. On the other hand, however, the fading away of the physical, which revealed itself in pain and torture, the end of the tremendous tumult that had erupted in the body, the reattainment of a hollow stability, is satisfying and soothing. There are even euphoric moments, in which the return of weak powers of reason is felt as an extraordinary happiness. The bundle of limbs that is slowly recovering human semblance feels the urge to articulate the experience intellectually, right away, on the spot, without losing the least bit of time, for a few hours afterward could already be too late.

Thinking is almost nothing else but a great astonishment. Astonishment at the fact that you had endured it, that the tumult had not immediately led also to an explosion of the body, that you still have a forehead that you can stroke with your shackled hands, an eye that can be opened and closed, a mouth that would show the usual lines if you could see it now in a mirror. What? you ask yourself—the same person who was gruff with his family because of a toothache was able to hang there by his dislocated arms and still live? The person who for hours was in a bad mood after slightly burning his finger with a cigarette was lacerated here with a horsewhip, and now that it is all over he hardly feels his wounds? Astonishment also at the fact that what happened to you yourself, by right was supposed to befall only those who had written about it in accusatory brochures: torture. A murder is committed, but it is part of the newspaper that reported on it. An airplane accident occurred, but that concerns the people who lost a relative in it. The Gestapo tortures. But that was a matter until now for the somebodies who were tortured and who displayed their scars at antifascist conferences. That suddenly you yourself are the Somebody, is grasped only with difficulty. That, too, is a kind of alienation.

If from the experience of torture any knowledge at all remains that goes beyond the plain nightmarish, it is that of a great amazement and a foreignness in the world that cannot be compensated by any sort of subsequent human communication. Amazed, the tortured person experienced that in this world there can be the other as absolute sovereign, and sovereignty revealed itself as the power to inflict suffering and to destroy.

The dominion of the torturer over his victim has nothing in common with the power exercised on the basis of social contracts, as we know it. It is not the power of the traffic policeman over the pedestrian, of the tax official over the taxpayer, of the first lieutenant over the second lieutenant. It is also not the sacral sovereignty of past absolute chieftains or kings; for even if they stirred fear, they were also objects of trust at the same time. The king could be terrible in his wrath, but also kind in his mercy; his autocracy was an exercise of authority. But the power of the torturer, under which the tortured moans, is nothing other than the triumph of the survivor over the one who is plunged from the world into agony and death.

Astonishment at the existence of the other, as he boundlessly asserts himself through torture, and astonishment at what one can become oneself: flesh and death. The tortured person never ceases to be amazed that all those things one may, according to inclination, call his soul, or his mind, or his consciousness, or his identity, are destroyed when there is that cracking and splintering in the shoulder joints. That life is fragile is a truism he has always known—and that it can be ended, as Shakespeare says, "with a little pin." But only through torture did he learn that a living person can be transformed so thoroughly into flesh and by that, while still alive, be partly made into a prey of death.

Whoever has succumbed to torture can no longer feel at home in the world. The shame of destruction cannot be erased. Trust in the world, which already collapsed in part at the first blow, but in the end, under torture, fully, will not be regained. That one's fellow man was experienced as the antiman remains in the tortured person as accumulated horror. It blocks the view into a world in which the principle of hope rules. One who was martyred is a defenseless prisoner of fear. It is *fear* that henceforth reigns over him. Fear—and also what is called resentments. They remain, and have scarcely a chance to concentrate into a seething, purifying thirst for revenge.

9. Elie Wiesel

Elie Wiesel was born in 1928 in the town of Sighet in Transylvania, a region that had shifted between Romanian and Hungarian control, but was under the latter's rule when German troops occupied their former ally in the spring of 1944. Adolf Eichmann's SS unit moved in with them, and soon began the mass deportation of hundreds of thousands of Hungarian Jews, chiefly from rural areas, to the deathcamp at Auschwitz. Among them were the 15,000 Jews of Sighet, including Wiesel, his parents, and three sisters. His mother and younger sister were sent directly to the gas chambers; along with his two older sisters (who survived the war), he and his father were chosen for work.

Wiesel, who writes in French, tells of his experiences in Auschwitz, and later in Buchenwald, in his most famous work, the autobiographical memoir *Night* (1958). An earlier and much longer Yiddish version, called *Un di velt hot geshvign* (*And the World Remained Silent,* 1956), appeared in Buenos Aires, but has never been published in English. Left alone after his father's death in Buchenwald, the young Wiesel was sent to France and eventually enrolled in the Sorbonne. As a correspondent for newspapers in Tel Aviv, Paris, and New York, he traveled widely, learned English, settled in New York, and became an American citizen in 1963. Since *Night,* he has written more than thirty volumes, including novels, essays, plays, and biblical commentaries reflecting his prewar study of Torah, Talmud, and the Hasidic masters.

Since 1976, Wiesel has been Andrew Mellon Professor in the Humanities at Boston University. He has received numerous literary prizes, and in 1986 was awarded the Nobel Peace Prize. "A Plea for the Dead," from *Legends of Our Time* (1968), illustrates Wiesel's lifelong resolve to protect the victims of Nazi persecution from misguided efforts to blame them for their own fate.

A Plea for the Dead

I was not quite fifteen when, for the first time, completely fascinated, I was present at a strange discussion about dignity and death and the possible relationship between the two.

People who were dead and did not know it yet were discussing the necessity, rather than the possibility, of meeting death with dignity.

The reality of certain words escaped me and the weight of that reality as well. The people around me were talking and I did not understand.

Now, I am more than twenty years older, and all paths leading to the cemetery are known to me. The discussion still goes on. Only the participants have changed. Those of twenty years ago have died and they know it now. As for me, I understand even less than before.

I had just stepped off into unreality. It must have been about midnight. Later, I learned that executioners are usually romantic types who like perfect productions: they find in darkness a stage setting and in night an ally.

Somewhere a dog began to howl, another echoed him, then a third. We were, it seemed, in the kingdom of dogs. One of the women went mad and let out a cry that no longer resembled anything human; it was more like barking; no doubt she wanted to become a dog herself. A pistol shot put an end to her hallucination; silence fell over us again. In the distance, red and yellow billows of fire, spewed out by immense smokestacks, rose toward the moonless sky, as if to set it aflame. A quarter of an hour before, or less, our train had stopped at a small suburban station. Standing at the grates, people read the name aloud: *Auschwitz.*

Someone asked: "We've arrived?"

Another answered: "I think so."

"Auschwitz, you know it?"

"No. Not at all."

The name evoked no memory, linked itself to no anguish. Ignorant in matters of geography, we supposed it was a small peaceful spot somewhere in Silesia. We did not yet know it had already made history with its populace of several million dead Jews. We learned it one minute later, when the train doors opened into an earsplitting din and when an army of inmates began to shout: "Last stop! Everybody off!"

Like conscientious tour guides, they described the surprises in store for us: "You're acquainted with Auschwitz? You're not? Too bad, you'll get to know it, it won't be long before you know it."

They sneered: "Auschwitz, you don't know Auschwitz? Really not? Too bad. Someone is waiting for you here. Who? Why, death, of course. Death is waiting for you. It waits only for you. Look and you will see."

And they pointed to the fire in the distance.

Later, many years later, I asked one of my friends: "What were your first impressions of Auschwitz?"

Somber, he answered me: "I found it a spectacle of terrifying beauty."

I found it neither beautiful nor terrifying. I was young and I simply refused to believe my eyes and ears. I thought: our guides are mocking us in order to scare us. It amuses them. We are living in the twentieth century, after all; Jews are not burned anymore. The civilized world would not allow it. My father walked alongside me, on my left, his head bowed. I asked him: "The Middle Ages are behind us, aren't they, Father, far behind us?"

He did not answer me.

I asked him: "I'm dreaming, Father. Am I not dreaming?"

He did not answer me.

We kept moving on toward the unknown. It was then, like a whisper, a feverish discussion went through the ranks. Some youths overcoming their stupor, grasping at their anger, called for a revolt. Without arms? Yes, without arms. Fingernails, fists, and a few penknives hidden in their clothes, that will suffice. But won't that mean certain death? Yes, so what? There is nothing more to lose and everything to gain, especially honor, that can still be gained, honor. To die as free men: that is what they advocated, those youths. There is defeat only in resignation.

But their fathers were opposed. They went on dreaming. And waiting. They invoked the Talmud: "God can intervene, even at the very last moment, when everything seems lost. We must not rush things, we must not lose faith or hope."

The argument won everyone over. I asked my father: "What do you think?"

This time he did answer me: "Thinking isn't much use anymore."

The human herd marched ahead, we did not know where our steps were leading us. No: we knew, our guides had told us. But we pretended not to know. And the discussions continued. The young were in favor of rebellion, their elders against. The former finally conceded, one must obey one's parents, the Bible says; their wishes must be respected.

And so the revolt did not take place.

In recent times, many people are beginning to raise questions about the problem of the incomprehensible if not enigmatic behavior of Jews in what was concentration-camp Europe. Why did they march into the night the way cattle go to the slaughterhouse? Important, if not essential, for it touches on timeless truth; this question torments men of good conscience who feel the need to be quickly reassured, to have the guilty parties named and their crimes defined, to have unraveled for them the meaning of a history which they have not experienced except through intermediaries. And so those millions of Jews, whom so-called civilized society had abandoned to despair and to agonize in silence and then in oblivion, suddenly are all brought back up to the surface to be drowned in a flood of words. And since we live at a time when small talk is king, the dead offer no resistance. The role of ghost is imposed upon them and they are bombarded with questions: "Well, now, what was it really like? How did you feel in Minsk and in Kiev and in Kolomea, when the earth, opening up before your eyes, swallowed up your sons and your prayers? What did you think when you saw blood—your own blood—gushing from the bowels of the earth, rising up to the sun? Tell us, speak up, we want to know, to suffer with you, we have a few tears in reserve, they pain us, we want to get rid of them."

One is sometimes reduced to regretting the good old days when this subject, still in the domain of sacred memory, was considered taboo, reserved for the initiates, who spoke of it only with hesitation and fear, always lowering their eyes, and always trembling with humility, knowing themselves unworthy and recognizing the limits of their language, spoken and unspoken.

Now in the name of objectivity, not to mention historical research, everyone takes up the subject without the slightest embarrassment. Accessible to every mind, to every intellect in search of stimulation, this has become the topic of fashionable conversation. Why not? It replaces Brecht, Kafka, and communism, which are now overdone, overworked. In intellectual, or pseudo-intellectual circles, in New York and elsewhere too, no cocktail party can really be called a success unless Auschwitz, sooner or later, figures in the discussion. Excellent remedy for boredom; a good way to ignite passions. Drop the names of a few recent works on this subject, and watch minds come alive, one more brilliant, more arrogant than the next. Psychiatrists, comedians, and novelists, all have their own ideas about the subject, all are clear, each is ready to provide all the answers, to explain all the mysteries: the cold cruelty of the executioner and the cry which strangled the victim, and even the fate that united them to play on the same stage, in the same cemetery. It is as simple as saying hello. As hunger, thirst, and hate. One need only understand history, sociology, politics, psychology, economics; one need know only how to add. And to accept the axiom that everywhere A + B = C. If the dead are dead, if so many dead are dead, that is because they desired their own death, they were lured, driven by their own instincts. Beyond the diversity of all the theories, the self-assurance of which cannot but arouse anger, all unanimously conclude that the victims, by participating in the executioner's game, in varying degrees shared responsibility.

The novelty of this view cannot fail to be striking. Until recently, Jews have been held responsible for everything under the sun, the death of Jesus, civil wars, famines, unemployment, and revolutions: they were thought to embody evil; now, they are held responsible for their own death: they embody that death. Thus, we see that the Jewish problem continues to be a kind of no-man's land of the mind where anyone can say anything in any way at all—a game in which everyone wins. Only the dead are the losers.

And in this game—it is really nothing else—it is quite easy to blame the dead, to accuse them of cowardice or complicity (in either the concrete or metaphysical sense of that term). Now, this game has a humiliating aspect. To insist on speaking in the name of the dead—and to say: these are their motivations, these the considerations that weakened their wills, to speak in their name—this is precisely to humiliate them. The dead have earned something other than this posthumous humiliation. I never before wholly understood why, in the Jewish faith, anything that touches corpses is impure. Now I begin to understand.

Let us leave them alone. We will not dig up those corpses without coffins. Leave them there where they must forever be and such as they must be: wounds, immeasurable pain at the very depth of our being. Be content they do not wake up, that they do not come back to the earth to judge the living. The day that they would begin to tell what they have seen and heard, and what they have taken most to heart, we will not know where to run, we will stop up our ears, so great will be our fear, so sharp our shame.

I could understand the desire to dissect history, the strong urge to close in on the past and the forces shaping it; nothing is more natural. No question is more important for our generation which is the generation of Auschwitz, or of Hiroshima, tomorrow's Hiroshima. The future frightens us, the past fills us with shame: and these two feelings, like those two events, are closely linked, like cause to effect. It is Auschwitz that will produce Hiroshima, and if the human race should perish by the nuclear bomb, this will be the punishment for Auschwitz, where, in the ashes, the hope of man was extinguished.

And Lot's apprehensive wife, she was right to want to look back and not be afraid to carry the burning of doomed hope. "Know where you come from," the sages of Israel said. But everything depends on the inner attitude of whoever looks back to the beginning: if he does so purely out of intellectual curiosity, his vision will make of him a statue in some salon. Unfortunately, we do not lack statues these days; and what is worse, they speak, as if from the top of the mountain.

And so I read and I listen to these eminent scholars and professors who, having read all the books and confronted all the theories, proclaim their erudition and their power to figure everything out, to explain everything, simply by performing an exercise in classification.

At times, especially at dawn, when I am awakened by the first cry I heard the first night behind barbed wire, a desire comes over me to say to all these illustrious writers who claim to go to the bottom of it all: "I admire you, for I myself stumble when I walk this road; you claim to know everything, there again I admire you: as for me, I know nothing.

What is to be done, I know I am still incapable of deciphering—for to do so would be to blaspheme—the frightened smile of that child torn away from his mother and transformed into a flaming torch; nor have I been able, nor will I ever be able, to grasp the shadow which, at that moment, invaded the mother's eyes. You can, undoubtedly. You are fortunate, I ought to envy you, but I do not. I prefer to stand on the side of the child and of the mother who died before they understood the formulas and phraseology which are the basis of your science."

Also, I prefer to take my place on the side of Job, who chose questions and not answers, silence and not speeches. Job never understood his own tragedy which, after all, was only that of an individual betrayed by God; to be betrayed by one's fellow men is much more serious. Yet, the silence of this man, alone and defeated, lasted for seven days and seven nights; only afterward, when he identified himself with his pain, did he feel he had earned the right to question God. Confronted with Job, our silence should extend beyond the centuries to come. And we dare speak on behalf of our knowledge? We dare say: *"I know"?* This is how and why victims were victims and executioners executioners? We dare interpret the agony and anguish, the self-sacrifice before the faith and the faith itself of six million human beings, all named Job? Who are we to judge them?

One of my friends, in the prime of life, spent a night studying accounts of the holocaust, especially the Warsaw Ghetto. In the morning he looked at himself in the mirror and saw a stranger: his hair had turned white. Another lost not his youth but his reason. He plunged back into the past and remains there still. From time to time I visit him in his hospital room; we look at one another and we are silent. One day, he shook himself and said to me: "Perhaps one should learn to cry."

I should envy those scholars and thinkers who pride themselves on understanding this tragedy in terms of an entire people; I myself have not yet succeeded in explaining the tragedy of a single one of its sons, no matter which.

I have nothing against questions: they are useful. What is more, they alone are. To turn away from them would be to fail in our duty, to lose our only chance to be able one day to lead an authentic life. It is against the answers that I protest, regardless of their basis. Answers: I say there are none. Each of these theories contains perhaps a fraction of truth, but their sum still remains beneath and outside what, in that night, was truth. The events obeyed no law and no law can be derived from them. The subject matter to be studied is made up of death and mystery, it slips away between our fingers, it runs faster than our perception: it is everywhere and nowhere. Answers only intensify the question: ideas and words must finally come up against a wall higher than the sky, a wall of human bodies extending to infinity.

For more than twenty years, I have been struggling with these questions. To find one answer or another, nothing is easier: language can mend anything. What the answers have in common is that they bear no relation to the questions. I cannot believe that an entire generation of fathers and sons could vanish into the abyss without creating, by their very disappearance, a mystery which exceeds and overwhelms us. I still do not understand what happened, or how, or why. All the words in all the mouths of the philosophers and psychologists are not worth the silent tears of that child and his mother, who live their own death twice. What can be done? In my calculations, all the figures always add up to the same number: six million.

Some time ago, in Jerusalem, I met by chance one of the three judges in the Eichmann trial. This wise and lucid man, of uncompromising character, is, to use an expression dear to Camus, at once a person and a personage. He is, in addition, a conscience.

He refused to discuss the technical or legal aspects of the trial. Having told him that side was of no interest to me, I asked him the following question:

"Given your role in this trial, you ought to know more about the scope of the holocaust than any living person, more even than those who lived through it in the flesh and in their memory. You have studied all the documents, read all the secret reports, interrogated all the witnesses. Now tell me: do you *understand* this fragment of the past, those few pages of history?"

He shuddered imperceptibly, then, in a soft voice, infinitely humble, he confessed:

"No, not at all. I know the facts and the events that served as their framework; I know how the tragedy unfolded minute by minute, but this knowledge, as if coming from outside, has nothing to do with understanding. There is in all this a portion which will always remain a mystery; a kind of forbidden zone, inaccessible to reason. Fortunately, as it happens. Without that . . ."

He broke off suddenly. Then, with a smile a bit timid, a bit sad, he added:

"Who knows, perhaps that's the gift which God, in a moment of grace, gave to man: it prevents him from understanding everything, thus saving him from madness, or from suicide."

In truth, Auschwitz signifies not only the failure of two thousand years of Christian civilization, but also the defeat of the intellect that wants to find a Meaning—with a capital *M*—in history. What Auschwitz embodied has none. The executioner killed for nothing, the victim died for nothing. No God ordered the one to prepare the stake, nor the other to mount it. During the Middle Ages, the Jews, when they chose death, were convinced that by their sacrifice they were glorifying and sanctifying God's name. At Auschwitz the sacrifices were without point, without faith, without divine inspiration. If the suffering of one human being has any meaning, that of six million has none. Numbers have their own importance; they prove, according to Piotr Rawicz, that God has gone mad.

I attended the Eichmann trial, I heard the prosecutor try to get the witnesses to talk by forcing them to expose themselves and to probe the innermost recesses of their being: Why didn't you resist? Why didn't you attack your assassins when you still outnumbered them?

Pale, embarrassed, ill at ease, the survivors all responded in the same way: "You cannot understand. Anyone who was not there cannot imagine it."

Well, I was there. And I still do not understand. I do not understand that child in the the Warsaw Ghetto who wrote in his diary: "I'm hungry, I'm cold; when I grow up I want to be a German, and then I won't be hungry anymore, and I won't be cold anymore."

I still do not understand why I did not throw myself on the Kapo who was beating my father before my very eyes. In Galicia, Jews dug their own graves and lined up, without any trace of panic, at the edge of the trench to await the machine-gun barrage. I do not understand their calm. And that woman, that mother, in the bunker somewhere in Poland, I do not understand her either; her companions smothered her child for fear its cries might betray their presence; that woman, that mother, having lived this scene of biblical intensity, did not go mad. I do not understand her: why, and by what right, and in the name of what, did she not go mad?

I do not know why, but I forbid us to ask her the question. The world kept silent while the Jews were being massacred, while they were being reduced to the state of objects good for the fire; let the world at least have the decency to keep silent now as well. Its questions come a bit late; they should have been addressed to the executioner. Do they trouble us? Do they keep us from sleeping in peace? So much the better. We want to know, to understand, so we can turn the page: is that not true? So we can say to ourselves: the matter is closed and everything is back in order. Do not wait for the dead to come to our rescue. Their silence will survive them.

We have questions? Very good. We do not want to put them to the executioner—who lives in happiness if not in glory at home in Germany—well then, pass them on to those who claim they never participated in the game, to those who became accomplices through their passivity. Their "ignorance" of the facts hardly excuses them, it was willful.

In London and in Washington, in Basel and in Stockholm, high officials had up-to-date information about every transport carrying its human cargo to the realm of ashes, to the kingdom of mist. In 1942–1943, they already possessed photographs documenting the reports; all were declared "confidential" and their publication prohibited.

Not many voices were raised to warn the executioner that the day of punishment is at hand; not many voices were raised to effectively console the victims: that there will be punishment and that the reign of night is only temporary.

Perhaps Eichmann was a small man after all. Hitler's Germany was full of small men like him, all carefully seeing to it that the extermination machine functioned well and efficiently. But, large and small, all were sure that in one regard, that of Nazi policy regarding Jews, they would have nothing to account for the day after their defeat: the fate of the Jews interested no one. Someday they would have to give back the occupied territories and eventually pay the victors for war damages, that is only normal. But the Jewish question would not weigh on them. The Allies could not have cared less about what the SS did with its Jews. In that area, the Eichmanns could act with impunity. It is only in this way one can understand how Heinrich Himmler, Grand Master of the death camps, could, toward the end of the war, have conceived the possibility of becoming the best negotiator for a separate peace with the western Allies; the fact that his successful direction of the annihilation of whole Jewish populations might disqualify him never even crossed his mind. And when, with feigned irony, Eichmann declared that no country was interested in saving Jews, he was telling the truth. Eichmann may have lied about his own role, but he did not lie about that of the Allies or of the neutral camp.

In fact, the Germans, known more for maniacal prudence rather than impulsiveness, developed their anti-Jewish policy step by step, gradually, stopping after each measure to catch their breath, after each move, to watch the reactions. There was always a respite between the different stages, between the Nuremberg laws and the Kristallnacht, between expropriation and deportation, between the ghettos and liquidation. After each infamy, the Germans expected a storm of outrage from the free world; they quickly became aware of their error: they were allowed to proceed. Of course, here and there, there were a few speeches, a few editorials, all indignant, but things stopped there. So, in Berlin, they knew what that meant. They said to themselves: since we have been given the green light, we can go on. Moreover, they were convinced—in all sincerity—that someday other peoples would be grateful to them for having done the job for them. Almost all the important Nazis expressed this idea in their writings; it also appeared in their speeches. They were killing the Jews for the good of the world, not only for the good of Germany. After all, the Germans should not be accused of thinking only of themselves.

I maintain that by forceful action, only once, by taking a stand without ambiguities, the free world would have been able to force the Germans to draw back, or at least to plan on a smaller scale. It is conceivable that for Berlin the absence of such action could only have meant a tacit agreement, unacknowledged, on the part of the Allied powers. One need only glance through the newspapers of the period to become disgusted with the human adventure on this earth: the phenomenon of the concentration camps, despite its horror and its overwhelming ramifications, took up less space, on the whole, than did ordinary traffic accidents.

It would be a mistake to believe the inmates of the camps were ignorant of this. Knowing themselves abandoned, excluded, rejected by the rest of humanity, their walk to death, as haughty as it was submissive, became an act of lucidity, of protest, and not of acceptance and weakness.

Yes: the transport of which I was a part did not rebel on the night of our arrival. What must be added is that the young men spoke also of the necessity of alerting the outside world: naïve, they still believed the Germans were doing their work secretly, like thieves, that the Allies knew nothing about it, for if they knew, the massacre would stop immediately. "We will fight," they said. "We will break this silence and the world will know that Auschwitz is a reality." I shall never forget the old man who, in a calm voice, terribly calm, answered them: "You are young and brave, my children; you still have a lot to learn. The world knows, no need to inform it. It knew before you did, but it doesn't care, it won't lose a minute thinking about our fate. Your revolt will have no bearing, no echo." The old man spoke without bitterness; he stated the facts. He was Polish and two years before had seen his family slaughtered: I do not know how he managed to escape and to slip across two frontiers before arriving among us, a refugee. "Save your strength for later," he told our young men. "Don't waste it." But they were persistent. "Even if you are right," they rejoined, "even if what you say is true, that doesn't change the situation. Let us prove our courage and our dignity, let us show these murderers and the world that Jews know how to die like free men, not like hunched-up invalids." "As a lesson, I like that," the old man's voice reached me. "But they don't deserve it."

Then we all held our heads up high and murmuring the words of the Kaddish we marched ahead, almost like conquerors, toward the gates of death where the elegant physician Dr. Josef Mengele—white gloves, monocle, and the rest—accomplished the sacred ritual of selection, of separating those who would live from those who would die.

The old man had seen things as they were. Had the Jews been able to think they had allies outside, men who did not look the other way, perhaps they might have acted differently. But the only people interested in the Jews were the Germans. The others preferred not to look, not to hear, not to know. The solitude of the Jews, caught in the clutches of the beast, has no precedent in history. It was total. Death guarded all the exits.

It was worse than the Middle Ages. Then, driven from Spain, the Jews were welcomed in Holland. Persecuted in one country, they were invited to another, given time to take heart again. But during the Hitler era the conspiracy against them seemed universal. The English closed off the gates of Palestine, the Swiss accepted only the rich—and later the children—while the poor and the adult, their right to life denied, were driven back into the darkness. "Even if I had been able to sell a million Jews, who would have bought them?" asked Eichmann, not without sarcasm, alluding to the Hungarian episode. Here again he was telling the truth. "What do you want us to do with a million Jews?" echoed the honorable Lord Moyne, British ambassador to Cairo. It is as though every country—

and not only Germany—had decided to see the Jew as a kind of subhuman species, an unnecessary being, not like others; his disappearance did not count, did not weigh on the conscience. He was a being to whom the concept of human brotherhood did not apply, a being whose death did not diminish us, a being with whom one did not identify. One could therefore do with him what one would, without violating the laws of the spirit; one could take away his freedom and joy without betraying the ideal of man. I often wonder what the world's reaction would have been had the Nazi machine ground up and burned day after day, not twenty thousand Jews, but twenty thousand Christians. It is better not to think about that too much.

If I dwell so long on the culpability of the world, it is not to lessen that of the Germans, nor to "explain" the behavior of their victims. We tend to forget.

The fact, for example, that in the spring of 1944 we, in Transylvania, knew nothing about what was happening in Germany is proof of the world's guilt. We listened to the short-wave radio, from London and Moscow: not a single broadcast warned us not to leave with the transports, not one disclosed the existence, not even the name of Auschwitz. In 1943, when she read three lines in a Hungarian newspaper concerning the Warsaw Ghetto uprising, my mother remarked: "But why did they do it? Why didn't they wait *peacefully* for the end of hostilities?" Had we known what was happening there, we might have been able to flee, to hide: the Russian front was only thirty kilometers away. But we were kept in the dark.

At the risk of offending, it must be emphasized that the victims suffered more, and more profoundly, from the indifference of the onlookers than from the brutality of the executioner. The cruelty of the enemy would have been incapable of breaking the prisoner; it was the silence of those he believed to be his friends—cruelty more cowardly, more subtle—which broke his heart.

There was no longer anyone on whom to count: even in the camps this became evident. *"From now on we shall live in the wilderness, in the void: blotted out of history."* It was this conviction which poisoned the desire to live. If this is the world we were born into, why cling to it? If this is the human society we come from—and are now abandoned by—why seek to return?

At Auschwitz, not only man died, but also the idea of man. To live in a world where there is nothing anymore, where the executioner acts as god, as judge—many wanted no part of it. It was its own heart the world incinerated at Auschwitz.

Let no one misinterpret. I speak without hatred, without bitterness. If at times I do not succeed in containing my anger, it is because I find it shocking if not indecent that one must plead to protect the dead. For that is the issue: they are being dug up in order to be pilloried. The questions asked of them are only reproaches. They are being blamed, these corpses, for having acted as they did: they should have played their roles differently, if only to reassure the living who might thus go on believing in the nobility of man. We do not like those men and women for whom the sky became a common grave. We speak of them without pity, without compassion, without love. We juggle their thousand ways of dying as if performing intellectual acrobatics: our heart is not in it. More than that: we despise them. For the sake of convenience, and also to satisfy our mania to classify and define everything, we make some distinctions: between the Germans and the Judenrat,* between the Kapos and the ghetto police, between the nameless victim and the victim who obtained a reprieve for a week, for a month. We judge them and we hand out certificates for good or bad conduct. We detest some more than others: we are on the other side of the wall, and we know exactly the degree of guilt of each of them. On the whole, they inspire our disgust rather than our anger.

That is what I reproach us for: our boundless arrogance in thinking we know everything. And that we have the right to pass judgment on an event which should, on the contrary, serve as proof that we are poor, and that our dreams are barren—when they are not bloody.

I plead for the dead and I do not say they are innocent; that is neither my intention nor aim. I say simply we have no right to judge them; to confer innocence upon them is already to judge them. I saw them die and if I feel the need to speak of guilt, it is always of my own that I speak. I saw them go away and I remained behind. Often I do not forgive myself for that.

Of course, in the camps I saw men conquered, weak, cruel. I do not hesitate to admit I hated them, they frightened me; for me, they represented a danger greater than the Germans. Yes, I have known sadistic Kapos; yes, I have seen Jews, a savage gleam burning in their eyes, whipping their own brothers. But, though they played the executioner's game, they died as victims. When I think about it, I am still astonished that so few souls were lost, so few hearts poisoned, in that kingdom of the night, where one breathed only hate, contempt, and self-disgust. What would have become of me had I stayed in the camps longer, five years, or seven, or twelve? I have been trying to answer that question for more than twenty years and at times, after a sleepless night, I am afraid of the answer. But many people are not afraid. These questions, which are discussed as one might discuss a theorem or a scientific problem, do not frighten us. For that, too, I reproach us.

Since the end of the nightmare I search the past, whose prisoner I shall no doubt forever remain. I am afraid, but I still pursue my quest. The further I go, the less I understand. Perhaps there is nothing to understand.

On the other hand, the further I go, the more I learn of the scope of the betrayal by the world of the living against the world of the dead. I take my head in my hands and I think: it is insanity, that is the explana-

*Jewish Council, the administrative body in most ghettoes.

tion, the only conceivable one. When so great a number of men carry their indifference to such an extreme, it becomes sickness, it resembles madness.

Otherwise how to explain the Roosevelts, the Churchills, the Eisenhowers, who never expressed their indignation? How to explain the silence of the Pope? How to explain the failure of certain attempts in London, in Washington, to obtain from the Allies an aerial bombardment of the death factories, or at least of the railway lines leading to them?

One of the saddest episodes of that war, not lacking in sad episodes, had as hero a Polish Jewish leader exiled in London: to protest the inaction of the Allies, and also to alert public opinion, Arthur Ziegelbaum, member of the "National Committee to Free Poland," put a bullet through his head in broad daylight in front of the entrance to the House of Commons. In his will he expressed his hope that his protest would be heard.

He was quickly forgotten, his death proved useless. Had he believed his refusal to live among men voluntarily blind would move them, he had been wrong. Ziegelbaum dead or Ziegelbaum living: to those hearts of stone it was all the same. For them he was only a Polish Jew talking about Jews and living their agony; for them he might just as well have perished over there, with the others. Arthur Ziegelbaum died for nothing. Life went on, so did the war: against the Axis powers, which continued their own war against the Jews. And the world stopped up its ears and lowered its eyes. Sometimes the newspapers printed a small item: the Ghetto of Lodz had been liquidated, the number of European Jews massacred already exceeded two or three million. This news was published as if these were normal events, almost without comment, without anguish. It seemed normal that Jews should be killed by the Nazis. Never had the Jewish people been so alone.

The more I search, the more reasons I find for losing hope. I am often afraid to reopen this Pandora's box, there are always the newly guilty to emerge from it. Is there no bottom to this evil box? No.

I repeat: hatred is no solution. There would be too many targets. The Hungarians put more passion than did the Germans into the persecution of Jews; the Rumanians displayed more savagery than the Germans; the Slovaks, the Poles, the Ukrainians: they hunted down Jews cunningly, as if with love. Perhaps I should hate them, it would cure me. But I am incapable. Were hatred a solution, the survivors, when they came out of the camps, would have had to burn down the whole world.

Now almost everywhere I am told: you mustn't bear a grudge against us, we didn't know, we didn't believe it, we were powerless to do anything. If these justifications suffice to assuage people's consciences, too bad for them. I could answer that they did not want to know, that they refused to believe, that they could have forced their governments to break the conspiracy of silence. But that would open the door to discussion. It is too late, in any case: the time for discussion is past.

Now, I shall simply ask: is it any surprise that the Jews did not choose resistance? And die fighting like soldiers for the victory of their cause? But what victory and what cause?

Let me reveal a secret, one among a thousand, about why Jews did not resist: to punish us, to prepare a vengeance for us for later. We were not worth their sacrifice. If, in every town and every village, in the Ukraine and in Galicia, in Hungary and in Czechoslovakia, Jews formed endless nightly processions and marched on to eternity as if carrying within themselves a pure joy, one which heralds the approach of ecstasy, it is precisely to reveal to us the ultimate truth about those who are sacrificed on the margins of history. In staying alive, at that price, we deserve neither salvation nor atonement. Nor do we even deserve that lesson of solemn dignity and lofty courage which, in spite of everything, in their own way, they gave us by making their way toward death, staring it full in the face, point blank, their heads high in the joy of bearing this strength, this pride within themselves.

Let us, therefore, not make an effort to understand, but rather to lower our eyes and not understand. Every rational explanation would be more esoteric than if it were mystical. Not to understand the dead is a way of paying them an ancient debt; it is the only way to ask their pardon.

I have before me a photograph, taken by a German officer fond of souvenirs, of a father who, an instant before the burst of rifle fire, was still speaking calmly to his son, while pointing to the sky. Sometimes I think I hear his dreamy voice: "You see, my son, we are going to die and the sky is beautiful. Do not forget there is a connection between these two facts." Or perhaps: "We are going to die, my son, yet the sky, so serene, is not collapsing in an end-of-the-world crash. Do you hear its silence? Listen to it, you must not forget it." It occurs to me that were I to ask him a question, any question, that same father would answer me. But I bury my eyes in what remains of him and I am silent.

Just as I am silent every time the image comes to my mind of the Rebbe in Warsaw who stood erect, unyielding, unconquerable, before a group of SS; they were amusing themselves by making him suffer, by humiliating him; he suffered, but did not let himself be humiliated. One of them, laughing, cut off his beard, but the Rebbe stared right into his eyes without flinching; there was pain in his expression, but also defiance, the expression of a man stronger than evil, even when evil is triumphant, stronger than death, even when death assumes the face of a comedian playing a farce—the expression of a man who owes nothing to anyone, not even to God.

I have long since carried that expression buried within me, I have not been able to part with it, I no longer want to part with it, as though wanting always to remember there are still, there will always be, somewhere in the world, expressions I will never understand. And when such an expression lights upon me, at the dinner table, at a concert, or beside a happy woman, I give myself up to it in silence.

For the older I grow, the more I know that we can do little for the dead; the least we can do is to leave them alone, not project our own guilt onto them. We like to think the dead have found eternal rest: let them be. It is dangerous to wake them. They, too, have questions, questions equal to our own.

My plea is coming to an end, but it would be incomplete if I said nothing about the armed assaults which, in spite of what the prosecution may think, Jews did carry out against the Germans. If I have difficulty understanding how multitudes went to their death without defending themselves, that difficulty becomes insurmountable when it comes to understanding those of their companions who chose to fight.

How, in the ghettos and camps, they were able to find the means to fight when the whole world was against them—that will always remain a mystery.

For those who claim that all the Jews submitted to their murderers, to fate, in common cowardice or common resignation, those people do not know what they are saying or—what is worse—knowingly falsify the facts only to illustrate a sociological theory, or to justify a morbid hatred which is always self-hatred.

In truth, there was among the victims an active elite of fighters composed of men and women and children who, with pitiful means, stood up against the Germans. They were a minority, granted. But is there any society where the active elite is not a minority? Such groups existed in Warsaw, in Bialystok, in Grodno, and—God alone knows how—even in Treblinka, in Sobibor, and in Auschwitz. Authenticated documents and eye-witness accounts do exist, relating the acts of war of those poor desperadoes; reading them, one does not know whether to rejoice with admiration or to weep with rage. One wonders: but how did they do it, those starving youngsters, those hunted men, those battered women, how were they able to confront, with weapons in hand, the Nazi army, which at that time seemed invincible, marching from victory to victory? Where did they take their sheer physical endurance, their moral strength? What was their secret and what is its name?

We say: weapons in hand. But what weapons? They had hardly any. They had to pay in pure gold for a single revolver. In Bialystok, the legendary Mordecai Tenenbaum-Tamaroff, leader of the ghetto resistance, describes in his journal—miraculously rediscovered—the moment he obtained the first rifle, the first ammunition: twenty-five bullets. "Tears came to my eyes. I felt my heart burst with joy." It was thus with one rifle and twenty-five bullets that he and his companions were going to contain the vast onslaught of the German army. It is easy to imagine what might have happened had every warrior in every ghetto obtained one rifle.

All the underground networks in the occupied countries received arms, money, and radio equipment from London, and secret agents came regularly to teach them the art of sabotage: they felt themselves organically linked to the outside world. In France or Norway a member of the resistance who was caught could comfort himself with the thought that somewhere in that town as well as on the other shore, there were people who feared for his life, who lived in anxiety because of him, who would move heaven and earth to save him: his acts registered somewhere, left traces, marks of sorrow, produced results. But the Jews were alone: only they were alone.

They alone did not receive help or encouragement; neither arms nor messages were sent them; they were not spoken to, no one was concerned with them; they did not exist. They cried for help, but the appeals they issued by radio or by mail fell on deaf ears. Cut off from the world, from the war itself, the Jewish fighters participated, fully aware they were not wanted, they had already been written off; they threw themselves into battle knowing they could count on no one, help would never arrive, they would receive no support, there would be no place to retreat. And yet, with their backs to the burning wall, they defied the Germans. Some battles are won even when they are lost.

Yes, competent elite existed even at Sobibor, where they organized an escape; at Treblinka, where they revolted; and at Auschwitz, where they blew up the crematoria. The Auschwitz insurgents attempted an escape, but in the struggle with the SS, who obviously had an advantage of superiority in weapons and men, all were killed. Later the Germans arrested the four young Jewish girls from Warsaw who had obtained the explosives for the insurgents. They were tortured, condemned to death, and hanged at a public ceremony. They died without fear. The oldest was sixteen, the youngest twelve.

We can only lower our heads and be silent. And end this sickening posthumous trial which intellectual acrobats everywhere are carrying on against those whose death numbs the mind. Do we want to understand? There is no longer anything to understand. Do we want to know? There is nothing to know anymore. It is not by playing with words and the dead that we will understand and know. Quite the contrary. As the ancients said: "Those who know do not speak; those who speak do not know."

But we prefer to speak and to judge. We wish to be strong and invulnerable. The lesson of the holocaust—if there is any—is that our strength is only illusory, and that in each of us is a victim who is afraid, who is cold, who is hungry. Who is also ashamed.

The Talmud teaches man never to judge his friend until he has been in his place. But, for the world, the Jews are not friends. They have never been. Because they had no friends they are dead.

So, learn to be silent.

II

Journals and Diaries

Although diaries of the deathcamp experience were still being dug up in the ruins of the gas chambers at Auschwitz as late as 1962, they were few in number, often fragmentary, and usually in poor condition. The most detailed records of Jewish existence during the Holocaust are to be found in archives and chronicles from such ghettos as Warsaw, Vilna, Lodz, and Kovno. Carefully hidden or buried at the time, they surfaced after the war to furnish vivid accounts of how an oppressed and deceived people managed the daily rhythms of life while the Germans contrived their relentless migration toward death.

Reading these narratives is a chastening task. More than any other kind of Holocaust writing, they force us to forgo the wisdom of hindsight and to tackle inquiries more modest than the larger question of why the victims did not foresee their fate and prepare to resist it. Retrospective vision seeks to integrate reality, even though all the evidence suggests that most ghetto residents faced a constant *dis*integration—through hunger, illness, or despair—of the family and community supports that ordinarily help us to thrive. Looking backward, as Tolstoy insisted in *War and Peace,* always makes history seem less random than it actually is.

If we enter the world of the ghetto through the eyes of its witnesses, many of whom left their written testimonies as their sole legacy to the future, we can begin to imagine how a constant wavering between expectation and futility drained the self of the strength that sustains the individual in more normal times. For example, among the papers rescued from the Lodz ghetto is the diary of an anonymous young girl who, in daily entries, noted the quantity of food she consumed each day, measured in decagrams (roughly equivalent to one-third ounces). Of course, she had other themes, but this was the focus of her being, exposing an inner gnawing that invaded everything else she wrote about:

> *Saturday, 7 March 1942*
> Beautiful sunny day today. When the sun shines, my mood is lighter. How sad life is. One yearns for a different life, better than this grey and sad one in the ghetto. When we look at the fence separating us from the

> rest of the world, our souls, like birds in a cage, yearn to be free. . . . How I envy the birds that fly to freedom. Longing breaks my heart, visions of the past come to me. Will I ever live in better times? After the war will I be with my parents and friends? Will I live to eat bread and rye flour until I'm full? Meanwhile, hunger is terrible. Again we have nothing to cook. I bought 1/4 kg. [about 1/2 pound] of rye flour for 11 1/2 RM [Reichsmarks, the currency at the time]. Everybody wants to live.

Another young diarist, who perished of tuberculosis in the ghetto at the age of nineteen, wrote: "Yesterday a student in our class died from hunger exhaustion. Because he looked awful, he was allowed to have as much soup in school as he wanted, but it didn't help. He was the third victim of starvation in the class." Such details give us a glimpse of the ceaseless if unequal struggle between everyone's desire to live and the firm resolve of a regime determined to see them die.

But the ghetto chronicles also address much more comprehensive issues, even though sometimes indirectly. We now know what the Jews in the ghettos did not: that the Germans planned to concentrate them there only as a stage in their eventual extermination. We learn much about the Jewish administration of the ghettos from these documents, less about the German officials pulling the strings from the wings. If Mordechai Chaim Rumkowski, the Elder of the Jews in Lodz, turns out to have been a pathetic megalomaniac, deceiving and self-deceived, and Adam Czerniaków, head of the Jewish Council in Warsaw, a helpless man of integrity, less deceiving than deceived, reading between the lines from the vantage point of today enables us to see that, despite their efforts to preserve life, both were little more than pawns in the hands of the Germans, who maneuvered them with lies until they were ready to strike.

The selections in this section give us insight into the minds and wills of men and women who want to live but have been deprived of the means. Once we realize that *nothing* Rumkowski or Czerniaków might have done could have saved the lives of the Jews of Lodz or Warsaw, we can perhaps approach judgment of their behavior with a more lenient stance. Rumkowski may have believed to the end that he had sacrificed the "few" to save the many: in this, he has his defenders and his foes. In Czerniaków's opinion, Rumkowski was "replete with self-praise, a conceited and witless man." Refusing to preside over the deportation of Warsaw's Jews to Treblinka, Czerniaków took poison and ended his life. This gesture, too, has earned both rebuke and admiration. The more we immerse ourselves in the daily ordeal of the ghetto residents, leaders and ordinary inhabitants alike, the more we see that they were all faced with a choice between impossibilities—no meaningful choice at all. Even the courageous Warsaw ghetto uprising was a venture in behalf of an impossibility: it could save almost no one. Its fighters were not eager to die; they preferred to live. The Germans would not allow it.

If the picture emerging from these ghetto documents resembles a moral labyrinth more than the serene spiritual vista we often hear about, this results not from something inherent in the victims' natures, but from the malicious intentions of their killers. Although some of the German officials in charge of the ghettos (prodded by the Labor Ministry) agreed to keep alive as long as possible those Jews producing uniforms and munitions for the military, they also cooperated with the SS as they systematically reduced the size of the ghettos by deporting less "useful" members of the community to the deathcamps. Offering false assurances to the beleaguered leaders, who passed them on half-heartedly to their desperate people, the oppressors preyed on their victims' instinct for hope. Rumors of impending improvement in these pages expel darker hints of doom, as the Jews cling to the shreds of life even as the fate of friends and family members gnaw at their will to survive.

Only extreme psychological naïveté might lead us to be surprised. Among the most hated groups in certain ghettos, after the SS, were the Jewish police, appointed by the administration and used to maintain order, but also to round up candidates for deportation. Although many Jews refused to serve in such units, many enrolled, and for obvious reasons: they and their families were temporarily exempt, higher up in the hierarchy of those who felt immune from persecution. In the end, predictably, along with Jewish Council members and Elders like Rumkowski himself, they, too, were shipped to their death. The Germans never planned to spare anyone. Why didn't the Jewish police decline to do the dirty work of the Germans, and volunteer to go first themselves? This too is a naïve question, born of our privileged sense of being well fed and unthreatened. The purpose of these excerpts is to carry us back into the reality of that milieu, governed by daily fear and hunger, when most human beings were stripped of the luxury of caring for anyone but one's own.

One of the bitterest truths we encounter in these ghetto records is how easily the concept of "one's own" shrank into the narrow identity of one's *self.* With ruthless, heartbreaking honesty, Abraham Lewin admits that he did not join his wife when she was rounded up for deportation because he was afraid: he was not prepared to submit to death. Should her misfortune be his, too? Those who criticize him from the safety of not having to make that choice themselves only confirm the abyss separating us from the moral maze of those days, when conventional ethical systems were challenged by an unprecedented external situation that left the victim bereft of guidance.

This does not mean that members of the Jewish Council in Kovno, Czerniaków in Warsaw, and even Rumkowski in Lodz did not argue or negotiate with their German masters to better conditions for their fellow Jews, often at considerable risk to themselves. They reasoned and pleaded to increase rations, reduce the number of deportees, improve housing and sanitation—and sometimes they succeeded. But German appeasement had nothing to do with pity; it was all part of a covert deception to mislead their victims and allay suspicion, drawing on the pliant human impulse to be beguiled when the issue is one's imminent death.

When 27,000 Jews from the Kovno ghetto are summoned to a massive "roll call" from which 10,000 would be selected for execution, we gain an unforgettable view of the miserable cruelty of survival. The Germans were interested in totals, not persons; if the figure was correct, the identity of the victims was trivial. In the chaos of the moment, much shifting from the "bad" side to the "good" side occurred. In addition, members of the Jewish Council were able to convince the SS to free some of those who had been selected if they were vital to the order of the community. But today, who can avert his face from the ruthless German algebra dictating that every life saved must be paid for with the death of someone else? And who is morally responsible for the feat of balancing this infernal equation?

After this mass selection of 10,000 doomed men, women, and children, the "reprieved" remnant returned to the ghetto. We know what they escaped *from;* what had they escaped *to?* Memory of their loss intervened to forestall relief at their survival, a rhythm prototypical of the Holocaust experience. Immediately, crowds thronged to the quarters of Dr. Elchanan Elkes, chairman of the Kovno Jewish Council, pleading with him to save those family members who had been taken away. He scrawled down as many names as he could, made his way to SS headquarters, and miraculously managed to gain permission to rescue 100 from among those who were on the way to their death. How was he to choose the 1 percent to be spared, while the other 99 percent perished? Armed with his approved petition, filled with despair and hope, he reached the ranks of the condemned, but his presence caused such turmoil among them that the Lithuanian guards, fearful of a rebellion, beat him to the ground with their rifle butts, and he was carried bleeding back into the ghetto.

Human beings should not have to ask themselves to make such choices—what I call choiceless choices—shorn of dignity and any of the spiritual renown we normally associate with moral effort. Nazi rule in the ghettos was designed to exploit and humiliate the ethical bias of its victims, even poisoning the value system implicit in language. If Elkes had succeeded in his mission, could he possibly have been considered "heroic"? Or would those he might have "saved" merely abide as hostages to the SS desire to nurture illusion among the living remainder? We know that the willingness of the SS to release 100 victims had nothing to do with their respect for the petitioner, who was trapped by a form of malice he was unable or unwilling to envision. What appeared to be courageous resistance from the Jewish point of view (and is, of course, to Elkes's credit) is nothing but contemptuous tolerance within the Nazi design of mass murder. Both choice and power were in the hands of the murderers, but failure to grant this caused some Jewish leaders to cling to a belief in the meaning of options whose results we are still trying to appraise.

The most notorious example of this dilemma is the "Give Me Your Children" speech delivered by Rumkowski in the Lodz ghetto on September 4, 1942. Evidently convinced that by surrendering the few he could preserve the many, Rumkowski conceded, "I must cut off limbs in order to save the body itself." But those "limbs" were all the children in the ghetto under the age of ten; did he worry that without its limbs, the body would lose much of its vital force? The questions that come so easily to us today—if the fear of death was so overwhelming then, what would the fear of *life* be like for those who survived the war, once they had time to reflect on the loss that had ensured *their* continuing existence?—may have seemed less urgent to a community robbed of its future, mired in the terror of the single day. Only a strenuous leap of the imagination, aided by the words of the chroniclers themselves, can lift us out of the flow of time and fix us in the arrested moment when Rumkowski's audience stood frozen in disbelief on that September afternoon.

By asking for their cooperation, did Rumkowski want to spare his people in Lodz the horror and brutality of the Warsaw ghetto, where even as he spoke the Germans were nearing the end of their vicious roundup and deportation of nearly 300,000 Jews to Treblinka? Was his rhetoric his own, or was he duped by Nazi promises? "So which is better?" he asks. "What do you want: that eighty to ninety thousand Jews remain, or, God forbid, that the whole population be annihilated?" This language is unmistakable; Rumkowski must have known the impending fate of the children and of the ill and incurable who were to join them (the Germans had demanded 24,000, but Rumkowski announced, and apparently believed, that he had bargained the figure down to "only" 20,000). Encouraged by his Nazi masters to trust in options, Rumkowski found his reasoning irreproachable: "[P]ut yourself in my place, think logically, and you'll reach the conclusion that I cannot proceed any other way. The part that can be saved is much larger than the part that must be given away."

If we shrink now from Rumkowski's request for the children, and from their parents' eventual consent to surrender them, we display a habit of mind that flourishes in an atmosphere free of the terror that prevailed then. Some survivors of the Lodz ghetto insist that they would be dead today if Rumkowski had not delayed the date of their deportation through such "arrangements." Czerniaków in Warsaw, however, committed suicide rather than ask his fellow Jews to give up their children for "resettlement." He was left without illusions and without hope, and that eroded his will to live. He apparently saw what Rumkowski balked at accepting: the part that was saved was *the same as* the part that was given away, all being candidates for destruction, some now, some later. Rumkowski's error was in assuming that sound premises would ensure stable conclusions. He refused to admit—and he was not alone in this—that the Third Reich buried logic as well as people.

The events described in these journals and diaries oblige us to forfeit the cherished idea that civilized men and women always achieve dignity by making moral choices and abiding by the results, whether they are beneficial or not. When the challenge to being is the minimal demand of staying alive, the test of character has little to do with spiritual stature. These texts cause us to revise our notion of what reasonable creatures are capable of under duress. The anonymous young diarist quoted earlier records her horror when one day she surprises her father in the act of furtively eating part of her mother's bread ration. She is dismayed, but less quick to condemn, because she knows how ceaseless hunger has infected her own will to live. The clash between family and community duties and the needs of the self was constant, remorseless, and ultimately beyond reconciling.

The appetite of the German death machine for more victims from the ghettos was unappeasable. The "lucky" ones died before deportation; we know the doom of the others. Choice had little to do with their fate, which took them beyond the frontiers of moral endeavor. If this is a harsh truth for us to absorb, how much more painful must it have been for the victims, who had to concede in their final misery that there were no limits to the evil their persecutors were capable of, while the good that might aid them was distressingly finite, drained by an ever-dwindling number of fragile chances to survive.

10. Abraham Lewin

Abraham Lewin was born in Warsaw in 1893. He came from an Orthodox Hasidic family. His father was a rabbi; his grandfather, a *shokhet* responsible for the ritual slaughtering that made food kosher. He attended Hebrew school as a child, and then studied at a yeshiva, or rabbinical academy. Lewin's father died when the boy was in his teens; by the time he was twenty, he had abandoned traditional Hasidic dress.

Forced to support his mother and sisters, Lewin took a job as teacher of Hebrew, biblical studies, and Jewish studies at a private Jewish secondary school for girls. Among the staff members was Emmanuel Ringelblum, who was to found the Oneg Shabbes archive, a secret record of the history of Warsaw Jewry under Nazi oppression, which included Abraham Lewin's diaries.

In 1928, Lewin married Luba Hotner, a teacher at the school. Her roundup and deportation is described with terrible desperation in "Diary of the Great Deportation," which is included here. In 1934, Lewin and his wife visited Palestine with their daughter, Ora, and considered emigrating, but Lewin's poor health forced them to return to Poland, where they remained until the German invasion in 1939 made further plans to leave impossible.

The Warsaw ghetto was established by the Germans in October 1940; the following month, they announced that it would be closed off from the rest of the city. Lewin probably began his diary entries a few months before this, though the surviving portions show an initial entry dated March 26, 1942. The Germans established a Judenrat, or Jewish Council, to govern the internal affairs of the Warsaw Jewish community. To head it they appointed Adam Czerniaków. Like Chaim Rumkowski in Lodz, Czerniaków worked tirelessly to improve conditions for his fellow Jews, but unlike Rumkowski he had fewer and fewer illusions about the impact of his negotiating powers. In July 1942, on the eve of the mass deportations to Treblinka, he committed suicide rather than preside over the presumable slaughter of the bulk of Warsaw's Jews.

Lewin's detailed description of this ordeal, which began on July 22 and lasted for fifty-four days, is nothing short of harrowing. Approximately 300,000 of the ghetto's Jews were sent to their deaths in the gas chambers, chiefly at Treblinka. At the end of the period, about 50,000 Jews remained. As the days drift by and Lewin meticulously records the names of his friends and associates who have vanished forever, one gets the sense of a nightmarish atmosphere of terror and despair that was slowly paralyzing and consuming an entire people. When the "action" finally paused in mid-September 1942—it was to resume, on a smaller scale, the following January—Lewin wrote in his diary: "Jewish Warsaw now has the air of a cemetery."

Lewin was not to live to see it totally reduced to ashes. His last entry is dated January 16, 1943, a few days before the beginning of the second "action," and one assumes that he and his daughter were caught in this roundup and sent to their deaths. But Lewin had not been naïve about the prospects for survival. On January 11, he wrote: "*[O]ver our heads hangs the perpetual threat of total annihilation.* It seems they have decided to exterminate the whole of European Jewry."

As a member of Ringelblum's Oneg Shabbes enterprise (a code name for the underground archive that literally means "Joy of the Sabbath" and refers to the custom of celebrating the end of a Sabbath service with light refreshments), Lewin shared the responsibility for chronicling for future generations all features of ghetto life. His devotion to this task must have been all-consuming, since it continued despite the loss of his wife, a personal tragedy to which he returned repeatedly in his diary. Together with other volumes of the archive, Lewin's diary was buried in milk cans and metal chests, from which some parts were recovered in 1946 and 1950. They included Lewin's entries from March 1942 to January 1943.

Before his own capture and execution, Ringleblum wrote of Lewin's work: "The clean and compressed style of the diary, its accuracy and precision in relating facts, and its grave contents qualify it as an important literary document which must be published as soon as possible after the War." The first part of Lewin's diary is written in Yiddish. The second, about the great deportation, is in Hebrew: portions were translated into Yiddish in the early 1950s. Both sections only appeared in English, as *A Cup of Tears: A Diary of the Warsaw Ghetto,* almost fifty years after the events they record.

Diary of the Great Deportation

A paper-shop at the corner of Nowy Świat,[1] Daughter of P., Cecylia, was shot.

At 14 Nowolipie Street, P. himself reappeared, miraculously back from the dead . . . This I heard from A.

28 Świętojerska Street—Monday—five in the afternoon eight in the evening, the house-porter at 13 Nalewki.[2]

Tuesday[3]—five in the morning—3 Niska Street, Smocza Street

The shopkeeper who opened for R. before five,

The policeman Ajzensztajn[4]

Dr. Sztajnk[5]

The arrests—in the Supply Office (ZZ)[6] and in the Jewish community offices.[7]

15 Chłodna Street—more than 10

The Day of Judgement—whence will come our help?

We are preparing ourselves for death. What will be our fate?

Karmelicka Street—round-up into vehicles.

There is talk of 20 dead since this morning

Szmul of the 'conquerers'—15 years.[8]

Someone called Rozen and his father and uncle were shot yesterday before ten o'clock going into the shop.

Wednesday, 22 July[9]—The Day Before Tishebov

A day of turmoil, chaos and fear: the news about the expulsion of Jews is spreading like lightning through the town, Jewish Warsaw has suddenly died, the shops are closed, Jews run by, in confusion, terrified. The Jewish streets are an appalling sight—the gloom is indescribable. There are dead bodies at several places. No one is counting them and no names are being given in this terrifying catastrophe. The expulsion is supposed to begin today from the hostels for the homeless,[10] and from the prisons. There is also talk of an evacuation of the hospital.[11] Beggar children are being rounded up into wagons. I am thinking about my aged mother—it would be better to put her to sleep than to hand her over to those murderers.[12]

Ora brings exaggerated stories from Sweden [that the war is coming to an end].[13]

Thursday, 23 July—Tishebov

Disaster after disaster, misfortune after misfortune. The small ghetto has been turned out on to the streets.[14] My nephew Uri arrived at half past seven.[15]

The people were driven out from 42–44 Muranowska Street during the night.

Garbatko, 300 women, 55 children.[16] Last Tuesday in the night. Rain has been falling all day. Weeping. The Jews are weeping. They are hoping for a miracle. The expulsion is continuing. Buildings are blockaded. 23 Twarda Street.[17] Terrible scenes. A woman with beautiful hair. A girl, 20 years old, pretty. They are weeping and tearing at their hair. What would Tolstoy have said to this?

On Zamenhof Street the Germans pulled people out of a tram,[18] and killed them on the spot (Muranowska Street).

Friday, 24 July, Six in the Morning

The turmoil is as it was during the days of the bombardment of Warsaw.[19] Jews are running as if insane, with children and bundles of bedding. Buildings on Karmelicka and Nowolipie Streets are being surrounded. Mothers and children wander around like lost sheep: where is my child? Weeping. Another wet day with heavy skies: rain is falling. The scenes on Nowolipie Street. The huge round-up on the streets. Old men and women, boys and girls are being dragged away. The police are carrying out the round-up, and officials of the Jewish community wearing white armbands are assisting them.[20]

The death of Czerniaków yesterday at half past eight in the Jewish community building.[21] As for the reasons: during the ceremony at Grzybowska Street,[22] he said: 'Szlag mnie i tak trafi, proszę pani' ['I'll die anyway, Madam'].

The round-up was halted at three o'clock. How Jews saved themselves: fictitious marriages with policemen. Guta's marriage to her husband's brother.[23] The savagery of the police during the round-up, the murderous brutality. They drag girls from the rickshaws, empty out flats, and leave the property strewn everywhere. A pogrom and a killing the like of which has never been seen.[24]

Merenlender's visit. She and her father were taken the first day. In what kind of train-wagons are the prisoner's kept? According to her they will not even last a night. Many buildings have received an order to present themselves on their own. The manager of 30 Świętojerska Street, Nadzia, gave himself up. People get attacks of hysteria; 11,000 people have been rounded up; 100 policemen held hostage. One of them let himself down on a rope, fell, and was badly wounded. The policeman Zakhajm has been shot. Terrifying rumours about the night. Will there be a pogrom?

Schultz is dismissing 100 Jews.[25] His explanation for his action. The great hunger in the ghetto. Someone saves his sister and a four-year-old child, passing her off as his wife. The child does not give the secret away. He cries out: 'Daddy!' I am trying to save my mother with a paper from the Jewish Self-help Organization [ŻTOS].[26]

Saturday, 25 July

Last night I couldn't sleep. It passed peacefully. Everything reminds one of September 1939. People rushing through the streets. The day is so long. Packages, mainly of pillows and bedclothes. Noisy movement. The never-ending questions: 'Meken do durkhgen?' ['Can one get through there'].[27] Disaster: Gucia has been thrown out of her flat. Five killed in Dzielna Street in the night. Terrible scenes in the streets. The police are carrying out elegant furniture from the homes of those who have been driven out. Umschlagplatz:[28] a policeman is crying. He is struck. 'Why are you crying?' 'Meine mutter, meine frau!' 'Frau, ja; Mutter, nicht' ['My mother, my wife!' 'Wife, yes; mother, no'].[29] A smuggler who threw himself out from the fourth floor, I saw him on his sick-bed.

How did Czerniaków die? 10,000![30] The Wajcblum family. The looting of property. Last night there were a lot of suicides.[31] Conditions at the Umschlagplatz. People are dying where they are being held. You can't go in or out. By yesterday 25,000 had been taken away, with today, 30,000.[32] With each day the calamity worsens. Many give themselves up voluntarily. It is supposed that hunger forces them into it.[33]

The new proclamation: non-productive elements are being sent to the East.[34] Vast numbers of dead among those being expelled. The German Jews are content to go. For them it is a long journey. The Jewish Self-help Organization is flooded with Jews begging for mercy, stretching out their hands for help—who is there to help them? Then every Jew would come and ask for papers from the organization.

Since Tuesday there has been no newspaper in the ghetto, apart from the very sketchy 'Jewish' paper *Gazeta Zydowska*.[35]

Sunday, 26 July

The 'action' continues.[36] The buildings at 10–12 Nowolipie Street are surrounded. Shouts and screams. Outside my window they are checking papers and arresting people. Human life is dependent on some little piece of paper. It's really enough to drive you insane. A lovely morning, the sky is wonderfully beautiful: 'the sun is shining, the acacia is blooming and the slaughterer is slaughtering.' The blockade of our courtyard. How it was carried out. Winnik's story.[37] *'Good news'* from Brześć.[38] The closing of the post office.[39] The seizing of an eight-year-old girl, prettily dressed. She screams: 'Mummy!' Libuszycki, Lejzerowicz.[40] The terrible hunger. Many give themselves up. They are not accepted, so great is the number that are going. Yet they still set up blockades so as to extort money. The terrible corruption of our police and their assistants. An outrage, an outrage![41]

6 Solna Street; 99 victims. Today 12,000 martyrs. The closing of the post office. A kilo of bread—50 zloty. Potatoes—20. The violence of the police. Warszawski's son,[42] an official of the Jewish community, was seized and ransomed for 250 zloty. Czudner [Meir Czudner, Hebrew poet, editor and translator, who died in the ghetto]. Kirzner's sister—seven people. The breakup of families—Mendrowski. Pola. It hurts so much. 37,000 martyrs today [till today, that is, in the first five days of the deportation]. The Jewish community and Jewish Self-help Organization workers are also not safe. Only the workers in the 'shops' seem to be still safe.[43]

A new leadership for the community. Lichtenbaum, chairman, the deputies: Wielikowski, Sztolcman, Orliański.[44] The shot in Brustman's window. Lola Kapelusz, the wife of a lawyer from Łódź. She goes twice with her daughter to give herself up because they are starving. 'We haven't eaten now for two days.' They send them away because of huge crowds of people giving themselves up. Confiscations of packages at the post office. First people were given a receipt confirming that the office had received the packages—then, the confiscation.[45]

Monday, 27 July

The 'action' still continuing at full strength. People are being rounded up. Victims on Smozca Street. People were dragged from the trams and shot. One hundred dead (old people and the sick) at the Umschlagplatz.[46] Huge numbers of dead at 29 Ogrodowa Street. The remaining occupants were taken out, no notice was taken of their papers. The cause—a piece of glass fell on to the street when there were Germans passing. Shooting all day. Dead on Pawia and other streets.

The terrible hunger. Bread—60 zloty, potatoes—20, meat—80. There are round-ups in the street. The commandant from Lublin is in Warsaw.[47] How high will the numbers of deported become? Opinions differ: 100,000, 200,000. Some will go even further: about 50,000 will be left and these will also be removed to Grochow or Pelcowizna.[48] Today the number of those deported will reach about 44,000. And according to Wielikowski there is no prospect of an end to the 'action'. A break for 48 hours (so people are saying). Auerswald has returned.[49] Perhaps things will get easier? Suicides in great numbers. The Cytryn family, mother and son embracing. The attitude of the Poles. Kalman weeping over the telephone. He calls for revenge. Neustadt has been murdered.[50]

Tuesday, 28 July

The 'action' continues relentlessly. There are many volunteers, two families from 8 Nowolipie left their flats and gave themselves up (10 + 5). The reason—the terrible hunger. Bialer—execution because he didn't remove his hat. The incident with Kirzner. Up to yesterday 45,000. Wealthy Jews have left Warsaw.[51] The Rozencwajgs on a wagon.[52] The seizure of Gutgold. Lazar: taken off a tram. Deaths on Smocza Street. How was a strong young man shot between the eyes? He tried to escape, was wounded in the arm. He begged for mercy, and was killed by two bullets in the head. Gruzalc's mother has been taken away.[53] He works at Többens'.[54]

Pessimism of Kon.[55] The Germans want to leave 60,000 Jews in the town. The fate of those who work for the Jewish Self-help Organization. Some say that their identity papers will only be recognized as valid for another two days.[56] This is what Szeryński is said to have announced.[57] A blockade on our building for the second time. The two Walfisz boys were taken away.

The sight of Nowolipie and Smocza Streets at midday—a hunt for wild animals in the forest. The world has never seen such scenes. People are thrown into wagons like dogs, old people and the sick are taken to the Jewish cemetery and murdered there. I heard that a smuggler who lives on our courtyard wanted to get rid of her old, sick mother: she handed her over to the butchers. Jewish policewomen.[58] The huge numbers of people and sewing-machines assembled in the courtyards of 44–46 Nowolipie Street.[59]

Wednesday, 29 July

The eighth day of the 'action' that is continuing at full strength. At the corner of Karmelicka Street—a 'wagon'. People are thrown up on to it. In the courtyard of 29 Nowolipie Street the furniture of the occupants who were thrown out of the buildings is still standing there. A Jew sleeps in the open air.

Kon recounts: a young woman who returning from work at a *placówka* told of the murder of two 19-year-old boys, shot dead. One was left dying for a whole hour. They were shot for no reason. Ilenman Igla, the daughter from the ZZ [Food Office] walks with her mother. Places of execution: Piaseczno, Pustelnik, Bełżec.[60] People standing at the windows are shot at. A Christian woman on Leszno Street, seeing the wagons with those who have been rounded up, curses the Germans. She presents her chest and is shot. On Nowy Świat a Christian woman stands defiantly, kneels on the pavement and prays to God to turn his sword against the executioners—she had seen how a gendarme killed a Jewish boy.

A meeting of Oneg Shabbes.[61] Its tragic character. They discuss the question of ownership and the transfer of the archive to America to the YIVO if we all die.[62]

The terrible news about the Germans' plans. It is being assumed that they intend to deport 250,000. So far, 53,000. The terrible pessimism of G. and K——n.[63] They talk of death as of something that will certainly come. Announcements in the streets: all those who present themselves voluntarily before the first of the next month will receive 3 kg of bread and 1 kg of jam.[64]

'Workshop-mania'.[65] Will that save people? The Germans thank the police for their 'productive efforts'. It is said that they are going to put the police to 'work' in other locations. How are the Jews listening to the loudspeakers?[66] . . . So far eight Jewish policemen have committed suicide. Conditions in the streets get worse every day. Many Jews with identity papers from the Jewish Self-help Organization have been arrested.

A bulldog that had been taught to attack only Jews with armbands in Warsaw-Praga. A Jew was seized by him.[67]

How do Jews hide? In couches, in beds, cellars, attics. The Rozencwajgs were set free for 500 zloty. A memorandum has been handed to the authorities, offering a ransom in return for the halting of the expulsion. No reply has yet been received.

No Germans appear until four in the afternoon. The Jews do everything in an orderly fashion. Each day about 1 per cent of those rounded up, between 60 and 70 people, are killed.[68] They throw loaves of bread into the wagons. Those at the front grab even two or three of them, those at the back get none at all. The savage round-ups in the streets will go on until 1 August. Then those who are not working will receive orders. Children will not be separated from their mothers. Someone called our policemen 'gangsters'.

The day after Czerniaków's death the German officer W. came and apologized,[69] justifying himself by saying he was not responsible for the death and giving his word of honour as a German officer that those being deported are not being killed.[70]

At the employment office there are lists of the community workers and employees of the Jewish Self-help Organization. For the moment they are being left alone. It is supposed that they are going to check them. For now they are sorting out the workshop employees.

Sometimes I am quite calm about my life and sometimes a little indifferent, but suddenly I am gripped by fear of death that drives me insane. Everything depends on the news coming in from the street. Blockades in the streets. On Nowolipie Street near the Jewish Self-help Organization they seized a girl aged 15 or 16 who was going with a basket to buy something. Her shouts and screams filled the air. In H. well-dressed women were found. About 95 per cent of the people are sent away without any kind of packing, of linen or clothes.

Thursday, 30 July

The ninth day of the 'action' that is continuing with all its fearfulness and terror. From five in the morning we hear through the window the whistles of Jewish police and the movement and the running of Jews looking for refuge. Opposite my window, in Nowy Zjazd Street, a policeman chases a young woman and catches her. Her cries and screams are heartbreaking. The blockade on our building. How was the Rajchner family saved? How did I save Mrs. Minc?

Today the post office was opened again. Brandstetter was seized yesterday afternoon by the Germans.[71] He was released at the Umschlagplatz. Dr. Fuswerg's wife was seized, as was Klima.[72] They were freed this morning.

From midday yesterday onwards the shooting has not stopped next to our building. A soldier stands at the corner of Zamenhof and Nowolipie Streets and abuses the passers-by. Terrifying rumours: the authorities have closed the Jewish Self-help Organization. Brandt expresses his condolences to the committee members on Cz.'s death.[73] Höfle defends himself by saying he was not responsible for the death.[74] The terrible appearance of Nalewki Street. A woman shot dead there yesterday when she came out of the courtyard and began to run. Workers were removed and deported from Többens' workshop at 6 Gęsia Street. Community officials were also seized and deported. All the workshops were emptied. Those who hid themselves or refused to go were shot. All the workers have been removed from the workshop of the second section of the Jewish Self-help Organization, which was in the process of being set up. At the corner of Karmelicka and Nowolipie Streets they used axes to break off the locks and open up the shops. By midday 4,000 people had been rounded up, among them 800 volunteers. By yesterday evening the total number passed 60,000.

The notice in the *Deutsche Allgemeine Zeitung:* they are broadcasting continually on London radio, 'SOS Save 400,000 Jews in the Warsaw ghetto whom the Germans are slaughtering. The "lie" of the English propaganda is clear to anyone who saw the film from the Warsaw ghetto.'

A letter from Bialystok that a Polish policeman brought, from a woman to her husband. She and her son are together with several other families and have to work hard in the fields, but they are receiving food. About 2,000 people have been removed from 27–29 Ogrodowa Street. Also a lot of children from the Pnimia children's home.[75] Today many employees of the community were seized (including teachers) and from the Jewish Self-help Organization. The story of Pigowski.

Friday, 31 July

The tenth day of the slaughter that has no parallel in our history. Yesterday a large number of officials were rounded up. The female director of Többens' workshop, Neufeld. At the corner of 11 Mylna Street they stop me and lead me up to an officer. The Jewish Self-help Organization identity-cards 'have no value any more'. *Odjazd*—ambiguous.[76] I was terrified.

They are driving out the old people from the old people's home at 52 Nowolipki Street. Those rounded up are divided up into: those fit for work (*arbeitsfähig*), those able to survive (*lebensfähig*) and those not fit to be transported (*transportunfähig*). The last group is killed on the spot.[77] About 2,000 people have been removed from the buildings at 27, 29 and 31 Ogrodowa Street. Also from the boarding-school at 27 Ogrodowa Street. The official from the centre,[78] Rozen, and his father taken away. A certain young man, Frydland, who has been working for several months at one of the *placówki* has been seized and deported. Rabbi Nisenbaum's wife was put into the wagon.[79] Mrs. Mławer was shot and wounded. They removed the caretaker and his family from Centos.[80]

Yesterday 1,500 reported voluntarily.[81] Today by midday, 750. Among them members of the intelligentsia. There is talk that part of the 'squad' has already left for Radom.[82] The remainder are leaving tomorrow. I have heard that they will be taking 500 Jewish policemen with them from here. At four o'clock they suddenly took those who had been rounded up out of the wagons and announced that the action was being suspended. The joy and hope that this brought forth. At six to half past six the blockades started again.

A woman called Mydlarska jumped up into the wagon after her husband had been taken. In our courtyard a woman threw herself from the third floor—she was starving. Today about 3,000 people were taken away from Waliców and Grzybowska Streets. No attention was paid to identity papers 'Zay gezunt! Zay gezunt!' ['Goodbye, goodbye!'] a young Jew shouts from the wagon.

The calamity of the 'dead souls'. 120,000 fictitious food-coupons (*bony*).[83]

Saturday, 1 August

'Outside there is destruction by the sword, and inside there is terror.' The 11th day of the 'action' that gets progressively more terrible and brutal. Germans are in the process of emptying whole buildings and sides of streets. They took about 5,000 people out of 20–22 and other buildings on Nowolipie Street. The turmoil and the terror is appalling. There is a general expulsion of all the occupants of Nowolipie Street between Karmelicka and Smocza Streets. The awful sight: people carrying packages of pillows and bedclothes. No one thinks of moving furniture. Fajnkind says to his sister-in-law: 'Hide yourself and your beautiful child! Into the cellar!'

The nightmare of this day surpasses that of all previous days. There is no escape and no refuge. The round-ups never cease, Sagan[84] and Chilinowicz,[85] Sztain,[86] Zołotow, Karcewicz, Prync,[87] Opoczynski have been seized.[88] Mothers lose their children. A weak old woman is carried on to the bus. The tragedies cannot be captured in words.

The rabbi from 17 Dzielna Street has been seized and apparently shot. Children walking in the street are seized. The property of those who have been expelled is grabbed by neighbours who are left, or by the new tenants, the 'shop'-workers.

Fifty of the customers, 10 staff were removed from the officials' kitchen at 30 Nowolipie Street.[89] People who have hidden are shot. I spent the whole day at 25 Nowolipki Street and didn't go to eat, so was saved.

Sunday, 2 August

I spent the night at my sister's, at 17 Dzielna Street. The 12th day of the 'action', which becomes more and more intense. From yesterday the parents of police have been excluded from the category of those protected.[90] Last night a lot of people were killed or wounded.

A new proclamation in the streets of the ghetto from the head of the Jewish police: the action will continue. All those who are not employed in organizations or by the German authorities have to report voluntarily on 2, 3 and 4 August and they will receive 3 kg of bread and 1 kg of jam. Families will not be split up.[91]

Today three people were taken away from the kitchen at 22 Nowolipie Street. It looks like they have stopped recognizing the identity papers of the Jewish Self-help Organization.[92] Yesterday evening a large group of hundreds of Jews who have been driven from their homes was taken into the Pawiak. Early today some of them were brought out of the Pawiak, among them old people, young women with small babies on pillows. They were led by Jewish policemen. Jehoszua Zegal has been seized. Among the tragedies: Karcewicz has been taken away; she left behind two children aged four and seven. Magidson. People murdered on Nowolipki Street. The 'action' continues. People are saying that there will be a break for three weeks from 5 August. Those who remain will be able to get themselves fixed up with work. When that period is up, they will return and take away and liquidate anyone who is not working. A large number of people—estimated at 15,000—have been taken from the small ghetto. Grandmother was killed by a single shot: she was standing at the window that looks over Sienna Street. Mother has gone to Gucia's.

Monday, 3 August

The 13th day of slaughter. A night of horrors. Shooting went on all night. I couldn't sleep. In the morning I went to L.'s sawmill.[93] A mass of people, men, women and children, were gathered in the courtyard and in the garden. They were trying to save themselves. Will they be saved? It is said that from those who were taken to the Umschlagplatz yesterday about 2,000 were freed who had various papers. People are consoling themselves with the thought that the savage round-up will stop tomorrow and it will be carried out in an orderly way.

Everyone was taken from TOZ [Society for the Protection of Health] who was found there.[94]

At the cemetery 56 Jewish prisoners were killed. A few days ago more than 100 people were murdered on Nowolipie Street. Today the Germans have surrounded the following streets: Gęsia, Smocza, Pawia, Lubiecka, and took away all the occupants. Yesterday the following were taken away: Kahanowicz,[95] Rusak, and Jehoszua Zegal's whole family.[96]

Tuesday, 4 August

The 14th day of the 'action' that is being continued at full speed. Today the blockades were set up at ten in the morning. The Germans work together with the Jewish police. The small ghetto was surrounded and also Gęsia and Zamenhof Streets. There are stories of terrible lootings and violence during the expulsions. They deport the people and loot and pillage their possessions. Shops are also broken open and the goods carried off. In this participate Jewish police, ordinary Jewish neighbours and Germans.

It was announced that 14 Jews were killed who had sought refuge at the cemetery, those who work in the cemetery organizations, in addition to the 56 sentenced to death, who were killed.[97]

I have heard the following: they found a woman who had recently given birth and her three-day-old child on Szczęśliwa Street. They shot dead both the mother and the child—this is a true story. They are expelling the occupants from the buildings once again: from 45 Nowolipie Street and other buildings. Even if someone is not seized and sent away to die there is no certainty that when they return home they will find a roof over their head.

Again there is talk that the savage round-up will stop today. But we have heard this before and nothing came of it. The workers from the kitchen at 8 Prosta Street have been removed. A Junak was crying: 'Szkoda tych Żydków' ['It's a shame about these Jews']. At 9 Nalewki Street a sick woman was murdered. The 'action' will continue until 17 August. Zegal's father has been taken away. How do they get the corpses out? Tozsa Apfel has been taken and sent away.[98] Our feelings have been numbed! We hear of great calamities happening to those closest to us and we do not react. A letter from Baranowicze. The writer is working as a farm-labourer. She asks for underwear. Life is cheap, 7 zloty for white bread, 1.80 for potatoes. It would be good if she could be sent underwear. The letter came by post.[99]

Wednesday, 5 August

The 'action' continues unabated. We have no more strength to suffer. There are many murders. They kill the sick who don't go down to the courtyards. Yesterday about 3,000 volunteers reported. Not all of them were taken. These they sent away. In the town they are rounding up people regardless of the papers they have. Whoever falls into the hands of the Germans or the Jewish police is seized. The Jewish policemen took away Hillel Cajtlin.[100] He was released. Balaban has been taken.[101] At 13 Dzielna Street they killed Mrs. Grun who was ill and a girl. Yesterday the 'actions' in Radom began.

Thursday, 6 August

The 16th day of the 'action', which is continuing. Yesterday they took away everyone from the offices of the Jewish Self-help Organization who were there at the time, about 60 or 70 people. Some of them (Dr. Bornsztajn,[102] Sztolcman, a girl) were freed. They are predicting a hot day today. Once again there are the theories that the action will be suspended tomorrow because the annihilation squads are about to leave for Radom. Conditions in Landau's sawmill. The expulsion of the occupants of 10–12, 23 and 25 Nowolipki Street. Redoubled savagery and the maltreatment of Jews. While flats are being emptied out and people come to save their belongings, the SS arrive and seize the occupants.

Kohn and Heller have been killed.[103] During a blockade by the

Germans the Jewish police storm into Zylberberg's building. They are terrified that they have found them and say: 'Hide and lock yourself in well!' It was then that the whole family of the Radomsko rebbe was killed.[104]

P——wer saw orders with regard to trains and their numbers that were sent to Treblinka (the place of execution?) [105] Starvation haunts the survivors more and more, a kilo of bread—45 zloty, a kilo of potatoes, 15 zloty. Today they have already taken about 5,000 Jews from the small ghetto. Tosza Apfel has been caught and deported.

Friday, 7 August

The 17th day of the massacres. Yesterday was a horrendous day with a great number of victims. People were brought out from the small ghetto in huge numbers. The number of victims is estimated at 15,000. They emptied Dr. Korczak's orphanage with the doctor at the head.[106] Two hundred orphans. In the evening they drove out the people from the flats in the square bounded by Dzielna, Zamenhof, Nowolipki and Karmelicka Streets. There are no words to describe the tragedies and disasters. Rozencwajg's two sisters were sent away. One with a child of six months, the other with a four-year-old. Mrs. Schweiger is not there.[107] How terrible. Today Germans and Ukrainians came to the sawmill of the L. brothers and they rounded up a large number of women with their children from among the factory-workers, and women who just happened to be there. Wasser,[108] Smolar,[109] with her child, Tintpulwer and others, many others.[110]

The workers turn on the intellectuals. A shocking experience. Many rabbis have been sent away. Mendel Alter from Kalisz,[111] and more, more. During the pogrom on Nowolipie Street about 360 people were killed. At number 30 more than 30 people were killed. Górny's mother has been killed by the Germans,[112] and he came to sell meal-coupons as if nothing had happened. So dulled have our feelings become.

The hunger presses in on us in a terrible way. Today I had no bread for breakfast. I ate pickled cucumber. Today a kilo of bread costs 55 zloty. A new order has been issued that if people report voluntarily for deportation from 7 to 14 August they will receive a kilo of bread and half a kilo of jam. From this it can be deduced that the 'action' will continue for at least a week. Stupnicki has been sent away.[113] Three thousand were brought from Otwock directly to the Umschlagplatz. The number who have fallen victim is enormous. The crematorium near Malkinia and Sokołów.[114] I have heard that Erlich (nicknamed Kapote) has disappeared.[115] The shooting and the killing flourishes. During the blockade on Leszno Street four people were killed. Hirszhorn has committed suicide.[116] Many kill themselves. It is a miracle that there are people still alive.

Saturday, 8 August

The 'action' continues. The 18th day. There are still reports of our cherished and loved ones who fell victim yesterday: our children. The children of our boarding-schools led away (to be killed): 12–14 Wolność Street,[117] about 1,200, 18 Mylna Street, Koninski with his wife and the children from the boarding-school.[118] They intend to eradicate the whole of Warsaw Jewry. I hear reports today that the Germans are blockading Żelazna and Leszno Streets. They are driving people out of all the buildings on Miła Street. We have lived through a shattering and terrifying day—30 Gęsia Street.[119]

The numbness of everyone is staggering. Górny loses his mother and sells meal-tickets. Smolar has lost his wife and daughter.[120] Tintpulwer—widowed—goes around in despair, a broken man, and tries to 'find' work so as to be involved in something, not to be superfluous. Terrifying reports from the town. At 64 Lubiecka Street victims, many victims on Miła Street. All the cows were taken away—about 120—from the 'farm'.[121] A loss of millions: there will no longer be the small amount of milk that was distributed to the children.

In the evening a pogrom in the streets. A great many killed at various locations: Smocza, Pawia, Miła, Zamenhof and others. I was on my way home at half past eight. Hela comes towards me. Luba and Ora are not there.[122] I am sure they have been seized. They come home at nine o'clock. During the blockade they had stayed in the boarding-school at 67 Dzielna Street.[123] What Luba recounted of the children (150) and the women teachers during the blockade. Their packages in their hands, ready to set off—to their deaths. Kon said yesterday: 'I am writing a testament about the events.' Chmielewski's parents were taken away yesterday and he comes to the factory and is still on his feet.

Sunday, 9 August

The 19th day of the 'action' of which human history has not seen the like. From yesterday the expulsion took on the character of a pogrom, or a simple massacre. They roam through the streets and murder people in their dozens, in their hundreds. Today they are pulling endless wagons full of corpses—uncovered—through the streets.

Everything that I have read about the events in 1918–19 pales in comparison with what we are living through now. It is clear to us that 99 per cent of those transported are being taken to their deaths. In addition to the atrocities, hunger haunts us. People who during the war were previously well-fed come to ask for a little soup at a factory kitchen. The 'elite' still get some, but the rabble don't even get that.

Twenty Ukrainians, Jewish policemen (a few dozen) and a small number of Germans lead a crowd of 3,000 Jews to the slaughter. One hears only of isolated cases of resistance.[124] One Jew took on a German and was shot on the spot. A second Jew fought with a Ukrainian and escaped after being wounded. And other cases of this kind. The Jews are going like lambs to the slaughter. Yesterday 23 Jews were killed in one flat.

I have heard that the 'action' in Radom is already over after a week and three days—7,000 victims.[125] That is the target they set in advance. And here we have no idea when they will say—enough. I have heard about letters that arrive from France telling of expulsions of Jews there. They also say that they will be brought to the Warsaw ghetto. It is a wonder that people can endure so much suffering, living the whole day on a knife-edge between life and death and clinging with all their might to life in the hope that they may be among the ten survivors.

Monday, 10 August

Yesterday was horrific in the full sense of the word. The slaughter went on from early morning until nine and half past nine at night. This was a pogrom with all the traits familiar from the Tsarist pogroms of the years 1905–6. A mixed crowd of soldiers of various nationalities, Ukrainians, Lithuanians and over them the Germans, stormed into flats and shops, looting and killing without mercy. I have heard that people are being slaughtered with bayonets. Yesterday there were vast numbers of deaths.

In the town, proclamations have been published ordering the occupants of the small ghetto to leave their homes today by six o'clock.[126] This is a further terrible calamity. Firstly for all those who have remained there. There is no possibility for them to take out with them a few of their possessions, clothes and bedding, because there is a danger of being seized while moving. And secondly, where can they move to? A great many streets and blocks of buildings have been emptied by German factory-owners. The number of buildings still in Jewish hands is very low.

The number of those deported out (read: murdered) is estimated at 150,000. Yesterday the Guzik family was seized. He was freed, his family not.[127]

It is reported that the community organization is to be dismantled[128] and a commissar appointed. I have heard that Gancwajch, whom they wanted to kill a few weeks ago, has climbed back to prominence.

I was unable to go home. We all spent the night at 30 Gęsia Street. It was a very difficult night. Until two we sat on chairs, from two to five—we lay on plywood boards. We were told that 2,500 officials of the Jewish community and two members of the Judenrat have to present themselves at the Umschlagplatz. Later this was denied. The embitterment of the workers against the unwelcome intelligentsia is growing continually.

They feel they have been wronged by them. The wife of the editor Wolkowicz has killed herself.[129]

The terrible hunger: bread, 88 zloty, potatoes, 30. The appalling appearance of the Jewish streets. The shops and flats stand open, the Jewish crowds have remained—looting. In a building on Leszno Street, where 150 people used to live, there are now 30 left. Of these eight were killed yesterday.

I have heard that in the course of the massacres yesterday the famous Warsaw singer Marisia Ajzensztat, the only daughter of her parents and a former pupil at the Yehudia School, was attacked and killed.[130] I have heard that yesterday Kohn's sister and her husband were killed. And she who had celebrated her recent marriage so exuberantly. A worker sat weeping, Jewish policemen had come to his home and taken away his 16-year-old son. What brutality!

Tuesday, 11 August

Things are deteriorating fast. Appalling, horrendous. The brutal expulsion from the small ghetto. Whole buildings have been emptied of their occupants and all their possessions left behind. Christians are already beginning to loot. 24 Sienna Street, 28 Śliska Street. Except for Jakub's family,[131] there is not even a single tenant remaining in the building; the house-porter is also gone. Aunt Chawa and Dora Fejga have been seized and deported. The destruction of families. Early this morning the Germans and the rioters spread through the ghetto. By the evening they were distributed throughout the ghetto and were seizing people. In the course of five minutes they drove out all the occupants on Gęsia Street between Zamenhof and Lubiecka Streets. They pay no attention to papers.

The Jewish community offices have moved to 19 Zamenhof Street, the post office building.[132] They have reduced their personnel by half. The number of victims has already risen above 150,000. Today they will complete three weeks since the beginning of the terrible massacre. In the night a large number of women who worked at Többens' were removed. It looks like there is a policy to liquidate women and children. Yesterday at Többens' three Jews died at their work. Blockades and murders in the streets that still belong to the ghetto. The heavy blockade on the entrance to the buildings of the Warschauer Union, with two killed and a vast number seized, nearly 100 women, children and men. The mortal terror that gripped us as we sat in the office.

Smolar rang Sokołów. He was told that those that are deported, or if they are to deported to Tr., are going to their 'death'.[133] The news that K. brought. In Warsaw there is a Jew by the name of Slawa who has brought reports of Treblinka. Fifteen kilometres before the station at Treblinka the Germans take over the train. When people get out of the train they are beaten viciously. Then they are driven into huge barracks. For five minutes heart-rending screams are heard, then silence. The bodies that are taken out are swollen horribly. One person cannot get their arms round one of these bodies, so distended are they. Young men from among the prisoners are the gravediggers, the next day they too are killed. What horror!

Wednesday, 12 August

Eclipse of the sun, universal blackness. My Luba was taken away during a blockade on 30 Gęsia Street. There is still a glimmer of hope in front of me. Perhaps she will be saved. And if, God forbid, she is not? My journey to the Umschlagplatz—the appearance of the streets—fills me with dread. To my anguish there is no prospect of rescuing her. It looks like she was taken directly into the train. Her fate is to be a victim of the Nazi bestiality, along with hundreds of thousands of Jews. I have no words to describe my desolation. I ought to go after her, to die. But I have no strength to take such a step. Ora—her calamity. A child who was so tied to her mother, and how she loved her.

The 'action' goes on in the town at full throttle. All the streets are being emptied of their occupants. Total chaos. Each German factory will be closed off in its block and the people will be locked in their building. Terror and blackness. And over all this disaster hangs my own private anguish.

Thursday, 13 August

The 23rd day of the slaughter of the Jews of Warsaw. Today about 3,600 people were removed from Többens' buildings, mainly women and children. Today is Ora's fifteenth birthday. What a black day in her life and in my life. I have never experienced such a day as this. Since yesterday I have not shed a single tear. In my pain I lay in the attic and could not sleep. Ora was talking in her sleep: 'mamo, mamusiu, nie odchódź beze mnie!' ['Mother, Mama, don't leave me']! Today I cried a lot, when Gucia came to visit me. I am being thrown out of the flat at 2 Mylna Street: they have already taken most of my things. Those who have survived are thieving and looting insatiably. Our lives have been turned upside down, a total and utter destruction in every sense of the word.

I will never be consoled as long as I live. If she had died a natural death, I would not have been so stricken, so broken. But to fall into the hands of such butchers! Have they already murdered her? She went out in a light dress, without stockings, with my leather briefcase. How tragic it is! A life together of over 21 years (I became close to her beginning in 1920) has met with such a tragic end.

Friday, 14 August

The last night that I will spend in my war-time flat at 2 Mylna Street. The sight of the streets: the pavements are fenced off, you walk in the middle of the road. Certain streets, such as Nowolipie (on both sides of Karmelicka), Mylna and others are completely closed off with fences and gates and you can't get in there. The impression is of cages. The whole of Jewish Warsaw has been thrown out of the buildings.[134] There is a full-scale relocation of all Jews who have not yet been rounded up and are still in the town. Whole streets that have been given over to the German firms: Müller, Többens, Schultz, Zimmerman, Brauer and others.[135] We have been sold as slaves to a load of German manufacturers. The living-conditions of those in the workshops: hunger and hard labour. Their ration: a quarter kilo of bread a day and a bowl of soup.

The 'action' continues—today is the 23rd day. Yesterday they took away from Többens' workshops about 3,000–4,000 men and women, mostly women and children. This morning the Jewish community-council posted a new announcement: all Jews who live in Biała, Elektoralna, Zielna, Orla, Solna, Leszno, odd numbers in Ogrodowa, Chłodna Streets have to leave their flats by tomorrow, 15 August. Yesterday and today, a huge number of people killed—victims of the blockades. I am moving my things over to Nacia's at 14 Pawia Street.

Setting up of blockades on Nowolipie and Karmelicka Streets. Further victims—there are more deaths today, and very many driven out. There is talk of 15,000. I have heard that measures decreed in the expulsion orders are directed mainly against women and children. The police commandant of the second district is trying to save his wife and children. A new raid on the Jewish Self-help Organization at 25 Nowolipki Street. Dr. Bornsztajn and his wife taken away, Elhonen Cajtlin with his son and others.[136] This was carried out by Jewish policemen without the Germans, that is, on their own initiative. Renja Sztajnwajs. I have heard that Yitshak Katznelson's wife and one of his children have been seized.[137] The second day that I am without Luba. I am now also without a place to live. I have nowhere to lay my head. The number rounded up has reached 190,000, just counting those expelled, excluding those who have been killed and those who have been sent to the *Dulag* at 109 Leszno Street.[138]

Every crime in history, like the burning of Rome by Nero, pales into insignificance in comparison with this. Kirzhner has been taken away from work and deported. Together with him they took away a further 28 people. All were aged 35 and over. The same thing has happened, I have learnt, in another *placówka:* 29 people were taken away and deported.

Saturday, 15 August

Today is the 25th day of the bloody 'action' carried out by the butchers. I spent the night at 17 Dzielna Street. The rain of shooting started at half past nine in the evening. Deaths in the street. The whole night incessant movement in and out of the Pawiak. Gutkowski sends his only son, three and a half years old, to the cemetery to have him taken to Czerniaków.[139]

I have nowhere to rest my head at night. Gucia is being thrown out of her flat. Nacia and Frume are not allowed to enter. All the orphanages have been emptied.[140] Korczak went at the head of his children. The pain because of the loss of L. is becoming more intense. My soul can find no peace, for not having gone after her when she was in danger, even though I could also have disappeared and Ora would have been left an orphan. The most terrible thing is that Landau and Sonszajn misled me by saying that Luba wasn't in the queue. Be that as it may, the anguish is terrible and it will never be dimmed.

Rumours about reports arriving from women who were deported from Biała-Podlaska and Białystok.

Today by eight o'clock there was a blockade on Miła, Gęsia, Zamenhof and other streets. 'Our spirit is weary of the killing.' How much longer? Yesterday a huge number of bodies were brought to the cemetery, victims of the blockade of Többens' workshops. Today they were also taking people from the 'shops'. It will soon be seven o'clock and the blockade on Gęsia is still continuing, around our factory. The Jewish police have been looting, breaking open flats, emptying cupboards, smashing crockery and destroying property, just for the fun of it. More people were killed today in the course of blockades. People killed during the blockade. Mirka Priwes, her mother and brother have been deported. Yitshak Katznelson's wife and two of his children have been seized and deported.

The desolation and chaos is greatest on the streets from Chłodna to Leszno Streets, all the Jewish possessions have been abandoned and Polish thugs with the Germans will loot everything. The whole of Jewish Warsaw has been laid waste. That which remains is a shadow of what was, a shadow that tells of death and ruin.

Sunday, 16 August

Today is the 26th day of the 'action', which is continuing with all its atrocities and animal savagery, a slaughter the like of which human history has not seen. Even in the legend of Pharaoh and his decree: every newborn boy will be thrown into the river.

People who have returned from the Umschlagplatz have told of women who were seized yesterday who were freed if they sacrificed their children. To our pain and sorrow many women saved themselves in this way—they were separated from their children, aged three to 12 to 14, and if they had identity papers, they were freed. Any woman carrying a child or with a child next to her was not freed. The Germans' lust for Jewish blood knows no bounds, it is a bottomless pit. Future generations will not believe it. But this is the unembellished truth, plain and simple. A bitter, horrifying truth.

The Jewish police have received an order that each one of them must bring five people to be transported. Since there are 2,000 police, they will have to find 10,000 victims. If they do not fulfil their quotas they are liable to the death-penalty. Some of them have already received confirmation that they have presented the required number. Since every Jew has some kind of documentation—in the main valid ones—they tear up every document they are shown and round up the passers-by. It is now dangerous for every Jew to go out on to the street. No one goes out.

Rumours have reached me again that letters have allegedly arrived from the deportees saying that they are working in the area of Siedlce and conditions are not bad. Lifschitz's son (my friend from elementary school) told me that his daughter herself had read one of these letters from an elderly couple.

As things are developing, a handful of Jews will be left, those of a designated age. Apart from this there will be no way for a Jew to survive: there will be nowhere to live and no bread. The position of the old is especially tragic: they have no way out. They can either give themselves up into the hands of the butchers, or take their lives themselves, or hide out and live in dark corners and cellars, which is also very difficult because of the general expulsions from the buildings and the upheaval of the residents. In those buildings that have been taken over by new occupants, no strangers are let in. It is easier for an animal to find a hiding place and a refuge in the forest than for a Jew to hide in the ghetto.

Now (four in the afternoon) I have heard that there are no Germans at all in the Umschlagplatz. There are only Jews there and they are carrying out the bloody and terrible operation. Today rumours are going round that an order has been issued that all wives and children of officials have to report at the Umschlagplatz. Josef Erlich and his family have been killed, so I have heard. According to certain reports, Czerniaków's place here with us—à la Rumkowski[141]—will be inherited by Gancwajch, the man they had been hunting and trying to kill. He is outside the ghetto at the moment.

Monday, 17 August

The 27th day of the annihilation. Yesterday I came to 14 Pawia Street very late at night by a round-about route (via Zamenhof) and was anguished to hear the terrible news about Jakub, Frume and Uri. A very great blow. There is still a faint hope that they can be saved, since there were no train-wagons yesterday and they weren't taken straight to the train. This morning I saw in the streets an announcement about a new reduction in size of the ghetto. Very many streets and sides of streets (the odd or the even numbers) must be vacated by the Jews by 20 August, at four in the afternoon. The ghetto will be a third or a quarter of its original size, if there are no further decrees of this kind. They are emptying those streets that had already been handed over to the German firms, and been fenced off, for example Mylna, Nowolipie, Dzielna Streets and many others. The enemy's claw is reaching out for us and it is still not sated.

Yesterday hundreds of officials of the community and of the Jewish Self-help Organization were taken away. The Gestapo commandant Brandt stood there and struck the detainees with his own hands. Jakub, Uri and Frume were hit. The 'action' is continuing today. There was a blockade on the cemetery. Ora, who works with the group from Hashomer Hatsair, was in great danger. The group was saved today thanks to the intervention of Commissar Hensel.[142] Jewish policemen round up people all day. It is said that they have received an order that each policeman must find six Jews. They abuse those who are rounded up, and smash and loot the empty flats. I have heard that a thousand policemen have received an order to report at the Umschlagplatz. This report turns out to be false—for the time being.

Harsh conditions at the factory. Before 80 people were employed there and now almost a thousand are registered there. Hundreds of people wander around bored with nothing to do. They sit around in dread of German blockades and many hide themselves in all kinds of dark corners.

The pain over the loss of L. is getting more and more intense. During the day I am often choked with tears. The fact there is no news about her suffering and torment, whether she is alive or dead, how she died—gives me no peace. If I knew that she was alive and that she was not suffering too much, I would be calm. And if I knew that she had died but did not suffer much at her death—then I would also be calm.

I have been told that Yitshak Katznelson shows great inner strength and endurance, keeping hold of himself after the terrible disaster that has befallen him.

The Ejduses have been seized. Every day there are killings. When Jakub, Frume and Uri were taken away, someone tried to escape. He was killed on the spot. For a week now we have had no news of the progress of the war. The last report was a few days ago of the heavy bombardment of Mainz. The story about the Jew Chunkis (one of the directors of Adriatika).

Frume and Uri have returned. What they have told me about what is going on at the Umschlagplatz. Hell, pure hell. The rich save themselves, if they are not shut into the wagons straight away. The tragic fate of the Taubers. He was killed on the spot, his wife and beautiful and charming son (with statuesque features)—Rapusz—were deported.

Tuesday, 18 August

Today marks four weeks or 28 days of this blood operation, which has no parallel in history. The Germans and the Jewish police have been carrying out further blockades. Disaster has struck our family once again. Gucia and Hela have been taken away by the Germans, who entered their building. This is a very heavy blow for me. She had been so concerned for us and helped us in the war-years. I have heard talk again about the new rise of Gancwajch. He will take over Lichtenbaum's place and become commissar of the Jewish community.

Today I went with three friends to collect up the books that are in the flats that our firm has been allocated on Miła Street. We set eyes on an appalling vision, all the doors broken open, all the goods and property smashed and scattered through the courtyards. Russian pogromists would have been unable to make a more thorough and shattering pogrom than that carried out by the Jewish police. This sight, which is everywhere to be seen, stunned us. The destruction and the annihilation of the greatest Jewish community in Europe.

New proclamations from the Judenrat have been hung up which have caused panic among the Jews. Jews who are not employed are not permitted south of Leszno Street. Those who are caught there will be shot.[143] The families of those working are no longer protected. In fact all those who are not working, even the families of those who are employed, have to report voluntarily at the Umschlagplatz. Otherwise their food-cards will be taken away and they will be driven out by force. We can see that the Germans are playing a game of cat and mouse with us. Those employed have protected their families, now the families are being deported (killed) and they want to leave behind the working slaves for the time being. *What horror!* They are preparing to destroy us utterly.

Wednesday, 19 August

The 29th day of the bloody action. Last evening ended in a massacre and with a large raid on the brush workshop on Franciszkańska and Świętojerska Streets.[144] About 1,600 people were removed. Eight were killed. Among those taken away were large numbers of well-known and cherished individuals such as Mrs. Mokarska,[145] Rabbi Huberband and others.[146] I have been informed that Nisenbaum's father and Szczeranski have been seized.[147] Rokhl Sztajn poisoned herself at the Platz.[148] Hillel Cajtlin's wife was taken. Last night the terrible news reached me that Mrs. Schweiger has been taken away.[149] Inka tried to poison herself, but they saved her.

There is no 'action' in Warsaw itself today. The squad has left for Otwock.[150] And there is an 'action' there, according to reports. The Jewish police is carrying out checks in the buildings in search of 'outsiders' who are certain to be hiding there hidden in dark corners and cellars. Large numbers of Jews commit suicide.[151] The number of victims of the expulsion has reached approximately a quarter of a million. Today is the seventh day since the great calamity that befell me. If only I could die and be free of the whole nightmare. But I am still tied to life and it is still difficult for me to take my own life.

The squad is running riot in Otwock. I have heard that they have emptied Brius and Sofjówka.[152] Who knows how many cherished and beloved victims we have lost today. And to think that many had gone to Otwock to find an escape from death.

Thursday, 20 August

There was no 'action' yesterday in Warsaw. However, it is reported that there was a hunt on the Aryan side for Jews who had fled there. The squad carried out the action in Otwock with the help of 500 Jewish police from Warsaw (so it is said). I have heard that those rounded up have been marched to Warsaw on foot.

My sister Gucia is lost and her daughter Hela killed. This new disaster adds still further to the weight of my gloom. She was the best of sisters and was very concerned for me during the war-years. It is such anguish and only death will end my suffering.

An order for the caretakers in the buildings has been issued: to collect up and gather in one place all the goods and possessions of those who have been driven out and hand them over to the community representative who will call. A great looting is being prepared, Nazi-style; the Germans are preparing to remove all the Jews' possessions.

Friday, 21 August

Yesterday evening after six the Jewish police moved into the buildings which were supposed to have been evacuated by the occupants. They drove out the occupants by force, broke into locked flats, robbed and looted and smashed whatever they found and at the same time seized women, especially those who had no papers (*Meldekarten*—population registration coupons). Where did the Jews get this brutality from? The Germans' spirit has passed into them. They also entered our workshop at eight and caused panic among the women. They were bribed and left. I was worried about Ora and walked with her late to Frume's.

At eleven in the evening the air-raid on Warsaw began. It was as bright as on a clear, moonlit night, because of the bombardment. Bombs were dropped. In spite of the danger we welcomed the raid with a feeling of great satisfaction.[153] Perhaps our salvation will come, perhaps this evil power will be broken. There is talk of a second front in France and Holland. If these things had happened four or five weeks ago, perhaps we would have been saved from the catastrophe.

I have heard there was an assassination attempt on the chief of police Szeryński.[154] He was wounded in the cheek. According to rumours he was wounded by a Pole from the Polish Socialist Party disguised as a Jewish policeman. Today leaflets were distributed against the Jewish police, who have helped to send 200,000 Jews to their deaths. The whole police force has been sentenced to death.[155]

What should I do with my mother? Old people have nowhere to turn. They have nowhere to hide. They have been driven out of their former flats and are not let into the new blocks. The entrances are closed and the caretakers will not admit strangers. There are those who go to the Umschlagplatz and hand themselves over to the butchers to die a martyr's death.

Saturday, 22 August

Yesterday there was no 'action', but the squad is active in the Warsaw area. Yesterday, it is said, it was in Mińsk Mazowiecki.[156] The heavy and gloomy scene that I had at Nacia's because of mother. The Tombecks will not let her sleep there.[157] She has nowhere to go, except to go and give herself up to the executioners. How horrifying. Ora is ill and she has nowhere to lie down and no medicine. We have no roof over our heads. I leave her at Nacia's. I am deeply worried and frightened.

Sunday, 23 August

Yesterday was the 33rd day of the bloody 'action' in Warsaw, which has not been discontinued; on the contrary, it is still continuing. Yesterday the Germans rounded up mainly women and children with rampant viciousness and savagery. I was told that a prettily dressed 10-year-old girl was seized. The girl screamed in anguish and cried out: 'Mr. Policeman!' but her pleas were of no avail. He was deaf to her screams and put her in a rickshaw to the Umschlagplatz. Yesterday the two Ostrowiec girls from Hashomer Hatsair who work at our office were seized. I have heard that the Jewish police have been ordered to round up 1,000 victims a day, and that the hunt will continue for another 14 days. In short, not one of us is sure to survive, especially women and children, who are left living with the threat of destruction, every day, every minute. The knowledge of this so preys on their nerves that many nearly go insane.

There is talk of an 'action' in Piotrków.[158] The group (from Hashomer

Hatsair) that works at the Jewish cemetery, and the 'business' that is carried on with the Christians. The Germans are looting Jewish property. Yesterday they came to the Tepicyn factory and took away the best suitcases and everything of value. They even pillage from the wagons in the street. Life in the ghetto has become quiet. The few who have survived wander the streets like shadows, like corpses, and their number diminishes from day to day.

R. told me something he had heard from some German woman that in the area around the station at Kosow you can hear the screams of them being tortured to death 3 km away.[159] The savagery of the Jewish police against their unfortunate victims: they beat viciously, they steal, and they loot and pillage like bandits in the forest. What degeneracy! Who has raised these bitter fruits among us?

I have heard that in Falenica all the Jews left their homes, dispersed in the nearby small towns and the woods, and not a single one remained behind. The executioners were left without prey. Apparently they massacred everyone in Otwock and didn't deport people. In Falenica the Jews tried to resist the Germans and they were slaughtered.[160] Anyone who was found was killed. The remainder ran off into the woods. Today the butchers are supposed to return to Warsaw and there are terrible rumours circulating because of this: that they intend to liquidate the whole of Jewish Warsaw and empty it of Jews. Horrific.

Fela has returned from the Umschlagplatz and told me that Suchowolska has been seized. A few hours later I learned the end of the story. The neighbours collected a little money, clothing and food . . . and sent the mother and her two children, boys aged 10–12. Asch's brother-in-law has died—Alberg.[161] Yesterday he was still walking around, and today he died because his strength gave out.

Today the Jewish police carried out the 'action' with savage brutality. They simply ran riot. There is a dread of tomorrow, of an 'action' with the participation of the Germans. The firm [Ostdeutsche Bautischlerei Werkstätte] has set up a confiscation-team, which drives around carts with beds, couches and other valuable goods. The whole society has been pillaged from top to bottom. God, is there any help or salvation for us? Will the survivors stay alive, or will our end be the same as that of the hundreds and thousands who have already died?

1. One of the main streets of Warsaw. These notes, made by Lewin on 20 July just before the start of the great deportation, were never fully written up and testify to his anguish of spirit in the face of the great catastrophe confronting Warsaw Jewry. It is at this stage that he switches from Yiddish to Hebrew.

2. The porter of this house was put to death because he would not reveal the supposed hiding place of a young man being pursued by a German gendarme (Chaim Kaplan, *The Warsaw Diary of Chaim A. Kaplan,* trans. and ed. Abraham Katsch [New York: Collier, 1973], p. 379).

3. 21 July.

4. A Jewish policeman who was killed on the same day. The murder is recorded in Kaplan, *Warsaw Diary*, pp. 378–79. According to the memoirs of an anonymous police officer, between 20 and 30 members of the Jewish police were murdered or deported for refusing to participate in the round-up (Yisrael Gutman, *The Jews of Warsaw, 1939–1943: Ghetto, Underground, Revolt*, trans. Ina Friedman [Bloomington: Indiana University Press, 1989], p. 447).

5. Apparently Dr. Zygmunt Steinkalk, a well-known Warsaw pediatrician; murdered by the Germans on 21 July 1942 at the entrance to 26 Karmelicka Street (Kaplan, *Warsaw Diary*, p. 378). Yitshak Katznelson was also an eyewitness to the murder.

6. Zaklad Zaopatrzenia (Supply Office), the supply institution of the ghetto, under the supervision of the Judenrat. Set up on 1 August 1941, its functions were the acquisition and distribution of food as well as the importation of other essential goods. In spite of its sensitive function it enjoyed more confidence among the ghetto population than any other institution of the Judenrat. This was mainly because of the effectiveness and impartiality of its head, the veteran businessman Abraham Gepner. On this, see Gutman, *The Jews of Warsaw*, pp. 85–86.

7. The same day some 60 members of the Judenrat and the intellectuals of the ghetto were arrested as hostages. The arrests were accompanied by murders in the streets.

8. Perhaps Shmuel Breslaw, a member of the Anti-Fascist Bloc and Hashomer Hatsair activist. Murdered by the Gestapo in the Pawiak.

9. The first day of the massive deportation. The orders for the deportation had been dictated in the morning at a meeting of the Judenrat by the SS officer who headed the deportation staff. This is described in Adam Czerniaków, *The Warsaw Diary of Adam Czerniaków: Prelude to Doom*, ed. Raul Hilberg, Stanislaw Staron, and Josef Kermish (New York: Stein & Day, 1979), p. 385.

10. A shelter for refugees, deportees from elsewhere who were resettled in the Warsaw ghetto: 'points', a literal translation of the Polish *punkty*, here meaning depot, assembly point. For the most part these were miserable refugees, with great numbers of sick and starving people concentrated at terrible density. The morality in such places was extremely high, in some cases as much as 30 per cent.

11. With the establishment of the ghetto, the hospital—one of the most impressive institutions of the Warsaw Jewish community—was transferred from its premises in the suburb of Czyste to Czysta Street near the Umschlagplatz. With the beginning of the deportation the hospitals were used as assembly points and the patients, 761 in number, were transferred to improvised hospitals. In the final stage of the deportation, on 12 September 1942, the hospital was completely destroyed and the patients and the medical team, some 1,000 people in all, were deported to Treblinka.

12. Even for those with exemptions from the deportation order, the exemption only covered wives and children and did not extend to aged parents. The problem facing the diarist and his family preoccupied many others.

13. Ora—the diarist's daughter—who was fourteen. The sentence is unclear. Literally it reads 'Ora—shotzia, fabricated stories'. 'Shotzia' is perhaps a reference to the false information coming from Sweden (Szwecja in Polish) to the effect that the end of the war was approaching.

14. The 'Aryan' Chłodna Street cut the ghetto into two parts, a northern sector (the 'main ghetto') and a southern sector (the 'small ghetto'), which occupied approximately 23 per cent of the area of the ghetto. Between the two ghettos

there is an alleyway, at the corner of Żelazna and Chłodna Streets. Eventually a wooden bridge was constructed over Chłodna Street. The small ghetto was liquidated completely during the deportation of August 1942.

15. Uri Tombeck, the diarist's nephew.

16. Even in the darkest days of the deportation the diarist continued to record the information reaching him on deportations and murders in other places. Garbatko—a resort near Radom.

17. The main 'action' in Warsaw was carried out in the form of a siege or 'blockade', first on individual houses and later on streets and entire neighbourhoods. In the initial stage, the deportation was carried out entirely by the ghetto police. The historians of the Warsaw ghetto have left us many descriptions of atrocities committed at the time of the blockades. On this, see Gutman, *The Jews of Warsaw*, pp. 203–18.

18. A tram still traversed the ghetto but the route was short: from Żelazna Street via Leszno and Karmelicka Street to Dzika Street. It stopped running on 15 August, when the deportation intensified.

19. A reference to the bombing at the time of the German siege in September 1939. The Jewish neighbourhoods in Warsaw were shelled heavily during the Jewish New Year.

20. The staff of the Judenrat and its associated bodies were ordered to assist the Jewish police in carrying out the deportation order. They were supplied with special armbands bearing the symbol of the Judenrat, a serial number and an appropriate caption in German. In contravention of this order, the community institutions forbade their staff to assist in the deportation.

21. On the second day of the deportation, 23 July, at 8:30 P.M., the chairman of the Judenrat killed himself by taking poison. He did this after the Germans had informed him that the daily deportation quota was to be increased.

The report of the Jewish underground (15 November 1942), which was sent to London, was very critical of Czerniaków for not warning the public and not making them aware of the Germans' plan to completely destroy the ghetto.

Mention should also be made of Kaplan's opinion that Czerniaków 'may not have lived his life with honour but he did die with honour. Some merit paradise by the deeds of an hour but President Adam Czerniaków earned his right to paradise in a single moment' (Kaplan, *Warsaw Diary*, p. 385). The writer mentions that Czerniaków refused to sign the deportation order, which was therefore signed, irregularly, merely 'The Jewish Council' (ibid., p. 384).

22. This is apparently a reference to a party held in the community-building in 26 Grzybowska Street some 10 days before the deportation to mark the birthday of Abraham Gepner, a man who had long been a leading figure in the community and was head of the supply institution of the ghetto (see n. 6).

A slightly different version, equally typical of the mood of the chairman at that time, was recorded by Perets Opoczynski. 'Last week Czerniaków stalked around his office, clearly in a rage. A woman approached him and advised him to put a scarf round his neck because he had a cold. Czerniaków responded, "What are you afraid of, madam, I'll die anyway" ' (unpublished diary in Yad Vashem archives. Entry for 12 July).

23. Guta, the diarist's young sister. Fictitious marriages between brothers and sisters were not unknown.

24. See Gutman, *The Jews of Warsaw*, pp. 207–11.

25. Fritz Schultz, a German industrialist, owner of one of the largest workshops in the Warsaw ghetto. The workshop owners made a great deal of money

from the plight of the Jews. They stole their possessions, exploited their labour, sold jobs in their factories at extortionate prices, and co-operated with the Gestapo in deporting them from the ghetto.

26. Żydowskie Towarzystwo Opieki Społecznej, the official name of the Jewish Self-help Organization. [The Jewish Self-help Organization was set up during the siege of Warsaw in September 1939 and reorganized in January 1942. It covered the whole area of the General-Government and divided the ghetto into five action districts. The diarist was active in the second district. On its activities, see Gutman, *The Jews of Warsaw,* pp. 40–45, 102–6.] Of all the institutions in the ghetto, it had most successfully maintained its independence, operated many social welfare programmes and served as a cover for many underground operations. In accordance with the deportation order of 22 July, its workers were exempt from deportation, and in fact in the first days of the deportation possession of work-papers from this institution did constitute some immunity.

27. That is, is there no danger of being ambushed here? Is the street blockaded?

28. Umschlagplatz: a square on the outskirts of the ghetto, on Stawki Street, by the railway sidings. In this square, which had previously been used for the handling of merchandise, and the adjacent buildings the Jews were assembled before being shipped off to Treblinka. Only a few of those brought to the Umschlagplatz ever succeeded, by trickery or bribery, in returning to the ghetto.

29. Even in the first days of the deportation, exemptions extended only to the wives and children of those who held them (including people in the Jewish police), not to parents.

30. The chairman of the Judenrat was required to deliver 7,000 Jews for deportation on 22 July, and on the following day 10,000. This was the reason for Czerniaków's suicide.

31. Among those who killed themselves were many members of the intelligentsia.

32. This number is consistent with the German figures, which state that in the first four days of the campaign 29,078 people were 'deported'.

33. The report of the Jewish underground mentions another factor: there were cases of people volunteering for deportation in order to be reunited with members of their family who had been deported earlier.

34. After Czerniaków's death the Judenrat issued a proclamation (on the basis of promises by Gestapo officers) refuting the rumours circulating in the ghetto and appealing to the population not to hide but to present themselves for deportation. The proclamation stressed that the evacuation was aimed only at those who were not productive, and that the evacuees were indeed being sent to the borderlands of the east.

In order to strengthen the belief in this promise, at the start of the deportation the policeman used to make a point of releasing those who were employed in workshops and institutions.

35. *Gazeta Zydowska,* a Polish-language 'Jewish' newspaper published in Kraków two or three times a week with the permission of the German authorities.

36. The diarist renders the German *Aktion* into Hebrew as 'מפעל' literally the 'undertaking' or 'project'. In reality it meant deportation to a death-camp.

37. Israel Winnik, active in the Bund after the war, one of the young writers of the ghetto, organizer of the underground archive of the Bund in the ghetto; prepared a survey of the *punktim,* the assembly points for refugees in the ghetto. Killed during the deportation.

38. Through a rumour originated by the Gestapo, word spread in the ghetto that letters were arriving from people from Warsaw deported to the east (Brześć, Białystok, Pinsk) with news of good conditions in the areas in which they had been resettled.

39. As of 23 July 1942, all postal links between the ghetto and the outside world ceased; the post office at 19 Zamenhof Street was closed the next day. On 29 July, however, the delivery of post addressed to the ghetto was renewed (as noted in the entry for 30 July), but the ban on sending letters from the ghetto continued.

40. Aharon Libuszycki, Jewish writer and teacher, came to Warsaw from Łódź; M. Lejzerowicz, one of the editors of the Yiddish-language daily *Haynt.* Both lost their lives in the deportation.

41. Other diarists voiced similar condemnations of the ghetto police, 'filling their pockets with gold', who released some of those arrested in return for money (Kaplan, *Warsaw Diary,* pp. 386, 389).

42. Yakir Warszawski, writer and Hebrew teacher, a friend of the diarist. Died in the ghetto. The author writes of him again, at length, on 30 November.

43. At the beginning of the deportation the Germans let it be known that the intention was to deport non-productive people, while those employed in a German *shop* (workshop) would be allowed to stay in the ghetto. Hence *shopomania*—the blind race for employment in a workshop as security against deportation; *shopovatzim*—people employed in workshops.

44. After Czerniaków's death, an engineer by the name of Marc Lichtenbaum was appointed head of the Judenrat. His deputies were Dr. Gustaw Wielikowski and Alfred Sztolcman. Collaborators with the Germans and staunch opponents of the underground, all three were murdered by the Germans when the uprising broke out in April 1943. Orliański was probably Mścisław Orliański, one of the senior Judenrat officials and head of the Industrial Department.

45. Throughout 1941 and in the first months of 1942, food parcels continued to reach the ghetto from provincial towns in the General-Government as well as from abroad. These parcels were an important source of sustenance for many families in the ghetto. In the summer of 1942 the German authorities frequently confiscated them, on the pretext of a war against smuggling, although before they did so they would make the clerks in the Jewish post office sign that they had received the parcel. On the eve of the massive deportation some 2,000 parcels were reaching the ghetto daily.

46. In the first stage of the deportation the Germans used to kill the old and the sick to reinforce the impression that the other deportees really were being taken to work in the east. The execution took place on the spot, in the Umschlagplatz or the Jewish cemetery (testimony of Yitshak Zuckerman at the Eichmann trial).

47. A reference to Odilo Globocnik, head of the SS and the police in the Lublin district; directed the operation to destroy the Jews in the General-Government.

48. Grochów, Pelcowizna—small towns near Warsaw on the east bank of the Vistula.

49. With the start of the deportation command over the ghetto passed to the SS. Auerswald, the ghetto commissar, retained nominal control and up to the last minute assured Czerniaków that rumours of an impending deportation were false. This, of course, was merely a ruse to facilitate the Nazis' murderous plans (Czerniaków, *Diary,* p. 384; Gutman, *The Jews of Warsaw,* pp. 201–4).

50. Leon Neustadt, one of the directors of the 'Joint' in Warsaw: as the holder

of a South American passport he presented himself at the Pawiak at the beginning of the deportation and was shot.

51. With the growing wave of rumours about the approaching deportation, some of the wealthier families left the ghetto for places in the vicinity of the city, mainly Otwock; within a short time the deportation orders reached these places too.

52. Apparently the family of Rozencwajg, a poet and Hebrew teacher.

53. The mother of Leib Gruzalc, a member of the Bund youth movement, Tsukunft, and a commander of one of the fighting groups during the uprising.

54. Walter Caspar Többens, a German industrialist, owner of the largest workshops in the Warsaw ghetto. Made a great deal of money from exploiting the cheap labour of his Jewish employees. After the revolt of January 1943 he was made responsible for the evacuation of the ghetto and the transfer of the factories, together with their equipment and workers, to the Poniatowo and Trawniki labour camps. He was prevented from fulfilling this task by the Jewish Fighting Organization.

55. Here and below the reference is apparently to Menahem-Mendel Kon, a public figure, one of the moving spirits behind the Jewish underground. A colleague of the diarist in the secret archive.

56. Even though in theory those working for the organization were exempt from deportation, the Jewish police were secretly instructed to ignore their papers.

57. Józef Szeryński, commander of the ghetto police. A convert to Catholicism. Before the war, a colonel in the Polish police. On Szeryński, see Gutman, *The Jews of Warsaw*, pp. 88–90.

58. The Jewish police in the ghetto included a women's unit which was headed by Mrs. Horowitz, a lawyer from Łódź. The policewomen were posted, together with their male colleagues, by the gates of the ghetto, and their job was to search the women coming through. Sometimes they also used to direct the traffic in the streets of the ghetto. The female warders in the Jewish prison in the ghetto belonged to the same unit.

59. Owners of sewing-machines tried to get themselves into the existing workshops or those that were set up at the time of the deportation.

60. Bełżec—a death-camp in the Lublin district. Operated from March 1942 to June 1943. Piaseczno, Pustelnik—settlements in the Warsaw district, close to the capital.

61. [A reference to the weekly meetings of the workers of the secret archive established by the historian Dr. Emanuel Ringelblum, which were held on Saturdays and referred to as the Oneg Shabbes group. Oneg Shabbes was directed by a scientific organizational advisory board and had many active supporters—public figures, teachers, writers, rabbis and ordinary people. Lewin was a member of the Oneg Shabbes directorate. The activity of the Oneg Shabbes is described in detail in Josef Kermish, ed., *To Live with Honor and Die with Honor: Selected Documents from the Warsaw Ghetto Underground Archives (OS) (Oneg Shabbath)* (Jerusalem: Yad Vashem, 1986).]

62. Yidishe Visenshaftlikher Organizatsie, the Jewish Scientific Institute, established in Vilno in 1925; after the outbreak of war its offices were transferred to New York.

63. Apparently a reference to two colleagues of the diarist at Oneg Shabbes. Eliyahu Gutkowski and Menahem-Mendel Kon. On Kon, see n. 55.

64. The commander of the ghetto police announced that people presenting

themselves at the Umschlagplatz of their own free will on 29, 30 and 31 July would receive 3 kg of bread and 1 kg of jam—a strong temptation to the starving inhabitants of the ghetto. The proclamation was issued again on 1 August and the offer extended for a further three days. See also the entry for 2 Aug.

65. Lewin uses the word 'Shopomania'–שפמניה– to refer to the panic to obtain work in one of the German workshops. The illusion that people employed in the workshops were immune from deportation was carefully maintained by the Germans.

66. Megaphones were installed in the streets of Warsaw to broadcast German propaganda. The inhabitants ridiculed these broadcasts and regarded the official pronouncements with great scepticism.

67. Apparently a reference to an event that occurred at a place of work outside the ghetto. The name 'Warsaw-Praga' refers to the railway station in Praga, to the east of Warsaw, across the Vistula; at that time the railway workshops were employing many Jewish slave-labourers.

68. That is, the victims murdered there were 1 per cent of all those deported in the 'action'.

69. Apparently Obersturmführer Witossek, responsible for Jewish Affairs in the Ministry of Defence.

70. On receiving this false promise the Judenrat issued a comforting pronouncement (see n. 34).

71. Michael Brandstetter, an experienced teacher and educator, son of the writer M. D. Brandstetter; before the war, headmaster of the Jewish secondary school for girls in Łódź. In the Warsaw ghetto worked in the social assistance organization and as a teacher at the underground Jewish secondary school. Shot in the first uprising (January 1943).

72. Klima (real name Bluma) Fuswerg, teacher, director of the puppet theatre in the ghetto; later served as the contact of the Jewish National Committee on the Aryan side; among the survivors.

73. Untersturmführer Karl Georg Brandt, SS officer, among those responsible (together with Witossek) in the Jewish Department IV-B-4 in the Warsaw Gestapo. Took an active part in the destruction of the Jews.

74. A reference to Herman Höfle, Globocnik's chief of staff. Commander of the deportation operation in Warsaw. Never brought to trial, as he committed suicide in a Vienna prison in 1964.

75. This deportation of the children of the Pnimia orphanage on Ogrodowa Street marked the beginning of a systematic campaign to destroy all the children's institutions in the ghetto.

76. This word has two meanings in Polish: 'Get out' and 'departure'.

77. A few of those fit for work were taken from the Umschlagplatz and sent to a transit camp (*Dulag-Durchgangslager*) and from there to places of work.

78. A reference to the headquarters of the Jewish Self-help Organization. The diarist worked in one of the branch offices.

79. The wife of Rabbi Yitshak Nisenbaum, one of the spiritual mentors of the Zionist movement and a leader of the religious-Zionist party Mizrachi; the rabbi lost his life.

80. An organization very active before the war in looking after abandoned children and orphans, which continued to operate in the ghetto.

81. That is, volunteers for deportation, who came to the Umschlagplaz of their own free will.

82. The mobile murder squads who carried out the deportations throughout the General-Government under 'Operation Reinhard'.

83. *Bona;* in the ghetto slang, a food coupon. The number of coupons, on the basis of which supply was allocated, exceeded the number of people living in the ghetto. The difference ('the dead people') was not, however, as large as the diarist claims.

84. Shakhne Sagan, one of the leaders of Po'alei Tsion-Left and one of the most active figures in the field of self-help in the ghetto. Among the founders of the anti-Fascist groups.

85. Ben-Tsion Chilinowicz, journalist, a contributor to the main Yiddish newspaper *Der Moment.* In the ghetto he was active on the Journalists' and Authors' Aid Committee.

86. Dr. Edmond (Menachem) Sztajn, historian, lecturer in the Jewish Pedagogical Institute in Warsaw. Organizer of an illegal high school in the Warsaw ghetto. Among those deported to Trawniki, he was murdered at the beginning of January 1943 when the camps near Lublin were evacuated.

87. Zołotow, Karcewicz, Prync—teachers at the Yehudia Secondary School.

88. Perets Opoczynski, journalist and writer. Active in Zionist-Socialist Po'alei Tsion in the Warsaw ghetto underground; one of the close assistants of Ringelblum in Oneg Shabbes; wrote many reports and articles about life in the ghetto. Some were published in Warsaw in 1954. His unpublished diary is in the Yad Vashem archives. On his activity, see Kermish, Introduction to *To Live with Honor and Die with Honor,* pp. xviii–xx.

89. That is, 50 among those who were eating.

90. The ghetto police had been promised that their parents and other relatives (in addition to wives and children) would also have protection. This privilege was now revoked.

91. The proclamation was issued on 1 August and called on people to present themselves voluntarily at the Umschlagplatz.

92. The original deportation order gave exemption to its staff. Branches of the organization existed in every district capital.

93. The carpentry works of the Landau brothers (who lived in the United States) was expropriated by the Germans. Renamed the Ostdeutsche Bautischłerei Werkstätte (OBW), it was managed by the youngest brother, Alexander Landau. During the 'action' it gave shelter to many of the people active in the parties and the youth movements, among them the diarist. On this workshop, see Gutman, *The Jews of Warsaw,* who describes it as 'the most important centre of the ghetto underground during the deportation' (p. 222).

94. Towarzystwo Ochrony Zdrowia (Society for the Protection of Health), a very active association for medical aid and preventive medicine among the Jewish masses in Poland; in the ghetto, one of the affiliates of the general organization for mutual assistance.

95. Apparently Attorney Kahanowicz, an established community figure who worked in the Jewish Self-help Organization together with the diarist.

96. The diarist's relatives on his wife's side. Luba Lewin was the granddaughter of the chief rabbi of Warsaw, Yehuda Segal.

97. Inmates of the ghetto prison who had been sentenced to death (mainly for escaping from the ghetto or smuggling) were executed in the Jewish cemetery.

98. A former pupil of Yehudia.

99. On the letters that ostensibly arrived from the eastern borders see n. 38. A proclamation of the Jewish Fighting Organization of December 1942 warned the

inhabitants of the ghetto not to believe the false rumours and promises of the Germans.

100. Hillel Cajtlin (1872–1942), writer and publicist, one of the outstanding figures of Polish Jewry. Sick and very weak, he spent some time in the Jewish hospital. With the evacuation of the hospital during the deportation, Hillel Cajtlin was among those sent to Treblinka. He wrote a great deal, even during the war—among other things he translated the Book of Psalms into Yiddish—but his literary legacy was lost.

101. Professor Majer Balaban, a historian and teacher who was active in many fields and the author of many works on the history of Polish Jewry. In the ghetto he worked as director of the community archive. This time he was spared deportation. He died in the Warsaw ghetto in December 1942.

102. Dr. Alfred Bornsztajn, economist and journalist. In the ghetto he associated himself with underground circles. He lost his life on the Aryan side when he was working in the Jewish Department of the Polish Government Delegation in the underground.

103. Moritz Kohn and Zelig Heller, Gestapo agents in the ghetto. On Kohn and Heller, see Gutman, *The Jews of Warsaw,* pp. 92–93. The Germans killed their informers if they no longer needed them. Many of them knew too much and could have endangered their masters.

104. The Admor (Adonenu veMorenu, our Master and Teacher) of Radomsko, Rabbi Shlomo Hanoh Rabinowitz, his wife Esther, their daughter and son-in-law were killed in their flat at 30 Nowolipki Street.

105. Apparently a reference to Popower, the chief accountant of the Jewish council, who maintained contact with the underground.

106. Janusz Korczak (the pen-name of Dr. Henryk Goldszmid), doctor, writer, renowned educator. He was an innovator in the field of education, particularly in the care of orphans and street urchins. In the ghetto he continued to head an orphanage. During the deportation he refused to stay in the ghetto and abandon his pupils.

107. Bat-Sheva (Stefania) Hertzberg-Schweiger, the headmistress of the Yehudia Secondary School.

108. Bluma Kirschenfeld-Wasser, active in Oneg Shabbes, the wife of Hirsch Wasser. A survivor.

109. The wife of the teacher Natan Smolar.

110. The long-time skilled employees of the sawmill, who lost their wives and children, stormed against the 'new workers' (intellectuals who had found refuge in the plant), whom they considered to be endangering them by their presence.

111. Rabbi Mendel Alter, the president of the Association of Rabbis in Poland, the brother of the Admor of Gur; lived before the war in Panevezys (Lithuania) and Kalisz.

112. Yehiel Górny—one of the participants in Oneg Shabbes, later a member of the Jewish Fighting Organization. His account of the second deportation, in January 1943, has survived.

113. Saul Stupnicki, journalist and publicist; before the war edited *Lubliner Tagblat,* later a member of the editorial board of *Der Moment.* Committed suicide by taking poison at the Umschlagplatz, at the threshold of the deportation wagon.

114. A reference to the death-camp at Treblinka, in the vicinity of these two locations.

115. A reference to Josef Erlich ('Yosele Kapote'), an underworld figure who, thanks to his links with the SD (Sicherheitsdienst, the German Intelligence Ser-

vice), reached a position of dominance in the ghetto, holding the office of inspector in the ghetto police. Erlich dealt with the holders of foreign passports, who used to present themselves at the Pawiak Prison in order to be exchanged for Germans in Allied countries. As the 'action' intensified he too presented himself, together with all his family, at the prison, foreign passports in hand. He was shot by the SD men whom he had served for so long.

116. Shmuel Hirszhorn, journalist, an editor of the Polish-language Jewish newspaper *Nasz Przegląd.* A contributor to the Yiddish newspaper *Der Moment.*

117. The house at 14 Wolność Street was the largest of the children's homes in the ghetto; some thousand children lived there after it had taken in children from other institutions during the deportation. Most of the workers of the home went to be deported together with the children.

118. Nathan (Aharon) Koniński, educator, head of a children's home in the ghetto. Did not abandon his pupils in the deportation but went with them to the Umschlagplatz and then to Treblinka.

119. Landau's carpentry shop, in which the author worked, was at 30 Gęsia Street.

120. Natan Smolar, a teacher in schools founded by Tsysho (the Bundist educational organization), headmaster of the Borokhov School; worked in the carpentry shop together with Lewin. Died in April 1943.

121. We have not found a satisfactory explanation for this sentence or identified the farm referred to.

122. Luba Lewin, née Hotner, the diarist's wife. Ora was their only daughter.

123. 67 Dzielna Street was an orphanage for young children; the entire staff, led by the headmistress, Sarah Janowska, went to the Umschlagplatz together with the children.

124. A report of the liquidation of the ghetto written on the initiative of the Polish underground soon after the deportation mentions an instance of brave and poignant self-defence. When the gendarmes came to seize Seweryn M[ajda], formerly director of the theatre, he threw a heavy ashtray at one of them and was shot on the spot.

125. In Radom the first stage of the deportation took place in August 1942. The result was 10,000 people deported to Treblinka. Some 20,000 Jews from Radom were deported on 16 and 17 August. All that now remained of the Jewish community was some 4,000 Jewish slave-labourers employed in German factories. On this, see Martin Gilbert, *The Holocaust: A History of the Jews of Europe During the Second World War* (New York: Holt, Rinehart and Winston, 1985).

126. On 10 August, some three weeks after the start of the deportation, a decree was issued giving the inhabitants of the small ghetto (see n. 14) a few hours to leave their homes and move to the large ghetto. After the evacuation the southern area ceased to be part of the ghetto. Only the workers of Többens' workshop remained there, in Prosta Street.

127. The wife, son, and daughter of Daniel David Guzik, a director of the 'Joint' in Poland. At the time of the ghetto Guzik was one of the important figures in the Jewish Self-help Organization, was connected with underground circles and supported the efforts of the pioneering movement. Died in a plane crash in March 1946.

128. That is, the Judenrat.

129. Shmuel Wolkowicz, journalist, an editor of the Polish-language Jewish newspaper *Nasz Przegląd.* Settled in Israel.

130. A talented and popular singer, the daughter of Dawid Ajzensztat, the conductor of the choir of the Great Synagogue in Warsaw. Finished her studies at the Yehudia Secondary School in 1939. Frequently appeared at ghetto theatre evenings and concerts. Known as 'the nightingale of the ghetto'. Also appeared at performances organized by the youth movements and the underground. Shot at the Umschlagplatz for resisting the SS.

131. Dr. Jakub Tombeck, the diarist's brother-in-law.

132. The liquidation of the small ghetto forced the Judenrat to leave its offices in the former community-building, at 26–28 Grzybowska Street. It now operated from 19 Zamenhof Street, the premises of the Jewish post office in the ghetto.

133. The quotation marks around 'death' ('למיחה') suggest that communication of the terrible news was based on the similarity in sound between the word for 'death' (*mot*) and the word for 'bed' in Hebrew (*mita*).

134. Decrees took effect in mid-August limiting the area of the ghetto and evacuating entire streets. These internal 'transfers' were designed to confuse the Jews, to distract their attention from the deportations and to facilitate the plunder of their property.

135. The workshops of Többens, Schultz and Zimmerman—among the largest in the ghetto—were considered to be safe and employed thousands of Jewish labourers in conditions of severe exploitation. Karl Heinz Müller's workshop in Mylna Street employed many painters and artists—the shop was liquidated during the deportation and all its workers died.

136. Elhonon Cajtlin, writer and well-known journalist, a contributor to the Yiddish daily *Unzer Express;* worked for the Jewish Self-help Organization of the ghetto; saved from the Umschlagplatz and died in the ghetto in December 1942.

David Cajtlin, his son, was taken to Treblinka in January 1943. He jumped off the train and returned to the ghetto; died a short time after the liberation, in one of the camps in Germany, at the age of 18.

137. Among the deportees on that day was Yitshak Katznelson's wife Hanna (née Rosenberg) and their two young sons, Ben-Tsion and Binyomin.

138. A few of those assembled in the Umschlagplatz would be sent to a transit camp *(Durchgangslager, Dulag)* in Leszno Street, and thence to labour camps; according to German statistics, in July, 1,612, in August, 7,403, and in September, 2,565; in total during the deportation, 11,580.

139. Czerniaków was a suburb of Warsaw where a Dror kibbutz training farm (*hakhshara*) still operated.

The Jewish cemetery on the edge of the ghetto served as a transit point in the smuggling of goods and people.

[Eliyahu Gutkowski, a history teacher from Łódź. During the war he taught history in the Dror Secondary School run by the Warsaw ghetto underground. He was among the organizers of the secret archive of the ghetto, one of its most important contributors and a member of its executive. Murdered in April 1943.]

140. During the deportation the relative immunity of the children's institutions was rescinded: in all, some 30 institutions of different sorts were destroyed, and some 4,000 children were deported to Treblinka.

141. Mordecai Haim Rumkowski, the Jewish Elder (Judenälteste), the sole ruler of the Jews in the ghetto of Łódź.

142. Members of Hashomer Hatsair in the Warsaw ghetto who were employed in the OBW (Ostdeutsche Bautischlerei Werkstätte) factory (formerly Landau's sawmill). They cultivated a plot of land in the cemetery. The firm provided cover

for this, and hence the involvement of Commissar Hensel, the German supervisor of the workshop.

143. This proclamation was published on 16 August 1942. It stated that the only Jews permitted south of Leszno Street were those working in factories in this part of the city. The second paragraph stated that henceforth the families of people with permanent places of work would also be subject to deportation.

144. The broom factory (*brushter-shop*) on Franciszkańska Street was one of the largest manufacturing establishments in the ghetto. It produced brooms for the German army. Many community figures and members of the pioneering youth movement were employed there.

145. Shulamit Mokarska, an active member of women's organizations in Lodz; in the Warsaw ghetto she was active in the Łódź refugee Landsmannschaften; head of the women's self-help organization.

146. Rabbi Shimon Huberband, one of the permanent workers of the secret archive; his writings from this period were discovered in the archive's milk churn. He was 33 years old when he was murdered.

147. Rav Yitshak Nisenbaum was one of the leaders of the religious-Zionists in Poland. Leib Szczeranski was an important figure in Mizrachi in Poland.

148. Before the war, Dr. Rokhl Sztajn had headed the women's organization of the Bund and been a member of the Warsaw city council; an active member of Centos, the organization for Jewish children.

149. Bat-Sheva (Stefania) Herzberg-Schweiger (see n. 107).

150. During the 'action' in Otwock, near Warsaw, some 10,000 people were deported to Treblinka. Many days after the deportation the Germans were still hunting down people who had gone into hiding. Anyone caught was shot on the spot. Their number reached some 2,000, among them not a few who had escaped from the Warsaw ghetto.

151. According to a report by the governor of the Warsaw district (12 October 1942) the number of suicides in August was 155, and in September, 60.

152. There were two Jewish medical institutions in Otwock: Briuś, for people with tuberculosis; and Sofjówka, for mental patients. A report drawn up close to the deportation by Adolph and Batya Berman stated that many of the doctors and staff of the two institutions committed suicide.

153. The Soviet bombardment was a source of encouragement to the population of the ghetto, allowing them to believe—if only for a short while—that this was a response to the terrors of the Umschlagplatz. One shell hit Schulz's workshop, killing some and wounding others.

154. An attempt was made on the life of the commander of the ghetto police not by the Polish underground but by Yisrael Kanal, a commander in the Jewish Fighting Organization and a member of the religious-Zionist youth movement Akiva. At the time of the incident, Kanal was wearing the uniform of a policeman. The attack is described in Gutman, *The Jews of Warsaw*, pp. 237–40. On Szeryński, see n. 57.

It is interesting to note that although Lewin was a member of the Oneg Shabbes and thus had links with the underground, and although he was working in the OBW factory, where many of the Jewish Fighting Organization's leaders were sheltering, he never suspected that they were involved in the attempted assassination of Szeryński.

155. The proclamation censuring the Jewish Ordnungsdienst for the role they played in the deportation was also issued by the Jewish Fighting Organization.

156. A town in the Warsaw district. The deportation there took place on 21

August; about 1,000 people were murdered on the spot; the members of the Judenrat were shot the following day; two work-groups remained. 'The gang'—the mobile extermination squad of 'Operation Reinhard'.

157. The diarist's sister and brother-in-law.

158. In the week of 14–21 October, more than 20,000 Jews were deported from Piotrków. Some 2,000 were allowed to remain in factories, and 2,000 managed to hide. On this, see Gilbert, *The Holocaust,* pp. 481–82.

159. This too is a reference to Treblinka, which was close to Kosow.

160. A resort between Warsaw and Otwock, which was very popular with the inhabitants of Warsaw; the 7,000 Jews of Falenica were deported to Treblinka on 20 August 1942; only the 100 working in the sawmill remained.

161. A reference to the brother-in-law of Dr. Natan Asch, the secretary of the General Zionist Party in Poland, Al Hamishmar. A secretary of the Jewish Self-help Organization.

11. Jozef Zelkowicz

The Lodz ghetto (called Litzmannstadt by the Germans) lasted longer than any other Jewish community in Poland. When it was finally liquidated in August 1944, all other cities and towns in the nation had already been "cleansed" of their Jews. Mordechai Chaim Rumkowski, the Elder of the Jews in the Lodz ghetto, established a ghetto archive to record for posterity the ordeal of the inhabitants during this period. One result was the *Daily Chronicle,* which sought to present a day-by-day account of Jewish existence under German rule. Lucjan Dobroszycki's edition, *The Chronicle of the Lódz Ghetto, 1941–1944* (1984), which contains about one-fourth of the material in the original multi-volume Polish version, makes available to English readers a representative sample of this work.

In addition to the *Daily Chronicle,* Rumkowski's staff of writers (which included the journalist Jozef Zelkowicz, author of the following selection) began (but never finished) an "Encyclopedia of the Ghetto." They also left behind numerous essays and reports, as well as private, unofficial diaries and jottings. Other ghetto residents added their own written memoirs or observations, making the Lodz ghetto (together with Warsaw) one of the best documented Jewish communities in Europe during World War II.

Through a heroic effort of research and inquiry, Alan Adelson and Robert Lapides have gathered in *Lódz Ghetto: Inside a Community Under Siege* (1989) the most comprehensive picture we have of Jewish ghetto life (and death) during the German occupation. The editors of this monumental volume describe it as "a collected consciousness." It includes writings by men and women, young and old, in the form of diaries, journals, and notebooks as well as fragments scrawled in margins of novels and on the back of old menus. Few of the authors survived; many remain anonymous to this day. But almost all were driven by a consuming need to record for the future the daily details of their hope and despair while their ranks were cruelly being thinned by hunger, disease, executions, and deportation "to the east" (which meant first the suffocation vans of Chelmno, and then the gas chambers of Auschwitz).

Among the many valuable materials in this collection are the visual documents, especially some very rare color photographs, including one of SS Reichsführer Heinrich Himmler being greeted upon his arrival in the ghetto by Rumkowski. Rumkowski's conviction—born of a combination of vanity and self-delusion—that by creating ghetto workshops to produce bullet casings, boots, and uniforms for the German war effort he could make some of his fellow Jews indispensable to their captors betrays his unwillingness or inability to accept what is plain to us today: a fundamental aim of German policy was the *total* destruction of European Jewry. Rumkowski was only delaying, not averting, the certain doom that awaited them all.

Of course, Rumkowski was not alone in failing to grasp this truth. Among the audience at his notorious if heartbreaking "Give Me Your Children" speech (portions of which are reproduced here) there must have been many who forced themselves to agree with his strategy: that by surrendering the "few," he would be able to save the many. In the course of time, this formula reversed itself: yielding the many might preserve the few. Such misguided hope may have been a necessary ingredient for psychological survival, though it is difficult to imagine how Rumkowski's listeners could have reconciled themselves to the price they were being asked to pay. Eventually, 60,000 people died in the ghetto, from various causes, while 130,000 were deported and murdered in Chelmno and Auschwitz. The Germans left more than 800 Jews behind to clean up the ghetto after the final liquidation in August 1944, and many of them were still alive when the Russians liberated the city on January 19, 1945. Including those who survived Auschwitz and other camps, as many as 10,000 former Lodz residents, according to some estimates, were still alive at the end of the war.

Narratives such as the following selection simultaneously record a reality and challenge our resistant imaginations (often in vain) to identify with it. "Days of Nightmare" by Jozef Zelkowicz tells of the time in early September 1942, when 20,000 Jews, including the elderly over sixty-five, the sick, and children under ten, were rounded up for deportation—which meant death. Zelkowicz, was born near Lodz in 1897. Although ordained as a rabbi, he never practiced his profession, but trained to become a teacher. After service in the Polish army, he turned to research and writing, chiefly in Yiddish. He published two important works: one examined the view of death in Jewish ethnography and folklore; the other, Jewish life in a Polish stetl. He died in Auschwitz in 1944.

Zelkowicz wrote many pieces for the ghetto archive and the *Daily Chronicle,* as well as private essays, often critical of the ghetto administration. His disenchantment with the role of the ghetto police is evident in this selection, though he is not entirely unsympathetic to their human dilemma. The Germans sabotaged morality by offering the Jews a "choiceless choice," not between right and wrong, but between the bad and the worse. If the ghetto police refused to cooperate in the roundup of Jews, their own children under the age of ten would be subject to deportation. If they agreed to cooperate, they had to victimize their fellow Jews. We have difficulty remembering today that most of them behaved not through inclination, but as a result of the macabre threat imposed on them by the Germans. If their "choice" seems to reflect a failure of moral heroism, the fault may lie more with our own romantic expectations than a genuine lapse in their commitment to Jewish community. Almost any decision that included acceding to a German request for more candidates for deportation left the individual in an impossible ethical situation.

One diarist of the Lodz ghetto enjoins us: "Listen and believe this, even though it happened here, even though it seems so old, so distant, and so strange." Contemporaneous accounts like these reverse time and thrust us into the midst of things, as epic poets were fond of doing. But the old, the distant, and the strange that we encounter in these pages bear no resemblance to the legends of Homer and Virgil, or to the Romantic poets, who also responded to their appeal. They reflect rather a vision of unredeemed anguish whose victims enter into the future only through our memory of their pain.

Days of Nightmare

Friday, September 4, 1942

The deportation of children and old people is a fact.

This morning the ghetto received a horrifying shock: What seemed improbable and incredible news yesterday has now become a dreadful fact. Children up to the age of ten are to be torn away from their parents, brothers and sisters, and deported. Old people over sixty-five are being robbed of their last life-saving plank, which they have been clutching with their last bits of strength—their four walls and their beds. They are being sent away like useless ballast.

If only they were really being "sent away," if only there were the slightest ray of hope that these "deportees" were being taken somewhere! That they were being settled and kept alive, even under the worst conditions, then the tragedy would not be so enormous. After all, every Jew has always been ready to migrate; Jewish life has always been based on a capacity for adjusting to the worst conditions; every Jew has always been prepared to fold up his tent at command, to leave his home and country, and all the more so here in the ghetto, where there's no wealth, no property, no peace of mind, and where he is not attached to anything. Jewish life has always relied only on faith in the ancient Jewish God, who, the Jew feels, has never abandoned him. "Somehow or other, everything will turn out all right. Somehow or other, we'll manage to survive, with our last bit of wretched life."

If there were the slightest assurance, the slightest ray of hope they were being sent somewhere, then the ghetto would not be in such a turmoil over this new and unwonted evil decree. There have already been so many new and unwonted evil decrees and we have had to put up with them and, whether or not we wanted to, we had to go on living, so that we might somehow or other swallow this one too. But the fact is that no one has the least doubt, we are all certain that the people now being deported from the ghetto are not being "sent" anywhere. They are being taken to nowhere, at least the old people. They are going to the scrap heap, as we say in the ghetto. How, then, can we be expected to make peace with this new evil decree? How can we be expected to go on living whether or not we want to?

There is simply no word, no power, no art able to transmit the moods, the laments, and the turmoil prevailing in the ghetto since early this morning.

To say that today the ghetto is swimming in tears would not be mere rhetoric. It would be simply a gross understatement, an inadequate utterance about the things you can see and hear in the ghetto of Litzmannstadt, no matter where you go or look or listen.

There is no house, no home, no family which is not affected by this dreadful edict. One person has a child, another an old father, a third an old mother. No one has patience, no one can remain at home with arms folded awaiting destiny. At home you feel forlorn, wretched, alone with your devouring cares. Just run into the street. Out there you don't feel so blind, you don't feel so abandoned. Animals, too, when they feel some sorrow, supposedly cling together, animals having mute tongues which cannot talk away their sorrows and their grief. How much more so human beings.

All hearts are icy, all hands are wrung, all eyes filled with despair. All faces are twisted, all heads bowed to the ground, all blood weeps.

Tears flow by themselves. They can't be held back. People know these tears are useless. Those who can help it refuse to see them, and those who see them—and they too shed useless tears—can't help themselves either. Worst of all, these tears bring no relief whatsoever. On the contrary. It's as though they were falling on, rather than from, our hearts. They only make our hearts heavier. Our hearts writhe and struggle in these tears like fish in poisoned waters. Our hearts drown in their own tears. But no one can help us in any way, no one can save us.

No one at all? No one in the whole ghetto who wants to, who can save us? Could there be someone after all? There must be someone who wants to, who's able to! Perhaps we still don't know who that someone is? Perhaps he's hiding somewhere, because he can't help everyone, he can't save everyone!

Maybe that's why people are scurrying all over the ghetto like poisoned mice. They're looking for that "someone." Maybe that's why the ghetto Jews are clutching at straws, maybe this straw is the "someone" they're looking for. Maybe that's him over here, maybe that's him over there. Everyone is looking to revive old acquaintanceship with those who can help, everyone is looking for pull. Perhaps God will help.

And the little children who don't yet understand, the little children who have no way of knowing about the Damoclean sword hanging over their innocent heads, perhaps subconsciously sense the enormous threat hovering over them, and their tiny hands cling tighter to the scrawny and shrunken breasts of their fathers and mothers.

Son of man, go out in the street. Look at all this, soak in the subconscious terror of the infants about to be slaughtered. And be strong and don't weep! Be strong and don't let your heart break, so that later on you can give a thoughtful and orderly description of just the barest essentials of what took place in the ghetto during the first few days of September in the year 1942.

Mothers run through the streets, one shoe on, one shoe off, their hair half combed, their shawls trailing on the ground. They are still holding on to their children. They can clasp them now tighter and closer to their emaciated breasts. They can still cover their bright little faces and eyes with kisses. But what will happen tomorrow, later on, in an hour?

People say: The children are to be taken from their parents as early as today. People say: The children are to be sent away as early as Monday. They are to be sent away—where?

To be sent away on Monday, to be taken away today. Meanwhile, for the moment, every mother clings to her child. Now she can still give her child everything, the very best thing in her possession—the last morsel of bread, all her love, the dearest and the best! Today the child doesn't have to wait for hours on end and cry until his father or mother figures out how much to give him from the half-pound of bread. Today they ask the child: "Darling, would you like a piece of bread now?" And today the piece of bread that the child gets isn't dry and tasteless like always. If there's just a bit of margarine left, they spread it on. If there's any sugar left, they sprinkle it on. The ghetto lives recklessly today. No one weighs or measures. No one hoards sugar or margarine to stretch it for a whole ten days until the next ration. Today in the ghetto no one lives for the future. Today one lives for the moment and now, for the moment, every mother still has her child with her; and wouldn't she, if she could, give it her own heart, her own soul? . . .

There are children who do indeed understand. In the ghetto, ten-year-old children are mature adults. They already know and understand what is in store for them. They may not as yet know why they are being torn away from their parents—they may not as yet have been told. For the moment it's enough for them to know that they are being torn away from their devoted guardians, their fathers and their loving and anxious mothers. It's hard to keep such children in one's arms or to take them by the hand. Such children go out into the streets on their own. Such children weep on their own, with their own tears. Their tears are so sharp and piercing that they fall upon all hearts like poisoned arrows. But hearts in the ghetto have turned to stone. They would rather burst but they can't, and this is probably the greatest, the harshest curse. . . .

The sorrow becomes greater, and the torture more senseless, when one tries to think rationally. Well, an old man is an old man. If he's lived his sixty-five years, he can convince himself, or others convince him, that he should utter something like: "Well, thank God, I've had my share of living, in joy and sorrow, weal and woe. That's life. Probably that's fate. And anyway, you don't live forever. So what's the difference if it's a few days, a few weeks, or even a few months sooner? Sooner or later, you've got to die, sooner or later, everything's over. That's life."

Maybe they can talk the old man into telling himself these things, maybe they can talk his family into telling themselves. But what about children who have only just been hatched, children who have only seen God's world in the ghetto, for whom a cow or a chicken is just a legendary creature, who have never in their lives so much as inhaled the fragrance of a flower, laid eyes upon an orange, tasted an apple or a pear, and who are now doomed to die? . . .

The sky above the ghetto, like yesterday and the day before, is unclouded. Like yesterday and the day before, the early autumn sun shines. It shines and smiles at our Jewish grief and agony, as though someone were merely stepping on vermin, as though someone had written a death-sentence for bedbugs, a Day of Judgment for rats which must be exterminated and wiped off the face of the earth.

There are nevertheless still enough people in the ghetto who doubt, still enough people in the ghetto who continue to live with faith. There are even those who reason logically:

"This ghetto, where eighty percent of the population performs useful work, is not one of your provincial towns which could have been made *judenrein,* free of Jews, in half an hour. Here people are necessary, are needed for work. It's not possible that they would take people from here and send them away."

And those who cannot argue rationally, who are just full of faith, they simply believe in miracles:

"Things like this have happened before. All through history, Jews have been threatened with bitter decrees, and deliverance has come at the last minute. There just was an air raid on Lodz for the first time since the war began. So maybe they'll withdraw the evil decree. Who can tell?" . . .

The files of the vital statistics records were sealed last night.

No one knows who did the sealing. The German authorities? No. They were sealed by the Jewish Resettlement Commission,[1] sealed so that none of the dates of birth could be falsified. The child is not to make himself older or the old man younger. And therefore the decree, in its full severity, remains in force. No miracles take place today; perhaps they did in the past, but the past is so old that none of the ghetto Jews remembers it. No miracle ever happened to a ghetto Jew. No ghetto Jew ever heard of a good rumor coming true. Always the other way around, the bad is always the brutal actuality. People also have been saying in the factories: "The Resettlement Commission has already been formed."

There are people in the ghetto who, compulsorily or merely of their own free will, have taken it upon themselves to act as a Great Sanhedrin and issue death sentences. A commission has been formed with Jakobson, Blemer, Rosenblat, Naftalin, and Greenberg, and they will be in charge of deporting the 20,000 old people and children. They are the ones who will have to sever living limbs from living bodies. They are the ones who will be halving Jewish families.

Certainly they've been charged, probably by the Chairman, to take care of this matter. Certainly there has to be such a body in charge of even a matter as halving Jewish families. But still no person with conscience would undertake to issue death sentences.

This commission and a whole staff of officials involved with them have been working all night long. They have drawn up a list of the population by streets and buildings.

The commission is operating in the Office of Vital Statistics at 4 Kościelna Street, the focal point where all eyes of the ghetto are turned. This is where people come to plead and weep for those near and dear to them. But so much chaos and turmoil are here, so much disorganization and confusion that none of the aforementioned gentlemen on the commission knew what was being asked of him, what his assignment was, or if he could do anything for anyone. None of the gentlemen on the commission knew anything definite about the substance and nature of the decree.

One of them said that the age in question was from one to ten years, so that children under one year would not be included, and another one claimed that the edict was inclusive, that it applied to children from one minute to nine years and 365 days old. The commission likewise had differences of opinion about old people. They didn't know whether the age limit began with people who were already past sixty-five or whether it applied to those who were still in their sixty-fifth year. . . .

It was rumored during the day that the deportations would take place without the involvement of the German authorities. The edict will be carried out by the Resettlement Commission with the help of the Jewish police, the Jewish fire department, and the like. As a reward for their efficient and loyal performance, they have been promised that their own families, i.e. their children and parents, will be exempt from the edict. The police will thus be able to "work" with untroubled heads and calm spirit. They will do a fine job—you can rely on them. They've already passed the test a few times.[2] The same deal obviously was promised to exempt the families of all the heads of the factories or offices. The reason is the same—so they can work with untroubled minds.

But the question is: At whose expense have they been privileged? If all the old people and children without exception amount to only 13,000 souls, and now, if they are going to make exceptions and exempt several thousand people in the families of the police, the fire department, and higher officials, then who will be deported in their place? . . .

At a quarter to five, accompanied by David Varshavsky[3] and S. Jakobson, the Chairman showed up. No one could fail to notice the tremendous change he underwent in the past few days or hours. His head was bowed as though he couldn't keep it on his shoulders, his eyes dull and lifeless. We were looking at an old man barely able to totter on his feet. He was an old man like all the old men who gathered on the square. Only his face was fuller and less emaciated than theirs. Only his body was garbed in fine and fancy wear instead of rags and patches like theirs. In the tangle of his white hair you could tell how fearfully he had lived through these last hours. In the spasms of his lips you could tell he had no word of comfort for the people gathered here, that there would be no cheerfulness today.

The Chairman barely dragged himself up to the platform and announced that Varshavsky had the floor. He's not much of a speaker, and he probably can't manage a personal conversation either. But then who wants to drink gall from a beautiful goblet? Poison from an ugly cup is as effective as from a beautiful cup. No one came today to hear and admire any feats or oratorical talents. Everyone came to hear the truth, and David Varshavsky told the truth in all its bitterness, in bitter and crude words:

"Yesterday the Chairman received an order to resettle some 20,000 people from among the children and the aged. How strange is human fate! We all know the Chairman. We all know how many years of his life, how much energy, how much work and effort he devoted to rearing the Jewish child. Now precisely from his hands is the sacrifice being demanded. Precisely from him who has reared more children than the present edict covers! But there's nothing we can do. This is a decree that cannot be annulled. They are asking for sacrifices and the sacrifices have to be offered. We understand and we feel the anguish. There is no place to hide it and there is nothing to hide from you. All children up to the age of ten and all old people have to be handed over. The decree cannot be annulled. We can only soften it perhaps by carrying it out quietly and peacefully. Also in Warsaw there was a similar decree. We all know how that was carried out, it's no secret. And it happened that way because the German authorities were in charge, not the Jewish community. We undertook to carry out the decree ourselves because we do not want, we cannot allow the execution of the decree to assume catastrophic proportions. Can I comfort you, set your minds at ease? Regrettably, I cannot. But there is one thing I can tell you, perhaps it will set your minds at ease, perhaps it will bring you some comfort. It seems, according to all probabilities, that after this decree we will be permitted to remain undisturbed. There is a war going on. Air-raid alarms are frequent. We often have to flee. At such time, children and old people are only a hindrance. They must, therefore, be sent away." . . .

The next to speak was the lawyer Jakobson.

"Residents of the ghetto! Yesterday an order arrived to deport over 20,000 people from the ghetto; they are to come from among the children, the aged, and the sick who have no chance of recovery. None of the previous resettlement decrees has been as difficult to carry out as this one. The present edict has been rendered even more difficult by our lack of means or possibilities to heal or even ease the wounds.

"We have had to assume this heavy duty and this great responsibility to carry out this edict ourselves. We have had to take it upon ourselves, because other cities have shown us what happens when the edicts are carried out by strangers and not by our own hands." (Loud weeping in the audience.)

"Weeping and wailing unfortunately won't help us. There were hints like: 'If you don't carry out the order yourselves, then we will carry it out.' There's nothing more to say. We already understand everything and much too much.

"The resettlement will involve some 3,000 people daily. The responsibility for complying with the edict has fallen upon the entire community. The whole ghetto is responsible that it be carried out." . . .

Neither speaker made anyone feel any easier. Both David Varshavsky and Jakobson explicitly pointed out there is nothing anyone can do. Varshavsky explicitly said he had no comfort to offer, and Jakobson explicitly said he had no means at his disposal for healing or even easing the wounds. The two speakers had only one virtue. Everyone learned the truth from them, the truth about the number to be sent away, about the categories to be sent away, and about the pedestrian embargo.[4] As bitter as these truths may be and as desperate, they are still ever so much better than the unuttered truths, the rumors, and various speculations. But everyone felt that not everything had been fully said. Everyone felt that the Chairman still had something to say. The crowd waited in great suspense and with bated breath in order better to hear the Chairman's words.

The Chairman wept.

This proud Jew, who till now had governed his realm highhandedly and with total despotism, this man who had never heeded anyone and always did everything on his own responsibility and his own authority, this same man stood before the crowd a shattered man. He could not control his tears. The Chairman wept like a child. We could see how stricken he was by the common woe, how deeply afflicted by the decree, even though he himself was neither directly nor indirectly subject to it. These were no artificial tears. These were Jewish tears flowing from a Jewish heart. These tears helped to score a great deal with the crowd.

As soon as he mastered himself, he began with the following words:

"The ghetto has been afflicted with a great sorrow. We are being asked to give up the best that we possess—children and old people. I was not privileged to have a child of my own, and so I devoted the best years of my life for the sake of the child. I have lived and breathed with the child. I would never have imagined that my hands would deliver the sacrifice to the altar. In my old age I must stretch forth my arms and beg: Brothers and sisters, yield them to me! Fathers and mothers, yield me your children." (Enormous and fearful weeping among the crowd.)

"I had a premonition that something threatened us. I was expecting something and I stood always like a sentry on guard to prevent this something from happening. But I couldn't, because I didn't know what was in store for us, I didn't know what awaited us.

"The removal of the sick from the hospitals was something I never expected at all. You have the best proof. I had my own nearest and dearest there and I could do nothing for them. I thought that with that it would stop. I thought that with that they would leave us in peace. But it turns out that instead of this peace which I crave so strongly, for which I have always worked and striven, something else is destined for us. That is, after all, the fate of Jews—always to suffer more and worse, especially in wartime.

"Yesterday afternoon, I was given an order to deport some 20,000 Jews from the ghetto. If not: 'We will do it.' And the question arose: Should we have taken it over and do it ourselves or leave it for others to carry out? Being guided by the thought not of how many would be going to their destruction, but rather of how many we could save, we, that is, I and my closest colleagues, concluded that however difficult it would be for us, we would have to take over the responsibility for carrying out the decree ourselves. I have to carry out this difficult and bloody operation. I have to cut off limbs in order to save the body! I have to take children, because otherwise—God forbid—others will be taken." (Terrible wails.)

"I have not come to comfort you. Nor have I come to set your hearts at ease, but to uncover your full grief and woe. I have come like a thief to take your dearest possession from your hearts. I left no stone unturned in my efforts to get the order annulled. But when this was impossible, I tried to mitigate it. I then gave orders for a registration of all nine-year-old children. I wanted at least to rescue all those aged nine to ten. But I could not get them to assent. I did succeed in one thing—saving all children past the age of ten. Let this be a comfort in our great sorrow.

"In the ghetto, we have a great number of tuberculars who have only a few days, perhaps a few weeks, left to live. I don't know, perhaps this plan is devilish, perhaps not, but I cannot hold back from uttering it: 'Give me these sick and in their place we can rescue the healthy.' I know how dearly each family, especially among Jews, cherishes its sick. But with such an edict, we have to weigh and measure: Who should, can, and may be saved? Common sense dictates that we should save those capable of being saved, those who have prospects of survival, and not those who can't be saved anyway.

"We live, after all, in the ghetto. Our life is so austere that we don't even have enough for the healthy, much less for the sick. Each of us keeps the sick man alive at the price of our own health. We give the sick man our bread. We give him our bit of sugar, our piece of meat, and the consequence is that not only does the sick man not become well, but we become sick. Naturally, such sacrifices are noble. But at a time when we must choose either to sacrifice the sick man, who not only has no chance of becoming well but is even likely to make others sick, or to rescue a well man. I could not mull over this problem for long, and I was forced to decide in favor of the well man. I have therefore given orders to the doctors and they will be compelled to turn over all the incurably ill in order to rescue in their stead all those who are well and who want, and are able, to live." (Terrible weeping.)

"I understand you, mothers, I see your tears. I can also feel your hearts, fathers, who, tomorrow, after your children have been taken from you, will be going to work, when just yesterday you had been playing with your dear little children. I know all this and I sympathize with it. Since 4

P.M. yesterday, upon hearing the decree, I have utterly collapsed. I live with your grief, and your sorrow torments me, and I don't know how and with what strength I can live through it. I must tell you a secret. They demanded 24,000 victims, 3,000 persons a day, for eight days, but I succeeded in getting them to reduce the number to 20,000, and perhaps even fewer than 20,000, but only on condition that these will be children to the age of ten. Children over ten are safe. Since children and old people add up only to 13,000, we will have to meet the quota by adding the sick as well.

"It is hard for me to speak. I have no strength. I will only utter the appeal I make to you: Help me carry out the action! I tremble. I am frightened at the thought that others, God forbid, might take it into their own hands.

"You see before you a broken man. Don't envy me. This is the most difficult order that I have ever had to carry out. I extend to you failing and trembling hands and I beg you: Give into my hands the victims, thereby to ensure against further victims, thereby to protect a community of a hundred thousand Jews." . . .

Saturday, September 5, 1942 . . .

It has begun.

It's only a few minutes after 7 A.M. now. All the people, practically the entire ghetto, are on the street. Whose nerves don't drive them out? Who can sit home? Who has peace of mind? Who can just sit with his arms folded? No one! . . .

Consequently, from early morning the streets of the ghetto are busier than ever. And what a strange busyness. A silent, lifeless busyness, if one can put it that way. People don't talk to one another, as though everyone had left his tongue at home or had forgotten how to speak. Acquaintances don't greet each other, as though they feel ashamed. Everyone is rigid in motion, rigid standing in the long lines at the distribution places and rigid in the enormous lines at the vegetable places, A dead silence dominates the ghetto. No one so much as sighs or moans. Today huge, heavy stones weigh on the hearts of the ghetto residents.

People run through the ghetto streets like transmigrant spirits, perhaps like sinful souls wandering through the world of chaos. With that same stubborn silence on their clenched lips, with that same dread in their eyes—that's the way those spirits must look. People stand in line, perhaps like prisoners condemned to death, standing and waiting until their turn comes to go to the gallows. Rigidity, terror, collapse, fear, dread—there is no word to describe all the feelings that swell and grow in these petrified hearts that can't even weep, can't even scream. There is no ear that can catch the silent scream that deafens with its rigidity and that rigidifies with its deafening silence.

They run over the three ghetto bridges, like a host of hundredheaded serpents surging back and forth. The host of serpents extends forward and back. These are people hurrying and hastening. The air is pregnant with oppressiveness. Macabre tidings are in its density. The sky keeps swelling, billowing, and will soon burst, and out of the void will tumble the full horror and the full reality.

It has begun!

No one knows what, no one knows where, no one knows how. Supposing that everyone keeps silent, supposing that no one looks at anyone else, supposing that everyone avoids everyone else the way the thief avoids his pursuer, then who was the first to utter these dreadful words: "It has begun!" . . .

No one. No one spoke them. No one uttered these macabre tidings. Only the heavens burst and its spilled guts dropped those words: "It has begun!" . . .

Where has it begun? People say: "They're already taking out all the residents of the old-age home on Dworska Street."

People say: "On Rybna Street there's already a truck, and they're loading it with old people and children."

Alas, all the stories are true: They're taking them from here, from there, from everywhere, and they're already loading on Rybna Street.

It has begun.

The Jewish police made the first start. They began as if they wanted to practice their work along the line of least resistance—the old-age home. There it was as easy as pie—they were just ready to be taken. And the people there are being taken wholesale, there's no selecting and rejecting. They're all old, and so all of them are to go to the scrap heap. It's really the line of least resistance. Who is going to speak up for them, who's going to waste words for these old people who have been living on the good graces of the community for weeks and months now? . . . They are being loaded on the trucks like lambs for slaughter and driven to the staging area. There they may get the condemned man's last meal consisting, supposedly, of a soup with lots of potatoes, cooked with horse bones, and later they'll be taken away from this staging area—

To the scrap heap. . . .

Over on Rybna Street the police have to take them out of apartments. There they are encountering resistance. There they have to cut living, palpitating limbs from bodies. There they wrench infants from their mothers' breasts. There they pull healthy molars out of mouths. On Rybna Street they tear grown children from under their parents' wings. They separate husband from wife, wife from husband, people who've been together for forty or fifty years, who've lived in sorrow and in joy, who've had children together, who've reared them together and lost them together. They've been with one another for forty or fifty years and become practically one body. . . .

The sick, too, are being taken there, sick people who at great risk escaped from the hospital, whom mortal terror gave the strength and courage to leap over barriers, sick people who were given someone's last crust of bread, last bit of sugar, last potato just to keep them alive one more day, one more week, one more month, because the war might end and then they could perhaps get back on their feet. Also the sick are being taken.

Living limbs are cut off. Healthy molars are extracted. Palpitating bodies are halved. The anguish is great. Let someone try to describe it, he won't be able to! Let someone try to depict it, he'll only collapse! Is it any wonder that people scream? . . .

People scream. And their screams are terrible and fearful and senseless, as terrible and fearful and senseless as the actions causing them. The ghetto is no longer rigid; it is now writhing in convulsions. The whole ghetto is one enormous spasm. The whole ghetto jumps out of its own skin and plunges back within its own barbed wires. Ah, if only a fire would come and consume everything! If only a bolt from heaven would strike and destroy us altogether! There is hardly anyone in the ghetto who hasn't gasped such a wish from his feeble lips, whether he is affected directly, indirectly, or altogether uninvolved in the events which were staged before his very eyes and ears. Everyone is ready to die; already now, at the very start, at this very moment, it is impossible to endure the terror and the horror. Already at this moment it is impossible to endure the screams of hundreds of thousands of bound cattle slaughtered but not yet killed; impossible to endure the twitching of the pierced but unsevered throats, which let them neither die nor live.

What has happened to the Jewish police, who undertook to do that piece of work? Have their brains atrophied? Have their hearts been torn out and replaced with stones? It's hard, very hard, to answer these questions. One thing is certain—they are not to be envied. And there are also all sorts of executioners. There is an executioner who for a worthless traitor's pay would raise his hand against his brother; another, besides getting his traitor's pay, also has to be gotten drunk, otherwise his ignoble hand will fumble. And there are executioners who do their bloody work for the sake of an idea. They were told: "So-and-so is not only useless to society, he's actually detrimental, he's got to be cleaned out." So they act for the good of society, they do the cleaning out.

The Jewish police have been bought. They have been intoxicated. They were given hashish—their children have been exempted from the order. They've been given three pounds of bread a day for their bloody bit of work—bread to gorge themselves on and an extra portion of sausage and sugar. They work for the sake of an idea, the Jewish police do. Thus our own hands, Jewish hands, extract the molars, cut off the limbs, slice up the bodies. . . .

No, they are not to be envied at all, the Jewish police!

The bloody page of this history should be inscribed with black letters for the so-called White Guard, the porters of Balut Market and of the

Food Supply Office. This rabble, fearful of losing their soup during pedestrian embargo, volunteered to help in the action on condition that they get the same as had been promised the police—bread, sausage, and sugar, and the exemption of their families. Their offer was accepted. They participated voluntarily in the action.

The bloody page of this history should be inscribed with black letters for all those officials who petitioned to have some role in this action, only in order to get bread and sausage rations instead of the soup they wouldn't have gotten sitting at home. . . .

The Seizures

O God, Jewish God, how defenseless Jewish blood has become!

Oh, God, God of all mankind, how defenseless human blood has become!

Blood flows in the streets. Blood flows over the yards. Blood flows in the buildings. Blood flows in the apartments. Not red, healthy blood. That doesn't exist in the ghetto. Three years of war, two and a half years of ghetto have devoured the red corpuscles. All the ghetto has is pus and streaming gall, that drips, flows, and gushes from the eyes, and inundates streets, yards, houses, apartments.

How can such blood satisfy the appetites of the beasts? It can only whet their appetites, nothing more!

It is no longer just a rumor, not just gossip; it is an established fact. The head of the ghetto administration, Biebow,[5] the man most interested in the ghetto's existence, in its survival, has put himself in charge of the action. He himself directs the "resettlement."

People are being seized. The Jewish police are seizing them with mercy, according to orders: children under ten, old people over sixty-five, and the sick whom doctors have diagnosed as incurable. . . .

The Jewish police have addresses. The Jewish police have Jewish concierges, and the concierges have house registers. The addresses inform that in such and such an apartment is a child who was born on such and such a date. The addresses inform that in such and such an apartment is an old man who was born so and so many years ago. A doctor comes into every apartment. He examines the occupants. He observes who is in good health and who just pretends to good health. He's had so much practice in the ghetto that a mere glance distinguishes the well from the mortally ill.

Nor does it avail the child to cling to its mother's neck with both little hands. Nor does it avail the mother to throw herself on the threshold and bellow like a slaughtered cow: "Only over my dead body will you take my child!" It does not avail the old man to clutch the cold walls with his bony fingers and plead: "Let me die here in peace." It does not avail the old woman to fall on her knees, kiss their boots, and plead: "I've already got grown-up grandchildren." It does not avail the sick man to bury his feverish head in the damp, sweaty pillow, and moan, and shed perhaps his last tears.

Nothing avails. The police have to supply their quota. They have to seize people. They cannot show pity. But when the Jewish police take people, they do so punctiliously. When they take people, they help them weep, they help them moan. When they take people, they try to comfort them with hoarse voices, to express their anguish. . . .

It's totally different when others come! [6]

They enter a yard and first off is a reckless shot of a revolver. Everyone loses courage. All blood stops coursing. All throats are stopped with hot lead. The lead freezes the gasps and sobs in the throat. You tremble. No! To tremble you have to have flowing blood, but your blood refuses to circulate. It has curdled, it is rigid like water in a frost. The Jews wait, benumbed, paralyzed, helpless. They wait for what happens next.

Next comes a harsh, terse, draconic order, yelled out, and then repeated by the Jewish police. "In two minutes everyone downstairs! No one is permitted to stay inside. All doors must be left open!"

Who can describe, depict the crazed and wild stampede on stairways and landings, the rigid and inanimate figures who hasten to obey the order on time? No one.

Old, rheumatic, twisted sclerotic legs stumble over crooked stairs and angular stones. Young, buoyant, deer legs fly with birdlike speed. The heavy and clumsy legs of the sick heaved from their beds are bent and bowed. Swollen legs of starvelings tap blindly along. All scurry, hurry, rush out into the courtyard.

Woe to the latecomer. He will never finish that last walk. He will have to swim in his own blood. Woe to him who stumbles and falls. He will never stand up again. He will slip and fall again in his own blood. Woe to the child who is so terrified that all he can do is scream "Mama!" He will never get past the first syllable. A reckless gunshot will sever the word in his throat. The second syllable will tumble down into his heart like a bird shot down in mid-flight. The experience of the last few hours proved this in its stark reality.

When the Jewish police take people, they take whomever they can, whoever is there. If someone has hidden and can't be taken, then he remains hidden. But when *they* come to take people, they take those who are there and those who are not there. If someone is not located, they take another in his place. If the missing man is found, he will not be seized; he will have to be carried out.

The Jewish police, further, can be bought. Not with ghetto marks of course, but with more valuable items. As long as it's hush-hush and no one can see or hear, then anyone who's got something can ransom his way out. But when *they* come to take you, you can buy your way out only with your rarest and most precious possession—your life. You can take the choice of not wishing to go, and you'll never have to go anywhere again. . . .

The sun sets bloodily in the west. All the west is bathed in blood. Ridiculous and grandiloquent to speak of the heavens reflecting the blood that was spilled in the ghetto today. The heavens are too far from earth. The screams and the shouts from the ghetto did not carry up there, not even an echo of their echo. The screaming was of no avail; the tears were shed for nothing and were lost. No one saw them; no one heard them.

To say that "the sun sets bloodily in the west" is therefore merely a datum that symbolizes today's bloody ghetto day. To say that "all the west is bathed in blood" is a datum that symbolizes this evening. But the day still strives with the evening. The day is not yet over, as if this day were not long enough. The measure is not yet full. The last tidings of Job have not yet been heard.

What iron strength this day has! It stretches as long as the Jewish Exile. It is as heavy as Jewish woe. What dark strength the ghetto people have! After three years of hunger, after three years of wretched enslavement, they can still endure such days as this day. . . .

Sunday, September 6, 1942

The clock says 12 noon. It is a full-blown summer's day. It's impossible to stay indoors. No one can stay at home. Perhaps because you must stay inside, perhaps because you're surrounded by your family and you have nothing to say to them. You yourself are indeed young. You yourself have your work certificate proving you're a useful citizen of the ghetto. Your wife is young too, and her papers are fully in order. Younger than both of you is your fourteen-year-old son, who already works. He is tall and slender and handsome, a fine figure of a lad. You have no reasons to be afraid, not for yourself, not for your wife, and not for your son. But you sit at home, listening to the sighs and shouts and screams of your neighbors, from whom pieces of flesh were torn away yesterday. You sit at home and hear yet another scream every minute from another neighbor, or mute sobs from people sick with sorrow. You sit at home and constantly hear yet another scream from a neighbor who, in his great despair over the children who were seized from him, tries to end his broken existence with a knife or by leaping from a high window. You sit at home, devoured by your own sorrow, your wife's sorrow, your son's sorrow, and the sufferings of all your neighbors and all Jews. Sitting at home like this, on and on, every time you glance at the mirror and it reflects a yellowed, sunken, confused countenance: "You, too, are a candidate for the scrap heap!" Sitting at home, casting furtive glances at your wife who has aged dozens of years in these last two days, then looking at your beautiful son, seeing his dark hollow face and the mortal fear lurking in his deep black eyes, then all the terror around you makes you fear for yourself, makes you fear for your wife, makes you fear for your trembling child. All of you are candidates!

1. The Resettlement Commission included leading officials of the Lodz ghetto administration: the head of the office of vital statistics (Naftalin), the Jewish police commandant (Leon Rosenblat), the prison commandant, the director of penal administration, the head of the office of investigations, and the chairman of the ghetto courts (Jakobson).

2. The ironic reference is to the role of the Jewish police during the deportations from Lodz in the first half of 1942.

3. David Varshavsky was head of the tailoring enterprises in the Lodz ghetto.

4. In order to carry out the deportations with a minimum of resistance, the Germans ordered Rumkowski to declare a general embargo on pedestrian traffic in the ghetto, beginning Saturday, September 5, 1942, at 5 A.M. Persons who wanted a pedestrian permit had to apply to Jewish police headquarters. All building concierges were held responsible to see that only bona-fide tenants of their buildings were domiciled there. The notice which Rumkowski issued about the pedestrian embargo warned that all unauthorized persons on the ghetto streets would be evacuated.

5. Hans Biebow was the German chief of the ghetto administration of Litzmannstadt. He was tried for war crimes in Poland, condemned to death, and executed in 1947.

6. The reference is to the SS, which had in fact taken over the deportations because the Jewish police had not proved sufficiently competent for the task.

12. Avraham Tory

Avraham Tory was born Avraham Golub (*golub* means "dove" in Russian; *tory* is the Hebrew equivalent for "turtledove") in Russian Lithuania in 1909. He graduated from the Hebrew high school in 1927 and began to study law at the University of Kovno. In 1930, he came to the United States, where he continued to study law at the University of Pittsburgh. After his father's unexpected death in 1931, he returned to Lithuania and the university.

Awarded his law degree in 1934, Tory eventually secured a job as assistant to one of the few Jewish law professors at the University of Lithuania. Because he was Jewish, Tory found it virtually impossible to obtain a license to practice law. During the 1930s, he was active in Zionist movements and was, in fact, attending an international Zionist conference in Geneva when the Germans invaded Poland on September 1, 1939. He decided to return to Lithuania, which by then had become a Soviet satellite.

Because of his Zionist activities, Tory was under surveillance by the Soviet secret police, so he left Kovno and went into hiding in Vilna. He was there when the Germans invaded Russia on June 22, 1941, but he soon returned to Kovno. Writing (and later dictating) in Yiddish, Tory made his first entry at midnight of that momentous invasion day. From then until he escaped from the ghetto in April 1944, he made regular entries detailing the fate of Kovno's 35,000 Jews. As deputy secretary to the Jewish Council of Elders (formed by German order), Tory had access to most German decrees, which he preserved along with his own chronicle. He also encouraged artists and photographers to keep a visual record; some of their work remains and is included in the English-language edition of Tory's diaries, *Surviving the Holocaust: The Kovno Ghetto Diary* (1990), from which the following selection, "Memoir," is taken.

Tory buried his materials in five crates before he escaped from the ghetto, but when he returned to rescue them after the Russian liberation, he was able to retrieve only three crates from the ruins. He ignored the orders of the Soviet secret police to turn over all documents to them, and after an arduous and risky journey through several East European countries (during which he was forced to leave some of his precious cargo in trustworthy Jewish hands for later delivery), he finally arrived in

Palestine in October 1947 and retrieved as much as he could. Tory, who now lives in Israel, estimates that he has in his possession about two-thirds of the original diary with its accompanying documents.

Unlike the terse entries in Adam Czerniaków's Warsaw ghetto diary, which attempt little portraiture of the Nazi administrators with whom the head of the Warsaw Jewish Council came into almost daily contact, Tory's descriptions of encounters with various German officials give us a complex glimpse at how ruthless these murderers were, disguising their intentions by assuming a restrained and sometimes civil demeanor when their victims bargained for more food, more fuel, more space, and more work. The leaders of the Jewish Council in Kovno showed none of the dogmatic arrogance displayed by Chaim Rumkowski in Lodz, but their efforts were in the end no more successful than his, not because their strategies were inept but because (unknown to them) their enemies were determined from the start to kill them all.

Among the most graphic narratives in Tory's diary are the reports of German "actions," when large numbers of Jews were rounded up and sent to the Ninth Fort outside Kovno for execution. One of the grimmest is his account of the burning of the Jewish hospital and orphanage on October 4, 1941; some of the patients and children were shot and buried in a pit in the courtyard, while others perished inside the buildings. But the most harrowing entry, which is included here, describes the day when most of Kovno's Jews were assembled in the ghetto square and 10,000 were selected and sent to their deaths. Tory's eyewitness account of this event, the notorious "great action," remains one of our most vivid testaments to the mass murder of European Jewry.

Memoir

October 28, 1941

On Friday afternoon, October 24, 1941,[1] a Gestapo car entered the Ghetto. It carried the Gestapo deputy chief, Captain Schmitz, and Master Sergeant Rauca. Their appearance filled all onlookers with fear. The Council was worried and ordered the Jewish Ghetto police to follow all their movements. Those movements were rather unusual. The two Ghetto rulers turned neither to the Council offices nor to the Jewish police, nor to the German labor office, nor even to the German commandant, as they used to in their visits to the Ghetto. Instead, they toured various places as if looking for something, tarried a while in Demokratu Square, looked it over, and left through the gate, leaving in their wake an ominously large question mark: what were they scheming to do?

The next day, Saturday afternoon, an urgent message was relayed from the Ghetto gate to the Council: Rauca, accompanied by a high-ranking Gestapo officer, was coming. As usual in such cases, all unauthorized persons were removed from the Council secretariat room and from the hallway, lest their presence invoke the wrath of the Nazi fiends.

The two Germans entered the offices of the Council. Rauca did not waste time. He opened with a major pronouncement: it is imperative to increase the size of the Jewish labor force in view of its importance for the German war effort—an allusion to the indispensability of Jewish labor to the Germans. Furthermore, he continued, the Gestapo is aware that food rations allotted to the Ghetto inmates do not provide proper nourishment to heavy-labor workers and, therefore, he intends to increase rations for both the workers and their families so that they will be able to achieve greater output for the Reich. The remaining Ghetto inmates, those not included in the Jewish labor force, would have to make do with the existing rations. To forestall competition and envy between them and the Jewish labor force, they would be separated from them and transferred to the small Ghetto. In this fashion, those contributing to the war effort would obtain more spacious and comfortable living quarters. To carry out this operation a roll call would take place. The Council was to issue an order in which all the Ghetto inmates, without exception, and irrespective of sex and age, were called to report to Demokratu Square on October 28, at 6 A.M. on the dot. In the square they should line up by families and by the workplace of the family head. When leaving for the roll call they were to leave their apartments, closets, and drawers open. Anybody found after 6 A.M. in his home would be shot on the spot.

The members of the Council were shaken and overcome by fear. This order boded very ill for the future of the Ghetto. But what did it mean? Dr. Elkes* attempted to get Rauca to divulge some information about the intention behind this roll call, but his efforts bore no fruit. Rauca refused to add another word to his communication and, accompanied by his associate, left the Council office and the Ghetto.

The members of the Council remained in a state of shock. What lay in wait for the Ghetto? What was the true purpose of the roll call? Why did Rauca order the Council to publish the order, rather than publish it himself? Was he planning to abuse the trust the Ghetto population had in the Jewish leadership? And if so, had the Council the right to comply with Rauca's order and publish it, thereby becoming an accomplice in an act which might spell disaster?

Some Council members proposed to disobey the Gestapo and not publish the order, even if this would mean putting the lives of the Council members at risk. Others feared that in the case of disobedience the archhenchmen would not be contented with punishing the Council alone, but would vent their wrath also on the Ghetto inmates, and that thousands of Jews were liable to pay with their lives for the impudence of their leaders. After all, no one could fathom the intentions of Rauca and his men; why, then, stir the beasts of prey into anger? Was the Council entitled to take responsibility for the outcome of not publishing the order? On the other hand, was the Council entitled to take upon itself the heavy burden of moral responsibility and go ahead with publishing the order?

The Council discussions continued for many hours without reaching a conclusion. In the meantime, the publication of the order was postponed and an attempt was made to inquire about Rauca's plans, using the contacts of Caspi-Serebrovitz[2] in the Gestapo. Zvi Levin, who was Caspi's fellow party member (they were both Revisionists),† was asked to leave for the city, to call on him and ask him what he knew about Rauca's plans, and to ask Rauca to grant an audience to Dr. Elkes. Levin found Caspi packing his bags. The latter was stunned to learn about the order and exclaimed spontaneously: "Aha, now I understand why Rauca is sending me to Vilna for three days just at this time. He wants to keep me away from Kovno, especially now."

Complying with Levin's request, Caspi set out to inform Rauca that disquiet prevailed in the Ghetto and that the Council chairman wished to see him that very evening. Rauca responded favorably.

The Council members agreed that the meeting with Rauca should take place in the modest apartment of Dr. Elkes, in order to keep the meeting

*Dr. Elchanen Elkes, head of the Jewish Council in Kovno, died in Dachau on July 25, 1944.

†Antisocialist party founded by Zeev Jabotinsky in 1925.

as secret as possible. At 6 P.M. Rauca arrived at Dr. Elkes's apartment. Yakov Goldberg, a member of the Council and head of the Council's labor office, was also present. Dr. Elkes began by saying that his responsibilities as leader of the community and as a human being obliged him to speak openly. He asked Rauca to understand his position and not to be angry with him. Then he revealed his and the Council members' fears that the decree spelled disaster for the Ghetto, since if the German authorities' intention was only to alter the food distribution arrangements, the Council was prepared to carry out the appropriate decrees faithfully and to the letter. Therefore, he went on to say, there is no need for roll call of the entire Ghetto population, including elderly people and babes in arms, since such a summons was likely to cause panic in the Ghetto. Moreover, the three roll calls which had taken place over the past three months had each ended in terrible "actions." Therefore, he, Dr. Elkes, pleaded with "Mr. Master Sergeant" to reveal the whole truth behind the roll call.

Rauca feigned amazement that any suspicion at all could have been harbored by the members of the Council. He repeated his promise that a purely administrative matter was involved and that no evil intentions lurked behind it. He added that at the beginning the Gestapo had, in fact, considered charging the Council with the distribution of the increased food rations for the Jewish labor force, but having given thought to the solidarity prevailing among the Jews, had suspected that the food distribution would not be carried out and that the food delivered to the Council would be distributed among all Ghetto residents—both workers and nonworkers—in equal rations. The Gestapo could not allow this to happen under the difficult conditions of the continuing war. Accordingly, the Gestapo had no choice but to divide the Ghetto population into two groups. The roll call was a purely administrative measure and nothing more.

Dr. Elkes attempted to appeal to the "conscience" of the Gestapo officer, hinting casually that every war, including the present one, was bound to end sooner or later, and that if Rauca would answer his questions openly, without concealing anything, the Jews would know how to repay him. The Council itself would know how to appreciate Rauca's humane approach. Thus, Dr. Elkes daringly intimated a possible defeat of Germany in the war, in which case Rauca would be able to save his skin with the help of the Jews. Rauca, however, remained unmoved: there was no hidden plan and no ill intention behind the decree. Having said this he left.

After this conversation, Dr. Elkes and Goldberg left for the Council offices, where the other Council members were waiting for them impatiently. Dr. Elkes's report of his conversation did not dispel the uncertainty and the grave fears. No one was prepared to believe Rauca's assertions that a purely administrative matter was involved. The question remained: why should the elderly and the infants, men and women, including the sick and feeble, be dragged out of their homes at dawn for a roll call by families and by workplace, if the purpose was simply the distribution of increased food rations to the workers? Even if the plan was just to transfer part of the Ghetto population to the small Ghetto—why was a total roll call needed? Was it not sufficient to announce that such-and-such residents must move into those living quarters within the small Ghetto which had been left empty after the liquidation of its residents and the burning of the hospital?

Even before Rauca ordered the Council to publish the decree, rumors originating in various Jewish workplaces in the city where there was contact with Lithuanians had it that in the Ninth Fort* large pits had been dug by Russian prisoners-of-war. Those rumors were being repeated by various Lithuanians and, naturally, they reached the Council. When Rauca announced the roll-call decree, the rumors and the roll call no longer seemed a coincidence.

As the rumors about digging of pits persisted, and the members of the Council failed to give any indication of their apprehension, an atmosphere of fear pervaded the Ghetto, growing heavier with each passing day. The very real apprehensions of the Council were compounded by the fear that any revelation of its suspicions and doubts might lead many Jews to acts of desperation—acts which were bound to bring disaster both on themselves and on many others in the Ghetto.

Since the members of the Council could not reach any decision, they resolved to seek the advice of Chief Rabbi Shapiro. At 11 P.M. Dr. Elkes, Garfunkel, Goldberg, and Levin set out for Rabbi Shapiro's house. The unexpected visit at such a late hour frightened the old and sick rabbi. He rose from his bed and, pale as a ghost, came out to his guests. He was trembling with emotion.

The members of the Council told Rabbi Shapiro about the two meetings with Rauca, and about the roll-call decree. They also told him about their fears and asked him to rule on the question of whether they, as public leaders responsible for the fate of the Jews in the Ghetto, were permitted or even duty bound to publish the decree.

The rabbi heaved a deep sigh. The question was complex and difficult: it called for weighty consideration. He asked them to come back to him at 6 A.M. the next day. Dr. Elkes and his colleagues replied by stressing the urgency of the matter, since the Council had been told that it must publish the decree before that time. Each further delay was liable to provoke the ire of the Gestapo. The rabbi promised that he would not close his eyes all night; that he would consult his learned books and give them an early reply.

When the Council members returned to the rabbi's house at 6 A.M. they found him poring over books which lay piled up on his desk. His face bore visible traces of the sleepless night and the great ordeal he had

*One of the nineteenth-century tsarist fortifications surrounding the city of Kovno, it was a site of mass executions of Jews.

gone through to find scriptural support for the ruling on the terrible question facing the Council. He lifted his head—adorned by white beard—and said that he had not yet found the answer. He asked them to come back in three hours' time. But at 9 o'clock he was still engrossed in study and put off his answer for another two hours. At last, at 11 o'clock, he came up with the answer. In studying and interpreting the sources, he had found that there had been situations in Jewish history which resembled the dilemma the Council was facing now. In such cases, he said, when an evil edict had imperiled an entire Jewish community and, by a certain act, a part of the community could be saved, communal leaders were bound to summon their courage, take the responsibility, and save as many lives as possible. According to this principle, it was incumbent on the Council to publish the decree. Other rabbis, and a number of public figures in the Ghetto, subsequently took issue with this ruling. They argued that it was forbidden for the Council to publish the decree, since by doing so it inadvertently became a collaborator with the oppressor in carrying out his design—a design which could bring disaster on the entire Ghetto. Those bereft of all hope added the argument that since the Ghetto was doomed to perdition anyway, the Council should have adopted the religious principle "yehareg u'bal yaavor" (to refuse compliance even on the pain of death), and refrained from publishing the decree.

Immediately after their visit to the chief rabbi, members of the Council convened for a special meeting and decided to publish the decree. So it was that on October 27, 1941, announcements in Yiddish and in German were posted by the Council throughout the Ghetto. Their text was as follows:

> The Council has been ordered by the authorities to publish the following official decree to the Ghetto inmates:
>
> All inmates of the Ghetto, without exception, including children and the sick, are to leave their homes on Tuesday, October 28, 1941, at 6 A.M., and to assemble in the square between the big blocks and the Demokratu Street, and to line up in accordance with police instructions.
>
> The Ghetto inmates are required to report by families, each family being headed by the worker who is the head of the family.
>
> It is forbidden to lock apartments, wardrobes, cupboards, desks, etc. . . .
>
> After 6 A.M. nobody may remain in his apartment.
>
> Anyone found in the apartments after 6 A.M. will be shot on sight.

The wording was chosen by the Council so that everyone would understand that it concerned a Gestapo order; that the Council had no part in it.

The Ghetto was agog. Until the publication of this order everyone had carried his fears in his own heart. Now those fears and forebodings broke out. The rumors about the digging of pits in the Ninth Fort, which had haunted people like a nightmare, now acquired tangible meaning. The

Ghetto remembered well the way the previous "actions"* had been prepared, in which some 2,800 people had met their deaths. An additional sign of the impending disaster was that on that very same day workers in various places were furnished with special papers issued by their German employers—military and paramilitary—certifying that their holders were employed on a permanent basis at such-and-such a German factory or workplace.

This category also included the airfield workers, who had been issued suitable cards and yellow armbands to be worn on their right sleeve, as well as members of the Jewish labor brigade which worked for the Gestapo. Workers in this brigade, headed by Lipzer,[3] were particularly conspicuous since, in addition to the documents certifying their employment by the Gestapo, they were provided by Lipzer with a sign bearing the word "Gestapo" as their workplace.

The overwhelming majority of the Ghetto inmates did not have in their possession such privileged documents, or Jordan certificates. People kept flocking to streetcorners and into courtyards, making inquiries, hoping to hear something which might put them at their ease. Everyone was busy interpreting each word in the decree. Particularly ominous was the threat that anyone found in his apartment after 6 A.M. would be shot. The Gestapo also announced that, immediately after this deadline, armed German policemen would be deployed in every house and courtyard and would kill anyone to be found there regardless of the reason.

No one in the Ghetto closed an eye on the night of October 27. Many wept bitterly, many others recited Psalms. There were also people who did the opposite: they decided to have a good time, to feast and gorge themselves on food, and use up their whole supply. Inmates whose apartments were stocked with wines and liquor drank all they could, and even invited neighbors and friends to the macabre drinking party "so as not to leave anything behind for the Germans."

There were also those who, despite everything, did not lose hope and kept themselves busy, hiding away money, jewelry, and other valuables in hiding places under floorboards or in door lintels, in pits they dug that night in their courtyards, and so on. Every unmarried woman looked for a family to adopt her, or for a bachelor who would present her as his wife. Widows with children also sought "husbands" for themselves and "fathers" for their children—all this in preparation for the roll call, in the hope of being able to save themselves.

I, too, adopted as my son an eleven-year-old boy, named Moishele Prusak, who was a remote relative of mine. His parents and the other members of his family were living in my native village of Lazdijai, whereas he lived alone in Slobodka, where he studied at the yeshiva.

Tuesday morning, October 28, was rainy. A heavy mist covered the sky and the whole Ghetto was shrouded in darkness. A fine sleet filled the air and covered the ground in a thin layer. From all directions, dragging

*_Aktion_ was the Nazi term for the round-up of victims for execution.

themselves heavily and falteringly, groups of men, women, and children, elderly and sick who leaned on the arms of their relatives or neighbors, babies carried in their mothers' arms, proceeded in long lines. They were all wrapped in winter coats, shawls, or blankets, so as to protect themselves from the cold and the damp. Many carried in their hands lanterns or candles, which cast a faint light, illuminating their way in the darkness.

Many families stepped along slowly, holding hands. They all made their way in the same direction—to Demokratu Square. It was a procession of mourners grieving over themselves. Some 30,000 people proceeded that morning into the unknown, toward a fate that could already have been sealed for them by the bloodthirsty rulers.

A deathlike silence pervaded this procession tens of thousands strong. Every person dragged himself along, absorbed in his own thoughts, pondering his own fate and the fate of his family whose lives hung by a thread. Thirty thousand lonely people, forgotten by God and by man, delivered to the whim of tyrants whose hands had already spilled the blood of many Jews.

All of them, especially heads of families, had equipped themselves with some sort of document, even a certificate of being employed by one of the Ghetto institutions, or a high school graduation diploma, or a German university diploma—some paper that might perhaps, perhaps, who knows, bring them an "indulgence" for the sin of being a Jew. Some dug out commendations issued by the Lithuanian Army; perhaps these might be of help.

The Ghetto houses were left empty, except for a handful of terminally ill persons who could not raise themselves from bed. In compliance with the instruction issued by the authorities, on every house in which a sick person had been left behind a note was posted giving his name and his illness. The Council offices, as well as the offices of the Jewish Ghetto police, the labor office, and the Ghetto workshops, were also left empty; doors of offices were left ajar, as well as closets, desks, and drawers, in compliance with Rauca's order, so that nobody could remain there in hiding. The storerooms containing the Ghetto's food supplies and raw materials were left unattended on that day, as was the private property of the Ghetto inmates.

As the Ghetto inmates were assembling in Demokratu Square, armed Lithuanian partisans raided the Ghetto houses, forcing their way into every apartment, every attic, every storage room, and every cellar, looking for who might be hiding Jews. Many of these Lithuanians took the opportunity to loot. Some carried with them suitcases stuffed with goods, but these looters were subsequently arrested by the German police officers, who disarmed them and removed them from the Ghetto.

The Ghetto fence was surrounded by machine guns and a heavy detachment of armed German policemen, commanded by Captain Tornbaum. He also had at his disposal battalions of armed Lithuanian partisans. A crowd of curious Lithuanian spectators had gathered on the hills overlooking the Ghetto. They followed the events taking place in the square with great interest, not devoid of delight, and did not leave for many hours.

The Ghetto inmates were lined up in columns according to the workplace of the family heads. The first column consisted of the Council members, followed by the column of the Jewish policemen and their families. On both sides and behind stood the workers in the Ghetto institutions, and many columns of the various Jewish labor brigades together with their families, since on that day the Ghetto was sealed off. No one was allowed to go out to work. The airfield workforce, which had left for work the previous day and had stayed there for an additional shift in compliance with General Geiling's order,[4] were returned to the Ghetto after the entire Ghetto population had already been assembled in the square. Having completed two shifts of hard labor, these people hurried to and fro among the columns, tired, hungry, and dirty, in an effort to locate their families.

In the meantime, dawn broke. The grayish light of a rainy day replaced the nocturnal darkness. Old people, and those too weak to remain standing on their feet for long hours, collapsed on the ground. Others, having learned from past experience, had brought with them a chair or a stool on which they could sit and rest. Some had even equipped themselves with food before coming to the roll call, but the great majority of the Ghetto inmates remained standing on their feet, hungry and tired, among them mothers and fathers with children in baby carriages or in their arms.

Three hours went by. The cold and the damp penetrated their bones. The endless waiting for the sentence had driven many people out of their minds. Religious Jews mumbled prayers and Psalms. The old and the sick whimpered. Babies cried aloud. In every eye the same horrible question stood out: "When will it begin?! When will it begin?!"

At 9 A.M. a Gestapo entourage appeared at the square: the deputy Gestapo-chief, Captain Schmitz, Master Sergeant Rauca, Captain Jordan, and Captain Tornbaum, accompanied by a squad of the German policemen and Lithuanian partisans.

The square was surrounded by machine-gun emplacements. Rauca positioned himself on top of a little mound from which he could watch the great crowd that waited in the square in tense and anxious anticipation. His glance ranged briefly over the column of the Council members and the Jewish Ghetto police, and by a movement of his hand he motioned them to the left, which, as it became clear later, was the "good" side. Then he signaled with the baton he held in his hand and ordered the remaining columns: "Forward!" The selection had begun.

The columns of employees of the Ghetto institutions and their families passed before Rauca, followed by other columns, one after another. The Gestapo man fixed his gaze on each pair of eyes and with a flick of the finger of his right hand passed sentence on individuals, families, or even whole groups. Elderly and sick persons, families with children, single women, and persons whose physique did not impress him in terms of labor power, were directed to the right. There, they immediately fell into the hands of the German policemen and the Lithuanian partisans, who showered them with shouts and blows and pushed them toward an opening especially made in the fence, where two Germans counted them and then reassembled them in a different place.

At first, nobody knew which was the "good" side. Many therefore rejoiced at finding themselves on the right. They began thanking Rauca, saying "Thank you kindly," or even "Thank you for your mercy." There were many men and women who, having been directed to the left, asked permission to move over to the right and join their relatives from whom they had been separated. Smiling sarcastically, Rauca gave his consent.

Those who tried to pass over from the right to the left, in order to join their families, or because they guessed—correctly, as it turned out—that that was the "good" side, immediately felt the pain of blows dealt by the hands and rifle butts of the policemen and the partisans, who brutally drove them back again to the right. By then everyone realized which side was the "good" and which the "bad" one.

When some old or sick person could not hold out any longer and collapsed on to the ground, the Lithuanians set upon him instantly, kicking him with their boots, beating him, and threatening to trample him underfoot if he did not get up at once. Drawing the last ounce of strength, he would rise to his feet—if he could—and try to catch up with his group. Those unable to get up were helped by their companions in trouble, who lifted them up, supported them, and helped them along to reach the assembly spot in the small Ghetto, to which they were marched under heavy guard.

In most cases these were old people, women, and children, frightened and in a state of shock, turned by screams and blows into a panic-stricken herd which felt it was being driven by a satanic, omnipotent force. It was a force which banished all thought and seemed to allow no hope of escape.

In especially shocking cases where members of a family were separated, when pleas and cries were heartrending, Dr. Elkes tried to come to the rescue, and at times he even succeeded in transferring whole families to the left. Among others, he intervened on behalf of a veteran public figure, the director of the hospital, a skillful artisan, and a number of activists of Zionist and non-Zionist underground circles. Unfortunately he did not succeed in transferring everyone to the left.[5]

The commander of the Jewish police, Kopelman, who stayed with Dr. Elkes near Rauca, also succeeded in saving Jews and whole families. The number of such survivors, throughout this bitter and hurried day, reached into the hundreds.

Rauca directed the job of selection composedly, with cynicism, and with the utmost speed, by mere movements of the finger of his right hand. When the meaning of the movement of his finger was not grasped instantly, he would roar: "To the right!" or "To the left!" And when people failed to obey at once he shouted at them: "To the right, you lousy curs!" Throughout the selection he did not exhibit any sign of fatigue or sensitivity at the wailing, pleas, and cries, or at the sight of the heartrending spectacles which took place before his eyes when children were separated from their parents, or parents from their children, or husbands and wives from each other—all those tragedies did not penetrate his heart at all.

From time to time, Rauca feasted on a sandwich—wrapped in wax paper lest his blood-stained hands get greasy—or enjoyed a cigarette, all the while performing his fiendish work without interruption.

When a column composed mostly of elderly people, or of women or children, appeared before him, he would command contemptuously: "All this trash to the right!" or "All this pile of garbage goes to the right!" To Dr. Elkes, when he tried to intervene in an attempt to save their lives, he would say: "Wait, you'll be grateful to me for having rid you of this burden."

Whenever Rauca condescended to respond favorably to Dr. Elkes's intercession, he would say carelessly: "Well, as far as I am concerned . . ." and then order the German policeman: "This fat one, or this short one, or this one with the glasses on, bring him back to me."[6]

Now and then Rauca would be handed a note with a number written on it, copied from the notebook kept by the German who diligently applied himself to the task of recording the number of Jews removed to the small Ghetto.

Rauca was quick to dispense "mercy" to those who, having found themselves on the left side, asked to be reunited with their families motioned to the right. In such cases he would say: "You want to be together—all right, everybody to the right!"

Everyone passing in front of Rauca would wave a document he held in his hand. This brought a scornful smile to Rauca's lips. He acted in accordance with his own criteria.

Members of the Jewish labor brigades working at Gestapo headquarters, headed by Benno Lipzer, Rauca sent to the "good" side, together with their parents and children. He also motioned to the left the entire brigade of workers employed at German military installations whose commanders had contacted Rauca earlier. He also treated benevolently single women, and even women with children or aged parents, when that woman called out that she was employed by a high-ranking German officer, or by the German police commandant or by the German Ghetto commandant and such like. Their employers also had contacted Rauca in advance of the selection.

In contrast, there were cases where he paid no heed to the Jordan certificates, regarded thitherto as secure life permits in the Ghetto, and sent their holders to the right.

The selection was a protracted affair. Hungry, thirsty, and dejected, thousands of people waited for their turn from dawn. Many had already undergone the selection process, yet the square still seemed full. No end to the torment seemed in sight. Many resigned themselves to their fate and sighed in despair: "Come what may, as long as this waiting comes to an end." Mothers clasped their little children to their broken hearts, hugging and kissing them as though aware they were doing so for the last time. As a matter of fact, it was not clear what would happen to those sent to the right, but it was clear that it was the bad side and in the Ghetto "bad" in most cases meant death.

Those who were weak—those who could not withstand the psychological tension and the bodily torment—collapsed and breathed their last even before their turn came to pass before Rauca.

Dr. Elkes stood there, his pale face bearing an expression of bottomless grief. Since 6 A.M. this sixty-five-year-old man[7] had been standing on his feet, refusing to sit on the stool that had been brought to him. Now and then, when he was overcome by a fit of weakness, those near him asked him to sit down to regain his strength, or offered him a piece of bread. He refused, murmuring: "Thank you, thank you, gentlemen; terrible things are happening here; I must remain standing on guard in case I can be of assistance." Whenever he succeeded in transferring someone from the bad to the good side, and the person saved would try to shake his hand, he would refuse, saying: "Leave me alone, leave me alone." Sometimes, when in his efforts to transfer somebody to the left side he would inadvertently step too close to the guard unit charged with keeping order at the dividing line, he would be showered with curses and threats from the Lithuanian partisans: "Get away, you old, stupid Zhid, or else you'll go together with them."

It was beginning to grow dark, yet thousands of people remained standing in the square. Captain Jordan now opened another selection place; he was assisted by Captain Tornbaum. Rauca could count on this pair without reservation.

Except for an occasional sarcastic smile, Rauca's face did not betray any emotion, whereas Jordan stood with a frozen, sullen expression on his face and watched with fear-inspiring eyes the Jews passing in front of him. At first he was a little hesitant, lest he transfer to the good side those "not deserving" it, but he regained his composure quickly and followed Rauca's example by motioning families and whole columns either to the left or to the right without hesitation.

There was, nevertheless, some difference between Jordan and Rauca. The former did not motion to the bad side anyone holding a Jordan certificate, not even family members of the permit holder, without taking into consideration their age and physical condition.

SS Captain Schmitz would show up between counts and whisper something in Rauca's ear, showing him a note he had just received from the crossing point to the small Ghetto. That note indicated the number of Jews who had already passed through the selection.

Time passed. The people waiting in the square suffered more and more; it seemed to them that the selection would never come to an end. Even those who emerged from the selection intact were in a state of shock as a result of all they had undergone in the recent days and during the long hours of waiting for their fate to be decided. There was no easing of tension even for these "fortunate" ones. Apart from that, there were many families of whom only part found themselves on the good side; either one or both parents, or a brother or a sister, or one or several children had been torn away from them. Those families succumbed to inconsolable grief.

The Jewish Ghetto policemen were instructed to keep order in that part of the square where those who had passed the selection were assembled. On that day the Jewish policemen displayed initiative, daring, and resourcefulness—they cheered up the dispirited, they lent a hand to those who collapsed on the ground or fainted, and gave them water. They even worked wonders: while lining up the survivors in a new column, they seized every opportunity of transferring individuals—and even whole families still waiting for their turn—over to the good side. They did this with cunning and deftness, they would signal by a wink, or a movement of the hand, to slip away, to jump quickly, or to crouch and crawl toward them without drawing the attention of the guards. Whenever such a person drew near them, the Jewish policemen would set upon him screaming and push him brutally to the good side, pretending that they were forcing him back to his correct place. But whenever they did not succeed in deceiving the guards, a hail of curses and blows would pour on the unfortunate policeman caught in the act.

The selection was completed only after nightfall, but not before Rauca made sure that the quota had been fulfilled and that some 10,000 people had been transferred to the small Ghetto. Only then were those who had passed through the selection, and had remained standing in the square, allowed to return to their homes.

About 17,000 out of some 27,000 people slowly left the vast square where they had been standing for more than twelve hours. Hungry, thirsty, crushed, and dejected, they returned home, most of them bereaved or orphaned, having been separated from a father, a mother, children, a brother or a sister, a grandfather or a grandmother, an uncle or an aunt. A deep mourning descended on the Ghetto. In every house there were now empty rooms, unoccupied beds, and the belongings of those who had not returned from the selection. One-third of the Ghetto population had been cut down. The sick people who had remained in their homes in the morning had all disappeared. They had been transferred to the Ninth Fort during the day.

The square was strewn with several dozen bodies of elderly and sick people who had died of exhaustion. Here and there stools, chairs, and empty baby carriages were lying about.

On his way back Dr. Elkes muttered: "It wasn't worthwhile living for more than sixty years in order to witness a day like this! Who can bear all this when you are being appealed to with heartrending cries and there is nothing much you can do? I can't bear it any longer!"

When we reached Dr. Elkes's house we found many people besieging his door. All of them wanted to know what had happened to the people who had been taken to the small Ghetto. Men and women implored him to save their parents, wives, children, brothers, sisters, or other relatives. Everyone had a moving story of his own to tell.

Deadly tired and crushed by the day's horrors as he was, Dr. Elkes listened to every one. In vain he tried to explain that he had no idea of the German plans regarding the people transferred to the small Ghetto and that he was powerless to get them out of there. Nonetheless he promised to do all he could and to intervene with the authorities to comply with their requests. He wrote down the names of those he had been asked to rescue, including details such as occupation or skill which might produce some result with the Germans.

Among the Jews transferred to the small Ghetto there were pessimists who felt that all was lost, whereas others refused to give up all hope. But everyone tried to keep his head above water in case a miracle might occur. As more and more people were transferred there, there was more conflict and even competition among them. Each family tried to take possession of a better apartment, to gather more wood for fuel, to get indispensable household utensils, and so on. Some people set about tidying up and improving apartments that had long ago been abandoned by their previous occupants, filling up holes in the wall and fixing windows to protect against wind and rain. The more industrious got together during the night to discuss how to organize their lives in the new quarters. Some even proposed to elect immediately a Council on the pattern of the Council in the large Ghetto. All night long they debated and haggled among themselves, but were unable to agree about the composition of the proposed Council, the distribution of apartments, and so on.

It was an autumnal, foggy, and gloomy dawn when German policemen and drunken Lithuanian partisans broke into the small Ghetto, like so many ferocious beasts, and began driving the Jews out of their homes. The assault was so unexpected and brutal that the wretched inmates did not have a single moment to grasp what was going on. The partisans barked out their orders to leave the houses and to line up in rows and columns. Each column was immediately surrounded by partisans, shouting "Forward march, you scum, forward march," and driving the people by rifle butts out of the small Ghetto toward the road leading to the Ninth Fort. It was in the same direction that the Jews had been led away in the "action" commanded by Kozlovski on September 26, 1941,[8] and in the "action" of the liquidation of the small Ghetto on October 4, 1941. The same uphill road led Jews in one direction alone—to a place from which no one returned.

It was a death procession. The cries of despair issuing from thousands of mouths were hovering above them. Bitter weeping could be heard from far off. Column after column, family after family, those sentenced to death passed by the fence of the large Ghetto. Some men, even a number of women, tried to break through the chain of guards and flee to the large

Ghetto, but were shot dead on the spot. One woman threw her child over the fence, but missed her aim and the child remained hanging on the barbed wire. Its screams were quickly silenced by bullets.

Thousands of inmates from the large Ghetto flocked to the fence and, with tearful eyes and frozen hearts, watched the horrible procession trudging slowly up the hill. Many recognized a brother, a sister, parents or children, relatives or friends, and called to them by name. They were thrust back brutally by the reinforced guard of Lithuanian partisans, who pointed to the signs posted on the fence. In German, Lithuanian, and Yiddish the signs warned: "Death zone! Whoever approaches within two meters of the fence will be shot on the spot without warning."

Dr. Elkes and the head of the Jewish Ghetto police, Kopelman, accompanied by their assistants, arrived at the fence. No sooner was Dr. Elkes seen than the cries and pleas went up from those being marched up the hill, as well as from those crowding near the fence, inside the large Ghetto: "Dr. Elkes, help!" The cries joined into one big outburst that rose up to heaven: "Save us!"

Dr. Elkes asked that Rauca be found with the utmost urgency. He was traced before long by Kopelman's men at the German Ghetto command. Elkes addressed him immediately, asking to "allow him to remove from the small Ghetto those people who had fallen victim to error during the selection." Rauca consented, but limited the number of men and women to be removed to 100. Dr. Elkes kept in his pocket a list of people whose relatives had pleaded with him during the night and morning to save them, but their number far exceeded the figure of 100.

Accompanied by two sentries from the German Ghetto Guard, Dr. Elkes passed into the area of the small Ghetto, where he was immediately assailed by a throng of people already lined up in a column ready for departure. They begged him to save their lives. One seized his hand, another took hold of the tail of his coat, while another clasped his neck and refused to let go. Those who surrounded him knew full well that by leaving the column they endangered their lives, but they kept crying: "It is better for us to be killed here. We won't let go of you, Doctor—save us!" Within seconds the entire column faltered. For a moment it seemed that those condemned to death had rebelled. The Lithuanian guards intervened immediately, and with blows and kicks pushed people back into their places and hurried the whole column toward the road to catch up with other columns that had moved ahead.

Dr. Elkes himself was ordered by the guards to clear out at once—if not, they threatened, they would take him too to the Ninth Fort. Dr. Elkes insisted on his right to remove 100 people, as permitted by Rauca. Thereupon the Lithuanian partisans pounced upon him, hitting him with their fists. One of them brought down the butt of his rifle upon his head. Dr. Elkes collapsed on the ground, unconscious and bleeding profusely.

Jewish policemen and other Ghetto inmates (myself included) rushed from behind the fence to help. We lifted him up and carried him on our shoulders to the large Ghetto across the road, and put him inside the first house near the fence. He lay there for several days. Physicians stitched his open head wounds and nursed him, until he was able to stand on his feet again and return to his home. His efforts to save a number of Jews from the small Ghetto had almost cost him his own life.

The procession, numbering some 10,000 people, and proceeding from the small Ghetto to the Ninth Fort, lasted from dawn until noon. Elderly people, and those who were sick, collapsed by the roadside and died. Warning shots were fired incessantly, all along the way, and around the large Ghetto. Thousands of curious Lithuanians flocked to both sides of the road to watch the spectacle, until the last of the victims was swallowed up by the Ninth Fort.

In the fort, the wretched people were immediately set upon by the Lithuanian killers, who stripped them of every valuable article—gold rings, earrings, bracelets. They forced them to strip naked, pushed them into pits which had been prepared in advance, and fired into each pit with machine guns which had been positioned there in advance. The murderers did not have time to shoot everybody in one batch before the next batch of Jews arrived. They were accorded the same treatment as those who had preceded them. They were pushed into the pit on top of the dead, the dying, and those still alive from the previous group. So it continued, batch after batch, until the 10,000 men, women, and children had been butchered.

Villagers living in the vicinity of the fort told stories of horrors they had seen from a distance, and of the heartrending cries that emanated from the fort and troubled their waking hours all day and night.

About 17,500 people were left in the Ghetto, most of them orphaned, or bereft of their children, or widowed. All of a sudden there seemed to be plenty of space in the Ghetto. In every house a void had been created, pervaded with a mute terror haunting the survivors: this fate awaits you too.

The homes, the furniture, and the belongings of the victims of October 28 seemed to exude the odor of death. Hardly anyone dared touch them or make use of them. In the courtyards, storerooms, and other places, beds, furniture, and various articles were piled up, with nobody claiming them. They were removed from the houses from which entire families had been murdered without leaving a single survivor.

On the day after October 28, the Gestapo called upon the Ghetto inmates to increase the labor force and to increase even more the output at the airfield, in digging peat in Palemonas,[9] and in other places of forced labor, since it was work alone that was a guarantee of survival.

The small Ghetto remained sealed off and surrounded by a heavy guard even after all the Jews had been removed to the Ninth Fort. For a whole week nobody was allowed to enter. On the afternoon of October 29, the Jewish Ghetto policemen were ordered to conduct a thorough search of every house and of every attic in the deserted small Ghetto to make sure no Jews remained in hiding there. The Germans gave this assignment to the Jewish police on purpose; their plan was to lead astray those Jews in hiding who, upon hearing the Jewish policemen speaking Yiddish, were supposed to be lulled into thinking they were coming to rescue them. These Jews would then leave their hideouts and fall into a trap. The Jewish policemen were alert to such possibility, especially as they were convinced that a handful of Jews had succeeded in hiding during those critical moments. Therefore they brought with them, unnoticed, Jewish police caps, armbands, and insignia, to give them to every Jew who might be found in a hideout, or might come out to meet them, in order to give him an outward appearance of being a member of the Jewish Ghetto police, and in this way enable him to accompany them and leave the small Ghetto as one of their number. Indeed, the ruse worked. Some twenty Jews were saved in this manner—they would have been killed upon being discovered by the Lithuanians or the Germans, who were waiting for them at the opening in the small Ghetto fence, but who could not distinguish them from the real Jewish policemen.[10]

After the "action" of October 28, fear and dread fell on the Ghetto. People no longer believed the Gestapo announcements that this was the last "action," that it was not to be repeated in the future, and that from now on everyone would remain unharmed, working and earning a living. Their feelings regarding the future resembled those of a man about to be called up for military service. For the time being he is free in his home, but the inevitable lurks around the corner.

Ten thousand people were driven out and put to death in this huge "action." Before that, on September 26 and October 4, as many as 2,500 people had been killed; altogether about 17,000 people remained in the Ghetto.

Rumors sprang up again about pits dug in the fort, in Aleksotas, in Palemonas, and in Alytus.[11]

1. Avraham Tory's original description of the "action" of October 28, 1941, was written immediately after the event and buried in the ground. When, in August 1944, Tory dug out what he had buried, he felt that this entry was too short, so he expanded it. This second version, published here, was thus written three years after the event. It contains details Tory had learned during the war in discussions with survivors of the "action."

2. Joseph Serebrovitz, a Jewish journalist, who wrote under the name Caspi. He had been imprisoned by the Soviet authorities in 1940 and released shortly after the German occupation. He at once offered the Germans his services against the Soviet Union. Later he tried to exert influence in the Ghetto because of his special relationship with the Gestapo. Permitted not to wear the yellow Star of David, and allowed to live in the town with his family, he was a frequent visitor to the Ghetto. In October 1941 he was sent to Vilna. He later returned to Kovno, but in July 1942 he was again transferred to Vilna. In June 1943 he once more appeared in the Kovno Ghetto; later in 1943 he was murdered by the Nazis, together with his wife and two daughters.

3. Benjamin (Benno) Lipzer, born in Grodno in 1896, in the Kovno Ghetto was head of the brigade (work group) of Jews employed by the Germans on various work details which served the headquarters building of the Gestapo in

Kovno. After he had developed contacts with the upper echelons of the Gestapo, he attempted to gain control of the Jewish police. In June 1942 the Ghetto authorized him to supervise the labor department of the Jewish Council. At the time of the liquidation of the Ghetto in July 1944, he hid in the Ghetto, but was driven from his hiding place by the Gestapo and killed.

4. General Geiling, supervisor of the construction of the large military airfield being built by the Germans in the suburb of Aleksotas.

5. In his book *Dem Goirel Antkegn (Facing Fate),* published in Johannesburg in 1952, Professor Aharon Peretz writes that he and his family were already on the "bad" side when all of a sudden, as a result of intervention of someone on the Council, he was transferred to the "good" side and saved, together with his family. This "someone" was Dr. Elkes, who pointed him out to Rauca as a specialist needed in the hospital.

6. Rauca did not mind transferring any Jew to the left, since, as was learned later on, a quota of 10,000 victims had been set for that day, and all he was concerned with was to know at any time how many had already been transferred to the right. Every so often he would turn to the German standing at his side and ask: "The number, give me the number, I want the exact number."

7. Dr. Elkes was born in 1879.

8. A month earlier, on September 26, 1941, Kozlovski, the German commander of the Ghetto Guard, had staged his own fake assassination, on orders from his superiors. That same day, 1,200 Jews had been taken to the Ninth Fort and shot.

9. Palemonas was ten kilometers from Kovno; a labor camp was set up there to dig turf in the vicinity. The work conditions and regime at the camp were so horrible that many Jews working there were killed, or died of hunger or exhaustion.

10. This paragraph ends the version of the "action" of October 28, 1941, which Avraham Tory wrote at the end of the war. The three following paragraphs are those with which he ended his initial account of the "Great Action," written immediately after it.

11. After the war, Tory commented on this sentence: "This time the pits were not designed for us. Although pits were made ready, they were designed for other Jews, Jews from Germany, Czechoslovakia, Austria. Those Jews were brought here from those distant places to spill their blood on the Lithuanian soil; to brand with the sign of Cain the Lithuanian people for their conscious and persistent collaboration with the Germans in the murder of the Jewish people."

III

Fiction

Reading Holocaust fiction is a venture into disorientation. It will never be otherwise. Earlier in the century, when readers first entered the narrative world of Faulkner's *The Sound and the Fury* or Joyce's *Ulysses,* they might have felt similar unease, but decades of critical commentary have illuminated the inner working of Benjy Compson's mind and explained the complex stylistic devices that first made some of Joyce's chapters so baffling. The patient reader who has done his or her homework can even master the verbal and mythical mysteries that inhabit *Finnegans Wake.*

Holocaust fiction presents difficulties of another sort. Like the event on which it is based, it leads away from rather than toward flashes of clarity, so that midway through the twenty-first century readers may grasp the import of such literature still less than we do today. Even with annotation, how will they regard a narrative like Tadeusz Borowski's "This Way for the Gas, Ladies and Gentlemen," which begins with the startling revelation, "All of us walked around naked"? When the dreadful history behind those puzzling words has fled from mental view, how will readers react to the news that in a place called Auschwitz (footnote), garments are being deloused (another footnote) with Cyclon B (yet another footnote), which kills lice in clothing—and humans in gas chambers (will that one day require a footnote, too)?

The following selections, disturbing in content and effect, awaken in the reader powerful impulses to flee behind psychological gestures of disavowal. When Jakov Lind begins a story ("Soul of Wood," not included here) with the bizarre nonsequitur that "those who had no papers entitling them to live lined up to die," the reader's natural instinct is to seize on the illogic of the line and assume that the subsequent narrative will lead into a domain of fantasy. But, in fact, many Jews (often only temporarily, to be sure) survived selection for mass execution because they possessed work permits that confirmed the essential labor they were performing for the Germans. Life by entitlement may be an idea hard to imagine today, but it was an article of faith in the Third Reich, whose own citizens required Aryan documents to validate their existence. Lind describes a reality, but its premises have grown so alien from what we know and accept that it seems to introduce us to a realm of dreams—or nightmares.

Dreams and nightmares evoke the atmosphere of Kafkaesque narrative, but there is a basic difference between the solemn, official labyrinths that fuel Kafka's vision and the literal atrocities that inspire Holocaust fiction. The threat at the heart of such literature is not the loss of spiritual harmony, which bewilders Kafka's characters, but the demolition of the physical self. Wandering through the corridors, tunnels, and bureaucratic havens of Kafka's universe, his creatures are vaguely oppressed by the confined spaces that surround them; but those neither resemble nor predict the ever-narrowing margins that remind us of the ruin of European Jewry: the enclosed ghetto, the hidden bunker, the sealed boxcar, the mass grave, the gas chamber, the crematorium and its chimney.

The spatial and temporal boundaries caging Holocaust fiction require an imaginative leap into a stifling region that often leaves us bereft of intellectual breathing room. Little is ever resolved. Trained by tragedy to meet a fate to be mastered, we greet instead a doom to be borne. Hoping to find stories of character braced by suffering, we exhume tales of men and women pruned by atrocity to their naked and vulnerable core. People chosen for death—this kind of death—do not behave like persons choosing life. Holocaust reality nurtures responses that seem disruptive only if we shun the ruthless logic that gives them birth.

The ruthless logic of Holocaust fiction emerges from a landscape that demands the willing suspension of so many of our beliefs that we enter it with a strong sense of peril. Like the narrator of Pierre Gascar's "The Season of the Dead," we can manage single burials with their consoling ceremonies; tradition supports our faltering feelings there. But when, in digging a new grave, he stumbles on the twisted limbs of a corpse cast into the earth like the withered branches of some wasted shrub, we are aghast and left emotionally drained. The bodies of anonymous Jews jumbled together in a mass grave conjure a soil fertilized by annihilation rather than remembrance. They occupy a void whose vital principle is destruction, not care. It is an antiworld to the one we comfortably dwell in.

Holocaust fiction makes vivid for us the discomforts that were part of the daily ordeal of its victims, who have lost their heroic roles. In their fear and agony, they are forced to forfeit the persona of protagonist. A hunted Jew in Gascar's novella spends his days hiding in a tree, his nights sleeping in an empty grave. He lives above and beneath the earth, but not on it, robbed by his tormentors of his human terrain. Like many of his fellow victims, he has shifted the focus of his existence, from choosing life to avoiding death. Nazi brutality has ousted the defiant response from his repertoire of dynamic gestures.

The frustration and paralysis of normal longings—children cheated of family bonds, as in *Tzili* by Aharon Appelfeld and "Bread" by Isaiah Spiegel, or parents unable to save the children they love, as in Ida Fink's "The

Key Game" and "A Spring Morning"—thrust us into a remote milieu ruled by dubious laws. The situational ethic that so often shapes conduct in this fiction will seem unfamiliar to many readers who have been educated by both literature and Scripture to believe that the sources of personal dignity lie in the privilege of choice. The sense of character in much Holocaust fiction—certainly in the examples presented here—reflects the German design to create a victim psychology that was reactive rather than active. First hunted by ruthless oppressors, then haunted by constant, unimaginable forms of dying, the victims who populate these pages, chiefly Jews, are trapped by circumstances no human being could have been expected to foresee. They face not a failure of some internal moral force, but the strength of an outer Nazi triumph that curbs the will to action. Thus the narrator in Sara Nomberg-Przytyk's stories "The Verdict" and "Friendly Meetings," does not ask how one could thwart death in Auschwitz, but whether or not it was better to know in advance that the moment of your murder had arrived. *This* was a major "ethical" issue in that dreadful place, where staying alive deposed spiritual poise from the throne of character.

The result may not be chronicles of shattered lives, but they are certainly fractured ones. Integrity of the self suffered a kind of geologic shift that splintered the quest for inner peace. Whether fictional persons represent the victims then or the survivors now or even, as in Alexander Kluge's "Lieutenant Boulanger," the perpetrators, whatever the guise or persona, they lead discontinuous lives, as past and present tug in opposite directions for their allegiance. Even if in the midst of their distress they dare to hope for refuge from death, as in "Infinity" by Arnošt Lustig, they know that those they leave behind will never free them to enjoy an unstained future.

Holocaust fiction is thus very much a literature of memory, whose impact is expressly mournful. Its details cancel the appealing dignity of a narrator like Camus's Dr. Rieux in *The Plague,* who *was* free to choose to be a healer throughout the calamity that ravaged his city and thus earned his role as protagonist. Quite different are the male prisoners in Lustig's story, who listen in anguish to the songs of the women as they enter the gas chamber, powerless to act to help the victims, or themselves. They become casualties to their memories of the deaths of others, and should they escape a similar doom, these memories will remain a legacy to shield them from the leisure of a truly liberated life.

Holocaust fiction, because of the history it reflects, is a lingering art. It leads only to an unreconciled understanding. Who is to purge Tzili of her misgivings in Appelfeld's novel or Borowski's narrator of his nausea as he implicates himself in the killing process to further his own survival? The closure we expect of narrative—in the form of insight, reconciliation, maturity, or moral triumph—never appears. Creatures of Holocaust fiction are exempt from these consolations, since there is no way of imagining them into an inner tranquility without deceiving ourselves or betraying their heritage. Hence the rhythms of this literature remain cropped, jagged, and unresolved, and its endings signify no arrival but merely another invitation to depart.

What this literature invites us to depart from is not only the eternally incomplete narrative of its content, which of its very nature eludes the spatial and temporal finality that we find in most finished works of art; but also the usual anticipation we bring to an encounter with fiction. Just as the characters in Holocaust stories find their expectations violated by the situations that assault them, so readers entering this imagined world must adjust to the strenuous demands it makes on *their* power to imagine. How do we salvage for art, to say nothing of our appreciation, that terrible moment in Spiegel's story when a starving father in the Lodz ghetto secretly eats a portion of his starving children's bread? Assailed by hunger, then smitten by shame, his "character" dissolves before our eyes, leaving the cringing shell of a man whom we are nonetheless summoned to pity, not scorn.

The major challenge of Holocaust fiction is to nurture a readiness to enter and accept a world that is valid in all its features despite the virtual absence of familiar analogies to our own. The hungry children in Spiegel's story peer through a crack in the wall of their room at the display window of a bakery just beyond the wire fence surrounding their ghetto, filled with yearning for the loaves and rolls that load its shelves. The arid distance between longing and fruition is the narrow space where the drama of Holocaust fiction unfolds, an area where the trapped self searches, often in vain, for some hidden source of dignity to soothe its grief or doom. The bread to appease the children's physical hunger is forever beyond their reach; the luxury of spiritual solace, though tempting, seems equally far.

What then are we left with? These writers know the limitations of their art, when the issue is mass murder. The evil they need to portray is so unlike Satan's, the suffering so remote from Job's, that the very categories inspiring their literary ancestors prove useless to them. No device can conceal the horror of the crime, which even when unnamed rumbles beneath the text until it explodes in the reader's brain, despite the authors' stylized efforts to control its disorder. The fate of the "converted" Jew in Lind's "Resurrection" mocks the conventions of disguise and concealment, to say nothing of the Christians whose faith he embraces to save his life. The tentacles of the German octopus reach everywhere, exposing the futility of old defenses. The narrator's ironic detachment in Borowski's story, a cherished means of shaping tone *and* response, proves to be another vain buffer. The chaos of the theme—the selection of newly arrived prisoners for gassing—erupts through the casual surface of the text, scattering more than symbolic ash. Even the impersonal question-and-answer pattern of Kluge's tale (which may owe a debt to a late chapter in Joyce's *Ulysses*) shatters its formal limits when the appalling nature of the subject's mission dawns on the reader.

The Holocaust fiction included here, like the reality on which it draws, casts a shadow over our natural longing for more fertile pastures. We would prefer to graze elsewhere. Camus's creed that "there is no sun without shadow, and it is essential to know the night" cannot be applied to the tensions in this literature. He meant it as a preface to lucidity, a modern restatement of the tragic view of experience. Launching hope from within its apparent gloom, his view is an invitation to renewal through art and creativity, a resolute defense of the human in the face of an indifferent universe. But the plague that destroyed European Jewry was more than a metaphor, having little to do with personal tragic conflict. It reflects a milieu cut off from the sources of ordinary solace, and imaginations stirred—and perhaps infected—by the details of this atrocity in our time.

A book, Kafka wrote, is an axe, to crack the frozen sea within us. These narratives are more like icebergs, to freeze our warm and coursing blood. They evoke an era that is fortunately no longer with us, so our blood slowly thaws once we have ended our literary encounter. But though the Holocaust is no longer with us, its memories wax within; the mere fact that it existed remains a troubling gadfly to our self-esteem and trust in civilization. Just as the possibilities of celebrating individual life by honoring individual death shrink for Gascar's grave-digging narrator in "The Season of the Dead" once he has uncovered the dismal fate of the Jews, so our own enthusiasm for human growth is curtailed by the vision unleashed in these fictions. We now live alongside the inhuman, and this promotes two adjacent worlds that embrace a life after death, called survival, and a life within death, for which we have no name, only the assurance, through testimony and art, that it happened. The passionate earnestness in the voice of one surviving victim frames the nature of that parallel existence:

> You sort of don't feel at home in this world anymore, because this experience—you can live with it, it's like constant pain: you never forget, you never get rid of it, but you learn to live with it. And that sets you apart from other people. Not that you can't enjoy yourself. On the contrary, when I am happy, I'm *so* happy, because I know how horribly unhappy I can be. I know the whole difference. But there is a certain—it's like a music in the background. It's that something is different.

The challenge to listen to that "different" music, its discords as well as its harmonies, is a summons that no one pledged to hear the strident echoes of twentieth-century reality can afford to ignore.

13. Ida Fink

Ida Fink was born in Poland in 1921. After the Nazi invasion, she lived in a ghetto through 1942, and then survived in hiding until the end of the war. She settled in Israel with her family in 1957. *A Scrap of Time*, from which the stories "The Key Game" and "A Spring Morning" are taken, won the first Anne Frank Prize for Literature. The piece called "The Table" in *A Scrap of Time* was written as a radio play; it has been broadcast and performed on television in Israel and Europe. Fink's novel, *The Journey*, which tells the story of the attempt by two Jewish sisters from Poland to use false papers to survive undetected within Germany, was published in the United States in 1992.

Ida Fink's personal experience in hiding plainly influenced the setting of many of her stories—confined places like barns, attics, tiny rooms, cellars, and even pigpens, where human beings manage to stay alive a little longer under minimal conditions. The two tales included here illustrate one of her recurrent themes: how fear of the German threat distracted and finally destroyed the power of normal family love to serve life. Fink's special genius is her mastery of spare, intense prose to portray the daily dilemma of numerous Jewish victims of this pitiless milieu: forced to exist with the terrible foreknowledge of their doom, but prevented by circumstance—with few exceptions—to do anything to thwart it.

The Key Game

They had just finished supper and the woman had cleared the table, carried the plates to the kitchen, and placed them in the sink. The kitchen was mottled with patches of dampness and had a dull, yellowish light, even gloomier than in the main room. They had been living here for two weeks. It was their third apartment since the start of the war; they had abandoned the other two in a hurry. The woman came back into the room and sat down again at the table. The three of them sat there: the woman, her husband, and their chubby, blue-eyed, three-year-old child. Lately they had been talking a lot about the boy's blue eyes and chubby cheeks.

The boy sat erect, his back straight, his eyes fixed on his father, but it was obvious that he was so sleepy he could barely sit up.

The man was smoking a cigarette. His eyes were bloodshot and he kept blinking in a funny way. This blinking had begun soon after they fled the second apartment.

It was late, past ten o'clock. The day had long since ended, and they could have gone to sleep, but first they had to play the game that they had been playing every day for two weeks and still had not got right. Even though the man tried his best and his movements were agile and quick, the fault was his and not the child's. The boy was marvelous. Seeing his father put out his cigarette, he shuddered and opened his blue eyes even wider. The woman, who didn't actually take part in the game, stroked the boy's hair.

"We'll play the key game just one more time, only today. Isn't that right?" she asked her husband.

He didn't answer because he was not sure if this really would be the last rehearsal. They were still two or three minutes off. He stood up and walked towards the bathroom door. Then the woman called out softly, "Ding-dong." She was imitating the doorbell and she did it beautifully. Her "ding-dong" was quite a soft, lovely bell.

At the sound of chimes ringing so musically from his mother's lips, the boy jumped up from his chair and ran to the front door, which was separated from the main room by a narrow strip of corridor.

"Who's there?" he asked.

The woman, who alone had remained in her chair, clenched her eyes shut as if she were feeling a sudden, sharp pain.

"I'll open up in a minute, I'm just looking for the keys," the child called out. Then he ran back to the main room, making a lot of noise with his feet. He ran in circles around the table, pulled out one of the sideboard drawers, and slammed it shut.

"Just a minute, I can't find them, I don't know where Mama put them," he yelled, then dragged the chair across the room, climbed onto it, and reached up to the top shelf of the étagère.

"I found them!" he shouted triumphantly. Then he got down from the chair, pushed it back to the table, and without looking at his mother, calmly walked to the door. A cold, musty draft blew in from the stairwell.

"Shut the door, darling," the woman said softly. "You were perfect. You really were."

The child didn't hear what she said. He stood in the middle of the room, staring at the closed bathroom door.

"Shut the door," the woman repeated in a tired, flat voice. Every evening she repeated the same words, and every evening he stared at the closed bathroom door.

At last it creaked. The man was pale and his clothes were streaked with lime and dust. He stood on the threshold and blinked in that funny way.

"Well? How did it go?" asked the woman.

"I still need more time. He has to look for them longer. I slip in sideways all right, but then . . . it's so tight in there that when I turn . . . And he's got to make more noise—he should stamp his feet louder."

The child didn't take his eyes off him.

"Say something to him," the woman whispered.

"You did a good job, little one, a good job," he said mechanically.

"That's right," the woman said, "you're really doing a wonderful job, darling—and you're not little at all. You act just like a grown-up, don't you? And you do know that if someone should really ring the doorbell someday when Mama is at work, everything will depend on you? Isn't that right? And what will you say when they ask you about your parents?"

"Mama's at work."

"And Papa?"

He was silent.

"And Papa?" the man screamed in terror.

The child turned pale.

"And Papa?" the man repeated more calmly.

"He's dead," the child answered and threw himself at his father, who was standing right beside him, blinking his eyes in that funny way, but who was already long dead to the people who would really ring the bell.

A Spring Morning

During the night there was a pouring rain, and in the morning when the first trucks drove across the bridge, the foaming Gniezna River was the dirty-yellow color of beer. At least that's how it was described by a man who was crossing this bridge—a first-class reinforced concrete bridge—with his wife and child for the last time in his life. The former secretary of the former town council heard these words with his own ears: he was standing right near the bridge and watching the Sunday procession attentively, full of concern and curiosity. As the possessor of an Aryan great-grandmother he could stand there calmly and watch them in peace. Thanks to him and to people like him, there have survived to this day shreds of sentences, echoes of final laments, shadows of the sighs of the participants in the *marches funèbres,* so common in those times.

"Listen to this," said the former secretary of the former town council, sitting with his friends in the restaurant at the railroad station—it was all over by then. "Listen to this: Here's a man facing death, and all he can think about is beer. I was speechless. And besides, how could he say that? I made a point of looking at it, the water was like water, just a little dirtier."

"Maybe the guy was just thirsty, you know?" the owner of the bar suggested, while he filled four large mugs until the foam ran over. The clock above the bar rattled and struck twelve. It was already quiet and empty in town. The rain had stopped and the sun had broken through the white puffs of clouds. The sizzle of frying meat could be heard from the kitchen. On Sunday, dinner should be as early as possible. It was clear that the SS shared that opinion. At twelve o'clock the ground in the meadow near the forest was trampled and dug up like a fresh wound. But all around it was quiet. Not even a bird called out.

When the first trucks rode across the bridge over the surging Gniezna, it was five in the morning and it was still completely dark, yet Aron could easily make out a dozen or so canvas-covered trucks. That night he must have slept soundly, deaf to everything, since he hadn't heard the rumbling of the trucks as they descended from the hills into the little town in the valley. As a rule, the rumbling of a single truck was enough to alert him in his sleep; today, the warning signals had failed him. Later, when he was already on his way, he remembered that he had been dreaming about a persistent fly, a buzzing fly, and he realized that that buzzing was the sound of the trucks riding along the high road above his house—the last house when one left the town, the first when one entered it.

They were close now, and with horrifying detachment he realized that his threshold would be the first they crossed. "In a few minutes," he thought, and slowly walked over to the bed to wake his wife and child.

The woman was no longer asleep—he met her gaze immediately, and was surprised at how large her eyes were. But the child was lying there peacefully, deep in sleep. He sat down on the edge of the bed, which sagged under his weight. He was still robust, though no longer so healthy looking as he used to be. Now he was pale and gray, and in that pallor and grayness was the mark of hunger and poverty. And terror, too, no doubt.

He sat on the dirty bedding, which hadn't been washed for a long time, and the child lay there quietly, round and large and rosy as an apple from sleep. Outside, in the street, the motors had fallen silent; it was as quiet as if poppy seeds had been sprinkled over everything.

"Mela," he whispered, "is this a dream?"

"You're not dreaming, Aron. Don't just sit there. Put something on, we'll go down to the storeroom. There's a stack of split wood there, we can hide behind it."

"The storeroom. What a joke. If I thought we could hide in the storeroom we'd have been there long ago. In the storeroom or in here, it'll make no difference."

He wanted to stand up and walk over to the window, but he was so heavy he couldn't. The darkness was already lifting. He wondered, are they waiting until it gets light? Why is it so quiet? Why doesn't it begin?

"Aron," the woman said.

Again her large eyes surprised him, and lying there on the bed in her clothing—she hadn't undressed for the night—she seemed younger, slimmer, different. Almost the way she was when he first met her, so many years ago. He stretched out his hand and timidly, gently, stroked hers. She wasn't surprised, although as a rule he was stingy with caresses, but neither did she smile. She took his hand and squeezed it firmly. He tried to look at her, but he turned away, for something strange was happening inside him. He was breathing more and more rapidly, and he knew that in a moment these rapid breaths would turn into sobs.

"If we had known," the woman said softly, "we wouldn't have had her. But how could we have known? Smarter people didn't know. She'll forgive us, Aron, won't she?"

He didn't answer. He was afraid of this rapid breathing; he wanted only to shut his eyes, put his fingers in his ears, and wait.

"Won't she, Aron?" she repeated.

Then it occurred to him that there wasn't much time left and that he had to answer quickly, that he had to answer everything and say everything that he wanted to say.

"We couldn't know," he said. "No, we wouldn't have had her, that's clear. I remember, you came to me and said, 'I'm going to have a child, maybe I should go to a doctor.' But I wanted a child, *I* wanted one. And I said, 'Don't be afraid, we'll manage it somehow. I won't be any worse than a young father.' *I* wanted her."

"If only we had a hiding place," she whispered, "if we had a hiding place everything would be different. Maybe we should hide in the wardrobe, or under the bed. No . . . it's better to just sit here."

"A shelter is often just a shelter, and not salvation. Do you remember how they took the Goldmans? All of them, the whole family. And they had a good bunker."

"They took the Goldmans, but other people managed to hide. If only we had a cellar here . . ."

"Mela," he said suddenly, "I have always loved you very much, and if you only knew—"

But he didn't finish, because the child woke up. The little girl sat there in bed, warm and sticky from her child's sleep, and rosy all over. Serious, unsmiling, she studied her parents' faces.

"Are those trucks coming for us, Papa?" she asked, and he could no longer hold back his tears. The child knew! Five years old! The age for teddy bears and blocks. Why did we have her? She'll never go to school, she'll never love. Another minute or two . . .

"Hush, darling," the woman answered, "lie still, as still as can be, like a mouse."

"So they won't hear?"

"So they won't hear."

"If they hear us, they'll kill us," said the child, and wrapped the quilt around herself so that only the tip of her nose stuck out.

How bright her eyes are, my God! Five years old! They should be shining at the thought of games, of fun. Five! She knows, and she's waiting just like us.

"Mela," he whispered, so the child wouldn't hear, "let's hide her. She's little, she'll fit in the coalbox. She's little, but she'll understand. We'll cover her with wood chips."

"No, don't torture yourself, Aron. It wouldn't help. And what would become of her then? Who would she go to? Who would take her? It will all end the same way, if not now, then the next time. It'll be easier for her with us. Do you hear them?"

He heard them clearly and he knew: time was up. He wasn't afraid. His fear left him, his hands stopped trembling. He stood there, large and solid—breathing as if he were carrying an enormous weight.

It was turning gray outside the window. Night was slipping away, though what was this new day but night, the blackest of black nights, cruel, and filled with torment.

They were walking in the direction of the railroad station, through the town, which had been washed clean by the night's pouring rain and was as quiet and peaceful as it always was on a Sunday morning.

They walked without speaking, already stripped of everything human. Even despair was mute; it lay like a death mask, frozen and silent, on the face of the crowd.

The man and his wife and child walked along the edge of the road by the sidewalk; he was carrying the little girl in his arms. The child was quiet; she looked around solemnly, with both arms wrapped around her father's neck. The man and his wife no longer spoke. They had said their last words in the house, when the door crashed open, kicked in by the boot of an SS-man. He had said then to the child, "Don't be afraid, I'll carry you in my arms." And to his wife he said, "Don't cry. Let's be calm. Let's be strong and endure this with dignity." Then they left the house for their last journey.

For three hours they stood in the square surrounded by a heavy escort. They didn't say one word. It was almost as if they had lost the power of speech. They were mute, they were deaf and blind. Once, a terrible feeling of regret tore through him when he remembered the dream, that buzzing fly, and he understood that he had overslept his life. But this, too, passed quickly; it was no longer important, it couldn't change anything. At ten o'clock they set out. His legs were tired, his hands were numb, but he didn't put the child down, not even for a minute. He knew it was only an hour or so till they reached the fields near the station—the flat green pastures, which had recently become the mass grave of the murdered. He also recalled that years ago he used to meet Mela there, before they were husband and wife. In the evenings there was usually a strong wind, and it smelled of thyme.

The child in his arms felt heavier and heavier, but not because of her weight. He turned his head slightly and brushed the little girl's cheek with his lips. A soft, warm cheek. In an hour, or two . . .

Suddenly his heart began to pound, and his temples were drenched with sweat.

He bent towards the child again, seeking the strength that flowed from her silky, warm, young body. He still didn't know what he would do, but he did know that he had to find some chink through which he could push his child back into the world of the living. Suddenly he was thinking very fast. He was surprised to see that the trees had turned green overnight and that the river had risen; it was flowing noisily, turbulently, eddying and churning; on that quiet spring morning, it was the only sign of nature's revolt. "The water is the color of beer," he said aloud, to no one in particular. He was gathering up the colors and smells of the world that he was losing forever. Hearing his voice, the child squirmed and looked him in the eye.

"Don't be afraid," he whispered, "do what Papa tells you. Over there, near the church, there are a lot of people, they are going to pray. They are standing on the sidewalk and in the yard in front of the church. When we get there, I'm going to put you down on the ground. You're little, no one will notice you. Then you'll ask somebody to take you to Marcysia, the milkmaid, outside of town. She'll take you in. Or maybe one of those people will take you home. Do you understand what Papa said?"

The little girl looked stunned; still, he knew she had understood.

"You'll wait for us. We'll come back after the war. From the camp," he added. "That's how it has to be, darling. It has to be this way," he whispered quickly, distractedly. "That's what you'll do, you have to obey Papa."

Everything swam before his eyes; the image of the world grew blurry. He saw only the crowd in the churchyard. The sidewalk beside him was full of people, he was brushing against them with his sleeve. It was only a few steps to the churchyard gate; the crush of people was greatest there, and salvation most likely.

"Go straight to the church," he whispered and put the child down on the ground. He didn't look back, he didn't see where she ran, he walked on stiffly, at attention, his gaze fixed on the pale spring sky in which the white threads of a cloud floated like a spider web. He walked on, whispering a kind of prayer, beseeching God and men. He was still whispering when the air was rent by a furious shriek:

"Ein jüdisches Kind!"

He was still whispering when the sound of a shot cracked like a stone hitting water. He felt his wife's fingers, trembling and sticky from sweat; she was seeking his hand like a blind woman. He heard her faint, whimpering moan. Then he fell silent and slowly turned around.

At the edge of the sidewalk lay a small, bloody rag. The smoke from the shot hung in the air—wispy, already blowing away. He walked over slowly, and those few steps seemed endless. He bent down, picked up the child, stroked the tangle of blond hair.

"Deine?"

He answered loud and clear, *"Ja, meine."* And then softly, to her, "Forgive me."

He stood there with the child in his arms and waited for a second shot. But all he heard was a shout and he understood that they would not kill him here, that he had to keep on walking, carrying his dead child.

"Don't be afraid, I'll carry you," he whispered. The procession moved on like a gloomy, gray river flowing out to sea.

14. Isaiah Spiegel

Isaiah Spiegel was born in Lodz, Poland, in 1906. He attended Jewish and public schools and studied at a teacher-training college. He taught Yiddish language and literature in Lodz until the Nazi invasion of Poland in 1939, and then worked in various capacities for the Lodz Jewish Council after the ghetto was established in May 1940. He remained in the ghetto until August 1944, when he left with the last deportations to Auschwitz, where the rest of his family perished. After Auschwitz was evacuated in January 1945, Spiegel himself spent the rest of the war in several labor camps.

Spiegel began publishing Yiddish poems in 1922. His first book of poetry, *Facing the Sun,* appeared in 1930. Many other unpublished manuscripts from the prewar period were lost during the Holocaust. While in the Lodz ghetto, he continued writing stories and poems, some of which he buried in a cellar and others he brought with him to Auschwitz (where, of course, they disappeared). Following his liberation, he reconstructed some of the missing works from memory, adding them to the surviving stories, which were published in Lodz in 1947 in a volume called *Ghetto Kingdom.* A second collection appeared in Paris in 1948 under the title *Stars Over the Ghetto.*

Spiegel resumed his teaching career, first in Lodz (1945–1948) and then in Warsaw (1948–1950). In 1951, he emigrated to Israel, where he worked for the Finance Ministry while remaining a prolific author of fiction and poetry. He died in Israel in 1991.

Like the story "Bread," which is included here, Spiegel's tales are deromanticized narratives of ghetto existence, when hunger and fear created an incurable daily misery for most of the inhabitants. His simple, laconic style highlights the horror of Jewish life under the inhuman demands of German rule, when parental instincts, shorn of the luxury of love, had to struggle with utter need to sustain the urge to nurture.

Bread

The little room where Mama Glikke has installed herself and her housekeeping, along with Shimmele and their two children, sits in the porch of the little house, its one narrow little window looking wistfully down at the small street that is already beyond the confines of the ghetto. The little street runs like a narrow tunnel outside the ghetto, next to the recently installed wire fences. The hovel itself—a wooden structure crumbling with age, sagging with the rains and snows of many generations—bows its roof ever closer to the top of the fence, like a person who laments: "Oh—oh, what I have lived to see. . . . Oh my, such a fate. . . ." None of the roofs of the neighboring stone houses has taken up the rain-song of the autumn night, the joyful caresses of the glittering snow—and the clear strolls of silvery moons—as has the old, crumbling roof on the old, sinking hovel.

From the narrow little window, "they" can always be seen below. For days, now, and months, the black, shiny military boots have been pacing the bridge, back and forth, day and night, without ceasing. The Kraut pacing here with his gun keeps on looking at the warped little house and at the tilted roof. None of those inside the house approach the little window. The window is open just a crack, because when the Kraut below sees an open window, or the shadow of a head—he shoots at it. As a matter of fact, at this moment the little window in Glikke's house is just the slightest bit ajar. It is a hot July day. Through the narrow opening in the window wafts a dry, hot wind.

The room itself is white, calcined, with a wooden pole that descends straight down to the stove. The paint is peeling from the walls, littering the ragged floor, which here and there is missing a few boards. The Gentile who had lived in the apartment before them, and who left before the Jews had been herded into the ghetto, took some of the floor with him when he left and took the stove apart too. When they moved in Mama Glikke had quickly set up a second stove, which now stands propped up on some red bricks emitting billows of smoke from the crevices every time Mama decides to cook a hot meal. Not a stick of furniture, no closet, no beds. When the family had fled here, Mama Glikke had even brought a decent cabinet with her from the old place in the city. Back there, at home, a pair of silver Sabbath candelabras used to stand in the cabinet, behind glass doors. But, as it happened, the cabinet fell apart on the journey, as they were rushing into the ghetto, and now the two Sabbath candelabras are lying on the floor in a corner near the window, in a pile of junk, among empty pots and torn clothes.

Set in the right wall of the room is a little wooden door, and beyond the door a tiny chamber, dark and narrow, where Mama has stored a bit of firewood, and also the sides of the collapsed cabinet. A cold gloom pervades that tiny chamber with its perpetual evening twilight; behind the slanted studs old green spiders are spinning their webs. The Gentile who used to live here kept pigeons in the tiny chamber. A triangular opening had been cut out of the front wall facing the street, and through this hole in the wall the Gentile's doves would fly in and out. You could still find little grains of oats in the soil of the black earthen floor of the little chamber. The door in the little chamber is always closed. Mama does not let the children, who are constantly digging holes in the earthen floor, into the room. It sometimes happens that in that chamber, in the cut-out opening, a pigeon will alight on the edge of a board. She has blundered here because of a familiar wind or cloud. She still can't divorce herself from the old home, from the old chamber under the crumbling roof. She bends her head into the empty, dark chamber. She stands trembling, frightened, with her red little feet on the board's edge, and with surprised, innocent eyes she soon flies away, beating her wings outside the window as if to bid a farewell: "Stay healthy! Who knows if we will see each other again?"

The mother and the father divide their day as follows:

At the crack of dawn, as soon as the sun appears on the eastern rim of the ghetto, Mama Glikke takes two huge pots and positions herself somewhere in a courtyard behind a fence. Mama stands there for hours, sometimes till late at night. There is a kitchen there where they cook dinners for many people. The boys who carry water from the pump to the kitchen know, by this time, that the desperate eyes of Mama Glikke are staring through a crack in the fence. After the soup has been portioned out, the boys take two pots from her and ladle out a bit of turbid, diluted liquid. Though hunger gnaws at Mama's bowels, as if someone were tearing her flesh with a pair of pliers, and although she has been standing at the fence staring through the crack all day, still Mama does not taste a drop of the soup until she has entered the little room. With two pots tucked under her scarf she runs across the ghetto like an athlete. She is terrified that, God forbid, she may spill a drop of soup, a spoonful of holy victuals. She runs, and the two corners of her scarf flutter behind her like the wings of a large, frightened bird.

The labors of the father, Shimmele, are quite another matter. Sickly, with a chronic, juicy cough and melting, running eyes, as if overflowing with tears, he, the father, sits all day in the room with the children, wrapped in his *tallis* and *tfillin.** He starts his prayers as soon as Glikke leaves, and does not take off the *tfillin* till just before evening. The children are afraid of the father. The last few days he just stands for hours on end next to the wall, wrapped in *tallis* and *tfillin,* without moving an inch. They imagine that father has been dead for a while, that out of the blue he has decided to give up his soul, that he is standing there with his face to the wall, stony, frosty, a dead man. And when he suddenly turns from the wall and gazes into the room after many hours of stony silence—the children see that their father's face is white as snow, his eyes soaked in tears. He sits down in the remotest corner of the room, covered with *tallis* and *tfillin,* and waits for the mother to walk through the door. And at his feet lie Avremele and Perele, terrified and lost.

Since they started rationing bread in the ghetto hunger has haunted the room. At the beginning everybody shared: mother, father, and children. When they picked up their ration of bread and laid it on the table, it would lie there for a long time till the fastidious and fluttering mother took the knife, said a blessing, and tremblingly sliced the bread. Father and the children would gather around her, and after everyone had received a portion sufficient to assuage the initial pangs of hunger, mother would wrap the bread in a white cloth and stash it away so that no one would be able to find it. The bread would have to last for a long, long time. As if out of spite, when bread entered the room hunger grew apace, tearing at the guts like a peverse imp. In dreams, fresh round loaves would swim into view and bring satisfaction. They knew that somewhere in the room a radiant treasure lay hidden. Oh, God, if only it were possible to cut slice after slice from the loaf and not be terrified at knowing that with every slice the loaf gets smaller and smaller, if only they could sit down and eat to their hearts' content. But that one could only dream. In the meantime they cursed every bite of bread that touched the palate. The bread did not satisfy. With each bite, hunger's curse sharpened. They examined each crumb of bread lying in their hands as if it were a diamond. And before putting it to the tongue they feasted their eyes on it, drawing a sad, sated joy from the sight of the Lord's bounty.

Once there was a stroke of misfortune.

That time, mother had denounced and cursed this wretched life all day. Something happened that almost caused father to hang himself out of shame in the little chamber. After that misfortune he spent whole days just lying on the floor, moaning. Who could have predicted such a thing? Mother just kept repeating: "Now there is nothing lower for him to stoop to, oh Master of the Universe, except to cut a chunk of meat out of the children and cook it."

They did not know how it came to pass, and even the father himself did not know how things had come to such misfortune and disgrace.

*Prayer shawl and phylacteries, leather thongs wound around the forehead and forearm during prayer.

There was no way that Shimmele could figure out what the power was that urged him to it. He just did not know.

At that moment a tornado had raged in his heart. It could not have been his very own hands that in the darkness of the chamber stuffed the children's bread into his own mouth. No . . . no. That wasn't the father any more. It wasn't Shimmele, but some kind of enchanted shadow that had separated itself from him, that had issued from his hands and feet, some kind of accursed *dybbuk** who used the father's hands and fingers to tear at the dark, chestnut-flour loaf and salivate over it dozens of times. Wet, half-chewed chunks fell from his stuffed mouth and lay in his palms. From there they were popped right back into the mouth that had spit them out in unnecessary haste. And when that was done, he remained standing, benighted and petrified, his head buried in his hands.

The same day the father ate the children's bread, Mama Glikke came home late. She found both children lying in a corner, famished. When she noticed that a quarter of a loaf was missing, she raised a racket, and it was only then that Shimmele crawled out of the little chamber. When she read the truth in his downcast eyes, she squawked like a slaughtered chicken:

"A father, eh? A fa—ther, is it? *Murderer!*"

And from that moment on the mother walked around worried and anxious, till she found a solution: she sewed little sacks out of shirts and stuffed the sliced portions of bread into the sacks; in that manner she traversed the streets, with both sacks hanging over her heart.

The mother calls Avremele: Umele. He is slight in stature, with a thin, narrow little head and protruding ears. He is the spitting image of his father. His eyes also have a moistness about them, and if you were to look closely, knowing his elderly father, you would see from Umele's pinched expression and pointed chin that he will turn out exactly like the old man. Though Umele has not yet reached his twelfth birthday there is already set within his countenance a trace of his father's agedness and brokenness. Like the old man, he always keeps his hand on his bowed forehead, thinking of something. A restless shadow hovers over his pale, transparent cheeks. The few ghetto months have completely transformed the children. They are no longer children, but ancients, on whose faces sit the ravages of heavy years.

Umele is sitting on the floor, and next to him, at his side, his little sister, Perele. Perele has a slight limp; her thin, sandy-colored hair cascades down her narrow shoulders. They are both sitting in the corner near the Sabbath candelabras, gazing across at the opposite corner, where the father is busy with something or other. The mother left the room very early in the morning, taking with her the little sacks of bread, and now

*In Jewish folklore, a wandering spirit capable of entering and taking possession of a living person.

they are hoping that the father will leave the room soon, so that they can go back into the little chamber.

But it is a long, long time before the father leaves. He just keeps on looking at an open book. Since that day when the great disgrace occurred he has not been able to look the children straight in the eye. Buried in the pages of his book, he seeks something there with his tiny pinpoint eyes, and every now and then he gives vent to the accumulated air in his lungs with a great, deep sigh, and remains lying with his face buried in the book as if his throat had been cut. He doesn't move for a long time, dozes off, then wakes up abruptly when he hears the crash of the book that has fallen out of his hands onto the floor.

Now the children watch as the father picks up the book, rummages around in his pocket looking for his key, and lets himself out the door.

When the two of them, Umele and Perele, see that they have been left alone, their eyes light with glee. Umele runs quickly to the door of the little chamber and pulls out the nail that is keeping the door fastened. The two of them run into the chamber, but as she is running Perele trips on the threshold. She gets up and, with her dragging foot, chases after Umele. By this time, Umele is already standing in the darkness of the chamber, near the wall that faces out onto the street. Umele had found a crevice in the wall.

In the wooden wall there are two crevices, one slightly removed from the other. At one of these crevices, Umele now stands and watches.

Down below, on the other side of the street, across from the wires stand a little store and the display window of a bakery. Umele's eye aims directly into the window. Brown shiny loaves and white roundish rolls are set out for display. Avremele had discovered that treasure yesterday, when the mother happened to leave the chamber door ajar. And what a treasure! If you really squeezed your eye right up close to the crevice you could see clearly: roundish fresh loaves and light wholesome rolls. There might be four loaves lying in the window and perhaps ten or more rolls. The sun, as it happens, is just opposite the window, shedding its light directly on the treasures. Umele is taking it all in with his left eye, while his tongue is swimming in streams of sweet saliva. When the left eye tires he switches to the right, then back to the left, and still later back to the right. Suddenly he sees a hand from inside the store removing one of the loaves. He lets out a sigh. Behind him stands Perele, tugging at his hand, and she keeps asking:

"Let me look, let me look too. I'm hungry too."

Umele doesn't budge.

Suddenly he cries out in a strange voice:

"Don't take it, don't take it, don't take it!"

Across the way the hand has once again swept two loaves and a whole pile of rolls. A single loaf of bread and two rolls remain in the window.

Umele still can't tear himself away from the crevice.

When Perele stubbornly grabs hold of his arm and will not relent, he flares up in anger and shouts:

"I don't want to and that's it. Leave me alone."

Through the crevice that Perele has been looking out of one can see a little shop. Laid out majestically in the display window is a pure white cheese, and nothing else. It is possible, by raising oneself a little higher on one's toes, to see part of a field, where a cow is grazing. From Umele's vantage-point there is only bread and rolls. Umele stands praying to God not to let the hand reappear. He murmurs a verse his father had taught him long before the ghetto. And that fragment has to accomplish everything. If you are ill, it can help; if you are very hungry and you say it with real feeling—your hunger disappears in the wink of an eye. Umele earnestly recites that holy verse now, and waits. The loaves and rolls are so close now, almost within reach. He can even see, at this point, two little holes in the bread. And he notices that a fly has alighted on one of the rolls, a large fly with large shimmering wings.

"Umele, would you like some cheese?" Perele does not stop her nagging. "Let me in, let me in. If you don't Mama will. . . ."

God forbid that Mama should find out about this. No, Mama must not know about this, because if she does, then the whole treasure is lost. It will no longer be possible to come into the little chamber and gorge oneself on rolls and loaves of white bread. Okay, he'll let her in just for a short while, only a minute.

"Not for a long stay, Perele, all right? I don't like cheese." He moves away slowly and peeps through the other crevice.

Through the crevices the eye can escape into a free, uncaged world. Just a single leap over the fence and you are free. You can go wherever you want: to the courtyard, from the courtyard to the open fields, from the fields to the forest, further and further. The childish eyes float out of the little chamber. First of all, across the road into the shops. Now Umele can see the flat white cheese, Perele the last shiny rolls. How those rolls laugh their way right up to the children; they come so close to the crevice, so close to the eye, that Perele actually licks her little lips. You can smell the sweetness of the rich, black poppyseeds. Perele's lips are already tasting the sweetness, and just look, she runs her tongue across her lips and really—she feels as if she had taken a lick of those sweetish poppyseeds.

Suddenly Perele lets out a scream. Across the street, the hand has just pulled the last of the bread and rolls out of the window . . . now the window is empty. Umele runs over to the crack—and really he sees that the window opposite has become a complete void.

Perele watches Umele sink to the ground, tears running down his cheeks. She is now standing at her own crevice, and in the field she sees a cow with black spots.

Perele turns her head back into the chamber. She feels something soft sliding around in her throat interfering with her breathing.

"Umele," she asks, "would you like to look at some cows?"

And when Umele fails to answer she sits down next to him and her eyes also start to overflow. The two of them sit there for a long while, mute, while with their fingers they dig into the dark earth of the chamber floor, till the ominous dusk settles on the little roof and the mother finds them nestled against each other, fast asleep.

That day the mother sealed the door with a nail, and from that moment on all was lost. And anyway, a couple of days later something happened that made the children forget all about the treasure they used to see through the window.

It happened in the morning, on a day of incessant, soaking rain. Swollen black clouds had settled over the ghetto. The mother was occupied with something in the corner of the room, when a neighbor came running and cried out to her in a panic: "Glikke, they've picked up Shimmele!"

At first the mother just stood there, not understanding a single word. But when the neighbor poured out the whole tale in a single breath, saying, "It has begun already," and explaining that today they started the first transports of Jews, picking up people in the streets and loading them into wagons—at that point the mother went limp, right where she was standing, giving a toss of her head and fainting dead away on the spot.

Now the panic really started. More than half the men who lived in the court, who had just happened to be out on the street, were missing. But that was just a prologue to the awesome black days that were about to descend on the ghetto.

That very same evening, Mama Glikke is to be seen going around with a wet headdress wrapped around her head. The window, draped in black, looks like a mirror covered to protect the soul of a corpse lying in the room. The father's *tallis* and *tfillin* hang in shame in a corner. Who needs them now? Who will now bind the thongs around the arm? The mother walks around all day, eyes swollen with weeping. The children drag around behind her, desperately hungry. Suddenly, the mother reminds herself of something. A ray of joy creeps into her swollen eyes. From a remote corner she takes out a whole loaf of bread and lays it on the table. That bread lights up the entire room, as if the sun has just peeped above the horizon. With deliberation she plunges the knife into the hard loaf of bread and, swallowing the new salty tears streaming down her cheeks, she keeps up an uninterrupted monologue:

"Eat, children, eat. It is your father's bread, your father's whole loaf of bread. . . ."

And for the first time in many, many months, the children joyfully eat to their hearts' content, as does Mama Glikke.

That night they all sleep peacefully and soundly, and father, Shimmele, appears only to Umele in a dream. Umele sees him praying over a large, thick book.

15. Bernard Gotfryd

Bernard Gotfryd was born in 1920 in Radom, Poland. During World War II, he was associated with the Polish underground, but he was eventually caught by the Germans and endured six concentration camps, including Majdanek and the stone quarry of Mauthausen. His parents were killed, but a brother and sister survived. In 1947, he emigrated to the United States, where for many years he was a staff photographer for *Newsweek*. *Anton, the Dove Fancier* (1990), from which "The Last Morning" is taken, is his first book.

Primo Levi praised Gotfryd as a writer "in constant search for goodness even in the most extreme situations." Like Ida Fink, Gotfryd draws on his own experiences, but unlike her he does not anchor our feelings to the moments of terror that temporarily estrange victims from their human selves. No matter how bizarre the circumstances he chronicles, he manages to salvage in his vignettes some fragment of dignity and even humor in the lives of his victims, and occasionally even in the lives of their tormentors. Whereas Fink stresses the fragility and anguish of existence under Nazi oppression, Gotfryd chronicles, in his own words, the "suffering and the endurance of the human spirit." Their divergent visions lead us along a spectrum from poignancy to dread, inviting us as readers to acknowledge the need for a point of view, and to develop one of our own.

The Last Morning

I very clearly remember the day I saw my mother for the last time. It was Sunday the sixteenth of August, 1942, a beautiful day with a clear blue sky and hardly a breeze. That morning she got up very early, earlier than usual, and quietly, so as not to wake us, she went out to the garden. I was already up. I watched her through the kitchen window. She sat down on the broken bench behind the lilac tree and cried. I always felt bad when I saw my mother cry, and this time it was even more painful.

My mother was going to be forty-four years old at the end of August. She never made a fuss over her birthday, as if it were her own secret, and so I never knew the exact date. She was of medium height, rather plump, with a most beautiful face. She had large brown eyes and long, dark brown hair sprinkled with gray, which she pulled back into a chignon. She smiled at people when she spoke and looked them straight in the eye.

When she came in from the garden she walked over to me and caressed my face as she used to do some years before the war, when I was a little boy. Now I was in my teens. Then she went over to the kitchen stove and started a fire. The wood was damp, and the kitchen filled with smoke. There was no more firewood left; this was the last of the broken-down fence from around our garden. She stood next to the stove fanning the smoke and asked me to open the door and the windows to let the smoke escape. Her eyes were red and teary, but when she turned to face me she smiled.

Soon the rest of the family was up, and Mother served a chicory brew with leftovers of sweet bread she had managed to bake some days earlier. There was even some margarine and jam, a great treat. We sat wherever we could, since the table was too small for the five of us. Because of limited table space my grandmother and my aunt ate their meals in their own room. None of us had much to say that morning. We just stared at one another as if to reaffirm our presence.

Suddenly my mother lifted her eyes and, looking at my father, asked him, "What are you thinking about?" My father, as if he had just wakened from a deep sleep, answered, "I stopped thinking, it's better not to think." We looked at him oddly. How could anyone stop thinking?

My mother got up from the table and started to tidy up the room. Then she asked me to go up to the attic and find her small brown suitcase for her. I found the suitcase, and, alone in the attic, I hugged it many times before I brought it to her.

The tension in the house nearly paralyzed me. It was stifling. I left in a hurry and, running all the way, went to investigate the ghetto square. It was still early in the morning, and clusters of people were congregating at street corners, pointing up at the utility poles. During the night the light bulbs had been replaced by huge reflectors. The ghetto police were out in force, preventing people from gathering. I noticed a poster reminding all inhabitants of the ghetto to deliver every sick or infirm member of their families to the only ghetto hospital. Noncompliance called for the death penalty.

My paternal grandmother was recovering from a stroke. She was able to walk with the help of a cane. I trembled at the thought of having to turn her in. The Nazis were preparing something devious. I knew the hospital wasn't big enough to absorb all the sick people in the ghetto.

My mother studied my face when I came back from the square. There was a frightened look in her eyes. She asked me what was happening out there, what people were saying, and I lied to her. I didn't mention the reflector bulbs, but I could tell that she knew what was coming.

She had her suitcase packed, and her neatly folded raincoat was laid out on the couch, as if she were going on an overnight trip the way she used to before the war. No one said much. We were communicating through our silence; our hearts were tense. My father took out the old family album and stood at the window, slowly turning the heavy pages. I looked over his shoulder and saw him examining his own wedding picture. He pulled it out of the album and put it inside his breast pocket. I pretended not to see.

My mother started preparing our lunch, and I helped her with the firewood. There was no more fence left, and somebody had just stolen our broken bench. I found an old tabletop that Father kept behind the house, covered with sheets of tar paper. It was dry and burned well. I didn't tell my mother where the wood had come from; I was afraid she might not like the idea of putting a good table to the fire.

It was past noon, and my mother was busy in the kitchen. She found some flour and potatoes she had managed to save and came up with a delicious soup, as well as potato pancakes sprinkled with fried onions. Was this to be our last meal together? I wondered.

Some friends and neighbors with scared expressions on their faces dropped in to confirm the rumors about the coming deportation and to say good-bye. The Zilber family came, and everybody cried. I couldn't bring myself to say good-bye to anybody; I feared that I would never see them again.

It was getting close to four o'clock in the afternoon when my grandmother, dressed in her best, came out of her room. She was ready, she said, if someone would escort her to the hospital. My brother and I volunteered. She insisted on walking alone, so we held her lightly by the arms in case she tripped. She walked erect, head high; from time to time she would look at one of us without saying a word. People passed us in bewilderment. They seemed like caged birds looking for an escape. An elderly man carrying a huge bundle on his shoulders stopped us and asked for the time. "Why do you need to know the time?" I inquired. He looked at me as if upset by my question and answered, "Soon it will be time for evening prayers, don't you know?" And he went on his way, talking to himself and balancing the awkward bundle on his shoulders.

When we reached the hospital gate my grandmother insisted we leave her there. She would continue alone. With a heavy heart I kissed her good-bye. She smiled and turned toward us, saying, "What does one say? Be well?" Then she disappeared behind the crumbling whitewashed gate of the hospital. I needed to cry but was ashamed to do so in front of my older brother. Determined to prove how tough I was, I held back my tears. We walked back in silence, each of us probably thinking the same thing.

I'll never forget coming back to the house after escorting Grandmother to the hospital. My mother was in the kitchen saying good-bye to one of her friends. I had never seen her cry as she was crying. When she saw us she fell upon us, and through her tears she begged us to go into hiding. She begged us to stay alive so that we could tell the world what had happened. Her friend was crying with her, and I felt my heart escaping.

A neighbor came in to tell us that the ghetto was surrounded by armed SS men, and it was official that the deportation was about to begin. The ghetto police were on full alert, and it was impossible to get any information out of them.

My brother and I turned and ran out of the house. Without stopping we ran the entire length of the ghetto until, dripping with sweat, we arrived at the fence. On the other side of the fence was a Nazi officers' club; farther off in the middle of a field stood a stable. By now the Ukrainian guards with their rifles were inside the ghetto. We scaled the fence behind their backs and made it across to the other side. We entered the stable through a side door. As far as I could tell, no one was there. The horses turned their heads and sized us up. My brother decided we should hide separately, so that if one of us was discovered, the other one would still have a chance. I climbed up on the rafters and onto a wooden platform wedged in between two massive beams. There was enough hay to cover myself with, and I stretched out on my stomach. Through the wide cracks between the boards of the platform I could scan the entire stable underneath me. I also found a crack in the wall that allowed me a wide view of the street across from the stable.

A mouse came out from under a pile of straw, stopped for a second, and ran back in. I lay there trying to make sense of every sound. As I turned on my side I felt something bulky inside my pocket. I reached for it and discovered a sandwich wrapped in brown paper. My mother must have put it there when my jacket was still hanging behind the kitchen door.

As I replaced the sandwich I heard the door open and saw a man enter. He walked to the other end of the stable and deposited a small parcel inside a crate. Then he started to tend to the horses while whistling an old Polish tune. He must be the caretaker, I thought. He appeared to be still young, even though I couldn't clearly see his face; he walked briskly and carried heavy bales of hay with ease. I feared the commotion he was causing might attract attention; he kept going in and out, filling the water bucket for the horses to drink. I was getting hungry. I was about to bite into the sandwich when on one of his trips he looked up at the spot where I was hiding. I froze. Could it be that he had heard me move? I couldn't imagine what had made him look up, and I broke out in a sweat. I held on to the sandwich but was too upset to eat it. Every time he opened the door it squeaked, and the spring attached to it caused it to shut with a loud bang. He spoke to the horses in Polish with a provincial accent and called each horse by its name. He lingered with some of them, slapped their backs or gently patted their necks. How I envied him. Why was he free while I had to hide?

I started to recall the events of the entire day. I realized I had run out of the house without saying good-bye to my parents. Seized with guilt, I started sobbing.

I must have fallen asleep. When I woke up I heard loud noises coming from behind the fence. I looked through the crack in the wall; it was dark outside. Suddenly a loud chorus of cries and screams rang out, intermingled with voices shouting commands in German. Rifle shots followed, and more voices calling out names pierced the darkness. The cries of little children made me shudder.

I imagined hearing the screaming of my four-year-old cousin, who was there with his mother; my aunt, her sister, with her two beautiful little daughters. They were all there, trapped, desperate, and helpless. I thought of our friend Mr. Gutman, who some years before had claimed that God was in exile. I wondered where he was and what he was saying now. I worried about my grandmother and what they were doing to her at the hospital. Frightened and burdened with my misgivings, I resolved to go on, not to give in.

I heard the squeak of the door and looked down to see the caretaker slipping out. He blocked the door with a rock to keep it open. The sounds coming in from the outside were getting louder; the horses became restless and started to neigh. Rifle shots were becoming more frequent and sounded much closer than before. All these noises went on for most of the night—it felt like an eternity.

I could picture my mother in that screaming, weeping crowd begging me to stay alive, and I could hear her crying for help. Was my father with her, I kept wondering, and where was my sister?

It was almost daybreak when the noises began to die down. The sun was rising; it looked like the beginning of a hot August day. Only occasional rifle shots could be heard, and a loud hum that sounded as if swarms of bees were flying overhead; it was the sound of thousands of feet shuffling against the pavement. Looking through the crack in the wall, I could see long columns of people being escorted by armed SS men with dogs on leashes. Most of the people carried knapsacks strapped to their backs; others carried in their arms what was left of their possessions. I focused on as many people as I could, hoping to recognize a face. I wanted to know if my mother was among them and kept straining my eyes until I couldn't see anymore. I wondered if my brother, at the other end of the stable, was able to see outside. As it was, we had no way to communicate.

I kept imagining the moving columns of people getting longer and wider until there was no more room for them to walk. As I pictured them they kept multiplying; soon they walked over one another like ants in huge anthills, and the SS men weren't able to control them any longer.

Suddenly I heard voices underneath me. Before I realized who was there I saw the caretaker climbing up toward my hiding place. I couldn't believe it. I stopped breathing. Two SS men wearing steel helmets and carrying rifles stood at the door watching the caretaker climb. He came close to the platform where I was lying and in a loud voice told me to get down. "They came to get you," he said. "I knew you were here hiding. You can't outsmart me." I was betrayed.

Next he walked right over to where my brother was hiding and called him out. The two of us took a terrible beating from the SS men before they escorted us back to the ghetto. The first thing I saw in the ghetto was a large horse-drawn cart on rubber wheels, loaded with dead, naked bodies. On one side, pressed against the boards, was my grandmother. She seemed to be looking straight at me.

No dictionary in the world could supply the words for what I saw next. My mother begged me to be a witness, however; all these years I've been talking and telling, and I'm not sure if anybody listens or understands me. I myself am not sure if I understand.

The following night my brother and I miraculously escaped the final deportation, only to be shipped off to the camps separately soon afterward. I never saw my mother again, nor was I ever able to find a picture of her. Whenever I want to remember her I close my eyes and think of that Sunday in August of 1942 when I saw her sitting in our ghetto garden, crying behind the lilac tree.

16. Sara Nomberg-Przytyk

Sara Nomberg-Przytyk was born in Lublin, Poland, in 1915. She came from a Hasidic background; her grandfather was a prominent Talmudist, principal of a yeshiva in Warsaw and rabbi of a town near Lublin. When the Germans invaded Poland in 1939, she fled east and in 1941 finally settled in Bialystok. In August 1943, she was deported from the Bialystok ghetto to Stutthof concentration camp, and from there to Auschwitz.

After liberation, she returned to Poland, married, and worked as a journalist. In 1968 she emigrated to Israel, and in 1975, settled in Canada near one of her sons. In an earlier volume, *Columny Samsona* (*The Pillars of Samson,* 1966), still untranslated, Nomberg-Przytyk describes events in the Bialystok ghetto before it was destroyed. The manuscript of *Auschwitz: True Tales from a Grotesque Land* is also dated 1966 and was scheduled to be published in Poland in 1967 when the outbreak of the Six-Day War led her publisher to insist that all references to Jews be removed from the narrative. She, of course, refused, taking the manuscript with her to Israel, where a copy was deposited in the archives of Yad Vashem in Jerusalem, where it was subsequently discovered by Eli Pfefferkorn.

In the following stories, "The Verdict" and "Friendly Meetings," taken from *Auschwitz: True Tales from a Grotesque Land* (1985), both reader and narrator learn that staying alive in Auschwitz requires tactics bearing little resemblance to those needed in the moral world we pride ourselves in inhabiting today. The tension between what is ethically desirable and what is necessary for physical survival remains the central challenging theme of Nomberg-Przytyk's art.

The Verdict

In October 1944 the whole hospital was moved to camp "C," the old gypsy camp. That is when I met Mrs. Helena. She had been doing the same job I was doing, except that she was a clerk in the infirmary for non-Jewish prisoners. In the new block, the separate infirmaries were liquidated and combined into one. The new infirmary was located in a separate barrack. In addition to the reception room there was a beautiful room containing three bunk beds. Five of the beds were occupied by the workers in the infirmary: Helena and I, the clerks; Mancy and Frieda, the two doctors; and nurse Marusia. The sixth bed was taken by Kwieta, who worked in the *Leichenkomando.* *

Mrs. Helena stuck out oddly in our group of five. Perhaps because she was older than we were, we felt very inhibited in her presence. She maintained a constant silence and seemed always to be steeped in her own thoughts. She lived her own life and said nothing to anyone. We did not even know how she had gotten to Auschwitz. She was slim, light-haired, and had an inscrutable face. She did not take part in our discussions, and she never judged anybody. She eavesdropped on our gossiping and seemed to be saying, "I would like to see how you would behave in a similar situation."

One evening, while we were discussing conscious and unconscious death, we were surprised to hear Helena break heatedly into our discussion:

"Listen to the story I am going to tell you about the death of 156 girls from Krakow, and then you can tell me what you think of the way I behaved." We all stopped talking, and complete silence descended on our cell.

"We were just finishing receiving the sick," Mrs. Helena started quietly. "While Mengele† was looking over the women who had been admitted to the area, we had but one thought in our minds: we hoped he would leave soon. I remember that it was a scorching July day. The atmosphere in the infirmary was almost unbearable. The last sick woman moved

*Corpse squad, the work detail responsible for transporting the bodies of the dead.

†Josef Mengele was the Auschwitz doctor notorious for his medical experiments with Jewish and Gypsy prisoners, especially identical twins.

through the line, passing in front of the German doctor. We heaved a sigh of relief. Mengele got up slowly, buttoned his uniform, stood facing me, and said: 'At fifteen hours the *Leichenauto** will come; I will come at fourteen hours.' We looked at one another in dumb amazement. Why the *Leichenauto* at fifteen hours? Usually, the car came to pick up the dead after darkness had fallen. What was Mengele planning to do here at fourteen hours? We couldn't speak. We were all sure that the *Leichenauto* was coming for us, to take us to the crematorium. We had to start cleaning up, but you can believe that everything kept dropping out of our hands, and that the hours dragged on without end. It's not easy to wait for the worst. After all, I don't have to tell you about that.

"At thirteen hours two young girls came to the infirmary, Poles from Krakow. They told us that the *blokowa*† had ordered them to report here because they had to leave for work in Germany and Mengele was going to examine them. I was so frightened by what I heard that I almost fainted. 'Is it only you the *blokowa* sent?' I asked in a quivering voice.

" 'Not just us,' they answered. 'There will be a lot more of us here. The rest will be coming soon. We came in first because we are in a hurry to join the transport that is leaving Auschwitz.'

"Quite a large number of women were now gathering in front of the infirmary, most of them young. They were happy to be leaving Auschwitz. They were talking loudly, laughing, never dreaming that they had been horribly deceived and that the *Leichenauto* was coming for them in about an hour. For us it was all clear; those Poles were condemned to death, and the sentence was going to be carried out in the infirmary. 'What to do?' I thought feverishly. Maybe I should tell them why they had been summoned here. Perhaps I should shout it out to them: 'Calm down! Don't laugh. You are living corpses, and in a few hours nothing will be left of you but ashes!' Then what? Then we attendants would go to the gas chambers and the women would die anyway. The women might run and scatter all over the camp, but in the end they would get caught. Their numbers have been recorded. There is no place for them to run.

"Believe me, we quietly took counsel, trying to decide what to do. We didn't tell them the terrible truth, not out of fear for our own lives, but because we truly did not know what would be the least painful way for the young women to die. Now they didn't know anything, they were carefree, and death would be upon them before they knew it. If we told them what was in store for them, then a struggle for life would ensue. In their attempt to run from death they would find only loneliness, because their friends, seeking to preserve their own lives, would refuse to help them. There were more than 150 women in front of the infirmary. They stood in rows of five, as at roll call, and waited for the doctor to examine them. Still, we did not know what to do. All our reasoning told us to say

*Vehicle for carrying corpses.
†Prisoner head of a women's barrack.

nothing. Today I know that it was fear for our own lives that made us reason this way, that induced us to believe that sudden, unexpected death is preferable to a death that makes itself known to your full and open consciousness.

"Precisely at fourteen hours, Mengele arrived, accompanied by an orderly named Kler. He looked at the lines of women standing there and then at us in such a way as to make us partners in the crimes that he was about to commit. At that moment I knew that we had made a mistake in not telling the young women what was awaiting them. Whether dying is supposed to be easy or difficult, I suppose every individual has to decide for herself. But it was our duty to inform the young women what awaited them.

" 'Bring them in for a checkup,' shouted Mengele. The first girl walked in, the one who was in a hurry to leave Auschwitz. She stood in front of me; I did not say anything. By filling out her hospital card, I was taking part in this deception that was making it easy for Mengele to execute his victim. She walked in without suspecting anything. Then I heard the crashing sound of a falling body; later, the second; then the third, the tenth, the twentieth. Always the same: the card, the squeaking of the door, the crash of a falling body. The corpses were thrown out into the waiting room, which was located behind the reception room. An SS man with a dog kept order in front of the infirmary. Calm and trusting, the women kept going in. I lowered my head so they wouldn't see my face. All I would see each time was a hand stretched out to receive a card. I really did not understand why they were so calm. Weren't they surprised not to see the other women coming out of the infirmary after they had been examined? I looked for some sign of anxiety in those stretched out hands, but to no avail. I had given out about a hundred cards when it started.

"One of the girls asked the SS man why the other women weren't coming out after having been examined. Instead of answering her he hit her over the head with his rifle butt. Then I heard one of the girls yell, 'We are not going in there. They will give us an injection of phenol.' A terrible outcry started. The girls really refused to enter the infirmary. When one of them tried to run away the SS man shot her. At the sound of a shot a whole troop of SS men and dogs ran in. The young women were completely surrounded. Each girl, having first been beaten, was dragged screaming, by two SS men, into the presence of Mengele. I didn't give out any more cards. It was no longer necessary.

"I jumped up from my seat and hid in a corner of the infirmary. The women did not want to die. They tore themselves out of the grip of the SS men and started to run away. Then the dogs were set on them. Their deaths were completely different from the deaths of the first batch of women who went to their deaths unknowing. Who knows which death was more difficult, but the first group seemed to die more peacefully.

"At fifteen hours the *Leichenauto* showed up, and an hour later the entire operation was completed. Up to the very last minute we were not certain that Mengele was not going to send us, the witnesses of that bloody happening, to the gas. Mengele left, calm, and with a smile he put down the sick card he had been holding. 'Herzanfall [heart attack],' he said."

Mrs. Helena finished her terrible tale. We did not utter a word. After a long pause she resumed: "I still don't know whether we should have told the women about the death that was waiting for them. What do you think?"

None of us said anything.

Friendly Meetings

A cold, penetrating rain had been falling for a few days. Such rains were not unusual in Auschwitz. I opened the gate of the infirmary very quietly so as not to disturb the performance and listened. "Plop, plop,"—the drops continued falling without a stop. Outside it was dark and quiet. The lights on the ramp of the station were out. It had been a few days since the last trainload of victims had arrived at Auschwitz. Perhaps, I thought, they would not bring any more victims here.

I sat down in the corner to watch the performance. It was Sunday. Since everything was at peace this day, Irena had organized a cultural evening in the infirmary. She had planned an evening of dancing—without men, of course. But then, it is possible to dance without men, too.

Irena was an actress. Although she was originally from Poland, she had lived in Paris some fourteen years before being shipped to Auschwitz. She was tall, strong, and straight. I remember that when I first met her it was hard for me to believe that she was an actress. Looking at her, you would absolutely never guess that she was an actress. The girls who knew her swore that once she got on the stage she changed so completely that you would never recognize her. She was particularly wonderful, they said, as a character actress.

In Auschwitz we often organized such friendly get-togethers. I remember that for the first few months of my stay here those get-togethers struck me as being indecent. How was it possible that we could sing while the sky above was red with the flames of the crematoria.

"How can you joke, dance, and tell stories," I asked, "when we are enveloped in a sea of suffering, pain, and tears?"

"You will get used to it. Then you will understand." So said the old prisoners.

One evening, as I was returning from the infirmary to the barracks for the night, I bumped into a group of girls from the *Leichenkomando,* whose job it was to load the dead into the trucks. One of them stood near a pile of corpses, the second near the truck, the third on a small stool, and the fourth on the platform of the truck. They were handing the dead to each other any old way: grabbing the corpse by the leg, or the arm, or the hair, and then swinging it onto the platform. I noted their indifference to the dead and tried to imagine what kind of women they had been a few years ago, when they loved and were loved in a world of normality. Every few minutes I could hear a sound—the thump of falling flesh and the cries of the women: "Hurry up. Why are you dawdling?"

One of the women started singing a song and immediately the rest of them joined in: "For a cup of flour, he kissed for an hour."

They sang to the melody of a German march. I descended on them, half choking:

"How can you sing a merry song in front of those skeletons?" I called out resentfully.

They looked at me in bewilderment, without the foggiest notion of what I was talking about.

"You'll get used to it," one of them said. Then, after a moment's silence, she added: "If you don't get used to it you'll drop dead."

I got used to it. After eight months in Auschwitz, I could look at the dead with indifference. When a corpse was lying across my path I did not go around it any more, I simply stepped over it, as if I were merely stepping over a piece of wood. I sang along with the others, and I laughed when I heard a good joke. I even told jokes myself. I even got used to the rats warming themselves in front of the stove like cats. I had imbibed all of the terrors of Auschwitz and lived. Then I really understood that my ability to adapt to just about anything was a most useful talent. Was this good or bad? It was difficult for me to know.

High on the ceiling a small light bulb was burning; we were sitting around, scattered all over the room. Some were sitting on the table, some were sitting on the floor. Marusia and Kwieta and all of the other Czechs were singing a beautiful youth song about those who "defy the wind." The words of the song said that only by swimming against the current could the strong achieve satisfaction. The French girls were singing French songs about Paris. I specialized in Russian songs. I sang without thinking about the red sky. Somebody recited a poem. Then, in hushed tones, we sang a prison hymn in German. The conductor of the orchestra, a Hungarian woman, was the main attraction. She had come to the camp with her violin; now she started playing Hungarian and gypsy melodies. We were sitting around, listening, as if bewitched. Suddenly the door of the infirmary opened with a loud crash, and there stood Hitler: moustache, hair, and a haughty, stupid expression. We all jumped up from our places, and Marusia even yelled "Achtung!" Hitler walked in with a long stride and an outstretched hand and kneeled in front of me. He set his hands beseechingly and whispered, "Maybe you would like to change places with me."

We all burst into joyous laughter. It was Irena. She was mocking my assertion that we had it better than Hitler because we had more of a chance of living through the war. Hitler would certainly not live through the war, but we might. She got up. We could not get over her impersonation. With the help of black shoe polish she had changed her face beyond recognition.

Now the girls started dancing to the accompaniment of the violin. Orli was standing next to me, pale and agitated.

"Has something happened?" I asked.

"Walk outside unobtrusively," Orli said, "so that nobody will notice."

We left the infirmary. It was raining without letup. The ramp was lighted up. The SS men were standing in front of the cars. We had been feeling happy because the transports were not arriving. Now the unloading of the people was starting again, but quietly, without the usual screaming. Nude men came out, so skinny that it was difficult to believe that those people were moving on their own power. They were, indeed, moving skeletons.

"Those are Russian prisoners," Orli whispered. "They were working someplace, and now, since they are incapable of working any more, they are being sent to the gas chambers."

They walked slowly under the cold rain. Some of them were swaying. They all went to the gas chambers. The lights on the station went out, the empty cars left, and we just stood there, outside.

"Let's not tell the girls anything. Let's not spoil their fun," Orli said quietly. We did not return to the infirmary. To sing and joke now was beyond our strength. After all, you could not get used to everything.

17. Aharon Appelfeld

Aharon Appelfeld was born in 1932 in Czernowitz (which is Paul Celan's birthplace as well) in the Bukovina region of Romania, which now belongs to Moldova in the former Soviet Union. When he was eight, the Germans killed his mother and sent his father to a labor camp. Although only a child, Appelfeld escaped from a camp and spent the next three years hiding in the woods. He was eleven when Russian troops found him; he became a kitchen helper for the Soviet army. His isolation during these years may be reflected in the lonely wanderings of Tzili in the short novel included in this section.

Appelfeld was thirteen when World War II ended. After living in a displaced persons' camp in Italy, with the help of a Jewish organization he emigrated to Palestine in 1946. He was only fourteen years old when he arrived and did not discover that his father was alive in Israel until 1960.

Appelfeld was educated in Palestine and Israel. Like a number of writers concerned with Holocaust themes (for example, Elie Wiesel, Piotr Rawicz, and Dan Pagis), he chooses to write in a language that is not his native tongue—in his case, like Pagis, in Hebrew. Currently a professor of modern Hebrew literature at Ben Gurion University in Beersheva, he has taught at several American universities, including Yale, Harvard, Brandeis, and Boston University. He has received the prestigious Israel Prize and many other awards for his work.

Appelfeld has written more than twenty books, including novels, short stories, and essays. It would be fair to say that he writes more "around" the Holocaust than about it, in the sense that most of his fiction concerned with this period is set before and/or after the war, leaving the atrocity itself as a great vacant canvas to be painted by the reader's imagination. This strategy supports his intuition that the destruction of European Jewry disrupted the inner life of its surviving victims as well as the smooth chronological flow of history. The mounting anxiety that broods over his landscapes, combined with vividly observed details of daily life, fuse naïveté and threat into narratives that reflect both his own Holocaust experience and the influence of Kafka.

Like many of the writers included in this collection, Appelfeld shatters the confining mold that would describe him merely as a "Holocaust writer." His characters are literally and figuratively uprooted creatures, orphaned by a mutilated age, sometimes as survivors, sometimes only as potential victims of the Holocaust, but also simply as human beings who have been deprived of their own culture or of the one into which they had been assimilated. A muted theme is the loss suffered by Christian society from the disappearance of the Jewish intelligentsia.

The narrative in *Tzili: The Story of a Life* (1983) has an internal shape of its own, resembling Appelfeld's personal Holocaust wanderings only in some superficial details. It reflects the ordeal of a guileless and solitary mind contending with the dilemma of the displaced and violated self. Tzili's life has been shaped, and continues to be haunted, by slogans that are totally irrelevant to her experience as a fugitive Jewish child: "If you want to you can"; "Women are lucky; they don't have to go to war"; "Without cigarettes there's no point in living"; "You have to forget; it's not a tragedy." But the main recurrent refrain, which becomes an ironic epitaph to millions of Tzili's fellow victims, is "Death is not as terrible as it seems. All you have to do is conquer your fear." One is reminded of the sinister chorus from Paul Celan's "Death Fugue": ". . . der Tod ist ein Meister aus Deutschland" (. . . death is a master from Germany). But Tzili has never read Celan or grasped, as he had, how atrocity had rotted the bond joining language to truth.

In *Tzili,* naïve consciousness rivals a dark history in a never-ending struggle to recover the impossible. Memory, oblivion, and hope collide frequently in the narrative: Shall we reclaim a ruined past, engage in drunken and forgetful revelry in the present, or set out in sober quest of a reborn future (in this instance, in Palestine)? Tzili's inference near the end of the novel that "in Palestine everything will be different" is charged with ambiguity: Is this one more bland slogan or an ominous warning? When another refugee cries out, "We've had enough words. No more words," she speaks not as a mouthpiece for Appelfeld, but for a point of view that has been rumbling beneath the surface of the text from the beginning. The Holocaust has taught us to mistrust not only what we hear, but also what we have heard. What appears to be a story of survival thus turns out to be equally a chronicle of loss and injury—to the past, the present, and the uncertain future, and the language that records them all. *Tzili* dramatizes the futility of trying to amputate hope from remembered cruelty or, even with the best of intentions, substitute reassurance for truth.

Tzili

1

Perhaps it would be better to leave the story of Tzili Kraus's life untold. Her fate was a cruel and inglorious one, and but for the fact that it actually happened we would never have been able to tell her story. We will tell it in all simplicity, and begin right away by saying: Tzili was not an only child; she had older brothers and sisters. The family was large, poor, and harassed, and Tzili grew up neglected among the abandoned objects in the yard.

Her father was an invalid and her mother busy all day long in their little shop. In the evening, sometimes without even thinking, one of her brothers or sisters would pick her out of the dirt and take her into the house. She was a quiet creature, devoid of charm and almost mute. Tzili would get up early in the morning and go to bed at night like a squirrel, without complaints or tears.

And thus she grew. Most of the summer and autumn she spent out of doors. In winter she snuggled into her pillows. Since she was small and skinny and didn't get in anyone's way, they ignored her existence. Every now and then her mother would remember her and cry: "Tzili, where are you?" "Here." The answer would not be long in coming, and the mother's sudden panic would pass.

When she was seven years old they sewed her a satchel, bought her two copybooks, and sent her to school. It was a country school, built of gray stone and covered with a tiled roof. In this building she studied for five years. Unlike other members of her race, Tzili did not shine at school. She was clumsy and somewhat withdrawn. The big letters on the blackboard made her head spin. At the end of the first term there was no longer any doubt: Tzili was dull-witted. The mother was busy and harassed but she gave vent to her anger nevertheless: "You must work harder. Why don't you work harder?" The sick father, hearing the mother's threats, sighed in his bed: What was to become of them?

Tzili would learn things by heart and immediately forget them again. Even the gentile children knew more than she did. She would get mixed up. A Jewish girl without any brains! They delighted in her misfortune. Tzili would promise herself not to get mixed up, but the moment she stood in front of the blackboard the words vanished and her hands froze.

For hours she sat and studied. But all her efforts didn't help her. In the fourth grade she still hadn't mastered the multiplication table and her handwriting was vague and confused. Sometimes her mother lost her temper and hit her. The sick father was no gentler than the mother. He would call her and ask: "Why don't you study?"

"I do study."

"Why don't you know anything?"

Tzili would hang her head.

"Why are you bringing this disgrace on your family?" He would grind his teeth.

The father's illness was fatal, but the dull presence of his youngest daughter hurt him more than his wound. Again and again he blamed her laziness, her unwillingness, but never her inability . "If you want to you can." This wasn't a judgment, but a faith. In this faith they were all united, the mother in the shop and her daughters at their books.

Tzili's brothers and sisters all worked with a will. They prepared for external examinations, registered for crash courses, devoured supplementary material. Tzili cooked, washed dishes, and weeded the garden. She was small and thin, and kneeling in the garden she looked like a servant girl.

But all her hard work did not save her from her disgrace. Again and again: "Why don't you know anything? Even the gentile children know more than you do." The riddle of Tzili's failure tortured everyone, but especially the mother. From time to time a deep groan burst from her chest, as if she were mourning a premature death.

In the winter evil rumors were already rife, but only echoes reached the remoter districts. The Kraus family labored like ants. They hoarded food, the daughters memorized dates, the younger son drew clumsy geometric figures on long sheets of paper. The examinations were imminent, and they cast their shadow over everyone. Heavy sighs emerged from the father's darkened room: "Study, children, study. Don't be lazy." The vestiges of a liturgical chant in his voice aroused his daughters' ire.

At home Tzili was sometimes forgotten, but at school, among all the gentile children, she was the butt of constant ridicule and scorn. Strange: she never cried or begged for mercy. Every day she went to her torture chamber and swallowed the dose of insults meted out to her.

Once a week a tutor came from the village to teach her her prayers. The family no longer observed the rituals of the Jewish religion, but her mother for some reason got it into her head that religious study would be good for Tzili, besides putting a little money the old man's way. The tutor came on different days of the week, in the afternoons. He never raised his voice to Tzili. For the first hour he would tell her stories from the Bible and for the second he would read the prayer book with her. At the end of the lesson she would make him a cup of tea. "How is the child progressing?" the mother would ask every now and then. "She's a good girl," the old man would say. He knew that the family did not keep the

Sabbath or pray, and he wondered why it had fallen to the lot of this dull child to keep the spark alive. Tzili did her best to please the old man, but as far as reading was concerned her progress left much to be desired. Among her brothers and sisters the old man's visits gave rise to indignation. He wore a white coat and shabby shoes, and his eyes glinted with the skepticism of a man whose scholarship had not helped him in his hour of need. His sons had emigrated to America, and he was left alone in the derelict old house. He knew that he was nothing but a lackey in the service of Tzili's family's hysteria, and that her brothers and sisters could not bear his presence in the house. He swallowed his humiliation quietly, but not without disgust.

At the end of the reading in the prayer book he would ask Tzili, in the traditional, unvarying formula:

"What is man?"

And Tzili would reply: "Dust and ashes."

"And before whom is he destined to stand in judgment?"

"Before the King of Kings, the Holy One blessed be He."

"And what must he do?"

"Pray and observe the commandments of the Torah."

"And where are the commandments of the Torah written?"

"In the Torah."

This set formula, spoken in a kind of lilt, would awaken loud echoes in Tzili's soul, and their reverberations spread throughout her body. Strange: Tzili was not afraid of the old man. His visits filled her with a kind of serenity which remained with her and protected her for many hours afterward. At night she would recite "Hear, O Israel" aloud, as he had instructed her, covering her face.

And thus she grew. But for the old man's visits her life would have been even more wretched. She learned to take up as little space as possible. She even went to the lavatory in secret, so as not to draw attention to herself. The old man, to tell the truth, felt no affection for her. From time to time he grew impatient and scolded her, but she liked listening to his voice and imagined that she heard tenderness in it.

2

When the war broke out they all ran away, leaving Tzili to look after the house. They thought nobody would harm a feeble-minded little girl, and until the storm had spent itself, she could take care of their property for them. Tzili heard their verdict without protest. They left in a panic, without time for second thoughts. "We'll come back for you later," said her brothers as they lifted their father onto the stretcher. And thus they parted from her.

That same night the soldiers invaded the town and destroyed it. A terrible wailing rose into the air. But Tzili, for some reason, escaped unharmed. Perhaps they didn't see her. She lay in the yard, among the barrels in the shed, covered with sacking. She knew that she had to look after the house, but her fear stopped her from doing so. Secretly she hoped for the sound of a familiar voice coming to call her. The air was full of loud screams, barks, and shots. In her fear she repeated the words she had been taught by the old man, over and over again. The mumbled words calmed her and she fell asleep.

She slept for a long time. When she woke it was night and everything was completely still. She poked her head out of the sacking, and the night sky appeared through the cracks in the roof of the shed. She lifted the upper half of her body, propping herself up on her elbows. Her feet were numb with cold. She passed both hands over the round columns of her legs and rubbed them. A pain shot through her feet.

For a long time she lay supporting herself on her elbows, looking at the sky. And while she lay listening, her lips parted and mumbled:

"Before whom is he destined to stand in judgment?"

"Before the King of Kings, the Holy One blessed be He."

The old man had insisted on the proper pronunciation of the words, and it was this insistence she remembered now.

But in the meantime the numbness left her legs, and she kicked away the sacking. She said to herself: "I must get up," and she stood up. The shed was much higher than she was. It was made of rough planks and used to store wood, barrels, an old bathtub, and a few earthenware pots. No one but Tzili paid any attention to this old shed, but for her it was a hiding place. Now she felt a kind of intimacy with the abandoned objects lying in it.

For the first time she found herself under the open night sky. When she was a baby they would close the shutters very early, and later on, when she grew up, they never let her go outside in the dark. For the first time she touched the darkness with her fingers.

She turned right, into the open fields. The sky suddenly grew taller, and she was small next to the standing corn. For a long time she walked without turning her head. Afterward she stopped and listened to the rustle of the leaves. A light breeze blew and the cool darkness assuaged her thirst a little.

On either side stretched crowded cornfields, one plot next to the other, with here and there a fence. Once or twice she stumbled and fell but she immediately rose to her feet again. In the end she hitched her dress up and tucked it into her belt, and this immediately liberated her legs. From now on she walked easily.

For some reason she began to run. A memory invaded her and frightened her. The memory was so dim that after she ran a little way it disappeared. She resumed her previous pace.

Her oldest sister, who was preparing for examinations, was the worst of them all. When she was swotting, she would chase Tzili away without even lifting her head from her books. Tzili loved her sister and the harsh words hurt her. Once her sister had said: "Get out of my sight. I never want to see you again. You make me nervous." Strange: these words rather than any others were the ones that seemed to carve themselves out of the darkness.

The darkness seeped slowly away. A few pale stripes appeared in the sky and turned a deeper pink. Tzili bent over to rub her feet and sat down. Unthinkingly she sank her teeth into a cornstalk. A stream of cool liquid washed her throat.

The light broke above her and poured onto her head. A few solitary animal cries drifted through the valley and a loud chorus of barks immediately rose to join them. She sat and listened. The distant sounds cradled her. Without thinking she fell asleep.

The sun warmed her body and she slept for many hours. When she woke she was bathed in sweat. She picked up her dress and shook off the grains of sand sticking to her skin. The sun caressed her limbs and for the first time she felt the sweet pain of being alone.

And while everything was still quiet and wrapped in shadows a shot pierced the air, followed by a sharp, interrupted scream. She bent down and covered her face. For a long time she did not lift her head. Now it seemed to her that something had happened to her body, in the region of the chest, but it was only a vague, hollow numbness after a day without eating.

The sun sank and Tzili saw her father lying on his bed. The last days at home, the rumors and the panic. Books and copybooks. No one showed any consideration for the feelings of his fellows. The examinations, which were to take place shortly in the distant town, threatened them all, especially her oldest sister. She tore out her hair in despair. Their mother too, in the shop, between one customer and the next, appeared to be repeating dates and formulas to herself. The truth was that she was angry. Only the sick father lay calmly in his bed. As if he had succeeded in steering the household onto the right course. He seemed to have forgotten his illness, perhaps even the dull presence of his youngest daughter. What he had failed to accomplish in his own life his children would accomplish for him: they would study. They would bring diplomas home.

And with these sights before her eyes she fell asleep.

3

When she woke, her memory was empty and weightless. She rose and left the cornfield and made for the outskirts of the forest. As if to spite her, another picture rose before her eyes, it too from the last days at home. Her youngest brother was adamant: he had to have a bicycle—all his friends, even the poorest, had bicycles. All his mother's pleas were in vain. She had no money. And what she had was not enough. Their father needed medicines. Tzili's seventeen-year-old brother shouted so loudly in the shop that strangers came in to quiet him. The mother wept with rage. And the older sister, who did not leave her books for a moment, shrieked: because of her family she would fail her exams. Tzili now remembered with great clarity her sister's white hand waving despairingly, as if she were drowning.

The day passed slowly, and visions of food no longer troubled her. She saw what was before her eyes: a thin forest and the golden calm of summer. All she had endured in the past days lost its terror. She was borne forward unthinkingly on a stream of light. Even when she washed her face in the river she felt no strangeness. As if it had always been her habit to do so.

And while she was standing there a rustle went through the field. At first she thought it was the rustle of the leaves, but she immediately realized her mistake: her nose picked up the scent of a man. Before she had time to recover she saw, right next to her, a man sitting on a little hillock.

"Who's there?" said the man, without raising his voice.

"Me," said Tzili, her usual reply to this question.

"Who do you belong to?" he asked, in the village way.

When she did not answer right away, the man raised his head and added: "What are you doing here?"

When she saw that the man was blind, she relaxed and said: "I came to see if the corn was ready for the harvesting." She had often heard these words spoken in the shop. Since the same sentence, with slight variations, was repeated every season, it had become part of her memory.

"The corn came up nicely this year," said the blind man, stroking his jacket. "Am I mistaken?"

"No father, you're not mistaken."

"How high is it?"

"As high as a man, or even higher."

"The rains were plentiful," said the blind man, and licked his lips.

His blind face went blank and he fell silent.

"What time is it?"

"Noon, father."

He was wearing a coarse linen jacket and he was barefoot. He sat at his ease. The years of labor were evident in his sturdy shoulders. Now he was looking for a word to say, but the word evaded him. He licked his lips.

"You're Maria's daughter, aren't you?" he said and chuckled.

"Yes," said Tzili, lowering her voice.

"So we're not strangers."

Maria's name was a household word throughout the district. She had many daughters, all bastards. Because they were all good-looking, like their mother, nobody harmed them. Young and old alike availed themselves of their favors. Even the Jews who came for the summer holidays. In Tzili's house Maria's name was never spoken directly.

A number of years before, Tzili's older brother had gotten one of Maria's daughters into trouble. Maria herself had appeared in the shop and created a scene. For days, the family had consulted in whispers, and in the end they had been obliged to hand over a tidy sum. The mother, worn out with work, had refused to forgive her son. She found frequent occasions to refer to his crime. Tzili had not, of course, grasped the details of the affair, but she sensed that it was something dark and sordid, not to be spoken of directly. Later on, their mother forgave her brother, because he began to study and also to excel.

"Sit down," said the blind man. "What's your hurry?"

She approached and seated herself wordlessly by his side. She was used to the blind. They would congregate outside the shop and sit there for hours at a time. Every now and then her mother would emerge and offer them a loaf of bread, and they would munch it noisily. Mostly they would sit in silence, but sometimes they would grow irritable and begin to quarrel. Her father would go out to restore order. Tzili would sit and watch them for hours. Their mute, upraised faces reminded her of people praying.

The blind man seemed to rouse himself. He groped for his satchel, took out a pear and said: "Here, take it."

Tzili took it and immediately sank her teeth into the fruit.

"I have some smoked meat too—will you have some?"

"I will."

He held the thick sandwich out in his big hand. Tzili looked at the big pale hand and took the sandwich. "Maria's daughters are all good-looking girls," he said and snickered. Now that he had straightened the upper half of his body he looked very strong. Even his white hands. "I don't like eating alone. Eating alone depresses me," he confessed. He chewed calmly and carefully, as blind men will, as if they were suspicious even of the food they put in their mouths.

As he ate he said: "They're killing the Jews. The pests. Let them go to America." But he didn't seem particularly concerned. He was more concerned with the coming harvest.

"Why are you so silent?" he said suddenly.

"What's there to say?"

"Maria's daughters are a cunning lot."

Tzili did not yet know that the notorious name of Maria would be her shield from danger. All her senses were concentrated on the thick sandwich the blind man had given her.

Once Maria had been a customer at the shop. She was a handsome, well-dressed woman and she used city words. They said that Maria had a soft spot for Jews, which did not add to her reputation. Her daughters too had inherited this fondness. And when the Jewish vacationers appeared, Maria would have a taste of what it meant to be indulged.

Tzili now remembered nothing but the heavy scent Maria left behind her in the shop. She liked breathing in this scent.

The blind man said casually: "Maria's daughters love the Jews, may God forgive them." And he snickered to himself again. Then he sat there quietly, as if he were a cow chewing the cud.

Now there was no sound but for the birds and the rustling of the leaves, and they too seemed muted. The blind man abandoned his full face to the sun and seemed about to fall asleep.

Suddenly he asked: "Is there anyone in the field?"

"No."

"And where did you come from?" The full face smiled.

"From the village square."

"And there's no one in the field?" he asked again, as if he wanted to hear the sound of his own voice.

"No one."

Upon hearing Tzili's reply he reached out and put his hand on her shoulder. Tzili's shoulder slumped under the weight of his hand.

"Why are you so skinny?" said the blind man, apparently encountering her narrow shoulder bones. "How old are you?"

"Thirteen."

"And so skinny." He clutched her with his other hand, too, the one he had been leaning on.

Tzili's body recoiled from the violence of the peasant's embrace, and he threw her onto the ground with no more ado.

A scream escaped her lips.

The blind man, apparently taken aback by this reaction, hurried to stop her mouth but his hand missed its aim and fell onto her neck. Her body writhed under the blind man's heavy hands.

"Quiet! What's the matter with you?" He tried to quiet her as if she were a restless animal. Tzili choked. She tried to wriggle out from under the weight.

"What has your mother been feeding you to make you choke like that?"

The blind man loosened his grip, apparently under the impression that Tzili was too stunned to move. With a swift, agile movement she slipped out of his hands.

"Where are you?" he said, spreading out his hands.

Tzili retreated on her hands and knees.

"Where are you?" He groped on the ground. And when there was no reply, he started waving his hands in the air and cursing. His voice, which had sounded soft a moment before, grew hoarse and angry.

For some reason Tzili did not run away. She crawled on all fours to the field. Evening fell and she curled up. The blind man's strong hands were still imprinted on her shoulders, but the pain faded as the darkness deepened.

Later the blind man's son came to take him home. As soon as he heard his son approaching, the blind man began to curse. The son said that one of the shafts had broken on the way and he had had to go back to the village to get another cart. The father was not convinced by this story and he said: "Why couldn't you walk?"

"Sorry, father, I didn't think of it. I didn't have the sense."

"But for the girls you've got sense enough."

"What girls, father?" said the son innocently.

"God damn your soul," said the blind man and spat.

4

By now Tzili's memories of home were blurred. They've all gone, she said blankly to herself. The little food she ate appeased her hunger. She was tired. A kind of hollowness, without even the shadow of a thought, plunged her into a deep sleep.

But her body had no rest that night. It seethed. Painful sensations woke her from time to time. What's happening to me? she asked herself, not without resentment. She feared her body, as if something alien had taken possession of it.

When she woke and rose to her feet it was still night. She felt her feet, and when she found nothing wrong with them she was reassured. She sat and listened attentively to her body. It was a cloudless and windless night. Above the bowed tops of the corn a dull flame gleamed. From below, the stalks looked like tall trees. She was astonished by the stillness.

And while she stood there listening she felt a liquid oozing from her body. She felt her belly, it was tight but dry. Her muscles throbbed rhythmically. "What's happening to me?" she said.

When dawn broke she saw that her dress was stained with a number of bright spots of blood. She lifted up her dress. There were a couple of spots on the ground too. "I'm going to die." The words escaped her lips.

A number of years before, her oldest sister had cut her finger on a kitchen knife. And by the time the male nurse came, the floor was covered with dark blood stains. When he finally arrived, he clapped his hands to his head in horror. And ever since they had spoken about Blanca's weak, wounded finger in solicitous tones.

"I'm going to die," she said, and all at once she rose to her feet. The sudden movement alarmed her even more. A chill ran down her spine and she shivered. The thought that soon she would be lying dead became more concrete to her than her own feet. She began to whimper like an animal. She knew that she must not scream, but fear made her reckless. "Mother, mother!" she wailed. She went on screaming for a long time. Her voice grew weaker and weaker and she fell to the ground with her arms spread out, as she imagined her body would lie in death.

When she had composed herself a little, she saw her sister sitting at the table. In the last year she had tortured herself with algebra. They had to bring a tutor from the neighboring town. The tutor turned out to be a harsh, strict man and Blanca was terrified of him. She wept, but no one paid any attention to her tears. The father too, from his sick bed, demanded the impossible of her. And she did it too. Although she did not complete the paper and obtained a low mark, she did not fail. Now Tzili saw her sister as she had never seen her before, struggling with both hands against the Angel of Death.

And as the light rose higher in the sky, Tzili heard the trudge of approaching feet. One of the blind man's daughters was leading her father to his place. He was grumbling. Cursing his wife and daughters. The girl did not reply. Tzili listened intently to the footsteps. When they reached his place on the hillock the girl said: "With your permission, father, I'll go back to the pasture now."

"Go!" He dismissed her, but immediately changed his mind and added: "That's the way you honor your father."

"What shall I do, father?" Her voice trembled.

"Tell your father the latest news in the village."

"They chased the Jews away and they killed them too."

"All of them?" he asked, with a dry kind of curiosity.

"Yes, father."

"And their houses? What happened to their houses?"

"The peasants are looting them," she said, lowering her voice as if she were repeating some scandalous piece of gossip.

"What do you say? Maybe you can find me a winter coat."

"I'll look for one, father."

"Don't forget."

"I won't forget."

Tzili took in this exchange, but not its terrible meaning. She was no longer afraid. She knew that the blind man would not move from his place.

5

Hours of silence came. Her oppression lifted. And after her weeping she felt a sense of release. "It's better now," she whispered, to banish the remnants of the fear still congealed inside her. She lay flat on her back. The late summer sunlight warmed her body from top to toe. The last words left her and the old hunger that had troubled her the day before came back.

When night fell she bandaged her loins with her shawl, and without thinking about where she was going she walked on. The night was clear, and delicate drops of light sparkled on the broad cornfields. The bandage pressing against her felt good, and she walked on. She came across a stream and bent down to cup the water in her hands and drink. Only now did she realize how thirsty she was. She sat calmly and watched the running water. The sights of home dissolved in the cool air. Her fear shrank. From time to time brief words or syllables escaped her lips, but they were only the sighs that come after long weeping.

She slept and woke and slept again and saw her old teacher. The look in his eye was neither kindly nor benign, but appraising, the way he looked at her when she was reading from the prayer book. It was a dispassionate, slightly mocking look. Strange, she tried to explain something to him but the words were muted in her mouth. In the end she succeeded in saying: I am setting out on a long journey. Give me your blessing, teacher. But she didn't really say it, she only imagined saying it. Her intention made no impression on the old man, as if it were just one more of her many mistakes.

Afterward she wandered in the outskirts of the forest. Her food was meager: a few wild cherries, apples, and various kinds of sour little fruits which quenched her thirst. The hunger for bread left her. From time to time she went down to the river and dipped her feet in the water. The cold water brought back memories of the winter, her sick father groaning and asking for another blanket. But these were only fleeting sensations. Day by day her body was detaching itself from home. The wound was fresh but not unhealthy. The seeds of oblivion had already been sown. She did not wash her body. She was afraid of removing the shawl from her loins. The sour smell grew worse.

"You must wash yourself," a voice whispered.

"I'm afraid."

"You must wash yourself," the voice repeated. In the afternoon, without taking off her dress, she stepped into the river. The water seeped into her until she felt it burn. And immediately drops of blood rose to the surface of the water and surrounded her. She gazed at them in astonishment. Afterward she lay on the ground.

The water was good for her, but not the fruit. In these early days she did not yet know how to distinguish between red and red, between black and black. She plucked whatever came to hand, blackberries and raspberries, strawberries and cherries. In the evening she had severe pain in her stomach and diarrhea. Her slender legs could not stand up to the pain and they gave way beneath her. "God, God." The words escaped her lips. Her voice disappeared into the lofty greenness. If she had had the strength, she would have crawled into the village and given herself up.

"What are you doing here?"

She was suddenly startled by a peasant's voice.

"I'm ill."

"Who do you belong to?"

"Maria."

The peasant stared at her in disgust, pursed his mouth, and turned away without another word.

6

Autumn was already at its height, and in the evenings the horizon was blue with cold. Tzili would find shelter for the night in deserted barns and stables. From time to time she would approach a farmhouse and ask for a piece of bread. Her clothes gave off a bad, moldy smell and her face was covered with a rash of little pimples.

She did not know how repulsive she looked. She roamed the outskirts of the forest and the peasants who crossed her path averted their eyes. When she approached farmhouses to beg for bread the housewives would chase her away as if she were a mangy dog. "Here's Maria's daughter," she would hear them say. Her ugly existence became a byword and a cautionary tale in the mouths of the local peasants, but the passing days were kind to her, molding her in secret, at first deadening and then quickening her with new life. The sick blood poured out of her. She learned to walk barefoot, to bathe in the icy water, to tell the edible berries from the poisonous ones, to climb the trees. The sun worked wonders with her. The visions of the night gradually left her. She saw only what was in front of her eyes, a tree, a puddle, the autumn leaves changing color.

For hours she would sit and gaze at the empty fields sinking slowly into grayness. In the orchards the leaves turned red. Her life seemed to fall away from her, she coiled in on herself like a cocoon. And at night she fell unconscious onto the straw.

One day she came across a hut on the fringes of the forest. Autumn was drawing to a close. It rained and hailed incessantly, and the frost ate into her bones. But she was no longer afraid of anyone, not even the wild dogs.

A woman opened the door and said: "Who are you?"

"Maria's daughter," said Tzili.

"Maria's daughter! Why are you standing there? Come inside!"

The woman seemed thunderstruck. "Maria's daughter, barefoot in this frost! Take off your clothes. I'll give you a gown."

Tzili took off her mildewed clothes and put on the gown. It was a fancy city gown, flowered and soaked in perfume. After many months of wandering, she had a roof over her head.

"Your mother and I were young together once, in the city. Fate must have brought you to my door."

Tzili looked at her from close up: a woman no longer young, with frizzy hair and prominent cheekbones.

"And what is your mother doing now?"

Tzili hesitated a moment and said: "She's at home."

"My name is Katerina," said the woman. "If you see your mother tell her you saw Katerina. She'll be very glad to hear it. We had a lot of good times together in the city, especially with the Jews."

Tzili trembled.

"The Jews are great lovers. Ours aren't a patch on them, I can tell you that—but we were fools then, we came back to the village to look for husbands. We were young and afraid of our fathers. Jewish lovers are worth their weight in gold. Let me give you some soup," said Katerina and hurried off to fetch a bowl of soup.

After many days of wandering, loneliness, and cold, she took in the hot liquid like a healing balm.

Katerina poured herself a drink and immediately embarked on reminiscences of her bygone days in the city, when she and Maria had queened it with the Jews, at first as chambermaids and later as mistresses. Her voice was full of longing.

"The Jews are gentle. The Jews are generous and kind. They know how to treat a woman properly. Not like our men, who don't know anything except how to beat us up." In the course of the years she had learned a little Yiddish, and she still remembered a few words—the word *dafka,* * for instance.

Tzili felt drawn into the charmed circle of Katerina's memories. "Thank you," she said.

"You don't have to thank me, girl," scolded Katerina. "Your mother and I were good friends once. We sat in the same cafés together, made love to the same man."

Katerina poured herself drink after drink. Her high cheekbones stuck out and her eyes peered into the distance with a birdlike sharpness. Suddenly she said: "The Jews, damn them, know how to give a woman what she needs. What does a woman need, after all? A little kindness, money, a box of chocolates every now and then, a bed to lie on. What more does a woman need? And what have I got now? You can see for yourself.

"Your mother and I were fools, stupid fools. What's there to be afraid of? I'm not afraid of hell. My late mother never stopped nagging me: Katerina, why don't you get married? All the other girls are getting married. And I like a fool listened to her. I'll never forgive her. And you." She turned to Tzili with a piercing look. "You don't get married, you hear me? And don't bring any little bastards into the world either. Only the Jews, only the Jews—they're the only ones who'll take you out to cafés, to restaurants, to the cinema. They'll always take you to a clean hotel, only the Jews."

Tzili no longer took in the words. The warmth and the scented gown cradled her: her head dropped and she fell asleep.

7

From the first day Tzili knew what was expected of her. She swept the floor and washed the dishes, she hurried to peel the potatoes. No work was too arduous for her. The months out of doors seemed to have taught her what it meant to serve others. She never left a job half done and she never got mixed up. And whenever it stopped raining she would take the skinny old cow out to graze.

*Untranslatable Yiddish (and Hebrew) word referring to a stubborn, willful refusal to comply. "Just because!" is a loose rendering.

Katerina lay in her bed wrapped in goat skins, coughing and sipping tea and vodka by turn. From time to time she rose and went to stand by the window. It was a poor house with a dilapidated stable beside it. And in the yard: a few pieces of wood, a gaping fence, and a neglected vegetable patch. These were the houses outside the village borders, where the lepers and the lunatics, the horse thieves and the prostitutes lived. For generations one had replaced the other here, without repairing the houses or cultivating the plots. The passing seasons would knead such places in their hands until they could not be told apart from abandoned forest clearings.

In the evening the softness would come back to her voice and she would speak again of the days in the city when she and Maria walked the streets together. What was left of all that now? She was here and they were there. In the city a thousand lights shone—and here she was surrounded by mud, madmen and lepers.

Sometimes she put on one of her old dresses, made up her face, stood by the window, and announced: "Tomorrow I'm leaving. I'm sick of this. I'm only forty. A woman of forty isn't ready for the rubbish dump yet. The Jews will take me as I am. They love me."

Of course, these were hallucinations. Nobody came to take her away. Her cough gave her no rest, and every now and then she would wake the sleeping Tzili and command her: "Make me some tea. I'm dying." At night, when a fit of coughing seized her, her face grew bitter and malign, and no one escaped the rough edge of her tongue, not even the Jews.

Once in a while an old customer appeared and breathed new life into the hut. Katerina would get dressed, make up her face, and douse herself with perfume. She liked the robust peasants, who clutched her round the waist and crushed her body to them. Her old voice would come back to her, very feminine. All of a sudden she would be transformed, laughing and joking, reminiscing about times gone by. And she would reprimand Tzili too, and instruct her: "That's not the way to offer a man a drink. A man likes his vodka first, bread later." Or: "Don't cut the sausage so thin."

But such evenings were few and far between. Katerina would wrap herself in blankets and whimper in a sick voice: "I'm cold. Why don't you make the fire properly? The wood's wet. This wetness is driving me out of my mind."

Tzili learned that Katerina was a bold, hot-tempered woman. Knives and axes had no fears for her. At the sight of an unsheathed knife all her beauty burst forth. With drunks she was gentle, speaking to them in a tender and maternal voice.

Although their houses were far apart, Katerina was at daggers drawn with her neighbors. Once a day the leper would emerge from his house and curse Katerina, yelling at the top of his voice. And when he started walking toward her door, Katerina would rush out to meet him like a maddened dog. He was a big peasant, his body pink all over from the disease.

Winter came, and snow. Tzili went far into the forest to gather firewood. When she came back in the evening with a bundle of twigs on her shoulders bigger than she was, Katerina was still not satisfied. She would grumble: "I'm cold. Why didn't you bring thicker branches? You're spoiled. You need a good hiding. I took you in like a mother and you're shirking your work. You're like your mother. She only looked out for herself. I'm going to give you a good beating."

Of Katerina's plans for her Tzili had no inkling. Her life was one of labor, oblivion, and uncomprehending delight. She delighted in the hut, the faded feminine objects, and the scents that frequently filled the air. She even delighted in the emaciated cow.

From time to time Katerina gave her significant looks: "Your breasts are growing. But you're still too skinny. You should eat more potatoes. How old are you? At your age I was already on the streets." Or sometimes in a maternal voice: "Why don't you comb your hair? People are coming and your hair's not combed."

Winter deepened and Katerina's cough never left her for an instant. She drank vodka and boiling-hot tea, but the cough would not go away. From night to night it grew harsher. She would wake Tzili up and scold her: "Why don't you bring me a glass of tea? Can't you hear me coughing?" Tzili would tear herself out of her sleep and hasten to get Katerina a glass of tea.

It was a long winter and Katerina never stopped grumbling and cursing her sisters, her father, and all the seekers of her favors who had devoured her body. Her face grew haggard. She could no longer stand on her feet. There were no more visitors. The only ones who still came were drunk or crazy. At first she tried to pretend, but now it was no longer possible to hide her illness. The men fled from the house. Katerina accompanied their flight with curses. But the worst of her rage she spent on Tzili. From time to time she threw a plate or a pot at her. Tzili absorbed the blows in silence. Once Katerina said to her: "At your age I was already keeping my father."

8

Spring came and Katerina felt better. Tzili made up a bed for her outside the door. Now too she kept up a constant stream of abuse, but to Tzili she spoke mildly: "Why don't you go and wash yourself? There's a mirror in the house. Go and comb your hair." And once she even offered her one of her scented creams. "A girl of your age should perfume her neck."

Tzili worked without a break from morning to night. She ate whatever she could lay her hands on: bread, milk, and vegetables from the garden. Her day was full to overflowing. And at night she fell onto her bed like a sack.

No one came to ask for Katerina's favors any more and her money ran out. Even the male nurse, who pulled out two of her teeth, failed to collect his due. Katerina would stand slumped in the doorway.

One evening she asked Tzili: "Have you ever been to bed with a man?"

"No." Tzili shuddered.

"And don't you feel the need? At your age," said Katerina, with almost maternal tenderness, "I had already known many men. I was even married."

"Did you have any children?" asked Tzili.

"I did, but I gave them away when they were babies."

Tzili asked no more. Katerina's face was angry and bitter. She realized that she shouldn't have asked.

Summer came and there was no end to Katerina's complaints. She would speak of her youth, of her lovers, of the city and of money. Now she hardened her heart toward her Jewish lovers too and abused them roundly. The accusations poured out of her in a vindictive stream, scrambled up with fantasies and wishes. Every now and then she would get up and hurl a plate across the room, and the walls would shake with the clatter and the curses. Tzili's movements grew more and more confined, and the old fear came back to her.

From time to time Katerina would berate her: "At your age I was already keeping my father, and you . . ."

"What do you want me to do, mother?"

"Do you have to ask? I didn't have to ask my father. I went to the city and sent him money every month. A daughter has to look after her parents. I let those guzzlers devour my body."

Tzili's heart was full of foreboding. She guessed that something bad was going to happen, but she didn't know what. Her happiest hours were the ones she spent in the meadows grazing the cow. The air and light kneaded her limbs with a firm and gentle touch. From time to time she would take off her clothes and bathe in the river.

Katerina watched her with an eagle eye: "Your health is improving every day and I'm being eaten up with illness." Her back was very bent, and without her front teeth her face had a ghoulish, nightmare look.

One evening an old client of Katerina's came to call, a burly middle-aged peasant. Katerina was lying in bed.

"What's the matter with you?" he asked in surprise.

"I'm resting. Can't a woman rest?"

"I just wanted to say hello," he said, retreating to the door.

"Why not stay a while and have a drink?"

"I've already had more than enough."

"Just one little drink."

"Thanks. I just dropped in to say hello."

Suddenly she raised herself on her elbows, smiled and said: "Why don't you take the little lass to bed? You won't be sorry."

The peasant turned his head with a dull, slow movement, like an animal, and an embarrassed smile appeared on his lips.

"She may be small but she's got plenty of flesh on her bones." Katerina coaxed him. "You can take my word for it."

Tzili was standing in the scullery. The words were quite clear. They sent a shiver down her spine.

"Come here," commanded Katerina. "Show him your thighs."

Tzili stood still.

"Pick up your dress," commanded Katerina.

Tzili picked up her dress.

"You see, I wasn't lying to you."

The peasant dropped his eyes. He examined Tzili's legs. "She's too young," he said.

"Don't be a fool."

Tzili stood holding her dress fearfully in her hands.

"I'll come on Sunday," said the peasant.

"She's got breasts already, can't you see?"

"I'll come on Sunday," repeated the peasant.

"Go then. You're a fool. Any other man would jump at the chance."

"I don't feel like it today. I'll come on Sunday."

But he lingered in the doorway, measuring the little girl with his eyes, and for a moment he seemed about to drag her into the scullery. In the end he recovered himself and repeated: "I'll come on Sunday."

"You fool," said Katerina with an offended air, as if she had offered him a tasty dish and he had refused to eat it. And to Tzili she said: "Don't stand there like a lump of wood."

Tzili dropped her dress.

For a moment longer Katerina surveyed the peasant with her bloodshot eyes. Then she picked up a wooden plate and threw it. The plate hit Tzili and she screamed. "What are you screaming about? At your age I was already keeping my father."

The peasant hesitated no longer. He picked up his heels and ran.

Now Katerina gave her tongue free rein, abusing and cursing everyone, especially Maria. Tzili's fears were concentrated on the sharp knife lying next to the bed. The knife sailed through the air and hit the door. Tzili fled.

9

The night was full and starless. Tzili walked along the paths she now knew by heart. For some reason she kept close to the river. On either side, the cornfields stretched, broad and dark. "I'll go on," she said, without knowing what she was saying.

She had learned many things during the past year: how to launder clothes, wash dishes, offer a man a drink, collect firewood, and pasture a cow, but above all she had learned the virtues of the wind and the water. She knew the north wind and the cold river water. They had kneaded her from within. She had grown taller and her arms had grown strong. The further she walked from Katerina's hut the more closely she felt her presence. As if she were still standing in the scullery. She felt no resentment toward her.

"I'll go on," she said, but her legs refused to move.

She remembered the long, cozy nights at Katerina's. Katerina lying in bed and weaving fantasies about her youth in the city, parties and lovers. Her face calm and a smile on her lips. When she spoke about the Jews her smile narrowed and grew more modest, as if she were revealing some great secret. It seemed then as if she acquiesced in everything, even in the disease devouring her body. Such was life.

Sometimes too she would speak of her beliefs, her fear of God and his Messiah, and at these moments a strange light seemed to touch her face. Her mother and father she could not forgive. And once she even said: "Pardon me for not being able to forgive you."

Tzili felt affection even for the old, used objects Katerina had collected over the years. Gilt powder boxes, bottles of eau de cologne, crumpled silk petticoats and dozens of lipsticks—these objects held an intimate kind of magic.

And she remembered too: "Have you ever been to bed with a man?"

"No."

"And don't you feel the need?"

Katerina's face grew cunning and wanton.

And on one of the last days Katerina asked: "You won't desert me?"

"No," promised Tzili.

"Swear by our Lord Saviour."

"I swear by our Lord Saviour."

Of the extent to which she had been changed by the months with Katerina, Tzili was unaware. Her feet had thickened and she now walked surely over the hard ground. And she had learned something else too: there were men and there were women and between them there was an eternal enmity. Women could not survive save by cunning.

Sometimes she said to herself: I'll go back to Katerina. She'll forgive me. But when she turned around her legs froze. It was not the knife itself she feared but the glitter of the blade.

Summer was at its height, and there was no rain. She lived on the fruit growing wild on the river banks. Sometimes she approached a farmhouse.

"Who are you?"

"Maria's daughter."

Maria's reputation had reached even these remote farmhouses. At the sound of her name, a look of loathing appeared on the faces of the farmers' wives. Sometimes they said in astonishment: "You're Maria's daughter!" The farmers themselves were less severe: in their youth they had availed themselves freely of Maria's favors, and in later years too they had occasionally climbed into her bed.

And one day, as she stood in a field, the old memory came back to confront her: her father lying on his sickbed, the sound of his sighs rending the air, her mother in the shop struggling with the violent peasants. Blanca as always, under the shadow of the impending examinations, a pile of books and papers on her table. And in the middle of the panic, the bustle, and the hysteria, the clear sound of her father's voice: "Where's Tzili?"

"Here I am."

"Come here. What mark did you get in the arithmetic test?"

"I failed, father."

"You failed again."

"This time Blanca helped me."

"And it didn't do any good. What will become of you?"

"I don't know."

"You must try harder."

Tzili shuddered at the clear vision that came to her in the middle of the field. For a moment she stood looking around her, and then she picked up her feet and began to run. Her panic-stricken flight blurred the vision and she fell spread-eagled onto the ground. The field stretched yellow-gray around her without a soul in sight.

"Katerina," she said, "I'm coming back to you." As soon as the words were out of her mouth she saw the burly peasant in front of her, examining her thighs as she lifted up her skirt. Now she was no longer afraid of him. She was afraid of the ancient sights pressing themselves upon her with a harsh kind of clarity.

10

In the autumn she found shelter with an old couple. They lived in a poor hut far from everything.

"Who are you?" asked the peasant.

"Maria's daughter."

"That whore," said his wife. "I don't want her daughter in the house."

"She'll help us," said the man.

"No bastard is going to bring us salvation," grumbled the wife.

"Quiet, woman." He cut her short.

And thus Tzili found a shelter. Unlike Katerina's place, there were no luxuries here. The hut was composed of one long room containing a stove, a rough wooden table, and two benches. In the corner, a couple of stools. And above the stools a Madonna carved in oak, as simple as the work of a child.

It was a long, gray autumn, and on the monotonous plains everything seemed made of mud and fog. Even the people seemed to be made of the same substance: rough and violent, their tongue that of the pitchfork and cattle prod. The wife would wake her while it was still dark and push her outside with grunts: go milk the cows, go take them to the meadow.

The long hours in the meadows were her own. Her imagination did not soar but the little she possessed warmed her like soft, pure wool. Katerina, of course. In this gray place her former life with Katerina seemed full of interest. Here there were only cows, cows and speechlessness. The man and his wife communicated in grunts. If they ran short of milk or wood for the fire, the wife never asked why but brandished the rope as a sign that something was amiss.

Here for the first time she felt the full strength of her arms. At Katerina's they had grown stronger. Now she lifted the pitchfork easily into the air. The columns of her legs too were full of muscles. She ate whatever she could lay her hands on, heartily. But life was not as simple as she imagined. One night she awoke to the touch of a hand on her leg. To her surprise it was the old man. The old woman climbed out of bed after him shouting: "Adulterer!" And he returned chastised to his bed.

This was all the old woman was waiting for. After that she spoke to Tzili like a stray mongrel dog.

It was the middle of winter and the days darkened. The snow piled up in the doorway and barred their way out. Tzili sat for hours in the stable with the cows. She sensed the thin pipes joining her to these dumb worlds. She did not know what one said to cows, but she felt the warmth emanating from their bodies seeping into her. Sometimes she saw her mother in the shop struggling with hooligans. A woman without fear. In this dark stable everything seemed so remote—was more like a previous incarnation than her own life.

Between one darkness and the next the old woman would beat Tzili. The bastard had to be beaten so that she would know who she was and what she had to do to mend her ways. The woman would beat her fervently, as if she were performing some secret religious duty.

When spring comes I'll run away, Tzili would say to herself on her bed at night. Or: Why did I ever leave Katerina? She was good to me. Now she felt a secret affection for Katerina's hut, as if it were not a miserable cottage but an enchanted palace.

Sometimes she would hear her voice saying, "The Jews are weak, but they're gentle too. A Jew would never strike a woman." This mystery seemed to melt into Tzili's body and flood it with sweetness. At times like these her mind would shrink to next to nothing and she would be given over entirely to sensations. When she heard Katerina's voice she would curl up and listen as if to music.

But the old man could not rest, and every now and then he would dart out of bed and try to reach her. And once, in his avidity, he bit her leg, but the old woman was too quick for him and dragged him off before he could go any further. "Adulterer!" she cried.

Sometimes he would put on an expression of injured innocence and say: "What harm have I done?"

"Your evil thoughts are driving you out of your mind."

"What have I done?"

"You can still ask!"

"I swear to you . . ." The old man would try to justify himself.

"Don't swear. You'll roast in hell!"

"Me?"

"You, you rascal."

The winter stretched out long and cold, and the grayness changed from one shade to another. There was nowhere to hide. It seemed that the whole universe was about to sink beneath the weight of the black snow. Once the old woman asked her: "How long is it since you saw your mother?"

"Many years."

"It was from her that you learned your wicked ways. Why are you silent? You can tell us. We know your mother only too well. Her and all her scandals. Even I had to watch my old man day and night. Not that it did me any good. Men are born adulterers. They'd find a way to cheat on their wives in hell itself."

Toward the end of winter the old woman lost control of herself. She beat Tzili indiscriminately. "If I don't make her mend her ways, who will?" She beat her devoutly with a wet rope so that the strokes would leave their mark on her back. Tzili screamed with pain, but her screams did not help her. The old woman beat her with extraordinary strength. And once, when the old man tried to intervene, she said: "You'd better shut up or I'll beat you too. You old lecher. God will thank me for it." And the old man, who usually gave back as good as he got, kept quiet. As if he had heard a warning voice from on high.

11

When the snow began to thaw she fled. The old woman guessed that she was about to escape and kept muttering to herself: "As long as she's here I'm going to teach her a lesson she'll never forget. Who knows what she's still capable of?"

Now Tzili was like a prisoner freed from chains. She ran. The heads of the mountains were still capped with snow, but in the black valleys below, the rivers flowed loud and torrential as waterfalls.

Her body was bruised and swollen. In the last days the old woman had whipped her mercilessly. She had whipped her as if it were her solemn duty to do so, until in the end Tzili too felt that she was only getting what she deserved.

But for the mud she would have walked by the riverside. She liked walking on the banks of the river. For some reason she believed that nothing bad would happen to her next to the water, but she was obliged to walk across the bare mountainside, washed by the melted snow. The valleys were full of mud.

She came to the edge of a forest. The fields spreading below it steamed in the sun. She sat down and fell asleep. When she woke the sun was on the other side of the horizon, low and cold.

She tried to remember. She no longer remembered anything. The long winter had annihilated even the little memory she possessed. Only her feet sensed the earth as they walked. She knew this piece of ground better than her own body. A strange, uncomprehending sorrow suddenly took hold of her.

She took the rags carefully off her feet and then bound them on again. She treated her feet with a curious solemnity. It did not occur to her to ask what would happen when darkness fell. The sun was sinking fast on the horizon. For some reason she remembered that Katerina had once said to her, in a rare moment of peace: "Women are lucky. They don't have to go to war."

Now she felt detached from everyone. She had felt the same thing before, but not in the same way. Sometimes she would imagine that someone was waiting for her, far away on the horizon. And she would feel herself drawn toward it. Now she seemed to understand instinctively that there was no point going on.

As she sat staring into space, a sudden dread descended on her. What is it? she said and rose to her feet. There was no sound but for the gurgle of the water. On the leafless trees in the distance a blue light flickered.

It occurred to her that this was her punishment. The old woman had said that many punishments were in store for her. "There's no salvation for bastards!" she would shriek.

"What have I done wrong?" Tzili once asked uncautiously.

"You were born in sin," said the old woman. "A woman born in sin has to be cleansed, she has to be purified."

"How is that done?" asked Tzili meekly.

"I'll help you," said the old woman.

That night she found shelter in an abandoned shed. It was cold and her body was sore, but she was content, like a lost animal whose neck has been freed from its yoke at last. She slept for hours on the damp straw. And in her dreams she saw Katerina, not the sick Katerina but the young Katerina. She was wearing a transparent dress, sitting by a dressing table, and powdering her face.

12

When she woke it was daylight. Scented vapors rose from the fields. And while she was sitting there a man seemed to come floating up from the depths of the earth. For a moment they measured each other with their eyes. She saw immediately: he was not a peasant. His city suit was faded and his face exhausted.

"Who are you?" he asked in the local dialect. His voice was weak but clear.

"Me?" she asked, startled.

"Where are you from?"

"The village."

This reply confused him. He turned his head slowly to see if anyone was there. There was no one. She smelled the stale odor of his mildewed clothes.

"And what are you doing here?"

She raised herself slightly on her hands and said: "Nothing."

The man made a gesture with his hand as if he was about to turn his back on her. But then he said: "And when are you going back there?"

"Me?"

Now it appeared that the conversation was over. But the man was not satisfied. He stroked his coat. He seemed about forty and his hands were a grayish white, like the hands of someone who had not known the shelter of a man-made roof for a long time.

Tzili rose to her feet. The man's appearance revolted her, but it did not frighten her. His soft flabbiness.

"Haven't you got any bread?" he asked.

"No."

"And no sausage either?"

"No."

"A pity. I would have given you money for them," he said and turned to go. But he changed his mind and said in a clear voice: "Haven't you got any parents?"

This question seemed to startle her. She took a step backward and said in a weak voice: "No."

Her reply appeared to excite the stranger, and he said with a kind of eagerness: "What do you say?" The trace of a crooked smile appeared on his gray-white face.

"So you're one of us."

There was something repulsive about his smile. Her body shrank and she recoiled. As if some loathsome reptile had crossed her path. "Tell me," he pressed her, standing his ground. "You're one of us, aren't you?"

For a moment she wanted to say no and run away, but her legs refused to move.

"So you're one of us," he said and took a few steps toward her. "Don't be afraid. My name's Mark. What's yours?"

He took off his hat, as if he wished to indicate with this gesture not only respect but also submission. His bald head was no different from his face, a pale gray.

"How long have you been here?"

Tzili couldn't open her mouth.

"I've lost everyone. I'd made up my mind to die tonight." Even this sentence, which was spoken with great emotion, did not move her. She stood frozen, as if she were caught up in an incomprehensible nightmare. "And you, where are you from? Have you been wandering for long?" he continued rapidly, in Tzili's mother tongue, a mixture of German and Yiddish, and with the very same accent.

"My name is Tzili," said Tzili.

The man seemed overcome. He sank onto his knees and said: "I'm glad. I'm very glad. Come with me. I have a little bread left."

Evening fell. The fruit trees on the hillside glowed with light. In the forest it was already dark.

"I've been here a month already," said the man, composing himself. "And in all that time I haven't seen a soul. What about you? Do you know anybody?" He spoke quickly, swallowing his words, getting out everything he had stored up in the long, cold days alone. She did not understand much, but one thing she understood: in all the countryside around them there were no Jews left.

"And your parents?" he asked.

Tzili shuddered. "I don't know, I don't know. Why do you ask?"

The stranger fell silent and asked no more.

In his hideout, it transpired, he had some crusts of bread, a few potatoes, and even a little vodka.

"Here," he said, and offered her a piece of bread.

Tzili took the bread and immediately sank her teeth into it.

The stranger looked at her for a long time, and a crooked smile spread over his face. He sat cross-legged on the ground. After a while he said: "I couldn't believe at first that you were Jewish. What did you do to change yourself?"

"Nothing."

"Nothing, what do you say? I will never be able to change. I'm too old to change, and to tell the truth I don't even know if I want to."

Later on he asked: "Why don't you say anything?" Tzili shivered. She was no longer accustomed to the old words, the words from home. She had never possessed an abundance of words, and the months she had spent in the company of the old peasants had cut them off at the roots. This stranger, who had brought the smell of home back to her senses, agitated her more than he frightened her.

When it grew dark he lit a fire. He explained: the entire area was surrounded by swamps. And now with the thawing of the snow it would be inaccessible to their enemies. It was a good thing that the winter was over. There was a practical note now in his voice. The suffering seemed to have vanished from his face, giving way to a businesslike expression. There was no anger or wonder in it.

13

When she woke there was light in the sky and the man was still sitting opposite her, in the same position. "You fell asleep," he said. He rose to his feet and his whole body was exposed: medium height, a worn-out face, and a crumpled suit, very faded at the knees. A few spots of grease. Swollen pockets.

"Ever since I escaped from the camp I haven't been able to sleep. I'm afraid of falling asleep. Are you afraid too?"

"No," said Tzili simply.

"I envy you."

The signs of spring were everywhere. Rivulets of melted snow wound their way down the slopes, dragging gray lumps of ice with them. There was not a soul to be seen, only the sound of the water growing louder and louder until it deafened them with its roar.

He looked at her and said: "If you hadn't told me, I'd never have guessed that you were Jewish. How did you do it?"

"I don't know. I didn't do anything."

"If I don't change they'll get me in the end. Nothing will save me. They won't let anyone escape. I once saw them with my own eyes hunting down a little Jewish child."

"And do they kill everyone?" Tzili asked.

"What do you think?" he said in an unpleasant tone of voice.

His face suddenly lost all its softness and a dry, bitter expression came over his lips. Her uncautious question had apparently angered him.

"And where were you all the time?" he demanded.

"With Katerina."

"A peasant woman?"

"Yes."

He dropped his head and muttered to himself. Apparently in anger, and also perhaps regret. His cheekbones projected, pulling the skin tight.

"And what did you do there?" He went on interrogating her.

"I worked."

"And did she know that you were Jewish?"

"No."

"Strange."

In the afternoon he grew restless and agitated. He ran from tree to tree, beating his head with his fists and reproaching himself: "Why did I run away? Why did I have to run away? I abandoned them all and ran away. God will never forgive me."

Tzili saw him in his despair and said nothing. The old words which had begun to stir in her retreated even further. In the end she said, for some reason: "Why are you crying?"

"I'm not crying. I'm angry with myself."

"Why?"

"Because I'm a criminal."

Tzili was sorry for asking and she said: "Forgive me."

"There's nothing to forgive."

Later on he told her. He had escaped and left his wife and two children behind in the camp. He had tried to drag them too through the narrow aperture he had dug with his bare hands, but they were afraid. She was, his wife.

And while he was talking it began to rain. They found a shelter under the branches. The man forgot his despair for a moment and spread a tattered blanket over the branches. The rain stopped.

"And did you too leave everyone behind?" he asked.

Tzili said nothing.

"Why don't you tell me?"

"Tell you what?"

"How you got away?"

"My parents left me behind to look after the house. They promised to come back. I waited for them."

"And ever since then you've been wandering?"

For some reason he tore off a lump of bread and offered her a piece.

She gnawed it without a word.

"The bread should be heated up. It's wet."

"It doesn't matter."

"Don't you suffer from pains in your stomach?"

"No."

"I suffer terribly from pains in my stomach."

The rain stopped and a blue-green light floated above the horizon. The gurgling of the water had given way to a steady flow. The man washed his face in the rivulet and said: "How good it is. Why don't you wash your face in the water too?"

Tzili took a handful of water and washed her face.

They sat silently by the little stream. Tzili felt that her life had led her to a new destination, it too unknown. The closeness of the man did not excite her, but his questions upset her. Now that he had stopped asking she felt better.

Suddenly he raised his eyes from the water and said: "Why don't you go down to the village and bring us something to eat? We have nothing to eat. The little we had is gone."

"All right, I'll go," she said.

"And you won't forget to come back?"

"I won't forget," she said, blushing.

Immediately he corrected himself and said: "You can buy whatever you want, it doesn't matter, as long as it's something to fill our bellies. I'd go myself, and willingly, but I'd be found out. It's a pity I haven't got any other clothes. You understand."

"I understand," said Tzili submissively.

"I'd go myself if I could," he said again, in a tone which was at once ingratiating and calculating. "You, how shall I put it, you've changed, you've changed for the better. Nobody would ever suspect you. You say your r's exactly like they do. Where do you get it all from?"

"I don't know."

Now there was something frightening in his appearance. As if he had risen from his despair another man, terrifyingly practical.

14

Early in the morning she set out. He stood watching her receding figure for a long time. Once again she was by herself. She knew that the stranger had done something to her, but what? She walked for hours, looking for ways around the melted snow, and in the end she found an open path, paved with stones.

A woman was standing next to one of the huts, and Tzili addressed her in the country dialect: "Have you any bread?"

"What will you give me for it?"

"Money."

"Show me."

Tzili showed her.

"And how much will I give you for it?"

"Two loaves."

The old peasant woman muttered a curse, went inside, and emerged immediately with two loaves in her hands. The transaction was over in a moment.

"Who do you belong to?" she remembered to ask.

"To Maria."

"Maria? *Tfu.*" The woman spat. "Get out of my sight."

Tzili clasped the bread in both hands. The bread was still warm, and it was only after she had walked for some distance that the tears gushed out of her eyes. For the first time in many days she saw the face of her mother, a face no longer young. Worn with work and suffering. Her feet froze on the ground, but as in days gone by she knew that she must not stand still, and she continued on her way.

The trees were putting out leaves. Tzili jumped over the puddles without getting wet. She knew the way and weaved between the paths, taking shortcuts and making detours like a creature native to the place. She walked very quickly and arrived before evening fell. Mark was sitting in his place. His tired, hungry eyes had a dull, indifferent look.

"I brought bread," she said.

Mark roused himself: "I thought you were lost." He fell on the bread and tore it to shreds with his teeth, without offering any to Tzili. She observed him for a moment: his eyes seemed to have come alive and all his senses concentrated on chewing.

"Won't you have some too?" he said when he was finished eating.

Tzili stretched out her hand and took a piece of bread. She wasn't hungry. The long walk had tired her into a stupor. Her tears too had dried up. She sat without moving.

Mark passed his right hand over his mouth and said: "A cigarette, if only I had a cigarette."

Tzili made no response.

He went on: "Without cigarettes there's no point in living." Then he dug his nails into the ground and began singing a strange song. Tzili remembered the melody but she couldn't understand the words.

Gradually his voice lost its lilt and the song trailed off into a mutter.

The evening was cold and Mark lit a fire. During the long days of his stay here he had learned to make fire from two pieces of flint and a thread of wool which he plucked from his coat. Tzili marveled momentarily at his dexterity. The agitation faded from his face and he asked in a practical tone of voice: "How did you get the bread? Fresh bread?"

Tzili answered him shortly.

"And they didn't suspect you?"

For a long time they sat by the little fire, which gave off a pleasant warmth.

"Why are you so silent?"

Tzili hung her head, and an involuntary smile curved her lips.

The craving for cigarettes did not leave him. The fresh bread had given him back his taste for life, but he lost it again immediately. For hours he sat nibbling blades of grass, chewing them up and spitting them to one side. He had a tense, bitter look. From time to time he cursed himself for being a slave to his addiction. Tzili was worn out and she fell asleep where she sat.

15

When she woke she kept her eyes closed. She felt Mark's eyes on her. She lay without moving. The fire had not gone out, which meant that Mark had not slept all night.

When she finally opened her eyes it was already morning. Mark asked: "Did you sleep?" The sun rose in the sky and the horizons opened out one after the other until the misty plains were revealed in the distance. Here and there they could see a peasant ploughing.

"It's a good place," said Mark. "You can see a long way from here." The agitation had faded from his face, and a kind of complacency that did not suit him had taken its place. Tzili imagined she could see in him one of the Jewish salesmen who used to drop into her mother's shop. Mark asked her: "Did you go to school?"

"Yes."

"A Jewish school?"

"No. There wasn't one. I studied Judaism with an old teacher. The Pentateuch and prayers."

"Funny," he said, "it sounds so far away. As if it never happened. And do you still remember anything?"

"Hear, O Israel."

"And do you recite it?"

"No," she said and hung her head.

"In my family we weren't observant any more," said Mark in a whisper. "Was your family religious?"

"No, I don't think so."

"You said they brought you a teacher of religion."

"It was only for me, because I didn't do well at school. My brothers and sisters were all good at school. They were going to take external examinations."

"Strange," said Mark.

"I had trouble learning."

"What does it matter now?" said Mark. "We're all doomed anyway."

Tzili did not understand the word but she sensed that it held something bad.

After a pause Mark said: "You've changed very nicely, you've done it very cleverly. I can't imagine a change like that taking place in me. Even the forests won't change me now."

"Why?" asked Tzili.

"Because everything about me gives me away—my appearance, from top to toe, my nose, my accent, the way I eat, sit, sleep, everything. Even though I've never had anything to do with what's called Jewish tradition. My late father used to call himself a free man. He was fond of that phrase, I remember, but here in this place I've discovered, looking at the peasants ploughing in the valley, their serenity, that I myself—I won't be able to change anymore. I'm a coward. All the Jews are cowards and I'm no different from them. You understand."

Tzili understood nothing of this outburst, but she felt the pain pouring out of the words and she said: "What do you want to do?"

"What do I want to do? I want to go down to the village and buy myself a packet of tobacco. That's all I want. I have no greater desire. I'm a nervous man and without cigarettes I'm an insect, less than an insect, I'm nothing."

"I'll buy it for you."

"Thank you," said Mark, ashamed. "Forgive me. I have no more money. I'll give you a coat. That's good, isn't it?"

"Yes, that's good," said Tzili. "That's very good."

In the tent of branches he had a haversack full of things. He spread them out now on the ground to dry. His clothes, his wife's and children's clothes. He spread them out slowly, like a merchant displaying his wares on the counter.

Tzili shuddered at the sight of the little garments spotted with food stains. Mark spread them out without any order and they steamed and gave off a stench of mildew and sour-sweet. "We must dry them," said Mark in a businesslike tone. "Otherwise they'll rot." He added: "I'll give you my coat. It's a good coat, pure wool. I bought it a year ago. I hope you'll be able to get me some cigarettes for it. Without cigarettes to smoke I get very nervous."

Strange, his nervousness was not apparent now. He stood next to the steaming clothes, turning them over one by one, as if they were pieces of meat on a fire. Tzili too did not take her eyes off the stained children's clothes shrinking in the sun.

Toward evening he gathered the clothes up carefully and folded them. The coat intended for selling he put aside. "For this, I hope, we'll be able to get some tobacco. It's a good coat, almost new," he muttered to himself.

That night Mark did not light a fire. He sat and sucked soft little twigs. Chewing the twigs seemed to blunt his craving for cigarettes. Tzili sat not far from him, staring into the darkness.

"I wanted to study medicine," Mark recalled, "but my parents didn't have the money to send me to Vienna. I sat for external matriculation exams and my marks weren't anything to write home about, only average. And then I married very young, too young I'd say. Of course, nothing came of my plans to study. A pity."

"What's your wife's name?" asked Tzili.

"Why do you ask?" said Mark in surprise.

"No reason."

"Blanca."

"How strange," said Tzili. "My sister's name is Blanca too."

Mark rose to his feet. Tzili's remark had abruptly stopped the flow of his memories. He put his hands in his trouser pocket, stuck out his chest, and said: "You must go to sleep. Tomorrow you have a long walk in front of you."

The strangeness of his voice frightened Tzili and she immediately got up and went to lie down on the pile of leaves.

16

She slept deeply, without feeling the wind. When she woke a mug of hot herb tea was waiting for her.

"I couldn't sleep," he said.

"Why can't you sleep?"

"I can't fall asleep without a cigarette."

Tzili put the coat into a sack and rose to her feet.

Mark sat in his place next to the fire. His dull eyes were bloodshot from lack of sleep. For some reason he touched the sack and said: "It's a good coat, almost new."

"I'll look after it," said Tzili without thinking, and set off.

I'll bring him cigarettes, he'll be happy if I bring him cigarettes. This thought immediately strengthened her legs. The summer was in full glory, and in the distant, yellow fields she could see the farmers cutting corn. She crossed the mountainside and when she came to the river she picked up her dress and waded across it. Light burst from every direction, bright and clear. She approached the plots of cultivated land without fear, as if she had known them all her life. With every step she felt the looseness of the fertile soil.

"Have you any tobacco?" she asked a peasant woman standing at the doorway of her hut.

"And what will you give me in exchange?"

"I have a coat," said Tzili and held it up with both hands.

"Where did you steal it?"

"I didn't steal it. I got it as a present."

Upon hearing this reply an old crone emerged from the hut and announced in a loud voice: "Leave the whore's little bastard alone." But the younger woman, who liked the look of the coat, said: "And what else do you want for it?"

"Bread and sausage."

Tzili knew how to bargain. And after an exchange of arguments, curses, and accusations, and after the coat had been turned inside out and felt all over, they agreed on two loaves of bread, two joints of meat, and a bundle of tobacco leaves.

"You'll catch it if the owner comes and demands his coat back. We'll kill you," the old crone said threateningly.

Tzili put the bread, meat, and tobacco into her sack and turned to go without saying a word. The old crone showed no signs of satisfaction at the transaction, but the young woman made no attempt to hide her delight in the city coat.

On the way back Tzili sat and paddled in the water. The sun shone and silence rose from the forest. She sat for an hour without moving from her place and in the end she said to herself: Mark is sad because he has no cigarettes. When he has cigarettes he'll be happy. This thought brought her to her feet and she started to run, taking shortcuts wherever she could.

Toward evening she arrived. Mark bowed his head as if she had brought him news of some great honor, an honor of which he was not unworthy. He took the bundle of tobacco leaves, stroking and sniffing them. Before long he had a cigarette rolled from newspaper. An awkward joy flooded him. In the camp people would fight over a cigarette stub more than over a piece of bread. He spoke of the camp now as if he were about to return to it.

That evening he lit a fire again. They ate and drank herb tea. Mark found a few dry logs and they burned steadily and gave off a pleasant warmth. The wind dropped too, and seemed gentler than before, the shadows it brought from the forest less menacing. Mark was apparently affected by these small changes. Without any warning he suddenly burst into tears.

"What's wrong?"

"I remembered."

"What?"

"Everything that's happened to me in the past year."

Tzili rose to her feet. She wanted to say something but the words would not come. In the end she said: "I'll bring you more tobacco."

"Thank you," he said. "I sit here eating and smoking and they're all over there. Who knows where they are by now." His gray face seemed to grow grayer, a yellow stain spread over his forehead.

"They'll all come back," said Tzili, without knowing what she was saying.

These words calmed him immediately. He asked about the way and the village, and how she had obtained the food and the tobacco, and in general what the peasants were saying.

"They don't say anything," said Tzili quietly.

"And they didn't say anything about the Jews?"

"No."

For a few minutes he sat without moving, wrapped up in himself. His dull, bloodshot eyes slowly closed. And suddenly he dropped to the ground and fell asleep.

17

Every week she went down to the plains to renew their supplies. She was quiet, like a person doing what had to be done without unnecessary words. She would bathe in the river, and when she returned her body gave off a smell of cool water.

She would tell him about her adventures on the plains: a drunken peasant woman had tried to hit her, a peasant had set his dog on her, a passerby had tried to rob her of the clothes she had taken to barter. She spoke simply, as if she were recounting everyday experiences.

And because the weather was fine, and the rains scattered, they would sit for hours by the fire eating, drinking herb tea, listening to the forest and hardly speaking. Mark stopped speaking of the camp and its horrors. He spoke now about the advantages of this high, remote place. And once he said: "The air here is very fresh. Can you feel how fresh it is?" He pronounced the word "fresh" very distinctly, with a secret happiness. Sometimes he used words that Tzili did not understand.

Once Tzili asked what the words "out of this world" meant.

"Don't you understand?"

"No."

"It's very simple: out of this world—out of the ordinary, very nice."

"From God?" she puzzled.

"Not necessarily."

But it wasn't always like this. Sometimes a suppressed rage welled up in him. "What happened to you? Why are you so late?" When he saw the supplies, he recovered his spirits. In the end he would ask her pardon. She, for her part, was no longer afraid of him.

Day by day he changed. He would sit for hours looking at the wild flowers growing in all the colors of the rainbow. Sometimes he would pluck a flower and whisper: "How lovely, how modest." Even the weeds moved him. And once he said, as if talking to himself: "In Jewish families there's never any time. Everyone's in a hurry, everyone's in a panic. What for?" There was a kind of music in his voice, a melancholy music.

The days went by one after the other and nothing happened to arouse their suspicions. On the contrary, the silence deepened. The corn was cut in one field after the other and the fruit was gathered in the orchards, and Mark, for some reason, decided to dig a bunker, in case of trouble. This thought came to him suddenly one afternoon, and he immediately set out to survey the terrain. Straight away he found a suitable place, next to a little mound covered with a tangle of thorns. In his haversack he had a simple kitchen knife. This domestic article, dull with use, fired the desire for activity in him. He set to work to make a spade. The hard, concentrated work changed his face; he stopped talking, as if he had found a purpose for his transitory life, a purpose in which he drowned himself completely.

Every week Tzili went down to the plains and brought back not only bread and sausages but also vodka, in exchange for the clothes which Mark gave her with an abstracted expression on his face. His outbursts did not cease, but they were only momentary flare-ups, few and far between. Activity, on the whole, made him agreeable.

Once he said to her: "My late father's love for the German language knew no bounds. He had a special fondness for irregular verbs. He knew them all. And with me he was very strict about the correct pronunciation. The German lessons with my father were like a nightmare. I always got mixed up and in his fanaticism he never overlooked my mistakes. He made me write them down over and over again. My mother knew German well but not perfectly, and my father would lose his temper and correct her in front of other people. A mistake in grammar would drive him out of his mind. In the provinces people are more fanatical about the German language than in the city."

"What are the provinces?" asked Tzili.

"Don't you know? Places without gymnasiums, without theaters." Suddenly he burst out laughing. "If my father knew what the products of his culture were up to now he would say, 'Impossible, impossible.' "

"Why impossible?" said Tzili.

"Because it's a word he used a lot."

After many days of slow, stubborn carving, Mark had a spade, a strong spade. The carved instrument brightened his eyes, and he couldn't stop touching it. He was in good spirits and he told her stories about all the peculiar tutors his father hired to teach him mathematics and Latin. Young Jewish vagabonds, for the most part, who had not completed their university degrees, who ended up by getting some girl, usually not Jewish, into trouble, and had to be sent packing in a hurry. Mark told these stories slowly, imitating his teachers' gestures and describing their various weaknesses, their fondness for alcohol, and so on. This language was easier for Tzili to understand. Sometimes she would ask him questions and he would reply in detail.

And then he started digging. He worked for hours at a stretch. Every now and then it started raining and the digging was disrupted. Mark would grow angry, but his anger did not last long. The backbreaking work gave him the look of a simple laborer. Tzili stopped asking questions and Mark stopped telling stories.

After a week of work the bunker was ready, dug firmly into the earth. And it was just what was needed for the cold autumn season, a shelter for the cold nights. Mark was sure that the Germans would never reach them, but it was better to be careful, just in case. Tzili noticed that Mark often used the word *careful* now. It was a word he had hardly ever used before.

He put the finishing touches to the bunker without excitement. A quiet happiness spread over his face and hands. Now she saw that his cheeks were tanned and his arms, which had seemed so weak and flabby, were full and firm. He looked like a laboring man who knew how to enjoy his labors.

What will happen when we've sold all the clothes? the thought crossed Tzili's mind. This thought did not appear to trouble Mark. He was so pleased with the bunker, he kept repeating: "It's a good bunker, a comfortable bunker. It will stand up well to the rain."

18

After this the days grew cold and cloudy and Mark drank a lot of vodka. The tan faded suddenly from his face. He would sit silently, and sometimes he would talk to himself, as if Tzili weren't there. On her return from the plains he would ask: "What did you bring?" If she had brought vodka he would say nothing. If she hadn't he would say: "Why didn't you bring vodka?"

At night the words would well up in him and come out in long, clumsy, half-swallowed sentences. Tzili could not understand, but she sensed: Mark was now living in another world, a world which was full of people. Day after day he sat and drank. His face grew lean. There was a kind of strength in this leanness. His days became confused with his nights. Sometimes he would fall asleep in the middle of the day and sometimes he would sit up until late at night. Once he turned to her in the middle of the night and said: "What are you doing here?"

"Nothing."

"Why don't you go down to the village and bring supplies? Our supplies are running out."

"It's night."

"In that case," he said, "we'll wait for the dawn."

He's sad, he's drunk, she would murmur to herself. If I bring him tobacco and vodka he'll feel better. She no longer dared to return without vodka. Sometimes she would sleep in the forest because she was afraid to come back without vodka.

At that time Mark said many strange and confused things. Tzili would sit at a distance and watch him. Alien hands seemed to be clutching at him and kneading him. Sometimes he would lie in his vomit like a hired hand on a drunken spree. His old face, the face of a healthy working man, was wiped away.

And once in his drunkenness he cried: "If only I'd studied medicine I wouldn't be here. I'd be in America." In his haversack, it transpired, were a couple of books which he had once used to prepare for the entrance exams to Vienna University. And once, when it seemed to her that he was calmer, he suddenly burst out in a loud cry: "Commerce has driven the Jews out of their minds. You can cheat people for one year, even for one hundred years, but not for two thousand years!" In his drunkenness he would shout, make speeches, tear sentences to shreds and piece them together again.

Tzili sensed that he was struggling with people who were far away and strangers to her, but nevertheless—she was afraid. His lean cheeks were full of strength. On her return from the plains she would hear his voice from a long way off, rending the silence.

And again, just when she thought that his agitation had died down, he fell on her without any warning: "Why didn't you learn French?"

"We didn't learn French at school, we learned German."

"Barbarous. Why didn't they teach you French? And it's not as if you know German either. What you speak is jargon. It drives me out of my mind. There's no culture without language. If only people learned languages at school the world would be a different place. Do you promise me that you'll learn French?"

"I promise."

Afterward it began to rain and Mark dragged himself to the bunker. A rough wind was blowing. Mark's words went on echoing in the air for a long time. And Tzili, without knowing what she was doing, went up to the bunker and called softly: "It's me, Tzili. Don't worry. Tomorrow I'll bring you vodka and sausage."

19

After this the autumn weather grew finer and a cold, clear sun shone on their temporary shelter. Mark's troubled spirit seemed to lighten too and he stopped cursing. He didn't stop drinking, but his drinking no longer put him in a rage. Now he would often say: "There was something I wanted to say, but it's slipped my mind." A weak smile would break through the clouds, darkening his face. Far-off, forgotten things continued to trouble him, but not in the same shocking way. Now he would speak softly of the need to study languages, acquire a liberal profession, escape from the provinces, but he no longer scolded Tzili.

He would speak of the approaching winter as a frontier beyond which lay life and hope. And Tzili sensed that Mark was now absorbed in listening to himself. Every now and then he would conclude aloud: "There's still hope. There's still hope."

And once he questioned her about her religious studies. Tzili's life at home now felt so remote and scattered that it didn't seem to belong to her. On the way to the plains she would wonder about Maria, whose name she had so unthinkingly adopted. The more she thought about her, the clearer her features grew. A tall, proud woman, she gave her body to anyone who wanted it, but not without getting a good price. And when her daughters grew up, they too adopted their mother's gestures, they too were bold.

She didn't tell him about Maria, just as she didn't tell him about Katerina. Her femininity blossomed within her, blind and sweet. Outwardly too she changed. The pimples didn't disappear from her face, but her limbs were full of strength. She walked easily, even when she had a heavy sack to carry.

"How old are you?" Mark had once asked her in the days of his drunkenness. Afterward he didn't ask again. Now he would beg her pardon for his drunken behavior; his face recovered its former mildness. Tzili's happiness knew no bounds. Mark had recovered and he would never shout at her again. For some reason she believed that the new drink, which the peasants called slivovitz, was responsible for this change.

It seemed to Tzili that the happy days of the summer were about to return, but she was wrong. Mark now craved a woman. This secret he was keeping even from himself. He would urge Tzili to go down to the plains even before it was necessary. Her blooming presence was driving him wild.

And while Tzili was busy pondering ways and means of getting hold of the new, calming drink, Mark suddenly said: "I love you."

Tzili's mouth fell open. His voice was familiar, but very different. She was surprised, but not altogether. The last few nights had been cold and they had both slept in the bunker. They had sat together until late at night, with a warm, dark intimacy between them.

Mark stretched out his arms and clasped her round the waist. Tzili's body shrank from his hands. "You don't love me," he mumbled. The tighter he held her, the more her body shrank. But he was determined, and he slid her dress up with nimble fingers. "No," she managed to murmur. But it was already too late.

Afterward he sat by her side and stroked her body. Strange words came tumbling out of his mouth. For some reason he began talking again about the advantages of the place, the beautiful marshes, the forests, and the fresh air. The words were external, and they brushed past her naked body like a cold wind.

From now on they stayed in the bunker. The rain poured down, but for the time being they were sheltered against it. Mark drank all the time, but never to excess. His happiness was a drunken happiness, and he wanted to cut it up into little pieces and make it last. From time to time he ventured out to confirm what he already knew—that outside it was cold, dark, and damp.

"Tell me about yourself. Why don't you tell me?" he would press her. The truth was that he only wanted to hear her voice. He showered many words on her during their days together in the bunker. His heart overflowed. Tzili, for her part, accepted her happiness quietly. Secretly she was glad that Mark loved her.

Their supplies ran short. Tzili put off going out from day to day. She liked it in this new darkness. She learned to drink the insidious drug, and the more she drank the more slothful her body became. "I'd go myself, but the peasants would betray me." Mark would excuse himself. And in the meantime the rain and cold hemmed them in. They snuggled up together and their small happiness knew no bounds.

Distant sights, hungry malevolent shadows invaded the bunker in dense crowds. Tzili did not know the bitter, emaciated people. Mark went outside and cut branches with his kitchen knife to block up the openings, hurling curses in all directions. For a moment or two it seemed that he had succeeded in chasing them off. But the harder the rain fell the more bitter the struggle became, and from day to day the shadows prevailed. In vain Tzili tried to calm him. His happiness was being attacked from every quarter. Tzili too seemed affected by the same secret poison.

"Enough," he announced, "I'm going down."

"No, I'll go," said Tzili.

The dark, rainy plains now drew Mark to them. "I have to go on a tour of inspection," he announced. It was no longer a caprice but a spell. The plains drew him like a magnet.

20

But in the meantime they put off the decision from day to day. They learned to go short and to share this frugality too. He would drink only once a day and smoke only twice, half a cigarette. The slight tremor came back to his fingers, like a man deprived of alcohol. But for the many shadows besieging their temporary shelter, their small happiness would have been complete.

From time to time, when the shadows deepened, he would go outside and shout: "Come inside, please. We have a wonderful bunker. It's a pity we haven't got any food. Otherwise we'd hold a banquet for you." These announcements would calm them, but not for long.

Afterward he said: "There's nothing else for it, we'll have to go down. Death isn't as terrible as it seems. A man, after all, is not an insect. All you have to do is overcome your fear." These words did not encourage

Tzili. The dark, muddy plains became more frightening from day to day. Now it seemed that not only the peasants lay in wait for her there but also her father, her mother, and her sisters.

And reality stole upon them unawares. Wetness began to seep through the walls of the bunker. At first only a slight dampness, but later real wetness. Mark worked without a pause to stop up the cracks. The work distracted him from the multitude of shadows lying in wait outside. From time to time he brandished his spade as if he were chasing away a troublesome flock of birds.

One evening, as they were lying in the darkness, snuggling up to each other for warmth, the storm broke in and a torrent of water flooded the bunker. Mark was sure that the multitudes of shadows waiting in the trees to trap him were to blame. He rushed outside, shouting at the top of his voice: "Criminals."

Now they stood next to the trees, looking down at the gray slopes shivering in the rain. And just when it seemed that the steady, penetrating drizzle would never stop, the clouds vanished and a round sun appeared in the sky.

"I knew it," said Mark.

If only Tzili had said, "I'll go down," he might have let her go. Perhaps he would have gone with her. But she didn't say anything. She was afraid of the plains. And since she was silent, Mark said: "I'm going down."

In the meantime they made a little fire and drank herb tea. Mark was very excited. He spoke in lofty, dramatic words about the need to change, to adapt to local conditions, and not to be afraid. Fear corrupts human dignity, he said. The resolution he had had while building the bunker came back to his face. Now he was even more resolute, determined to go down to the plains and not to be afraid.

"Don't go," said Tzili.

"I must go down. Inspection of the terrain has become imperative—if only from the point of view of general security needs. Who knows what the villagers have got up their sleeves? They may be getting ready for a surprise attack. I can't allow them to take us by surprise."

Tzili could not understand what he was talking about, but the lofty, resolute words, which at first had given her a sense of security, began to hurt her, and the more he talked the more they stung. He spoke of reassessment and reappraisal, of diversion and camouflage. Tzili understood none of his many words, but this she understood: he was talking of another world.

"Don't go." She clung to him.

"You have to understand," he said in a gentle voice. "Once you conquer your fear everything looks different. I'm happy now that I've conquered my fear. All my life fear has tortured me shamefully, you understand, shamefully. Now I'm a free man."

Afterward they sat together for a long time. But although Tzili now said, "I'll go down. They know me, they won't hurt me," Mark had made up his mind: "This time I'm going down." And he went down.

21

Mark receded rapidly and in a few minutes he was gone. She sat still and felt the silence deepening around her. The sky changed color and a shudder passed over the mountainside.

Tzili rose to her feet and went into the bunker. It was dark and warm inside the bunker. The haversack lay to one side. For the past few days Mark had refused to go into the bunker. "A man is not a mole. This lying about is shameful." He used the word *shameful* often, pronouncing it in a foreign accent, apparently German.

The daylight hours crept slowly by, and Tzili concentrated her thoughts on Mark's progress across the mountainside. She imagined him going up and down the same paths that she herself had taken. She saw him pass by the hut where she had bartered a garment for a sausage. She saw it all so clearly that she felt as if she herself were there with him.

In the afternoon she lit a fire and said: "I'll make Mark some herb tea. He likes herb tea."

Mark was late.

"Don't worry, he'll come back," a voice from home said in her ear. But when twilight fell and Mark did not return anxiety began dripping into her soul. She went down to the river and washed the mugs. The cold water banished the anxiety for a moment. For some reason she spread a cloth on the ground.

Darkness fell. The days she had spent with Mark had blunted her fear of the night. Now she was alone again. Mark's voice came to her and she heard: "A man is not an insect. Death isn't as terrible as it seems." Now these words were accompanied by the music of a military band. Like in her childhood, on the Day of Independence, when the army held parades and the bugles played. The military voice gave her back a kind of confidence.

Mark was late.

Now she felt that the domestic smells that had enveloped the place were fading away. Fresh, cold air blew in their place. It occurred to her that if she took the clothes out of the haversack and spread them around, the homely smells would come back to fill the air, and perhaps Mark would sense them. Immediately she took the haversack out of the bunker and spread the clothes on the ground. The brightly colored clothes, all damp and crumpled, gave off a confined, moldy smell.

He's lost, he must be lost. She clung to this sentence like an anchor. She fell to her knees by the clothes. They were children's clothes, small and shrunken with the damp, spotted with food stains and a little torn.

Afterward she turned aside to listen. Apart from an occasional rustle or murmur there was nothing to be heard. From the distant huts scattered between the swamps, isolated barks reached her ears.

After midnight a thin drizzle began to fall and she put the things back into the bunker. This small activity revived an old scene in her memory. She remembered the first days, before the bunker, when she had brought him the tobacco. The way he had rolled the shredded leaves in a piece of newspaper, the way he had recovered his looks, his smile, and the light on his face.

The rain stopped but the wind grew stronger, bending the trees with broad, sweeping movements. Tzili went into the bunker. It was warm and full of the smell of tobacco. She breathed in the smell.

She sat in the dark and for some reason she thought about Mark's wife. Mark seldom spoke of her. Once she had even sensed a note of resentment against her. She imagined her as a tall, thin woman sheltering her children under her coat. Strange, she felt a kind of kinship with her.

22

The next day Mark still did not return. She stood on the edge of the plateau exposed to the wind. The downward slope drew her too. The slope was not steep and it glittered with puddles of water. Now she felt that something had been taken from her, something that belonged to her youth. She covered her face in shame.

For hours she sat and practiced the words, so that she would be ready for him when he came. "Where were you Mark? I was very worried. Here is some herb tea for you. You must be thirsty." She did not prepare many words, and the few she did prepare, she repeated over and over again in a voice which had a formal ring in her ears. Repeating the words put her to sleep. She would wake up in alarm and go to the bunker. The walls of the bunker had collapsed, the flimsy roof had caved in, and the floor was covered by a spreading gray puddle. There was an alien spirit in it, but it was the only place she could go to. Everywhere else was even more alien.

The days dragged out long and heavy. Tzili did not stir. And once a voice burst out from within her: "Mark." The voice slid down the mountainside, echoing as it went. No one answered.

Overnight the winds changed and the winter winds came, thin and sharp as knives. The fire burned but it did not warm her. Low, dark clouds covered the somber sky. She prayed often. This was the prayer which she repeated over and over: "God, bring Mark back. If you bring Mark back to me, I'll go down to the plains and I won't be lazy."

How many days had Mark been gone? At first she kept track, but then she lost count. Sometimes she saw Mark struggling with the peasants and hurling pointed sticks at them, like the ones he had made for the walls of the bunker. Sometimes he looked tired and crushed. Like the first time she had seen him, pale and gray. Man is not an insect, she remembered and made an effort to get up and stand erect.

For days she had had nothing to eat. Here and there she still found a few withered wild apples, but for the most part she now lived off roots. The roots were sweet and juicy. "I'll go on," she said, but she didn't move. For hours she sat and gazed at the mountainside sloping down to the plains, the two marshes, the shelter, and the haversack. Sometimes she took out the clothes and spread them on the ground, but Mark did not respond to her call.

The moment she decided to leave she would imagine that she heard footsteps approaching. A little longer, she would say to herself. Death is not as terrible as it seems.

Sometimes the cold would envelop her in sweetness. She would close her eyes and curl up tightly and wait for a hand to come and take her away. But none came. Winter winds tore across the hillside, cruel and cutting. "I'll go on," she said, and lifted the haversack onto her shoulders. The haversack was soaked through and heavy, with every step she felt that the burden was too heavy to bear.

"Did you see a man pass by?" she asked a peasant woman standing at the doorway of her hut.

"There's no man here. They've all been conscripted. Who do you belong to?"

"Maria."

"Which Maria?"

And when she did not reply the peasant woman understood which Maria she meant, snickered aloud, and said: "Be off with you, wretch! Get out of my sight."

One by one Tzili gave the little garments away in exchange for bread. "If I meet Mark I'll tell him that I was hungry. He won't be angry." The haversack on her back grew more burdensome from day to day but she didn't take it off. The damp warmth stuck to her back. She went from tree to tree. She believed that next to one of the trees she would find him.

23

It began to snow and she was obliged to look for work. The long tramp had weakened her. Overnight she lost her freedom and became a serf.

At this time the Germans were on the retreat, but here it was the middle of winter and the snow fell without a break. The peasants drove her mercilessly. She cleaned the cow shed, milked the cows, peeled potatoes, washed dishes, brought firewood from the forest. At night the peasant's wife would mutter: "You know who your mother is. You must pay for your sins. Your mother has corrupted whole villages. If you follow in her footsteps I'll beat you black and blue."

Sometimes she went out at night and lay down in the snow. For some reason the snow refused to absorb her. She would return to her sufferings, meek and submissive. One evening on her way back from the forest she heard a voice. "Tzili," called the voice.

"I'm Tzili," said Tzili. "Who are you?"

"I'm Mark," said the voice. "Have you forgotten me?"

"No," said Tzili, frightened. "I'm waiting for you. Where are you?"

"Not far," said the voice, "but I can't come out of hiding. Death is not as terrible as it seems. All you have to do is conquer your fear."

She woke up. Her feet were frozen.

From then on Mark appeared often. He would surprise her at every turn, especially his voice. It seemed to her that he was hovering nearby, unchanged but thinner and unable to emerge from his hiding place. And once she heard quite clearly: "Don't be afraid. The transition is easy in the end." These apparitions filled her with a kind of warmth. And at night, when the stick or the rope fell on her back, she would say to herself, "Never mind. Mark will come to rescue you in the spring."

And in the middle of the hard, grim winter she sensed that her belly had changed and was slightly swollen. At first it seemed an insignificant change. But it did not take long for her to understand: Mark was inside her. This discovery frightened her. She remembered the time when her sister Yetty fell in love with a young officer from Moravia, and everyone became angry with her. Not because she had fallen in love with a gentile but because the intimate relations between them were likely to get her into trouble. And indeed, in the end it came out that the officer was an immoral drunkard, and but for the fact that his regiment was transferred the affair would certainly have ended badly. It remained as a wound in her sister's heart, and at home it came up among other unfortunate affairs in whispers, in veiled words. And Tzili, it transpired, although she was very young at the time, had known how to put the pieces together and make a picture, albeit incomplete.

There was no more possibility of doubt: she was pregnant. The peasant woman for whom she slaved soon noticed that something was amiss. "Pregnant," she hissed. "I knew what you were the minute I set eyes on you."

Tzili herself, when the first fear had passed, suddenly felt a new strength in her body. She worked till late at night, no work was too hard for them to burden her with, but she did not weaken. She drew strength from the air, from the fresh milk, and from the hope that one day she would be able to tell Mark that she was bearing his child. The complications, of course, were beyond her grasp.

And in the meantime the peasant woman beat her constantly. She was old but strong, and she beat Tzili religiously. Not in anger but in righteousness. Ever since her discovery that Tzili was pregnant her blows had grown more violent, as if she wanted to tear the embryo from her belly.

Heaven and hell merged into one. When she went to graze the cow or gather wood in the forest she felt Mark close by her side, even closer than in the days when they had slept together in the bunker. She spoke to him simply, as if she were chatting to a companion while she worked. The work did not stop her from hearing his voice. His words too were clear and simple. "I'll come in the spring," he said. "In the spring the war will end and everyone will return."

Once she dared to ask him: "Won't your wife be angry with me?

"My wife," said Mark, "is a very forgiving woman."

"As for me," said Tzili, "I love your children as if they were my own."

"In that case," said Mark in a practical tone of voice, "all we have to do is wait for the war to end."

But at night when she returned to the hut reality showed itself in all its nakedness. The peasant's wife beat her as if she were a rebellious animal, in a passion of rage and fury. At first Tzili screamed and bit her lips. Later she stopped screaming. She absorbed the blows with her eyes closed, as if she knew that this was her lot in life.

One night she snatched the rope from the woman and said: "No, you won't. I'm not an animal. I'm a woman." The peasant's wife, apparently startled by Tzili's resolution, stood rooted to the spot, but she immediately recovered, snatched the rope from Tzili's hand, and began to beat her with her fists.

It was the height of winter and there was nowhere to escape to. She worked, and the work strengthened her. The thought that Mark would come for her in the spring was no longer a hope but a certainty.

Once the peasant's wife asked her: "Who made you pregnant?"

"A man."

"What man?"

"A good man."

"And what will you do with the baby when it's born?"

"I'll bring it up."

"And who will provide for you?"

"I'll work, but not for you." The words came out of her mouth directly and quietly.

The peasant's wife ranted and raved.

The next day she said to Tzili: "Take your things and get out of my sight. I never want to see you again."

Tzili took up the haversack and left.

24

Once more she had won her freedom. At that time the great battlefronts were collapsing, and the first refugees were groping their way across the broad fields of snow. Against the vast whiteness they looked like swarms of insects. Tzili was drawn toward them as if she realized that her fate was no different from theirs.

Strange, precisely now, at the hour of her newfound freedom, Mark stopped speaking to her. "Where are you and why don't you speak to me?" she would ask in despair. Nothing stirred in the silence, and but for her own voice no other voice was heard.

In one of the bunkers she came across three men. They were wrapped from top to toe in heavy, tattered coats. Their bloodshot eyes peeped through their rags, alert and sardonic.

"Who are you?"

"My name is Tzili."

"So you're one of us. Where have you left everyone?"

"I," said Tzili, "have lost everyone."

"In that case why don't you come with us? What have you got in that haversack?"

"Clothes."

"And haven't you got any bread?" one of them said in an unpleasant voice.

"Who are you?" she asked.

"Can't you see? We're partisans. Haven't you got any bread in that haversack?"

"No I haven't," she said and turned to go.

"Where are you going?"

"I'm going to Mark."

"We know the whole area. There's no one here. You'd better stay with us. We'll keep you amused."

"I," said Tzili opening her coat, "am a pregnant woman."

"Leave the haversack with us. We'll look after it for you."

"The haversack isn't mine. It belongs to Mark. He left it in my care."

"Don't boast. You should learn to be more modest."

"I'm not afraid. Death is not as terrible as it seems."

"Cheeky brat," said the man and rose to his feet.

Tzili stared at him.

"Where did you learn that?" said the man, taking a step backward.

Tzili stood still. A strength not hers was in her eyes.

"Go then, bitch," said the man and went back to the bunker.

From then on the snow stretched before her white and empty. Tzili felt a kind of warmth spreading through her. She walked along a row of trees, which now seemed rootless, stuck into the snow like pegs.

From time to time a harassed survivor appeared, asked the way, and disappeared again. Tzili knew that her fate was no different from the survivors, but she kept away from them as if they were brothers who might say: "We told you so."

And while she was walking without knowing where her feet would lead her, the walls of snow began to shudder. It was the month of March and new winds invaded the landscape. On the mountain slopes the first stripes of brown earth appeared. Not long afterward the brown stripes widened.

And suddenly she saw what she had not seen before: the mountain, undistinguished and not particularly lofty, the mountain where she and Mark had spent the summer, and not far from where she was standing the foot of the slope, and next to it the valley leading to Katerina's house. As if the whole world had narrowed down to a piece of land which she could feel with her hands.

She stood for a moment as if she were trying to absorb all these painful places into her body. She herself felt no pain.

And while she was standing there sunk into herself a refugee approached her and he said: "Jewish girl, where are you from?"

"From here."

"And you weren't in the camps?"

"No."

"I lost everyone. What shall I do?"

"In the spring they'll all come back. I'm sure of it."

"How do you know?"

"I'm quite sure. You can believe me." There was strength in her voice. And the man stood rooted to the spot.

"Thank you," he said, as if he had been given a great gift.

"Don't mention it," said Tzili, as she had been taught to say at home.

Without asking for further details, the man vanished as abruptly as he had appeared.

Evening drew near and the last rays of the sun fell golden on the hillside. "I lived here and now I'm leaving," said Tzili, and she felt a slight twinge in her chest. The embryo throbbed gently in her belly. Her vision narrowed even further. Now she could picture to herself the paths lying underneath the carpet of snow. There was no resentment in her heart, only longing, longing for the earth on which she stood. Everything beyond this little corner of the world seemed alien and remote to her.

For days she had not tasted food. She would sit for hours sucking the snow. The melted snow assuaged her hunger. The liquids refreshed her. Now she felt a faint anxiety.

And while she was standing transfixed by what she saw, Mark rose up before her.

"Mark," the word burst from her throat.

Mark seemed surprised. He stood still. And then he asked: "Why are you going to the refugees? Don't you know how bad they are?"

"I was looking for you."

"You won't find me there. I keep as far away as possible from them."

"Where are you?"

"Setting sail."

"Where to?"

All at once a flock of birds rose into the air and crossed the darkening horizon, and Tzili understood that he had only called her in order to take his leave.

She put the haversack down on the snow. Her eyes opened and she said: "My search was in vain." It was her own voice which had come back to her.

"So you're abandoning me to the refugees, those bad, wicked people." A voice that had been locked up inside her broke out and rose into the air. "I'm asking you a question. Answer me. If you don't want me, tell me. I'm not complaining. I love you anyway. I'll make you herb tea if you like. The mountain is unoccupied. It's waiting for us. There's no one there; we can go back to it. It's a good mountain, you said so yourself. I'll go to the village and bring back supplies, I won't be lazy. You can believe me. Don't you believe me?"

After this she sat for hours next to the haversack. She woke up and fell asleep again, and when she finally rose to her feet brown vapors were already rising from the valleys. Here and there a peasant stretched himself as if after a long sleep. There were no refugees to be seen.

"So you're abandoning me," the old anger flickered up in her again. It wasn't really anger, but only a weak echo of anger. She was with herself as if after a long hunger. She narrowed her eyes and stroked her belly, and then she said to herself: "Mark's probably not allowed to leave, for the time being. Later on, they'll let him go. He's not his own boss, after all."

The snow thawed and the first convoys of refugees poured down the hillsides. Strange, thought Tzili, the war's over and I didn't know. Mark must have known before me; he must be happy now. Suddenly she felt that her life was moving toward some other destination, where the colors were different. She heard the voices of the refugees, and the sound was so familiar that it hurt her.

She thought that she would go to the high mountain where she had first met Mark. She set her steps in that direction but the road was covered with mud and she gave up. Later, she said to herself.

Afterward the commotion died down and the lips of the land could be heard quietly sucking all over the plains. The liquids were being absorbed into the earth.

"Thank you, Maria," said Tzili. "Thanks to you I'm still alive. But for you I would already be in another world. Thanks to you I'm still here. Isn't it strange that thanks to you I'm still here? I'm grateful to you, Maria." Tzili was surprised by the words that came out of her mouth.

She let the memory of Maria flow over her for a moment, and as she did so she saw her figure emerge from the mist and stand solidly before her eyes. A tall, strong woman dressed with simple elegance. When she came into her mother's shop she filled it with the breath of city streets, cafés, and theaters. She never hid her opinions. She would often say that she was fond of the Jews, although not the religious ones. Those she had always hated, but the freethinking Jews who lived in the city were men after her own heart. They knew what civilization meant; they knew how to get the most out of life in the city. Of course, her appearances would always be accompanied by a certain fear, because of her connections with the provincial officials, the tax collectors, the police, and the hospitals. And when her daughters grew up, the circle of her acquaintances was enlarged. Not without scandals, of course.

And when Tzili's brother got one of her daughters into trouble, Maria's tone changed. She threatened them brutally, and in the end she extorted a tidy sum. Tzili remembered this episode too, but she felt no resentment. A proud woman, she concluded to herself.

Katerina, with whom she had spent two whole seasons, had often used this combination of words—"a proud woman." Katerina too, Tzili now remembered with affection.

And while Tzili was standing there in a kind of trance, the refugees streamed toward her in a hungry swarm. She wanted to run for her life, but it was too late. They surrounded her on all sides: "Who are you?"

"I'm from here." At last she found the words.

"And where were you during the war?"

"Here."

"Can't you see?" One of the refugees interrupted. "She's afraid."

"And you weren't in any of the camps?"

"No."

"And no one gave you away?"

"Can't you see?" The same man intervened again. "She doesn't look Jewish. She looks healthy."

"A miracle," said the questioner and turned aside.

The news spread from one to the other but it made no impression on the refugees.

Later on Tzili asked: "Did you see Mark?"

"What's his last name?" asked a woman.

Tzili hung her head. She did not know.

The cold spring sun exposed them like moles. A motley crew of men, women, and children. The cold light showed up their ragged clothes. The convoy turned south and Tzili went with them. No one asked, "Where are you from?" or "Where are you going?" From time to time a supply-laden cart appeared and the people swarmed over it like ants. The familiar words from home now sounded wild and foreign to her. The refugees did not appear contented with anything. They argued, laughed, and fought, and at night they fell to the ground like sacks.

"What am I doing here?" Tzili asked herself. "I prefer the mountains and the rivers. Mark himself told me not to go with them. If I go too far, who knows if I'll ever find him?"

From here she could still see the mountain where Mark had revealed himself to her, illuminated in the cold evening light. Now the bunker was ruined and the wind blew through it. Her voice broke with longing. No memory stirred in her, only a thin stream of longing flowing out of her toward the distant mountaintop. The calm of evening fell upon the deserted ranges and she fell asleep.

25

They made their way southward. The peasants stood by the roadside displaying their wares: bread, vodka, and smoked meat. But the refugees walked past without buying or bartering. Suffering had made them indifferent. But Tzili was hungry. She sold a garment and received bread and smoked meat in exchange. "Look," said one of the survivors, "she's eating."

Now she saw them from close up: thin, speechless, and withdrawn. The terror had not yet faded from their faces.

The sun sank in the sky and the crust of the earth dried up. The first ploughmen appeared on the mountainsides next to the plains. There were no clouds to darken the sky, only the trees, and the quietness.

They moved slowly through the landscape, looking around them as they walked. They slept a lot. Hardly a word was spoken. A kind of secret veiled their faces. Tzili feared this secret more than the dark nights in the forest.

A convoy of prisoners was led past in chains. From time to time a soldier fired a shot into the air and the prisoners all bent their heads at once. No one looked at them. The survivors were sunk into themselves.

A man came up to Tzili and asked: "Where are you from?" It wasn't the man himself who asked the question, but something inside him, as in a nightmare.

Tzili felt as if her eyes had been opened. She heard words which she had not heard for years, and they lapped against her ears with their whispers. "If I meet my mother, what will I say to her?" She did not know what everyone else already knew: apart from this handful of survivors, there were no Jews left.

The sun opened out. The people unbuttoned their damp clothes and sprawled on the river bank and slept. The long, damp years of the war steamed out of their moldy bodies. Even at night the smell did not disappear. Only Tzili did not sleep. The way the people slept filled her with wonder. A warm breeze touched them gently in their deep sleep. Are they happy? Tzili asked herself. They slept in a heap, defenseless bodies suddenly abandoned by danger.

The next day too no one woke up. "What do they do in their sleep?" she asked without knowing what she was asking. "I'll go on," she said. "No one will notice my absence. I'll work for the peasants like I did before. If I work hard they'll give me bread. What more do I need?" Her thoughts flowed as of their own accord. All the years of the war, in the forest and on the roads, even when she and Mark were together, she had not thought. Now the thoughts seemed to come floating up to the surface of her mind.

For a moment she thought of getting up and leaving the sleeping people and returning to the mountain where she had first met Mark. The mountain itself had disappeared from view, but she could still see the swamps below it. They shone like two polished mirrors. Her longings were deep and charged with heavy feelings. They drew her like a magnet, but as soon as she rose to her feet she felt that her body had lost its lightness. Not only her belly was swollen but also her legs. The light, strong columns which had borne her like the wind were no longer what they had been.

Now she knew that she would never go back to that enchanted mountain; everything that had happened there would remain buried inside her. She would wander far and wide, but she would never see the mountain again. Her fate would be the fate of these refugees sleeping beside her.

She wanted to weep but the tears remained locked inside her. She sat without moving and felt the sleep of the refugees invading her body. And soon she too was deep in sleep.

26

Their sleep lasted a number of days. From time to time one of them opened his eyes and stretched his arms as if he were trying to wake up. All in vain. He too, like everyone else, was stuck to the ground.

Tzili opened the haversack and spread the clothes out to dry. Two long dresses, a petticoat, children's trousers, the kitchen knife which Mark had used to make the bunker, and two books—this is what was left.

From the size of the garments Tzili understood that Mark's wife was a tall, slender woman and the children were about five years old, thin like their mother. And she noticed too that the dresses buttoned up to the neck, which meant that Mark's wife was from a traditional family. The petticoat was plain, without any flowers. There were two yellow stains on it, apparently from the damp.

She sat looking at the inanimate objects as if she were trying to make them speak. From time to time she stroked them. The silence all around, as in the wake of every war, was profound.

Whenever she felt hunger gnawing at her stomach she would take a garment from the haversack and offer it in exchange for food. At first she had asked Mark to forgive her, although then too, she had not given the matter too much thought. Later she had stopped asking. She was often hungry and she bartered one garment after the other. The haversack had emptied fast, and now this was all that was left.

These things I won't sell, she said to herself, although she knew that the first time she felt hungry she would have to sell them. She would often feel a voracious greed for food, a greed she could not overcome. Mark will understand, she said to herself, it's not my fault.

She sat and listened to the pulsing of the embryo inside her. It floated quietly in her womb, and from time to time it kicked. It's alive, she told herself, and she was glad.

The next day spring burst forth in a profusion of flowers. And the sleepers awoke. It was not an easy awakening. For hours they went on lying, stuck to the ground. Not as many as they had seemed at first—about thirty people all told.

In the afternoon, as the heat of the sun increased, a few of them rose to their feet. In the light of the sun they looked thin and somewhat transparent. Someone approached her and said: "Where are you from?" He spoke in German Jewish. He looked like Mark, only taller and younger.

"From here," said Tzili.

"I don't understand," said the man. "You weren't born here, were you?"

"Yes," said Tzili.

"And what did you speak at home?"

"We tried to speak German."

"That's funny, so did we," said the stranger, opening his eyes wide. "My grandmother and grandfather still spoke Yiddish. I like the way they talked."

Tzili had never seen her grandfather. This grandfather, her father's father, a rabbi in a remote village in the Carpathian mountains, had lived to a ripe old age and had never forgiven his son for abandoning the faith of his fathers. His name was never mentioned at home. Her mother's parents had died young.

"Where are we going?" the man asked.

"I don't know."

"I have to get there soon. My engineering studies were interrupted in the middle. I've missed enough already. If I don't arrive in time I may be too late to register. A person starts a course of study and all of a sudden a war comes and messes everything up."

"Where were you during the war?" asked Tzili.

"Why do you ask? With everyone else, of course. Can't you see?" he said and stretched out his arm. There was a number there, tattooed in dark blue on his skin. "But I don't want to talk about it. If I start talking about it, I'll never stop. I've made up my mind that from now on I'm starting my life again. And for me that means studying. Completing my studies, to be precise."

This logic astounded Tzili. Now she saw: the man spoke quietly enough, but his right hand waved jerkily as he spoke and fell abruptly to his side, as if it had been cut off in midair.

He added: "I've always been an outstanding student. My average was ninety. And that's no joke. Of course, it made the others jealous. But what of it? I was only doing what I was supposed to do. I like engineering. I've always liked it."

Tzili was enchanted by his eloquence. It was a long time since she had heard such an uninterrupted flow of words. It was the way Blanca and Yetty and her brothers used to talk. Exams, exams always around the corner. Now the words momentarily warmed her frozen memory.

After a pause he said: "There were two exams I didn't take, through no fault of my own. I won't let them get away with it. It wasn't my fault."

"Never mind," said Tzili, for some reason.

"I won't let them get away with it. It wasn't my fault."

And for a moment it seemed that they were sitting, not in an open field in the spring after the war, but in a salon where coffee and cheese-cake were being served. The hostess asks: "Who else wants coffee?" A student on vacation speaks of his achievements. Tzili now remembered her own home, her sister Blanca, sulkily hunching her shoulder, her books piled on the table.

The man rose to his feet and said: "I'm not hanging around here. I haven't got any time to waste. These people are sleeping as if time lasts forever."

"They're tired," said Tzili.

"I don't accept that," said the man, with a peculiar gravity. "There's a limit to what a person can afford to miss. I've made up my mind to finish. I'm not going to leave my studies broken off in the middle. I have to get there in time. If I arrive in time I'll be able to register for the second semester."

Tzili asked no more. His eloquence stunned her. And as he spoke, scene after scene of a drama not unfamiliar to her unfolded before her eyes: a race whose demanding pace had not been softened even by the years of war.

He looked around him and said: "I'm going. There's nothing for me to do here."

Tzili remembered that Mark too had stood on the mountainside and announced firmly that he was going. If she had said to him then, "Don't go," perhaps he would not have gone.

"Mark," she said.

The man turned his head and said, "My name isn't Mark. My name's Max, Max Engelbaum. Remember it."

"Don't go," said Tzili.

"Thank you," said the man, "but I haven't any time to waste. I have no intention of spending my time sleeping. And in general, if you understand me, I don't want to spend any more time in the company of these people." He made a funny little half bow, like a clerk rising from his desk, and abruptly said: "Adieu."

Tzili noticed that he walked away the way people had walked toward the railway station in former days, with brisk, purposeful steps which from a distance looked slightly ridiculous.

"Adieu," he called again, as if he were about to step onto the carriage stair.

The awakening lasted a number of days. It was a slow, wordless awakening. The refugees sat on the banks of the river and gazed at the water. The water was very clear now and a kind of radiance shone on its surface. No one went down to bathe. From time to time a word or phrase rose into the air. They were struggling with the coils of their sleep, which were still lying on the ground.

Tzili felt that she had come a very long way. And if she stayed with these people she would go even further away. Where was Mark? Was he too following her, or was he perhaps still waiting, imprisoned in the same place? Perhaps he did not know that the war was over.

And while she was sitting and staring, a woman came up to her and said: "You need milk."

"I have none," said Tzili apologetically.

"You need milk, I said." The woman was no longer young. Her face was haggard and there was a kind of fury in the set of her mouth.

"I'll see to it," said Tzili, in order to appease the woman's wrath.

"Do it straight away. A pregnant woman needs milk. It's as necessary to her as the air she breathes, and you sit here doing nothing."

Tzili said no more. When she did not respond, the woman grew angry and said: "A woman should look after her body. A woman is not an insect. And by the way, where's the bastard who did this to you?"

"His name is Mark," said Tzili softly.

"In that case, let him take care of it."

"He's not here."

"Where is he?"

Tzili sat looking at her without resentment. No one interfered. They were sitting sunk into themselves. The woman turned away and went to sit on the river bank.

That night cool spring winds blew, bringing with them shadows from the mountains. Quiet shadows that clung soundlessly to the trees but that nevertheless caused a commotion. At first people tried to chase them away as if they were birds, but for some reason the shadows clung to the trees and refused to go.

And as if to spite them, the night was very bright, and they could see the shadows clearly, breathing fearfully.

"Go away, leave us alone!" The shouts arose from every side. And when the shadows refused to go, people began to beat them.

The shadows did not react. Their stubborn resistance infuriated the people and they cast off all restraint.

All night long the battle lasted. Bodies and shadows fought each other in silence, violently. The only sound was the thud of their blows.

When day broke the shadows fled.

The survivors were not happy. A kind of sadness darkened their daylight hours. Tzili did not stir from her corner. She too was affected by the sadness. Now she understood what she had not understood before: everything was gone, gone forever. She would remain alone, alone forever. Even the fetus inside her, because it was inside her, would be as lonely as she. No one would ever ask again: "Where were you and what happened to you?" And if someone did ask, she would not reply. She loved Mark now more than ever, but she loved his wife and children too.

The woman who had grown angry with her before on account of the milk now sat wrapped up in herself. A kind of tenderness shone from her eyes, as if she were, not a woman who had lost herself and all she possessed, but a woman with children, whose love for her children was too much for her to bear.

27

Spring was now at its height, its light was everywhere. Some of the people could not bear the silence and left. The rest sat on the ground and played cards. The old madness, buried for years, broke out: cards and gambling. All at once they shook off their damp, rotting rags and put on carefree expressions, laughing and teasing each other. Tzili did not yet know that a new way of life was unconsciously coming into being here.

The holiday atmosphere reminded Tzili of her parents. When she was still small they had spent their summer vacations in a pension on the banks of the Danube. Her parents were short of money, but they had spared no effort in order to be in the company, if only for two weeks, of speakers of correct German. As if to spite them, however, most of the people there spoke Yiddish. This annoyed her father greatly, and he said: "You can't get away from them. They creep in everywhere." Afterward he fell ill, and they stayed at home and spent their money on doctors and medicine.

No one spoke of the war anymore. The card games devoured their time. A few of them went to buy supplies, but as soon as they got back they joined enthusiastically in the game. Every now and then someone would remember to say: "What will become of us?" But the question was not serious. It was only part of the game. "What's wrong with staying right here? We've got plenty of coffee, cigarettes—we can stay here for the rest of our lives"—someone would nevertheless take the trouble to reply.

Not far from where they sat the troops passed by, a vigorous army liberated from the siege, invading the countryside on fresh young horses. They all admired the Russians, the volunteers and the partisans, but it was not an admiration which entailed a desire for action. "Let the soldiers fight, let them avenge us."

Tzili was with herself and the tiny fetus in her womb. Words which Mark had spoken to her on the mountain rang in her ears. Scenes from the mountain days passed before her eyes like vivid, ritual tableaus. Mark no longer appeared to her. For hours she sat and waited for him to reveal himself. He's dead—the thought flashed through her mind and immediately disappeared.

One evening a few more Jewish survivors appeared, bringing a new commotion. And one of them, a youthful-looking man, spoke of the coming salvation. He spoke of the cleansing of sins, the purification of the soul. He spoke eloquently, in a pleasant voice. His appearance was not ravaged. Thin, but not horrifyingly thin. Some of them recognized him and remembered him from the camp as a quiet young man, working and suffering in silence. They had never imagined that he had so much to say.

Tzili liked the look of him and she drew near to hear him speak. He spoke patiently, imploringly, without raising his voice. As if he were speaking of things that were self-evident. And for a moment it seemed that he was not speaking, but singing.

The people were absorbed in their card game, and the young man's eloquence disturbed them. At first they asked him to leave them alone and go somewhere else. The young man begged their pardon and said that he had only come to tell them what he himself had been told. And if what he had been told was true, he could not be silent.

It was obvious that he was a well-brought-up young man. He spoke politely in a correct German Jewish, and wished no one any harm. But his apologies were to no avail. They ordered him to leave, or at any rate to shut up. The young man seemed about to depart, but something inside him, something compulsive, stopped him, and he stood his ground and went on talking. One of the card players, who had been losing and was in a bad mood, stood up and hit him.

To everyone's surprise, the young man burst out crying.

It was more like wailing than crying. The whole night long he sat and wailed. Through his wailing the history of his life emerged. He was an architect. Like his father and forefathers, he was remote from Jewish affairs, busy trying to set up an independent studio. The war took him completely by surprise. In the camp something had happened to him. His workmate in the forced labor gang, something of a Jewish scholar although not a believer, had taught him a little Bible, Mishna, and the Sayings of the Fathers. After the war he had begun to hear voices, clear, unconfused voices, and one evening the cry had burst from his throat: "Jews repent, return to your Father in Heaven."

From then on he never stopped talking, explaining, and calling on the Jews to repent. And when people refused to listen or hit him, he fell to the ground and wept.

The next day one of the card players found a way to get rid of him. He approached the young man and said to him in his own language, in a whisper: "Why waste your time on these stubborn Jews? Down below, not far from here, there are plenty of survivors, gentle people like you. They're waiting for someone to come and show them the way. You'll do it. You're just the right person. Believe me."

Strange, these words had an immediate effect. He rose to his feet and asked the way, and without another word he set out.

Tzili felt sorry for the young man who had been led astray. She covered her face with her hands. The others too seemed unhappy. They returned to their card playing as if it were not a game, but an urgent duty.

28

After this the weather was fine and mild, without wind or rain. The grass grew thick and wild and the people sat about drinking coffee and playing cards. There were no quarrels, and for a while it seemed as if things would go on like this forever.

From time to time peasant women would appear, spread out their wares on flowered cloths, and offer the survivors apples, smoked meat, and black bread. The survivors bartered clothes for food. Some of them had gold coins too, old watches, and all kinds of trinkets they had kept with them through the years of the war. They gave these things away for food without haggling about their worth.

Tzili too sold a dress. In exchange she received a joint of smoked meat, two loaves of fresh bread, and a piece of cheese. She remembered the woman's anger and asked for milk, but they had no milk. Tzili sat on the ground and ate heartily.

Apart from the card game nobody took any interest in anything. The woman who had scolded Tzili for not providing herself with milk played avidly. Tzili sat and watched them for hours at a time. Their faces reminded her of people from home, but nevertheless they looked like strangers. Perhaps because of the smell, the wet rot of years which clung to them still.

And while they were all absorbed in their eager game, a sudden fear fell on Tzili. What would she do if they all came back? What would she say, and how would she explain? She would say that she loved Mark. She now feared the questions she would be asked more than she feared the strangers. She curled up and closed her eyes. The fear which came from far away invaded her sleep too. She saw her mother looking at her through a very narrow slit. Her face was blurred but her question was clear: Who was this seducer, who was this Mark?

And Tzili's fears were not in vain. One evening everything exploded. One of the card players, a quiet man with the face of a clerk, gentle-mannered and seemingly content, suddenly threw his cards down and said: "What am I doing here?"

At first this sentence seemed part of the game, annoyance at some little loss, a provocative remark. The game went on for some time longer, without anyone sensing the dynamite about to explode.

Suddenly the man rose to his feet and said: "What am I doing here?"

"What do you mean, what are you doing here?" they said. "You're playing cards."

"I'm a murderer," he said, not in anger, but with a kind of quiet deliberation, as if the scream in his throat had turned, within a short space of time, to a clear admission of guilt.

"Don't talk like that," they said.

"You know it better than I do," he said. "You'll be my witnesses when the time comes."

"Of course we'll be your witnesses. Of course we will."

"You'll say that Zigi Baum is a murderer."

"That you can't expect of us."

"I, for one, don't intend hiding anything."

This exchange, proceeding without anger, in a matter-of-fact tone, turned gradually into a menacing confrontation.

"You won't tell the truth, then?"

"Of course we'll tell the truth."

"A man abandons his wife and children, his father and his mother. What is he if not a murderer?" He raised his head and a smile broke out on his face. Now he looked like a man who had done what had to be done and was about to take up his practical duties again. He took off his coat, sat down on the ground, and looked around him. He showed no signs of agitation.

For a moment it seemed as if he were about to ask a question. All eyes were on him. He bowed his head. They averted their eyes.

"It's not a big thing to ask, I think," he said to himself. "I didn't want to ask you to do it, I don't know if I should have asked you. The day of judgment will come in the end. If not in this world then in the next. I can't imagine life without justice."

He did not seem confused. There was a straightforward kind of matter of factness in his look. As if he wanted to bring a certain matter up for discussion, a matter which had become a little complicated, but not to such an extent that it could not be discussed with people who were close to him.

He took his tobacco out of his pocket, rolled himself a cigarette, lit it and inhaled the smoke.

Everyone breathed a sigh of relief. He said: "This is good tobacco. It's got the right degree of moisture. You remember how we used to fight over cigarette stubs? We lost our human image. Pardon me—do you say human image or divine image?"

"Neither," said a voice from behind.

This remark was apparently not to his liking. He clamped his teeth on the cigarette and passed his hand over his hair. Now you could see how old he was: not more than thirty-five. His cheeks were slightly lined, his nose was straight, and his ears were set close to his head. There was a concentrated look in his eyes.

"How much do I owe?" he asked one of the others. "I lost, I think."

"It's all written down. You'll pay us back later."

"I don't like being in debt. How much do I owe?"

There was no response. He inhaled and blew the smoke out downward. "Strange," he said. "The war is over. I never imagined it would end like this."

Darkness fell and the tension relaxed. Zigi looked slightly ashamed of the scandal he had caused.

And while they were all sitting there, Zigi rose to his feet, stretched his arms, and raised his knees as if he were about to run a race. In the camp too he had been in the habit of taking short runs, in order to warm himself up. They had saved him then from depression.

Now it seemed as if he were about to take a run, as in the old days. One, two, he said, and set out. He ran six full rounds, and on the seventh he rose into the air and with a broad, slow movement cast himself into the water.

For a moment they all stood rooted to the spot. Then they all rushed together to the single hurricane lamp and stood waving it in the air. "Zigi, Zigi," they cried. A few of them jumped into the river.

All night long they labored in the icy water. Some of them swam far out, but they did not find Zigi.

And when morning broke the river was smooth and placid. A greenish-blue light shone on its surface. No one spoke. They spread their clothes out to dry and the old moldy smell, which seemed to have gone away, rose once more into the air.

Afterward they lit a fire and sat down to eat. Their hunger was voracious. The loaves of bread disappeared one after the other.

Tzili forgot herself for a moment. Zigi's athletic run went on flashing past her eyes, with great rapidity. It seemed to her that he would soon rise from the river, shake the water off his body, and announce: "The river's fine for swimming."

In the afternoon the place suddenly seemed confined and threatening, the light oppressive. The peasant women came and spread their wares on their flowered cloths, but no one bought anything. The women sat and looked at them with watchful eyes. One of them asked: "Why aren't you buying today? We have bread and smoked meat. Fresh milk too."

"Let's go," someone said, and immediately they all stood up. Tzili too raised her heavy body from the ground. No one asked: "Where to?" A dumb wonder stared from their faces, as after enduring grief. Tzili was glad that the haversack was empty, and now she had nothing but her own body to carry.

29

They walked along the riverside, toward the south. The sun shone on the green fields. Now it seemed that Zigi Baum was floating on the current, his arms outspread. Every now and then his image was reflected on the surface of the water. No one stopped to gaze at this shining reflection. The current widened as it approached the dam, a mighty torrent of water.

Later on a few people turned off to the right. They turned off together, without asking any questions or saying good-by. Tzili watched them walk away. They showed no signs of anger or of happiness. They went on walking at the same pace—for some reason, in another direction.

Tzili, it appeared, was already in the sixth month of her pregnancy. Her belly was taut and heavy but her legs, despite the difficulties of the road, walked without stumbling. When the refugees stopped to rest, they ate in silence. The strange disappearance of Zigi Baum had infected them with a subtle terror, unlike anything they had experienced before.

Tzili was happy. Not a happiness which had any outward manifestations: the fetus stirring inside her gave her an appetite and a lust for life. Not so the others: death clung even to their clothes. They tried to shake it off by walking.

From time to time they quickened their pace and Tzili fell behind. They were as absorbed in themselves now as they had been before in their card game. No one asked: "Where is she?" but nevertheless Tzili felt that their closeness to her was stronger than their distraction.

She no longer thought much about Mark. As if he had set out on a long journey from which it would take a long time to return. He appeared to her now as a tiny figure on the distant horizon, beyond the reach of her voice. She still loved him, but with a different kind of love. A love which had no real taste. From time to time a kind of awe descended on her and she knew: it was Mark, watching her—not uncritically—from afar.

She would say: Mark is inside me, but she didn't really feel it. The fetus was now hers, a secret which no one but she could touch.

Once, when they had stopped to rest, a woman asked her: "Isn't it hard for you?"

"No," said Tzili simply.

"And do you want the baby?"

"Yes."

The woman was surprised by Tzili's reply. She looked at her as if she were some stupid, senseless creature. Then she was sorry and her expression changed to one of wonder and pity: "How will you bring it up?"

"I'll keep it with me all the time," said Tzili simply.

Tzili too wanted to ask: "Where are you from?" But she had learned not to ask. On their last halt a quarrel had broken out between two women as a result of a tactless question. People were very tense and questions brought their repressed anger seething to the surface.

"How old are you?" asked the woman.

"Fifteen."

"So young." Wonder softened the woman's face.

Tzili offered her a piece of bread and she said, "Thank you."

"I," said the woman, "have lost my children. It seems to me that I did everything I could, but they were lost anyway. The oldest was nine and the youngest seven. And I am alive, as you see, even eating. Me they didn't harm. I must be made of iron."

A pain shot through Tzili's diaphragm and she closed her eyes.

"Don't you feel well?" asked the woman.

"It'll pass," said Tzili.

"Give me your mug and I'll fetch you some water."

When the woman returned Tzili was already sitting calmly on the ground. The woman raised the cup to Tzili's mouth and Tzili drank. The woman now wanted more than anything to help Tzili, but she did not know how. Tzili, in spite of everything, had more food than she did.

Straight after this night fell and the woman sank to the ground and slept. She shrank to the size of a child of six. Tzili wanted to cover the woman with her tattered coat, but she immediately suppressed this impulse. She did not want to frighten her.

The others were awake but passive. The isolated words which fluttered in the air were as inward as a conversation between two lovers, no longer young.

The night was warm and fine and Tzili remembered the little yard at home, where she had spent so many hours. Every now and then her mother would call, "Tzili," and Tzili would reply, "Here I am." Of her entire childhood, only this was left. All the rest was shrouded in a heavy mist. She was seized by longing for the little yard. As if it were the misty edge of the Garden of Eden.

"I have to eat." She banished the vision and immediately put her hand into the haversack and tore off a piece of bread. The bread was dry. A few grains of coal were embedded in its bottom crust. She liked the taste of the bread. Afterward she ate a little smoked meat. With every bite she felt her hunger dulled.

30

The summer took them by surprise, hot and broad, filling them with a will to live. The paths all flowed together into green creeks, bordered by tall trees. Refugees streamed from all directions, and for some reason the sight recalled summer holidays, youth movements, seasonal vacations, all kinds of forgotten youthful pleasures. Words from the old lexicon floated in the air. Only their clothes, like an eternal disgrace, went on steaming.

Tzili sat still, this happiness made her anxious. Soon it would give way to screams, pain, and despair.

That night they made a fire, sang and danced, and drank. And as after every catastrophe: embraces, couplings, and despondency in their wake. Tall women with the traces of an old elegance still clinging to them lay sunbathing shamelessly next to the lake.

"What does it matter—there's no point in living anymore anyway," a woman who had apparently run wild all night confessed. She was strong and healthy, fit to bring many more children into the world.

"And you won't go to Palestine?" asked her friend.

"No," said the woman decisively.

"Why not?"

"I want to go to hell."

From this conversation Tzili absorbed the word "Palestine." Once when her sister Yetty had become involved with the Moravian officer, there had been talk of sending her to Palestine. At first Yetty had refused, but then she changed her mind and wanted to go. But by then they didn't have the money to send her. Now Tzili thought often of her sister Yetty. Where was she now?

Tzili's fears were not in vain. The calamities came thick and fast: one woman threw herself into the lake and another swallowed poison. The marvelous oblivion was gone in an instant and the same healthy woman, the one who had refused to go to Palestine, announced: "Death will follow us all our lives, wherever we go. There'll be no more peace for us."

In the afternoon the body was recovered from the lake and the funerals took place one after the other. One of the men, who had the look of a public official even in his rags, spoke at length about the great obligations which were now facing them all. He spoke about memory, the long memory of the Jewish people, the eternal life of the tribe, and the historic necessity of the return to the motherland. Many wept.

After the funeral there was a big argument and the words of the official were heard again. It appeared that the woman who had taken poison had taken it because of a broken promise: someone who wanted to sleep with her had promised to marry her, and the next day he had changed his mind. The woman, who in all the years of suffering had kept the poison hidden in the lining of her coat without using it, had used it now. And something else: before taking the poison the woman had announced her intention of taking it, but no one had believed her.

Now there was nothing left but to say: Because of one night in bed a person commits suicide? So what if he slept with her? So what if he promised her? What do we have left but for the little pleasures of life? Do we have to give those up too?

Tzili took in the words with her eyes shut. She understood the words now, but she did not justify any of them in her heart. She sensed only one thing: the grief which had washed through her too had now become empty and pointless.

31

Now they streamed with the sun toward the sea. And at night they grilled silver fish, fresh from the river, on glowing coals. The nights were warm and clear, bringing to mind a life in which pleasures were real.

There was no lack of quarrels in this mixture. The summer sun worked its magic. As if the years in the camps had vanished without a trace. A forgetfulness which was not without humor. Like, for example, the woman who performed night after night, singing, reciting, and exposing her thighs. No one reminded her of her sins in the labor camp. She was now their carnival queen.

Now too there were those who could not stand the merriment and left. There was no lack of prosecutors, accusers, stirrers up of the past, and spoilsports. At this time too, the first visionaries appeared: short, ardent men who spoke about the salvation of the soul with extraordinary passion. You couldn't get away from them. But the desire to forget was stronger than all these. They ate and drank until late at night.

"What are you doing here?" A man would accost her from time to time, but on seeing that she was pregnant he would withdraw at once and leave her alone.

Tzili was very weak now. The long march had worn her out. From time to time a pain would pierce her and afterward she would feel giddy. Her legs swelled up too, but she bit her lips and said nothing. She was proud that her legs bore her and her baby. For some reason she believed that if her legs were healthy no harm would befall her.

And her life narrowed down to little worries. She forgot everyone and if she remembered them it was casually and absentmindedly. She was with herself, or rather with her body, which kept her occupied day and night. Sometimes someone offered her a piece of fish or bread. When she was very hungry she would stretch out her hand and beg. She wasn't ashamed to beg.

Without anyone noticing, the green creeks turned into a green plain dotted with little lakes. The landscape was so lovely that it hurt, but people were so obsessed with their merrymaking that they took no notice of the change. After a night of drinking they would sleep.

The convoy proceeded slowly and at a ragged pace. Sometimes a sudden panic took hold of them and made them run. Tzili limped after them with the last of her strength. They traipsed from place to place as if they were at the mercy of their changing moods. At this time fate presented Tzili with a moment of peace. Everything was full of joy—the light and the water and her body bearing her baby within it—but not for long.

During one of the panic flights she felt she could not go on. She tried to get up but immediately collapsed again. But for the fat woman, the one who sang and recited and bared her thighs—but for her and the fact that she noticed Tzili's absence and immediately cried: "We've left the child behind"—they would have gone on without her. At first no one paid any attention to her cry, but she was determined to be heard. She called out again, with a kind of authority, like a woman used to raising her voice, and the convoy drew to a halt.

No one knew what to do. During the years of the war they had learned to run and to stop for no one. The fat woman made them stop. "Man is not an insect. This time no one will shirk his duty." A sudden shame covered their faces.

There was no doctor among them, but there was a man who had been a merchant in peacetime and claimed that he had once taken a course in first aid, and he said: "We'll have to carry her on a stretcher." Strange: the words did their work at once. One of them went to fetch wood and another rope, and the skinny merchant, who never opened his mouth, knelt down and with movements that were almost prayerful he joined and he knotted. And they produced a sheet too, and a ragged blanket, and even some pins and some hooks. By nightfall the merchant could survey his handiwork and say: "She'll be quite comfortable on this."

And the next day when the stretcher bearers lifted the stretcher onto their shoulders and set out at the head of the convoy, a mighty song burst from their throats. A rousing sound, like pent up water bursting from a dam. "We are the torch bearers," roared the stretcher bearers, and everyone else joined in.

They carried the stretcher along the creeks and sang. The summer, the glorious summer, turned every corner golden. Tzili herself closed her eyes and tried to make the giddiness go away. The merchant urged the stretcher bearers on: "Run, boys, run. The child needs a doctor." All his anxieties gathered together in his face. And when they stopped he would sit next to her and feed her. He bought whatever he could lay his hands on, but to Tzili he gave only milk products and fruit. Tzili had lost her appetite.

"Thank you," said Tzili.

"There's no reason to thank me."

"Why not?"

"What else have I got to do?" His eyes opened and in the white of the left eye a yellow stain glittered. His despair was naked.

"You're helping me."

"What of it?"

And Tzili stopped thanking him.

At night he would fold his legs and sleep at her side. And Tzili was suddenly freed of the burden of her survival. The stretcher bearers took turns carrying her from place to place. There was not a village or a town to be seen, only here and there a house, here and there a farmer.

"Where are you from?" asked Tzili.

The merchant told her, unwillingly and without going into detail, but he did tell her about Palestine. In his youth he had wanted to go to Palestine. He had spent some time on a Zionist training farm, and he even had a certificate, but his late father had fallen ill and his illness had lasted for years. After that he had married and had children.

There was nothing captivating in the way he spoke. It was evident that he wanted to cut things short in everything concerning himself, like a merchant who put his trust in practical affairs and knew that they took precedence over emotions. Tzili asked no further. He himself left the stretcher only to fetch milk for her. Tzili drank the milk in spite of herself, so that he would not worry.

He never asked: "Where are you from?" or "What happened to you?" He would sit by her side as dumb as an animal. His face was ageless. Sometimes he looked old and clumsy and sometimes as agile as a man of thirty.

Once Tzili tried to get off the stretcher. He scolded her roundly. On no account was she to get off the stretcher until she saw a doctor. He knew that this was so from the first aid course.

And the fat woman who had saved Tzili started entertaining them again at night. She would sing and recite and expose her fat thighs. The merchant raised Tzili's head and she saw everything. She felt no affection for any of them, but they were carrying her, taking turns to carry her, from place to place. Between one pain and the next she wanted to say a kind word to the merchant, but she was afraid of offending him. He for his part walked by her side like a man doing his duty, without any exaggeration. Tzili grew accustomed to him, as if he were an irritating brother.

And thus they reached Zagreb. Zagreb was in turmoil. In the yard of the Joint Distribution Committee people were distributing biscuits, canned goods, and colored socks from America. In the courtyard they all mingled freely: visionaries, merchants, moneychangers, and sick people. No one knew what to do in the strange, half-ruined city. Someone shouted loudly: "If you want to get to Palestine, you'd better go to Naples. Here they're nothing but a bunch of money-grubbing profiteers and crooks."

The stretcher bearers put the stretcher down in a shady corner and said: "From now on somebody else can take over." The merchant was alarmed by this announcement and he implored them: "You've done great things, why not carry on?" But they no longer took any notice of him. The sight of the city had apparently confused them. Suddenly they looked tall and ungainly. In vain the merchant pleaded with them. They stood their ground: "From now on it's not our job." The merchant stood helplessly in the middle of the courtyard. There was no doctor present, and the officials of the Joint Committee were busy defending themselves from the survivors, who assailed their caged counters with great force.

If only the merchant had said, "I can't go on anymore," it would have been easier for Tzili. His desperate scurrying about hurt her. But he did not abandon her. He kept on charging into the crowd and asking: "Is there a doctor here? Is there a doctor here?"

People came and went and in the big courtyard, enclosed in a wall of medium height, men and women slept by day and by night. Every now and then an official would emerge and threaten the sleepers or the people besieging the doors. The official's neat appearance recalled other days, but not his voice.

And there was a visionary there too, thin and vacant-faced, who wandered through the crowds muttering: "Repent, repent." People would throw him a coin on condition that he shut up. And he would accept the condition, but not for long.

Pain assailed Tzili from every quarter. Her feet were frozen. The merchant ran from place to place, drugged with the little mission he had taken upon himself. No one came to his aid. When night fell, he put his head between his knees and wept.

In the end a military ambulance came and took her away. The merchant begged them: "Take me, take me too. The child has no one in the whole world." The driver ignored his despairing cries and drove away.

Tzili's pains were very bad, and the sight of the imploring merchant running after the ambulance made them worse. She wanted to scream, but she didn't have the strength.

32

It was a makeshift hospital housed in an army barracks partitioned with blankets. Soldiers and partisans, women and children, lay crowded together. Screams rose from every side. Tzili was placed on a big bed, apparently requisitioned from one of the bombed houses.

For days she had not heard the throbbing of the fetus. Now it seemed to her that it was stirring again. The nurse sponged her down with a warm, wet cloth and asked: "Where are you from?" And Tzili told her. The broad, placid face of the gentile nurse brought her a sudden serenity. It was evident that the young nurse came from a good home. She did her work quietly, without superfluous gestures.

Tzili asked wonderingly: "Where are you from?" "From here," said the nurse. A disinterested light shone from her blue eyes. The nurse told her that every day more soldiers and refugees were brought to the hospital. There were no beds and no doctors. The few doctors there were torn between the hospitals scattered throughout the ruined city.

Later Tzili fell asleep. She slept deeply. She saw Mark and he looked like the merchant who had taken care of her. Tzili told him that she had been obliged to sell all the clothes in the haversack and in the commotion she had lost the haversack too. Perhaps it was with the merchant. "The merchant?" asked Mark in surprise. "Who is this merchant?" Tzili was alarmed by Mark's astonished face. She told him, at length, of all that had happened to her since leaving the mountain. Mark bowed his head and said: "It's not my business anymore." There was a note of criticism in his voice. Tzili made haste to appease him. Her voice choked and she woke up.

The next day the doctor came and examined her. He spoke German. Tzili answered his hurried questions quietly. He told the nurse that she had to be taken to the surgical ward that same night. Tzili saw the morning light darken next to the window. The bars reminded her of home.

They took her to the surgical ward while it was still light. There was a queue and the gentile nurse, who spoke to her in broken German mixed with Slavic words, held her hand. From her Tzili learned that the fetus inside her was dead, and that soon it would be removed from her womb. The anesthetist was a short man wearing a Balaklava hat. Tzili screamed once and that was all.

Then it was night. A long night, carved out of stone, which lasted for three days. Several times they tried to wake her. Medics and soldiers rushed frantically about carrying stretchers. Tzili wandered in a dark stone tunnel, strangers and acquaintances passing before her eyes, clear and unblurred. I'm going back, she said to herself and clung tightly to the wooden handle.

When she woke the nurse was standing beside her. Tzili asked, for some reason, if the merchant too had been hurt. The nurse told her that the operation had not taken long, the doctors were satisfied, and now she must rest. She held a spoon to her mouth.

"Was I good?" asked Tzili.

"You were very good."

"Why did I scream?" she wondered.

"You didn't scream, you didn't make a sound."

In the evening the nurse told her that she had not stirred from the hospital for a whole week. Every day they brought more soldiers and refugees, some of them badly hurt, and she could not leave. Her fiancé was probably angry with her. Her round face looked worried.

"He'll take you back," said Tzili.

"He's not an easy man," confessed the nurse.

"Tell him that you love him."

"He wants to sleep with me," the nurse whispered in her ear.

Tzili laughed. The thin gruel and the conversation distracted her from her pain. Her mind was empty of thought or sorrow. And the pain too grew duller. All she wanted was to sleep. Sleep drew her like a magnet.

33

She fell asleep again. In the meantime the soldiers and refugees crammed the hut until there was no room to move. The medics pushed the beds together and they moved Tzili's bed into the doorway. She slept. Someone strange and far away ordered her not to dream, and she obeyed him and stopped dreaming. She floated on the surface of a vacant sleep for a few days, and when she woke her memory was emptier than ever.

The hut stretched lengthwise before her, full of men, women, and children. The torn partitions no longer hid anything. "Don't shout," grumbled the medics, "it won't do you any good." They were tired of the commotion and of the suffering. The nurses were more tolerant, and at night they would cuddle with the medics or the ambulant patients.

Tzili lay awake. Of all her scattered life it seemed to her that nothing was left. Even her body was no longer hers. A jumble of sounds and shapes flowed into her without touching her.

"Are you back from your leave?" she remembered to ask the nurse.

"I quarreled with my fiancé."

"Why?"

"He's jealous of me. He hit me. I swore never to see him again." Her big peasant hands expressed more than her face.

"And you, did you love him?" she asked Tzili without looking at her.

"Who?"

"Your fiancé."

"Yes," said Tzili, quickly.

"With Jews, perhaps, it's different."

Bitter lines had appeared overnight on her peasant's face. Tzili now felt a kind of solidarity with this country girl whose fiancé had beaten her with his hard fists.

At night the hut was full of screams. One of the medics attacked a refugee and called him a Jewish crook. A sudden dread ran through Tzili's body.

The next day, when she stood up, she realized for the first time that she had lost her sense of balance too. She stood leaning against the wall, and for a moment it seemed to her that she would never again be able to stand upright without support.

"Haven't you seen a haversack anywhere?" she asked one of the medics.

"There's disinfection here. We burn everything."

Women who were no longer young stood next to the lavatories and smeared creams on their faces. They spoke to each other in whispers and laughed provocatively. The years of suffering had bowed their bodies but had not destroyed their will to live. One of the women sat on a bench and massaged her swollen legs with pulling, clutching movements.

Later the medics brought in a lot of new patients. They reclassified the patients and put the ones who were getting better out in the yard.

They put Tzili's bed out too. All the gentile nurses' pleading was in vain.

The next day officials from the Joint Committee came to the yard and distributed dresses and shoes and flowered petticoats. There was a rush on the boxes, and the officials who had come to give things to the women had to beat them off instead. Tzili received a red dress, a petticoat, and a pair of high-heeled shoes. A heavy smell of perfume still clung to the crumpled goods.

"What are you fighting for?" an official asked accusingly.

"For a pretty dress," one of the women answered boldly.

"You people were in the camps weren't you? From you we expect something different," said someone in an American accent.

Later the gentile nurse came and spoke encouragingly to Tzili. "You must be strong and hold your head high. Don't give yourself away and don't show any feelings. What happened to you could have happened to anyone. You have to forget. It's not a tragedy. You're young and pretty. Don't think about the past. Think about the future. And don't get married."

She spoke to her like a loyal friend, or an older sister. Tzili felt the external words spoken by the gentile nurse strengthening her. She wanted to thank her and she didn't know how. She gave her the petticoat she had just received from the Joint Committee. The nurse took it and put it into the big pocket in her apron.

Early in the morning they chased everyone out of the yard.

34

Now everyone streamed to the beach. Fishermen stood by little booths and sold grilled fish. The smell of the fires spread a homely cheerfulness around. Before the war the place had evidently been a jolly seaside promenade. A few traces of the old life still clung to the peeling walls.

Beyond the walls lay the beach, white and spotted with oil stains, here and there an old signpost, a few shacks and boats. Tzili was weak and hungry. There was no familiar face to which she could turn, only strange refugees with swollen packs on their backs and hunger and urgency on their faces. They streamed over the sand to the sea.

Tzili sat down and watched. The old desire to watch came back to her. At night the people lit fires and sang rousing Zionist songs. No one knew how long they would be there. They had food. Tzili too went down to the sea and sat among the refugees. The wound in her stomach was apparently healing. The pain was bad but not unendurable.

"These fish are excellent."

"Fish is good for you."

"I'm going up to buy another one."

These sentences for some reason penetrated into Tzili's head, and she marveled at them.

Somewhere a quarrel broke out. A hefty man shouted at the top of his voice: "No one's going to kill me anymore." Somewhere else people were dancing the hora. One of the refugees sitting next to Tzili remarked: "Palestine's not the place for me."

"Why not?" his friend asked him teasingly.

"I'm tired."

"But you're still strong."

"Yes, but there's no more faith in me."

"And what are you going to do instead?"

"I don't know."

Someone lit an oil lamp and illuminated the darkness. The voice of the refugee died down.

And while Tzili sat watching a fat woman approached her and said: "Aren't you Tzili?"

"Yes," she said. "My name is Tzili."

It was the fat woman who had entertained them on their way to Zagreb, singing and reciting and baring her fleshy thighs.

"I'm glad you're here. They've all abandoned me," she said and lowered her heavy body to the ground. "With all the pretty shiksas here, what do they need me for?"

"And where are you going to go?" said Tzili carefully.

"What choice do I have?" The woman's reply was not slow in coming.

For a moment they sat together in silence.

"And you?" asked the woman.

Tzili told her. The fat woman stared at her, devouring every detail. All the great troubles inhabiting her great body seemed to make way for a moment for Tzili's secret.

"I too have nobody left in the world. At first I didn't understand, now I understand. There's the world, and there's Linda. And Linda has nobody in the whole wide world."

One of the officials got onto a box. He spoke in grand, thunderous words. As if he had a loudspeaker stuck to his mouth. He spoke of Palestine, land of liberty.

"Where can a person buy a grilled fish?" said Linda. "I'm going to buy a grilled fish. The hunger's driving me out of my mind. I'll be right back. Don't you leave me too."

Tzili was captivated for a moment by the speaker's voice. He thundered about the need for renewal and dedication. No one interrupted him. It was evident that the words had been pent up in him for a long time. Now their hour had come.

Linda brought two grilled fish. "Linda has to eat. Linda's hungry." She spoke about herself in the third person. She held a fish in a cardboard wrapper out to Tzili.

Tzili tasted and said: "It tastes good."

"Before the war I was a cabaret singer. My parents disapproved of my way of life," Linda suddenly confessed.

"They've forgiven you," said Tzili.

"No one forgives Linda. Linda doesn't forgive herself."

"In Palestine everything will be different," said Tzili, repeating the speaker's words.

Linda chewed the fish and said nothing.

Tzili felt a warm intimacy with this fat woman who spoke about herself in the third person.

All night the speakers spoke. Loud words flooded the dark beach. A thin man spoke of the agonies of rebirth in Palestine. Linda did not find these voices to her taste. In the end she could no longer restrain herself and she called out: "We've had enough words. No more words." And when the speaker took no notice of her threats she went and stood next to the box and announced: "This is fat Linda here. Don't anyone dare come near this box. I'm declaring a cease-words. It's time for silence now." She went back and sat down. No one reacted. People were tired, they huddled in their coats. After a few moments she said to herself: "Phooey. This rebirth makes me sick."

That same night they were taken aboard the ship. It was a small ship with a bare mast and a chimney. Two projectors illuminated the shore.

"What I'd like now," said Tzili for some reason, "is a pear."

"Linda hasn't got a pear. What a pity that Linda hasn't got a pear."

"I feel ashamed," said Tzili.

"Why do you feel ashamed?"

"Because that's what came into my head."

"I have every respect for such little wishes. Linda herself is all one little wish."

For the time being the sight was not an inspiring one. People climbed over ropes and tarpaulins. Someone shouted: "There's a queue here, no one will get in without waiting in the queue."

The crush was bad and Tzili felt that pain was about to engulf her again. Linda no longer waited for favors and in a thunderous voice she cried: "Make way for the girl. The girl has undergone an operation." No one moved. Linda shouted again, and when no one paid any attention she spread out her arms and swept a couple of young men from their places on a bench.

"Now, in the name of justice, she'll sit down. Her name is Tzili."

Later on, when the commotion had died down and some of the people had gone down to the cabins below and a wind began to blow on the deck, Tzili said: "Thank you."

"What for?"

"For finding me a place."

"Don't thank me. It's your place."

Afterward shouts were heard from below. People were apparently beating the informers and collaborators in the dark, and the latter were screaming at the tops of their voices. Up on the deck, too, there was no peace. In vain the officials tried to restore order.

Between one scream and the next Linda told Tzili what had happened to her during the war. She had a lover, a gentile estate owner who had hidden her in his granaries. She moved from one granary to another. At first she had a wonderful time, she was very happy. But later she came to realize that her lover was a goy in every sense of the word, drunk and violent. She was forced to flee, and in the end she fled to a camp. She didn't like the Jews, but she liked them better than the gentiles. Jews were sloppy but not cruel. She was in the camp for a full year. She learned Yiddish there, and every night she performed for the inmates. She had no regrets. There was a kind of cruel honesty in her brown eyes.

The little ship strained its engines to cross the stormy sea. Up on the deck they did not feel it rock. Most of the day the passengers slept in the striped coats they had been given by the Joint Committee. From time to time the ship sounded its horn.

Linda managed to get hold of a bottle of brandy at last, and her joy knew no bounds. She hugged the bottle and spoke to it in Hungarian. She started drinking right away, and when her heart was glad with brandy she began to sing. The songs she sang were old Hungarian lullabies.

18. Tadeusz Borowski

Tadeusz Borowski was born in Zhitomir in the Russian Ukraine in 1922. His parents were Polish. When Borowski was only four, his father was exiled to a harsh labor camp beyond the Arctic Circle, and four years later, his mother was deported to Siberia. Borowski was raised by an aunt. By 1934, both parents had been repatriated, and the family was reunited in Warsaw.

During World War II, Borowski continued his education secretly, worked with various underground presses, and published a first volume of poems in a small edition. His fiancée was arrested, and while searching for her, Borowski was trapped and seized by the Gestapo. He and his fiancée were sent to Auschwitz in April 1943. As non-Jews, they were not candidates for the gas chamber. Later, Borowski was transported to Dachau; both he and his fiancée survived. Borowski's experiences in the camp formed the basis for the stories in *This Way for the Gas, Ladies and Gentlemen.* These stories were selected from two volumes published in Poland in 1948: *Farewell to Maria* and *A World of Stone.*

Borowski later turned to Communist party journalism, though it is not clear why he chose to serve the state instead of his own talents. He seems never to have recovered from his Auschwitz experience and the subsequent disillusionment with history and politics. In 1951, three days after the birth of his first daughter, he turned on the gas in his kitchen and committed suicide. He was not yet thirty.

Just as the neutral style of "This Way for the Gas, Ladies and Gentlemen" is designed to shock more than to report, so the first-person narrative is intended as more than mere autobiography. Borowski's confessional mode conceals as much as it reveals, drawing the reader into a complicity with events along with the narrator, who finds his detachment foiled by the episodes he describes. Staying alive in Auschwitz sabotaged the moral codes that allowed the victims—and, through a kind of imaginative osmosis, the readers of these tales—to distance themselves from its evil. Borowski's relentless pursuit of the pristine truth of the ordeal strips the façade of piety from us all, leaving us facing the naked world of Holocaust atrocity defenseless and unconsoled.

This Way for the Gas, Ladies and Gentlemen

All of us walk around naked. The delousing is finally over, and our striped suits are back from the tanks of Cyclone B solution, an efficient killer of lice in clothing and of men in gas chambers. Only the inmates in the blocks cut off from ours by the 'Spanish goats'[1] still have nothing to wear. But all the same, all of us walk around naked: the heat is unbearable. The camp has been sealed off tight. Not a single prisoner, not only solitary louse, can sneak through the gate. The labour Kommandos have stopped working. All day, thousands of naked men shuffle up and down the roads, cluster around the squares, or lie against the walls and on top of the roofs. We have been sleeping on plain boards, since our mattresses and blankets are still being disinfected. From the rear blockhouses we have a view of the F.K.L.—*Frauen Konzentration Lager;** there too the delousing is in full swing. Twenty-eight thousand women have been stripped naked and driven out of the barracks. Now they swarm around the large yard between the blockhouses.

The heat rises, the hours are endless. We are without even our usual diversion: the wide roads leading to the crematoria are empty. For several days now, no new transports have come in. Part of 'Canada'[2] has been liquidated and detailed to a labour Kommando—one of the very toughest—at Harmenz. For there exists in the camp a special brand of justice based on envy: when the rich and mighty fall, their friends see to it that they fall to the very bottom. And Canada, our Canada, which smells not of maple forests but of French perfume, has amassed great fortunes in diamonds and currency from all over Europe.

Several of us sit on the top bunk, our legs dangling over the edge. We slice the neat loaves of crisp, crunchy bread. It is a bit coarse to the taste, the kind that stays fresh for days. Sent all the way from Warsaw—only a week ago my mother held this white loaf in her hands . . . dear Lord, dear Lord . . .

We unwrap the bacon, the onion, we open a can of evaporated milk. Henri, the fat Frenchman, dreams aloud of the French wine brought by the transports from Strasbourg, Paris, Marseille . . . Sweat streams down his body.

*The women's camp barracks.

'Listen, *mon ami,* next time we go up on the loading ramp, I'll bring you real champagne. You haven't tried it before, eh?'

'No. But you'll never be able to smuggle it through the gate, so stop teasing. Why not try and "organize" some shoes for me instead—you know, the perforated kind, with a double sole, and what about that shirt you promised me long ago?'

'*Patience, patience.* When the new transports come, I'll bring all you want. We'll be going on the ramp again!'

'And what if there aren't any more "cremo" transports?' I say spitefully. 'Can't you see how much easier life is becoming around here: no limit on packages, no more beatings? You even write letters home . . . One hears all kind of talk, and, dammit, they'll run out of people!'

'Stop talking nonsense.' Henri's serious fat face moves rhythmically, his mouth is full of sardines. We have been friends for a long time, but I do not even know his last name. 'Stop talking nonsense,' he repeats, swallowing with effort. 'They can't run out of people, or we'll starve to death in this blasted camp. All of us live on what they bring.'

'All? We have our packages . . .'

'Sure, you and your friend, and ten other friends of yours. Some of you Poles get packages. But what about us, and the Jews, and the Russkis? And what if we had no food, no "organization" from the transports, do you think you'd be eating those packages of yours in peace? We wouldn't let you!'

'You would, you'd starve to death like the Greeks. Around here, whoever has grub, has power."

'Anyway, you have enough, we have enough, so why argue?'

Right, why argue? They have enough, I have enough, we eat together and we sleep on the same bunks. Henri slices the bread, he makes a tomato salad. It tastes good with the commissary mustard.

Below us, naked, sweat-drenched men crowd the narrow barracks aisles or lie packed in eights and tens in the lower bunks. Their nude, withered bodies stink of sweat and excrement; their cheeks are hollow. Directly beneath me, in the bottom bunk, lies a rabbi. He has covered his head with a piece of rag torn off a blanket and reads from a Hebrew prayer book (there is no shortage of this type of literature at the camp), wailing loudly, monotonously.

'Can't somebody shut him up? He's been raving as if he'd caught God himself by the feet.'

'I don't feel like moving. Let him rave. They'll take him to the oven that much sooner.'

'Religion is the opium of the people,' Henri, who is a Communist and a *rentier,** says sententiously. 'If they didn't believe in God and eternal life, they'd have smashed the crematoria long ago.'

"Why haven't you done it then?'

*Stockholder or annuitant, although here probably used ironically for a person of independent means.

The question is rhetorical; the Frenchman ignores it.

'Idiot,' he says simply, and stuffs a tomato in his mouth.

Just as we finish our snack, there is a sudden commotion at the door. The Muslims[3] scurry in fright to the safety of their bunks, a messenger runs into the Block Elder's shack. The Elder, his face solemn, steps out at once.

'Canada! *Antreten!** But fast! There's a transport coming!'

'Great God!' yells Henri, jumping off the bunk. He swallows the rest of his tomato, snatches his coat, screams *'Raus'* at the men below, and in a flash is at the door. We can hear a scramble in the other bunks. Canada is leaving for the ramp.

'Henri, the shoes!' I call after him.

'Keine Angst!'† he shouts back, already outside.

I proceed to put away the food. I tie a piece of rope around the suitcase where the onions and the tomatoes from my father's garden in Warsaw mingle with Portuguese sardines, bacon from Lublin (that's from my brother), and authentic sweetmeats from Salonica. I tie it all up, pull on my trousers, and slide off the bunk.

'Platz!' I yell, pushing my way through the Greeks. They step aside. At the door I bump into Henri.

'Was ist los?'‡

'Want to come with us on the ramp?'

'Sure, why not?'

'Come along then, grab your coat! We're short of a few men. I've already told the Kapo,' and he shoves me out of the barracks door.

We line up. Someone has marked down our numbers, someone up ahead yells, 'March, march,' and now we are running towards the gate, accompanied by the shouts of a multilingual throng that is already being pushed back to the barracks. Not everybody is lucky enough to be going on the ramp . . . We have almost reached the gate. *Links, zwei, drei, vier! Mützen ab!*§ Erect, arms stretched stiffly along our hips, we march past the gate briskly, smartly, almost gracefully. A sleepy S.S. man with a large pad in his hand checks us off, waving us ahead in groups of five.

'Hundert!' he calls after we have all passed.

'Stimmt!'‖ comes a hoarse answer from out front.

We march fast, almost at a run. There are guards all around, young men with automatics. We pass camp II B, then some deserted barracks and a clump of unfamiliar green—apple and pear trees. We cross the circle of watchtowers and, running, burst on to the highway. We have arrived. Just a few more yards. There, surrounded by trees, is the ramp.

A cheerful little station, very much like any other provincial railway stop: a small square framed by tall chestnuts and paved with yellow

*"Line up!"
†"Don't worry!"
‡"What's the matter?"
§"Left, two, three, four. Caps off!"
‖"One hundred!" "Correct!"

gravel. Not far off, beside the road, squats a tiny wooden shed, uglier and more flimsy then the ugliest and flimsiest railway shack; farther along lie stacks of old rails, heaps of wooden beams, barracks parts, bricks, paving stones. This is where they load freight for Birkenau: supplies for the construction of the camp, and people for the gas chambers. Trucks drive around, load up lumber, cement, people—a regular daily routine.

And now the guards are being posted along the rails, across the beams, in the green shade of the Silesian chestnuts, to form a tight circle around the ramp. They wipe the sweat from their faces and sip out of their canteens. It is unbearably hot; the sun stands motionless at its zenith.

'Fall out!'

We sit down in the narrow streaks of shade along the stacked rails. The hungry Greeks (several of them managed to come along, God only knows how) rummage underneath the rails. One of them finds some pieces of mildewed bread, another a few half-rotten sardines. They eat.

'*Schweinedreck,*'* spits a young, tall guard with corn-coloured hair and dreamy blue eyes. 'For God's sake, any minute you'll have so much food to stuff down your guts, you'll bust!' He adjusts his gun, wipes his face with a handkerchief.

'Hey you, fatso!' His boot lightly touches Henri's shoulder. '*Pass mal auf,*† want a drink?'

'Sure, but I haven't got any marks,' replies the Frenchman with a professional air.

'*Schade,* too bad.'

'Come, come, Herr Posten, isn't my word good enough any more? Haven't we done business before? How much?'

'One hundred. *Gemacht?*'

'*Gemacht.*'‡

We drink the water, lukewarm and tasteless. It will be paid for by the people who have not yet arrived.

'Now you be careful,' says Henri, turning to me. He tosses away the empty bottle. It strikes the rails and bursts into tiny fragments. 'Don't take any money, they might be checking. Anyway, who the hell needs money? You've got enough to eat. Don't take suits, either, or they'll think you're planning to escape. Just get a shirt, silk only, with a collar. And a vest. And if you find something to drink, don't bother calling me. I know how to shift for myself, but you watch your step or they'll let you have it.'

'Do they beat you up here?'

'Naturally. You've got to have eyes in your ass. *Arschaugen.*'

Around us sit the Greeks, their jaws working greedily, like huge human insects. They munch on stale lumps of bread. They are restless, wondering what will happen next. The sight of the large beams and the stacks of rails has them worried. They dislike carrying heavy loads.

*"Filthy pig!"
†"Pay attention"
‡"Done?" "Done."

*'Was wir arbeiten?'** they ask.

'Niks. Transport kommen, alles Krematorium, compris?'†

'Alles verstehen,'‡ they answer in crematorium Esperanto. All is well—they will not have to move the heavy rails or carry the beams.

In the meantime, the ramp has become increasingly alive with activity, increasingly noisy. The crews are being divided into those who will open and unload the arriving cattle cars and those who will be posted by the wooden steps. They receive instructions on how to proceed most efficiently. Motor cycles drive up, delivering S.S. officers, bemedalled, glittering with brass, beefy men with highly polished boots and shiny, brutal faces. Some have brought their briefcases, others hold thin, flexible whips. This gives them an air of military readiness and agility. They walk in and out of the commissary—for the miserable little shack by the road serves as their commissary, where in the summertime they drink mineral water, *Studentenquelle,* and where in winter they can warm up with a glass of hot wine. They greet each other in the state-approved way, raising an arm Roman fashion, then shake hands cordially, exchange warm smiles, discuss mail from home, their children, their families. Some stroll majestically on the ramp. The silver squares on their collars glitter, the gravel crunches under their boots, their bamboo whips snap impatiently.

We lie against the rails in the narrow streaks of shade, breathe unevenly, occasionally exchange a few words in our various tongues, and gaze listlessly at the majestic men in green uniforms, at the green trees, and at the church steeple of a distant village.

'The transport is coming,' somebody says. We spring to our feet, all eyes turn in one direction. Around the bend, one after another, the cattle cars begin rolling in. The train backs into the station, a conductor leans out, waves his hand, blows a whistle. The locomotive whistles back with a shrieking noise, puffs, the train rolls slowly alongside the ramp. In the tiny barred windows appear pale, wilted, exhausted human faces, terror-stricken women with tangled hair, unshaven men. They gaze at the station in silence. And then, suddenly, there is a stir inside the cars and a pounding against the wooden boards.

'Water! Air!'—weary, desperate cries.

Heads push through the windows, mouths gasp frantically for air. They draw a few breaths, then disappear; others come in their place, then also disappear. The cries and moans grow louder.

A man in a green uniform covered with more glitter than any of the others jerks his head impatiently, his lips twist in annoyance. He inhales deeply, then with a rapid gesture throws his cigarette away and signals to the guard. The guard removes the automatic from his shoulder, aims, sends a series of shots along the train. All is quiet now. Meanwhile, the trucks have arrived, steps are being drawn up, and the Canada men stand

*"What we work?"

†"Nothin'. Transport comes, everything Crematorium, understand?"

‡"Everything understand."

ready at their posts by the train doors. The S.S. officer with the briefcase raises his hand.

'Whoever takes gold, or anything at all besides food, will be shot for stealing Reich property. Understand? *Verstanden?*'

'Jawohl!' we answer eagerly.

'Also los! Begin!'

The bolts crack, the doors fall open. A wave of fresh air rushes inside the train. People . . . inhumanly crammed, buried under incredible heaps of luggage, suitcases, trunks, packages, crates, bundles of every description (everything that had been their past and was to start their future). Monstrously squeezed together, they have fainted from heat, suffocated, crushed one another. Now they push towards the opened doors, breathing like fish cast out on the sand.

'Attention! Out, and take your luggage with you! Take out everything. Pile all your stuff near the exits. Yes, your coats too. It is summer. March to the left. Understand?'

'Sir, what's going to happen to us?' They jump from the train on to the gravel, anxious, worn-out.

'Where are you people from?'

'Sosnowiec-Będzin. Sir, what's going to happen to us?' They repeat the question stubbornly, gazing into our tired eyes.

'I don't know, I don't understand Polish.'

It is the camp law: people going to their death must be deceived to the very end. This is the only permissible form of charity. The heat is tremendous. The sun hangs directly over our heads, the white, hot sky quivers, the air vibrates, an occasional breeze feels like a sizzling blast from a furnace. Our lips are parched, the mouth fills with the salty taste of blood, the body is weak and heavy from lying in the sun. Water!

A huge, multicoloured wave of people loaded down with luggage pours from the train like a blind, mad river trying to find a new bed. But before they have a chance to recover, before they can draw a breath of fresh air and look at the sky, bundles are snatched from their hands, coats ripped off their backs, their purses and umbrellas taken away.

'But please, sir, it's for the sun, I cannot . . .'

*'Verboten!'** one of us barks through clenched teeth. There is an S.S. man standing behind your back, calm, efficient, watchful.

'Meine Herrschaften, this way, ladies and gentlemen, try not to throw your things around, please. Show some goodwill,' he says courteously, his restless hands playing with the slender whip.

'Of course, of course,' they answer as they pass, and now they walk alongside the train somewhat more cheerfully. A woman reaches down quickly to pick up her handbag. The whip flies, the woman screams, stumbles, and falls under the feet of the surging crowd. Behind her, a child cries in a thin little voice 'Mamele!'—a very small girl with tangled black curls.

*"Forbidden!"

The heaps grow. Suitcases, bundles, blankets, coats, handbags that open as they fall, spilling coins, gold, watches; mountains of bread pile up at the exits, heaps of marmalade, jams, masses of meat, sausages; sugar spills on the gravel. Trucks, loaded with people, start up with a deafening roar and drive off amidst the wailing and screaming of the women separated from their children, and the stupefied silence of the men left behind. They are the ones who had been ordered to step to the right—the healthy and the young who will go to the camp. In the end, they too will not escape death, but first they must work.

Trucks leave and return, without interruption, as on a monstrous conveyor belt. A Red Cross van drives back and forth, back and forth, incessantly: it transports the gas that will kill these people. The enormous cross on the hood, red as blood, seems to dissolve in the sun.

The Canada men at the trucks cannot stop for a single moment, even to catch their breath. They shove the people up the steps, pack them in tightly, sixty per truck, more or less. Near by stands a young, cleanshaven 'gentleman', an S.S. officer with a notebook in his hand. For each departing truck he enters a mark; sixteen gone means one thousand people, more or less. The gentleman is calm, precise. No truck can leave without a signal from him, or a mark in his notebook: *Ordnung muss sein.** The marks swell into thousands, the thousands into whole transports, which afterwards we shall simply call 'from Salonica', 'from Strasbourg', 'from Rotterdam'. This one will be called 'Sosnowiec-Będzin'. The new prisoners from Sosnowiec-Będzin will receive serial numbers 131–2—thousand, of course, though afterwards we shall simply say 131–2, for short.

The transports swell into weeks, months, years. When the war is over, they will count up the marks in their notebooks—all four and a half million of them. The bloodiest battle of the war, the greatest victory of the strong, united Germany. *Ein Reich, ein Volk, ein Führer*† and four crematoria.

The train has been emptied. A thin, pock-marked S.S. man peers inside, shakes his head in disgust and motions to our group, pointing his finger at the door.

'*Rein.* Clean it up!'

We climb inside. In the corners amid human excrement and abandoned wrist-watches lie squashed, trampled infants, naked little monsters with enormous heads and bloated bellies. We carry them out like chickens, holding several in each hand.

'Don't take them to the trucks, pass them on to the women,' says the S.S. man, lighting a cigarette. His cigarette lighter is not working properly; he examines it carefully.

'Take them, for God's sake!' I explode as the women run from me in horror, covering their eyes.

The name of God sounds strangely pointless, since the women and the

*"Everything in an orderly manner."

†"One nation, one people, one leader." Nazi slogan.

infants will go on the trucks, every one of them, without exception. We all know what this means, and we look at each other with hate and horror.

'What, you don't want to take them?' asks the pockmarked S.S. man with a note of surprise and reproach in his voice, and reaches for his revolver.

'You mustn't shoot, I'll carry them.' A tall, grey-haired woman takes the little corpses out of my hands and for an instant gazes straight into my eyes.

'My poor boy,' she whispers and smiles at me. Then she walks away, staggering along the path. I lean against the side of the train. I am terribly tired. Someone pulls at my sleeve.

'*En avant,* to the rails, come on!'

I look up, but the face swims before my eyes, dissolves, huge and transparent, melts into the motionless trees and the sea of people . . . I blink rapidly: Henri.

'Listen, Henri, are we good people?'

'That's stupid. Why do you ask?'

'You see, my friend, you see, I don't know why, but I am furious, simply furious with these people—furious because I must be here because of them. I feel no pity. I am not sorry they're going to the gas chamber. Damn them all! I could throw myself at them, beat them with my fists. It must be pathological, I just can't understand . . .'

'Ah, on the contrary, it is natural, predictable, calculated. The ramp exhausts you, you rebel—and the easiest way to relieve your hate is to turn against someone weaker. Why, I'd even call it healthy. It's simple logic, *compris?*' He props himself up comfortably against the heap of rails. 'Look at the Greeks, they know how to make the best of it! They stuff their bellies with anything they find. One of them has just devoured a full jar of marmalade.'

'Pigs! Tomorrow half of them will die of the shits.'

'Pigs? You've been hungry.'

'Pigs!' I repeat furiously. I close my eyes. The air is filled with ghastly cries, the earth trembles beneath me, I can feel sticky moisture on my eyelids. My throat is completely dry.

The morbid procession streams on and on—trucks growl like mad dogs. I shut my eyes tight, but I can still see corpses dragged from the train, trampled infants, cripples piled on top of the dead, wave after wave . . . freight cars roll in, the heaps of clothing, suitcases and bundles grow, people climb out, look at the sun, take a few breaths, beg for water, get into the trucks, drive away. And again freight cars roll in, again people . . . The scenes become confused in my mind—I am not sure if all of this is actually happening, or if I am dreaming. There is a humming inside my head; I feel that I must vomit.

Henri tugs at my arm.

'Don't sleep, we're off to load up the loot.'

All the people are gone. In the distance, the last few trucks roll along the road in clouds of dust, the train has left, several S.S. officers promenade up and down the ramp. The silver glitters on their collars. Their boots shine, their red, beefy faces shine. Among them there is a woman—only now I realize she has been here all along—withered, flat-chested, bony, her thin, colourless hair pulled back and tied in a 'Nordic' knot; her hands are in the pockets of her wide skirt. With a rat-like, resolute smile glued on her thin lips she sniffs around the corners of the ramp. She detests feminine beauty with the hatred of a woman who is herself repulsive, and knows it. Yes, I have seen her many times before and I know her well: she is the commandant of the F.K.L. She has come to look over the new crop of women, for some of them, instead of going on the trucks, will go on foot—to the concentration camp. There our boys, the barbers from Zauna, will shave their heads and will have a good laugh at their 'outside world' modesty.

We proceed to load the loot. We lift huge trunks, heave them on to the trucks. There they are arranged in stacks, packed tightly. Occasionally somebody slashes one open with a knife, for pleasure or in search of vodka and perfume. One of the crates falls open; suits, shirts, books drop out on the ground . . . I pick up a small, heavy package. I unwrap it—gold, about two handfuls, bracelets, rings, brooches, diamonds . . .

*'Gib hier,'** an S.S. man says calmly, holding up his briefcase already full of gold and colourful foreign currency. He locks the case, hands it to an officer, takes another, an empty one, and stands by the next truck, waiting. The gold will go to the Reich.

It is hot, terribly hot. Our throats are dry, each word hurts. Anything for a sip of water! Faster, faster, so that it is over, so that we may rest. At last we are done, all the trucks have gone. Now we swiftly clean up the remaining dirt: there must be 'no trace left of the *Schweinerei*'. But just as the last truck disappears behind the trees and we walk, finally, to rest in the shade, a shrill whistle sounds around the bend. Slowly, terribly slowly, a train rolls in, the engine whistles back with a deafening shriek. Again weary, pale faces at the windows, flat as though cut out of paper, with huge, feverishly burning eyes. Already trucks are pulling up, already the composed gentleman with the notebook is at his post, and the S.S. men emerge from the commissary carrying briefcases for the gold and money. We unseal the train doors.

It is impossible to control oneself any longer. Brutally we tear suitcases from their hands, impatiently pull off their coats. Go on, go on, vanish! They go, they vanish. Men, women, children. Some of them know.

Here is a woman—she walks quickly, but tries to appear calm. A small child with a pink cherub's face runs after her and, unable to keep up, stretches out his little arms and cries: 'Mama! Mama!'

'Pick up your child, woman!'

'It's not mine, sir, not mine!' she shouts hysterically and runs on, covering her face with her hands. She wants to hide, then she wants to reach

*"Give it to me."

those who will not ride the trucks, those who will go on foot, those who will stay alive. She is young, healthy, good-looking, she wants to live.

But the child runs after her, wailing loudly: 'Mama, mama, don't leave me!'

'It's not mine, not mine, no!'

Andrei, a sailor from Sevastopol, grabs hold of her. His eyes are glassy from vodka and the heat. With one powerful blow he knocks her off her feet, then, as she falls, takes her by the hair and pulls her up again. His face twitches with rage.

'Ah, you bloody Jewess! So you're running from your own child! I'll show you, you whore!' His huge hand chokes her, he lifts her in the air and heaves her on to the truck like a heavy sack of grain.

'Here! And take this with you, bitch!' and he throws the child at her feet.

'*Gut gemacht*, good work. That's the way to deal with degenerate mothers,' says the S.S. man standing at the foot of the truck. '*Gut, gut, Russki.*'

'Shut your mouth,' growls Andrei through clenched teeth, and walks away. From under a pile of rags he pulls out a canteen, unscrews the cork, takes a few deep swallows, passes it to me. The strong vodka burns the throat. My head swims, my legs are shaky, again I feel like throwing up.

And suddenly, above the teeming crowd pushing forward like a river driven by an unseen power, a girl appears. She descends lightly from the train, hops on to the gravel, looks around inquiringly, as if somewhat surprised. Her soft, blonde hair has fallen on her shoulders in a torrent, she throws it back impatiently. With a natural gesture she runs her hands down her blouse, casually straightens her skirt. She stands like this for an instant, gazing at the crowd, then turns and with a gliding look examines our faces, as though searching for someone. Unknowingly, I continue to stare at her, until our eyes meet.

'Listen, tell me, where are they taking us?'

I look at her without saying a word. Here, standing before me, is a girl, a girl with enchanting blonde hair, with beautiful breasts, wearing a little cotton blouse, a girl with a wise, mature look in her eyes. Here she stands, gazing straight into my face, waiting. And over there is the gas chamber: communal death, disgusting and ugly. And over in the other direction is the concentration camp: the shaved heads, the heavy Soviet trousers in sweltering heat, the sickening, stale odour of dirty, damp female bodies, the animal hunger, the inhuman labour, and later the same gas chamber, only an even more hideous, more terrible death . . .

Why did she bring it? I think to myself, noticing a lovely gold watch on her delicate wrist. They'll take it away from her anyway.

'Listen, tell me,' she repeats.

I remain silent. Her lips tighten.

'I know,' she says with a shade of proud contempt in her voice, tossing her head. She walks off resolutely in the direction of the trucks. Someone tries to stop her; she boldly pushes him aside and runs up the steps.

In the distance I can only catch a glimpse of her blonde hair flying in the breeze.

I go back inside the train; I carry out dead infants; I unload luggage. I touch corpses, but I cannot overcome the mounting, uncontrollable terror. I try to escape from the corpses, but they are everywhere: lined up on the gravel, on the cement edge of the ramp, inside the cattle cars. Babies, hideous naked women, men twisted by convulsions. I run off as far as I can go, but immediately a whip slashes across my back. Out of the corner of my eye I see an S.S. man, swearing profusely. I stagger forward and run, lose myself in the Canada group. Now, at last, I can once more rest against the stack of rails. The sun has leaned low over the horizon and illuminates the ramp with a reddish glow; the shadows of the trees have become elongated, ghostlike. In the silence that settles over nature at this time of day, the human cries seem to rise all the way to the sky.

Only from this distance does one have a full view of the inferno on the teeming ramp. I see a pair of human beings who have fallen to the ground locked in a last desperate embrace. The man has dug his fingers into the woman's flesh and has caught her clothing with his teeth. She screams hysterically, swears, cries, until at last a large boot comes down over her throat and she is silent. They are pulled apart and dragged like cattle to the truck. I see four Canada men lugging a corpse: a huge, swollen female corpse. Cursing, dripping wet from the strain, they kick out of their way some stray children who have been running all over the ramp, howling like dogs. The men pick them up by the collars, heads, arms, and toss them inside the trucks, on top of the heaps. The four men have trouble lifting the fat corpse on to the car, they call others for help, and all together they hoist up the mound of meat. Big, swollen, puffed-up corpses are being collected from all over the ramp; on top of them are piled the invalids, the smothered, the sick, the unconscious. The heap seethes, howls, groans. The driver starts the motor, the truck begins rolling.

'Halt! Halt!' an S.S. man yells after them. 'Stop, damn you!'

They are dragging to the truck an old man wearing tails and a band around his arm. His head knocks against the gravel and pavement; he moans and wails in an uninterrupted monotone: *'Ich will mit dem Herrn Kommandanten sprechen*—I wish to speak with the commandant . . .' With senile stubbornness he keeps repeating these words all the way. Thrown on the truck, trampled by others, choked, he still wails: *'Ich will mit dem . . .'*

'Look here, old man!' a young S.S. man calls, laughing jovially. 'In half an hour you'll be talking with the top commandant! Only don't forget to greet him with a *Heil Hitler!'*

Several other men are carrying a small girl with only one leg. They hold her by the arms and the one leg. Tears are running down her face and she whispers faintly: 'Sir, it hurts, it hurts . . .' They throw her on the truck on top of the corpses. She will burn alive along with them.

The evening has come, cool and clear. The stars are out. We lie against the rails. It is incredibly quiet. Anaemic bulbs hang from the top of the high lamp-posts; beyond the circle of light stretches an impenetrable darkness. Just one step, and a man could vanish for ever. But the guards are watching, their automatics ready.

'Did you get the shoes?' asks Henri.

'No.'

'Why?'

'My God, man, I am finished, absolutely finished!'

'So soon? After only two transports? Just look at me, I . . . since Christmas, at least a million people have passed through my hands. The worst of all are the transports from around Paris—one is always bumping into friends.'

'And what do you say to them?'

'That first they will have a bath, and later we'll meet at the camp. What would you say?'

I do not answer. We drink coffee with vodka; somebody opens a tin of cocoa and mixes it with sugar. We scoop it up by the handful, the cocoa sticks to the lips. Again coffee, again vodka.

'Henri, what are we waiting for?'

'There'll be another transport.'

'I'm not going to unload it! I can't take any more.'

'So, it's got you down? Canada is nice, eh?' Henri grins indulgently and disappears into the darkness. In a moment he is back again.

'All right. Just sit here quietly and don't let an S.S. man see you. I'll try to find you your shoes.'

'Just leave me alone. Never mind the shoes.' I want to sleep. It is very late.

Another whistle, another transport. Freight cars emerge out of the darkness, pass under the lamp-posts, and again vanish in the night. The ramp is small, but the circle of lights is smaller. The unloading will have to be done gradually. Somewhere the trucks are growling. They back up against the steps, black, ghostlike, their searchlights flash across the trees. *Wasser! Luft!** The same all over again, like a late showing of the same film: a volley of shots, the train falls silent. Only this time a little girl pushes herself halfway through the small window and, losing her balance, falls out on to the gravel. Stunned, she lies still for a moment, then stands up and begins walking around in a circle, faster and faster, waving her rigid arms in the air, breathing loudly and spasmodically, whining in a faint voice. Her mind has given way in the inferno inside the train. The whining is hard on the nerves: an S.S. man approaches calmly, his heavy boot strikes between her shoulders. She falls. Holding her down with his foot, he draws his revolver, fires once, then again. She remains face down, kicking the gravel with her feet, until she stiffens. They proceed to unseal the train.

*"Water! Air!"

I am back on the ramp, standing by the doors. A warm, sickening smell gushes from inside. The mountain of people filling the car almost halfway up to the ceiling is motionless, horribly tangled, but still steaming.

'Ausladen!' comes the command. An S.S. man steps out from the darkness. Across his chest hangs a portable searchlight. He throws a stream of light inside.

'Why are you standing about like sheep? Start unloading!' His whip flies and falls across our backs. I seize a corpse by the hand; the fingers close tightly around mine. I pull back with a shriek and stagger away. My heart pounds, jumps up to my throat. I can no longer control the nausea. Hunched under the train I begin to vomit. Then, like a drunk, I weave over to the stack of rails.

I lie against the cool, kind metal and dream about returning to the camp, about my bunk, on which there is no mattress, about sleep among comrades who are not going to the gas tonight. Suddenly I see the camp as a haven of peace. It is true, others may be dying, but one is somehow still alive, one has enough food, enough strength to work . . .

The lights on the ramp flicker with a spectral glow, the wave of people—feverish, agitated, stupefied people—flows on and on, endlessly. They think that now they will have to face a new life in the camp, and they prepare themselves emotionally for the hard struggle ahead. They do not know that in just a few moments they will die, that the gold, money, and diamonds which they have so prudently hidden in their clothing and on their bodies are now useless to them. Experienced professionals will probe into every recess of their flesh, will pull the gold from under the tongue and the diamonds from the uterus and the colon. They will rip out gold teeth. In tightly sealed crates they will ship them to Berlin.

The S.S. men's black figures move about, dignified, businesslike. The gentleman with the notebook puts down his final marks, rounds out the figures: fifteen thousand.

Many, very many, trucks have been driven to the crematoria today.

It is almost over. The dead are being cleared off the ramp and piled into the last truck. The Canada men, weighed down under a load of bread, marmalade and sugar, and smelling of perfume and fresh linen, line up to go. For several days the entire camp will live off this transport. For several days the entire camp will talk about 'Sosnowiec-Będzin'. 'Sosnowiec-Będzin' was a good, rich transport.

The stars are already beginning to pale as we walk back to the camp. The sky grows translucent and opens high above our heads—it is getting light.

Great columns of smoke rise from the crematoria and merge up above into a huge black river which very slowly floats across the sky over Birkenau and disappears beyond the forests in the direction of Trzebinia. The 'Sosnowiec-Będzin' transport is already burning.

We pass a heavily armed S.S. detachment on its way to change guard. The men march briskly, in step, shoulder to shoulder, one mass, one will.

'Und morgen die ganze Welt . . .'* they sing at the top of their lungs.

'Rechts ran! To the right march!' snaps a command from up front. We move out of their way.

*"And tomorrow the whole world." From a well-known Nazi anthem: "Today Germany Is Ours: Tomorrow the Whole World."

1. Crossed wooden beams wrapped in barbed wire.

2. 'Canada' designated wealth and well-being in the camp. More specifically, it referred to the members of the labour gang, or Kommando, who helped to unload the incoming transports of people destined for the gas chambers.

3. 'Muslim' was the camp name for a prisoner who had been destroyed physically and spiritually, and who had neither the strength nor the will to go on living—a man ripe for the gas chamber.

19. Arnošt Lustig

Arnošt Lustig was born in Prague in 1926. Together with many other members of the Prague Jewish community, he and his parents were sent to Terezín (Theresienstadt) concentration camp and then to Auschwitz, where his father was killed. He was in Buchenwald when American troops liberated the camp in April 1945.

After the war, Lustig returned to Prague, where he became a journalist and screenwriter, and began to create the stories and novels that form the basis of his reputation today. Several were made into celebrated European films, including *Night and Hope, Darkness Casts No Shadow* (from the story "Diamonds of the Night"), and *Dita Saxova.* After the Soviets occupied his country in the summer of 1968, Lustig went into exile and eventually made his way to the United States in 1970. Today he is a United States citizen and teaches literature and writing at American University in Washington, D.C.

Virtually all of Lustig's novels and volumes of short stories are concerned with the moral and psychological burden of staying alive in ghettos and camps under Nazi oppression. His most celebrated novel is *A Prayer for Katerina Horovitzova* (1987). Other titles include *The Unloved: From the Diary of Perla S.* (1985) and *Street of Lost Brothers* (1990), from which the following story, "Infinity," is taken. The ambivalence of its title haunts us like a derisive refrain. Infinity has many contents, among them, the never-ending deaths in a place like Auschwitz. When Lustig's narrator tries to translate the women's singing into a hopeful sign, his efforts are met by the cynical counterpoint of his lice-ridden bunkmates. The shifting viewpoints draw the reader into the turmoil of their ordeal, leaving no escape from chaos but the trial of interpretation.

Infinity

During the night I dreamed that I was an SS officer and was in charge of the gas chamber selection. In the dream, the lager commandant asked me how I was doing. I answered that I felt as if every day I trampled underfoot thousands—tens of thousands—of ants on the flat stone, but it didn't seem to lead anywhere. I cannot destroy the breed of ants. The ants will go on living even when I trample a hundred times more under my feet. Then the commandant changed into a voice. It could speak, and listen, and it was omnipresent. I have had similar dreams in the camps many times already. Once I dreamed an octopus was reaching for me and every tentacled arm pulled me toward its mouth and I tried to resist. But it had the ability to speak and announced to me that the average life span in Auschwitz-Birkenau is fifteen minutes. During the course of the day I forget the dreams, but as I prepare for each night of sleep I fear what I will dream of.

We got the upper bunk for the three of us: Harry Cohen, Ervin Portman, and myself. We did not have blankets. They were taken to the laundry in the Delousing Station because they were so full of lice and infested with typhus bacteria. We pressed against each other to make the most of our bodies' warmth. Portman was quietly repeating the number tattooed on his forearm, as he usually did before going to sleep, so that he would remember it if someone woke him up in the middle of the night. He preferred to lie on his left side so that he could see the number, in case Rottenführer* Schiese-Dietz came to make his night selection.

Before sounding the taps, Rottenführer Schiese-Dietz had made the rounds with his whip. It had a long handle like the whips used by coachmen driving teams of horses. The whip itself, twice the length of the handle, was braided leather made from human skin, fastened at the end with a double golden ring (cast from gold teeth) into which a Jewish jeweler in the camp workshop had engraved his initial. You could never be sure when Rottenführer Schiese-Dietz would get the idea to come to the barracks. He'd learned to crack the whip like a lion tamer. It sounded like gunfire. If the tip of the whip happened to touch someone it would slice the skin. A little while ago, satisfied that the barracks were quiet and

*Corporal.

that he'd seen nothing unusual anywhere, he had returned to the SS quarters and played the piano. He could play Bach, Schumann, or Mozart from memory, proud that he remembered so well everything he had learned. He had a different whip before, from somewhere in Saudi Arabia or from some German enterprise exporting whips to the near Orient, to make camels go faster. He was told that a camel never forgets a blow, even a baby camel, and that they sometimes run away in the middle of the desert. This couldn't happen here. People never escaped. There was nowhere to go. He had worn out his camel switch by using it too often and too hard, to beat out of people their habit of asking futile questions about boredom, or why they were living or why they were born or, most often, why their lives were so miserable. It was up to his whip to show them why they were still alive, and what had to be accomplished and why they were dissatisfied with their lives.

Somewhere in the night, a German voice called out: *"Laufschritt! Laufschritt!"* On the double! On the double! And again silence spread over the camp. Everywhere was mud. It was good at least to be under a roof.

A half hour later, Portman asked me, "Why aren't you asleep?"

"I'm cold," I said.

"Quit lying. What are you waiting for?" asked Portman.

"They'll begin soon," I blurted out.

"Get some sense into your head," Portman reprimanded me.

He must have known what I was waiting for, and he was right. I was waiting. I knew that he must have been waiting too, if he was not asleep. Maybe he waited for the same reasons I did and maybe at the same time it was because he wanted to refute what I wanted to verify for myself once it began. I felt a secret shame and didn't know exactly why. It encompassed many other shames, and all were present in one question: how and why were we born as we were born? And probably Harry Cohen was also waiting, although he said, "You shouldn't wait for it. It gets on your nerves. It isn't half as encouraging as it would seem."

"You'd better sleep," said Portman. "Be glad that we can sleep and they're not driving us to do some loading or unloading. In the mud it's obnoxious. It goes right through my rags." He trembled. "You keep wiggling. Who can put up with that?"

I did not answer. I did not want to be talking to Portman when they began. There were the usual noises in the bunks. Someone was quietly praying and somebody else told him to shut up and leave him alone and let him sleep in peace. And close to them, someone had a dispute with his Creator. The Creator didn't answer but the man had madness in his voice.

The wrangle finally stopped. Harry Cohen got a bit of gossip from his neighbor: that morning, the Grüppenführer* in the office of the Gestapo had been found dead, shot by his own hand.

Harry Cohen put his arm under his head. "That Chinaman who once said that the more a man knows, the luckier he is, was wrong," he

* Major general.

whispered. "It's just the opposite. The more you know, the more your world is filled with misfortune. Shit. I think I'm already dead. Don't let me oversleep. Do you know how long it's been since I've had a dream?"

"It's better to wait for the swallow to come back in the spring than to wait for *them,*" said Portman.

He stretched out his arms. He had long arms. He believed that was a sign of luck, and that, when necessary, he could reach farther and easier than Harry Cohen or I. His ears stuck out and he was convinced that this was a sign of someone who was satisfied with his lot; he was always sleepy and tired, but at the same time he had great willpower, although not the best talent for choosing friends. He did not have much patience. It was good that Harry Cohen did. In his former existence, Harry Cohen had had a lot of good luck, both in cards and in love, and was very successful in business undertakings. Portman considered my kind of patience morbid. He could not understand how I could go on waiting.

But Portman still wasn't asleep. Maybe he could still hear the Rottenführer cracking his whip. Maybe he could imagine that golden ring, cast from teeth that were knocked out of corpses and sometimes, just for the gold, out of the living. Most of the gold went into sealed railroad cars to the underground safes of the Berlin State Bank, but quite a bit of it slipped through the fingers of the SS men. Sometimes he dreamed about his little sister, up the chimney four months ago already. Her skin was very fair, with a touch of pink or peach; she was so white he could still see in his mind the blue veins beneath her skin. Her soft skin smelled like spices, maybe like flowers. Her fingers—not rough or damaged by manual labor—were long for a child of eight, and she also had long, dainty feet. Perhaps Portman envied the dead just as you would envy the living. Yesterday he had mumbled something in his sleep about Samson, and no one wanted to remind him of it. On Monday he dreamed that he had gotten typhus, on Tuesday that he had diarrhea and couldn't stop it, and Wednesday morning he realized that it had been both a dream and a reality. It wasn't hard to figure out what brought Samson's name to his lips. Samson was definitely the last resort.

"The rabbis know what they're talking about when they say that Noah was wrong to send out a pair of doves to bring the news that the flood had subsided. Maybe it drove him mad," Harry Cohen said. "To go mad—that's not the worst there is. The worst is when you know it. This afternoon when we were coming back from the soccer field, they were picking out women who had no shoes and those who were sick. The deaf and dumb are gone already and so is that shipment of war invalids from Vienna."

"Did you think they would be feeding them white bread and milk here?" asked Portman.

"Half of the people that they added to make up the count for the transport were healthy and strong—something was going on," added Harry Cohen. "They're in a hurry now. But why do they want to get rid of the women first?"

"The Nazis don't like women," said Portman. "They're dead set in their beliefs that women are the source of all evil, just as we are. I heard Rottenführer shouting at the women in the laundry that they like to screw, especially when they were menstruating, so that they could infect everyone with syphilis and other infectious diseases. He yelled at them that here they would lose all their bad blood. He told them that he knew a woman could not wash laundry properly when she had her period and then he threatened to send them all to the bath and burn them with their dirty rags. And then he said point blank that he would shoot down any woman who got closer than three steps to him or dared to touch him."

"There are places where the men believe that if you glance at a menstruating woman your bones will go soft and you'll lose the ability to have children," said Harry Cohen.

"I hope it happens fast," I added. "At least without having to stand in the snow and mud." I liked Harry for never complaining about anything. Whenever they beat him he never moaned a minute after. Maybe he believed human dignity lies in never speaking of pain, especially afterward. In Prague he left an Aryan girl, Maruschka. Sometimes at night, when he was looking at the stars, his big lips would have a tender smile like the Mona Lisa's. Once the Rottenführer caught him with that tender smile and beat him to chase it away. Cohen opened his mouth, but didn't complain.

"Why don't you both go to sleep?" Portman said.

I couldn't sleep. I waited for them to begin. Maybe they wouldn't.

When the killing came so close that he couldn't pretend not to see it, Portman would always get nervous. The blood would rise to his brain, and he would take it out on me. I could imagine that he was putting the blame on both the living and the dead, on people who had already gone through it, as well as those who were still waiting. I did not confide in him that for the last few nights I had been seized by a vision of crashing stars that in a fraction of a second would crush the whole camp and the planet on which the Germans had built this camp with our hands. It was a strange wish and it actually had something to do with Samson—only it was not just a matter of a few columns holding up the roof of an ancient palace. My vision encompassed the destruction of everything and everybody: the crushing of the earth, down to the very last pebble. It embraced the transformation of the planet into stardust—together with all and with everything, down to the last crow and the last ant. It was a very disturbing yet comforting obsession, and I knew it would depart with the first sleep or when the women from the adjoining camp would begin. It was safe as long as the Germans only killed new transports and picked out the sick and those people who were guilty of being old, of having gray hair or wearing glasses. The old-timers still held on to some hope or illusion, which turned into a new truth when the Germans started killing even the healthiest and strongest, those who could still work for them.

I did not want to miss it when the women from the adjoining camp began. It was always the beginning of something else as well. The beginning and the continuation of something that had no end even when it was over.

There were two women's camps. One was for Jewish women from Hungary and Slovakia and the other was for both Jewish and non-Jewish women from other occupied countries. They worked in the hospital ward, in the showers where water actually flowed from the sprinklers, and in the delousing station where they exterminated lice and insects with Cyclon B. The latter came from the same cans as those which were used in the underground showers next to the disrobing stations and the crematoria from number one to number five, and was used in fourteen-hour cycles, necessary for the proper extermination of vermin. The women also worked in cleaning stations, laundries and kitchens, and in the warehouses where shoes, gems, underwear, hair, orthopedic devices, and costume jewelry were stored. These women also had opportunities to work through all seasons as performers in the whorehouse or in the concert hall. Polish and Ukrainian henchmen and guards and German criminals wearing purple triangles took their opportunities in the brothel, although they had to pay two Reichmarks from their wages or other remuneration for the services. One mark was for the whore, and the other went into a special account which the commander of Auschwitz I, II, and III had been ordered from above to keep aside.

In the family camp B2b, which was burnt to ashes during the night of March eighth, there were kindergartens and nursery schools, and their teachers, before being killed, had rehearsed theatrical productions and gymnastic performances which the SS would come to watch.

Would they begin? They really ought to begin. If they were planning to, they should begin now.

As far as Portman was concerned, I had not been behaving like a normal person for the last few days. It was not the first time he had caught me waiting for them to begin. He knew why I had stared at the wooden ceiling with my eyes open, why I stared at the open, glassless window under the roof where I could see the stars or the moon or snow or smoke, waiting all the time. Sometimes my teeth would chatter with the cold and Portman would hear it and it would upset him because we lay so close to get heat from each other. He would blame me for not sleeping because I waited for the women to begin. He was mad at the women for what the camp had done to them instead of being mad at the camp. He thought that the women were crazy. Their faces had become coarse and some had hair growing on their cheeks, which made them look old.

Sometimes Portman would get furious, though at the same time he was afraid to vent his fury. That made him stutter, although maybe it was because he could not speak out loud. Yesterday, for some unknown reason, he had broken a tooth and thought he would die of it. Harry Cohen mentioned that people who had such widely spaced teeth as Portman did would always have to look for their good fortune far away from home. Well, as long as he was here, he was far enough from home, but what good fortune could he have when his teeth were crumbling, even though they had not sent him to the showers yet? Sometimes he became speechless when I waited for the women, but not now. He also feared he had tuberculosis. He believed that he would pull through if he could make it through the next month, unless Rottenführer Schiese-Dietz would spoil everything with his next selection, needing one more for the showers. It was already the twenty-seventh of October. He told himself that October was not a good month for those who have TB, and he fixed his mind on November. Portman sometimes blamed it all on his mother, whom he had never known because she left when Portman was born. Sometimes the thought came to him that it would be rather ironic if his mother had not been Jewish. He also believed that he could get rid of his suspicion of tuberculosis if he could take one gulp of milk.

It was strange that the image of women was associated in my mind with the image of milk, but I preferred not to mention this to Portman. My back drew the heat from Portman's belly and I pressed my stomach and chest against Harry Cohen's back. That afternoon I had given Portman a piece of my bread and blood sausage because I knew what he was afraid of. He gulped it down at once. Now he had the hiccups.

Everything was permeated by the smoke that rose toward the sky during the day and during the night, smoke from the dead who would not have a grave. I remembered what Rabbi Gans once told us in the Vinohrady synagogue: that when they were in the Sinai desert, Moses had ordered his successor, Joshua, to bury him in the ground so that no one would find his grave. The Germans were now doing it on a grand scale.

Portman hiccupped again. "I washed my rages in the latrine and the Rottenführer came there and whipped me out of the hole. He yelled at me in his Bavarian German: *'Höre doch auf, Mensch!'* He told me to stop it or he'd take the handle of his whip and press me out through the hole in the planks to see if I could swim."

"They should have started," I said.

Portman yawned. "It's better if they stay quiet. It won't help the dead. And the living should stick with common sense. You don't want them to run into bayonets, do you?" Then he added, "It brings them bad luck, just like the night air. Don't they know that? It's making their brains soft. Or they've gone crazy already. They sound like men. I don't miss it. And most of them don't have periods anymore even if they're sixteen."

Harry Cohen was feeling his chest, his body, the last thing that he owned. He was at rock bottom. He had used up his last bit of willpower and self-control to pretend otherwise. He did not say that here in Auschwitz-Birkenau everything, even breathing the wind, a draft, or still air, brought people bad luck. The good luck of one always meant the bad luck of another. Everyone who lived, lived at the expense of someone who lived no more. It was not his fault. It was the Germans' organized way of killing. There were batches that were sent up the chimney for punishment, and the rest were picked to feed the crematoria according to numbers. And so there would be no cheating, these numbers were tattooed on each person's forearm. People could live in the camp only until it was their turn to go to the showers or into the furnaces and crematoria, because even the best German engineers could not figure out how to burn them all at once. The Germans loved order and kept better discipline than the Frenchmen, Englishmen, Italians, Czechs, Belgians, and other nationalities which the Germans deported to this camp for liquidation. Even those Germans who had no more than a seventh-grade education, like Rottenführer Schiese-Dietz, managed to command order, even when the situations got very confusing.

Harry Cohen lay with his mouth shut not only to avoid wasting energy talking, but in order to breathe through his nose so as not to inhale the germs of typhus and tuberculosis. Neither his former nor his future elegance were of any use to him here. Every memory of the past that could be uplifting was at the same time depressing. Everything that could heal could also wound. Once in Theresienstadt, Faiga Tannenbaum-Novakova ruminated about the omnipotence of the devil. The devil now seemed like an amateur compared to what the German Nazis had dreamt up in Auschwitz-Birkenau and its subsidiaries. Harry Cohen felt the skin in the middle of his chest with the balls of his fingers, trying to ascertain how much there was; it was thinner than wrapping paper, and he wondered how long it could last. He was probably thinking of all those things the tired, frightened minds were thinking, all those who hadn't yet managed to fall asleep.

"They're burning the old Hungarian women who came Monday," said Portman. "The Polish and Romanian railway engineers who brought them were completely drunk because the women stank so much. Somebody said he'd never seen women so full of lice. When the old women stripped at last, because they did not understand what they were supposed to do—since most of them did not understand a word of German, just like the Italians or Frenchmen—the barbers asked for double disinfection before they started cutting their hair."

His words became lost in the thick air of the barracks. Words were the first thing to become silent and lose their meaning. There were no innocent or inexperienced people among us. We had all been there too long, even a single day or a single night. None of the people believed any longer that they were going to take a bath, even if they actually went to baths to be deloused and disinfected before the journey. Nor did they believe that the Germans would send them to work in one of their dominions because they were experienced tailors, goldsmiths, blacksmiths, or automobile mechanics—or that they could find work in the German armaments industry. Everyone knew where he was going if the order was to take the road to the left or turn on the road to the right. The road to the right led out of the camp, into a room which had doors without door handles, and they knew that no one who entered the room came out alive. No one was fooled by the notices about cleanliness and health, by the cursing and light banter of the Nazis.

"The dental technicians had a hell of a job prying the gold teeth out of those old women," Portman added. "As usual, they searched everywhere, up and down, back and front, to see if they didn't have some hidden gold, bank notes, or diamonds. Most of them had used them already at home. A couple of them were beaten just because nothing could be found on them in any of the places that women tend to hide things."

The words faded again. They were lost in the rotation of the earth—which did not fall between the two crashing stars as I had wished—and in the thickness and smell of emaciated bodies, covered with sweat, but at the same time getting numb, because it was already the end of October and it was cold, made more so by the wind. Most of the men were quiet. This silence sounded like the dried-up language of the dumb, like the very last silence of the earth as it will one day be heard by the last human generation, even without Auschwitz-Birkenau, when the sun will approach with its enormous ball of fire, and all that is human will turn into fire, ashes and ice, before the planet earth turns into an infinite night, like this night, but much more desolate, much more silent, and much more cold. In my imagination it came sooner.

For the third time Portman said, "I'll bet anything, including your bread and blood sausage, that they will beat it out of them five minutes after they've begun."

Harry Cohen did not say anything. He was, as always in the rare moments of quiet, concerned with the number of possibilities every situation seemed to offer, while really offering only one. It encompassed all possible endings.

"It's not worth it," Portman repeated. "They should be glad that they had time to undress. They wouldn't even have to bother dressing again."

Women over forty, just like girls under fifteen, usually went into the furnace right away, directly from the ramp, or by the first or second selection. They had to wait only if the dressing rooms and the crematoria did not have time to process them all. Although it was cold now—not like during the summer heat—the Germans were afraid that the corpses might decompose and epidemics would start, and so the old women had waited since Monday afternoon. Their lice-infested underwear, clothes, and shoes were burned. And so they had to wait naked, some for hours, some for days and nights, in the wind, rain, and snow, and no one bothered to give them water or anything to eat. About two hundred thousand Hungarian Jewish women had come here since the summer. It was said that in Hungary there remained, until the beginning of 1944, about eight hundred thousand Jewish men, women, and children, and the Germans had now decided to liquidate them quickly, just in case Hungary changed sides in the war.

In a few hours the Hungarian women stopped asking what happened to their husbands, their children, and their families. They wanted to go into the showers and get it over with. When the women were undressed and their belongings taken away, they had no poison they could take, nor knives with which to slit their wrists.

"What's today? It must be Friday. Right, it's Friday," said Portman.

"Or maybe it's already Saturday," said Harry Cohen. "And it will begin all over again. But actually, why should it?"

No one had a watch. A watch was the first thing the Germans stole from you in Auschwitz-Birkenau. The first day here, Harry Cohen made up his mind that he should lie and tell them that he was a trained watchmaker in order to get a cushy job, but fortunately he did not do it.

"You were probably born like a rabbit, with open eyes," Portman grumbled in my direction. "Or do you sleep with your eyes open?"

I didn't say anything. I was waiting for the women to begin. Portman probably debated with himself whether he should continue his talk about food. He knew as well as anyone that talking about food made one hungry. Someone would tell him to shut up. Once, when Portman served as Rottenführer Schiese-Dietz's orderly, he got outside the camp and carried back an image of deep and silent woods that only here and there were fenced with barbed wire. These were endless woods, pines, firs, and sometimes deciduous trees, full of game and silence. The SS men went out to hunt, but they did not shoot deer, does, and hares—they shot people.

Outside, the flames cut through the night and the snow. The sparks flew out of the darkness. The wind howled. Now and then you could hear the crows across the evening sky. They were evil omens. They were flying to the left. That, according to Portman, was the worst. If they flew to the right, that meant he should be cautious. And he could make a wish, as when you see a falling star. All sorts of superstitions came back, and many of them resembled crows. Crows were always an omen of death, of the worst there is. In the night I could imagine the curved flight path of the crows. Portman believed that they talked to each other. But according to him, they talked only because some farmer and some village children had slit open their tongues. The crows held court and they sentenced individuals from their ranks; and a male and a female carried out the sentence together, pecking the culprit to death either in flight or on the ground. But Portman would reverse this sometimes and insist that crows are capable of helping the weakest ones in the flock, the young ones and the weary. A squadron of crows will send out a patrol from its midst to save the weak, the falling or ailing ones. And they know how to warn others of danger.

Already on Monday, before it got dark, Portman, Harry Cohen, and I saw the old Jewish women who spoke Hungarian or Yiddish. We knew what awaited them. We hoped it would be quick. But our wishes did not matter. I also wished that the crows would fly somewhere else with their cawing and hoped they would fly in a straight line without detours. Their sounds reminded me of the worst things: of humiliation, hunger, cold, of illness, weakness, and helplessness. I envied them—their life, their flight, their freedom. I couldn't understand how they could live a hundred and fifty years. But it occurred to me that they could collect endless secrets in that time. I thought of all that they had lived through and wondered where they could take it in their flight.

Portman talked only about the living. He pretended that he had never seen the dead in the camp, as if there weren't any dead here, not to mention the dying. He did not see them. He did not look. Or maybe he looked elsewhere. He forced his hungry red eyes not to see, his brain not to comprehend, as if he had heard about the dead only secondhand.

I wasn't even thinking about the selection that Rottenführer Schiese-Dietz had come to perform. He made the prisoners walk over the long stable-chimney that ran horizontally about knee-high all the way through the barracks ever since the days of the Austro-Hungarian empire when the cavalry was garrisoned here. The prisoners had to walk over the chimney bricks naked, so that the overseers could quickly discern their healthy or sickly condition. Men whose penises had turned black, who were skin and bone, or those with exalted feverish eyes went at a trot down the whole length of the chimney; and only the healthy ones, when the thumbs-up sign was given, could jump off and go back to their bunks. The others who did not get the thumbs-up were selected to continue to the end of the chimney and then through the back entrance down a plank to be loaded onto a truck that took them to the showers. If someone's penis got hard as he was running down the square chimney, either from excitement, from fear or cold, or from some nervous disorder, he had to run to the end of the chimney down the plank, into a car, and into the disrobing rooms, where Germans would make short shrift of his Jewish lust as well as of him, in the surest possible way.

In the women's barracks there had been today—like every day including Saturday and Sunday—two, three, perhaps even five routine selections. For the Germans, it was both a necessity and a pastime. Naked women, in the presence of two or three well-built and elegant SS men, would run or walk along the chimney as if on a stage, waiting for the sign of the thumb. Women with sagging or dried-up breasts, the skinniest ones who had turned old overnight or all of a sudden, women weakened by menopause and with fear of illness or nervousness in their eyes, women without husbands and children, swollen or again thinner than dying mares, women with male traits and beards which they had no time to pluck out, women who were bleeding in the groin but could not clean themselves up, or just women who were splattered with mud and dust—they all waited for the sign of the thumb. Up, horizontally, down. Down and horizontally. The SS would take turns. They had enough women so that among the three hundred and thirty-three, each one could choose his own figures. Each time it was ten or twenty women who got the sign to walk or run to the back entrance at the back of the truck.

"Now," said Harry Cohen all of a sudden, just one second before I wanted to come out with it.

"I know," I said. "I can hear them."

"They're like swans," said Portman. "Women are always faithful beyond the grave to someone, just like swans. Or maybe like geese. But they're wrong if they think they're laying golden eggs in the darkness. How will this help them? I bet they haven't eaten. Are you telling me that you both knew they'd begin, as if nothing had broken them?" Sometimes, after the selection, the women would sing. They would wait for darkness to cover the camp, when everything except the fires had died down. No one could guess how long it would go on. Sometimes the guards sicced dogs on them, and sometimes the guards would fire a few volleys from a machine gun from the door to silence them. Sometimes a few shots were enough, sometimes a few magazines. It was always only a question of time. But before that there was the question whether they would begin, or whether they would give up even this, the very last thing they had. And now they began. It evoked a horror that most people could no longer sense. It gave a new birth to something that had perished long ago. They filled the void that spreads from space, like waves over a calm surface rippled by a wind out of nowhere. It was like a stone that had been given a voice.

"The heat from the chimneys gets on some people's brains," said Portman. "They must all be mad, or they wouldn't be asking for it. Do they want to drive everybody crazy?"

I could not yet identify the words of the melody, but I felt how they filled the space of the camp and of the whole night, the space of the world and beyond, extending into infinity, the incomprehensible void, which they filled with their voices, with their melody that was only beginning to take shape. I imagined a labyrinth, bodies, voices, wooden bunks in the dark, a maze from which the women tried to find a way out—like all the blind, deaf and dumb, drugged or poisoned, tottering from blows to the head, driven nearly out of their minds.

"All the rats have left the women's barracks," said Portman.

The women sang for all those who had lost someone, someone selected at daylight to be sent up the chimney and now no longer among the living; they sang to distract the bereaved from thoughts of the fires and of those who were alive only a few hours ago. The women who sang belonged to the Cleaning Outfit, which scrubbed even the trucks, scrubbing them clean of the dirt left behind by the condemned ones. The singing filled the night with double voices. Double darkness. Double eyes. Double blood. Double snow, and far above the snow clouds, in a double star-filled sky. Double memory. Double hope and illusion. It filled it with all the things that man possesses only once in his life and loses.

"They make me nervous as a cat," said Portman.

You could not see the flames, you could only hear them. The women who had lost someone already—yesterday, the day before, last week or last month—sang for the mothers who had lost a daughter, for the daughters who had lost a mother, for the sisters without sisters, and for friends without friends and acquaintances who had lost someone they knew. At first the singing was low, then louder, and finally quite loud. They would sing for a short time and sometimes longer, although never too long, and their singing was joined by those who were afraid at first. Finally they were joined by those that were stricken, until in the end, everybody was singing.

"If you sing at night or in bed, that's bound to bring you bad luck," Portman said again. "If you sing at night, you'll be crying before dawn. If you're singing because you don't have a reason for doing it, someone will give you hell for it."

The whole month Portman had argued that even here there were good days and bad days. And that only children and old people could not survive their uselessness.

I waited, as did all the men in the camps, for the first tune—just as you wait for the first evening star, making believe that it has some special significance for you. I tried to imagine the faces or figures of the women from Norway, Belgium, and Holland, the women and girls from Rome, Warsaw, or Sofia, women and girls from Berlin, Paris, or the island of Corfu who had come here by boat and then by train, half dead with thirst. That which they were singing was, at least for the moment, beyond the reach of the hierarchy of the ruling and the humiliated, those who condemned them while they still had the chance to do the condemning. It reminded one of a fortress crumbling invisibly, even though its ramparts were strong and remained tall and unassailable. The singing came from the darkness distorted, as if from a great distance. I felt something that no one could understand. It contained everything that I had ever waited for, everything I was afraid of and which filled me with mystery and fear, as well as with the wonder of life, because death had become simple, comprehensible, and ordinary. This is the world in which the killers were born together with us, but by killing would live without us; this was the world in which each one of us was the last in the world, before he disappeared and left behind a sliver of ash in the museum of an extinct race, remembered as someone who happened to appear for a couple of thousand years on the surface of this earth. It filled me with something resembling chloroform, which knocks you out like a fragment of a dream.

I waited to hear it interrupted by the rattle of the train on the ramp or a volley from a machine gun or a revolver. At the same time I hoped it would not stop. I wished that the women would sing, just as I wished to see the only thing that would make up for the destruction of justice: my image of two stars flying toward each other before they crash and crush the world that had culminated in this camp and in the killing of innocence in the name of an idea. It embraced something that I never could explain and which I probably would never be able to explain. Everything that man is and is not.

In those women's voices I heard all that is insignificant as well as all that is great, that which is pitiful and full of a silent glory, that which is comprehensible and incomprehensible, like every man and everything he has gone through and still must go through in the future. I understood that man in his smallness and misery is part of something greater, something that has had and will have many names, something that is unfathomable and great like the sea and the earth, the clouds and trillions of stars or galaxies, and at the same time lost like a grain of sand in the desert, or a fish or drop of water or salt in the ocean; that for which man wants to live, even in a place like this, permeated with death and killing, just as the sky is permeated with stars or a snowy night with snowflakes. I understood why man has the strength to die when it is his turn, so that someone else will not have to die in his place; and what he will share, when he has nothing left but his body heat and a spoon carved from an alder branch so that he can eat his soup and not lap it up like some dog, wolf, or rat.

I expected to hear the barking of dogs from the darkness of the night. I felt a different kind of fear and anxiety, different from what I experienced in the afternoon while I looked at the barbed wire, at the German uniforms, at the whip with the golden ring in the hand of the Rottenführer Schiese-Dietz. When I closed my eyes, I had the feeling that I was witnessing the birth of a new planet which was yet to be peopled, a feeling of going way back into time, into the very oldest times when the cooled-off planet earth had just become inhabited. I felt a new infinity, that infinity upon which man trespasses now and again with every breath, word, act or even the blinking of an eye. Something closer than closeness and more distant than distance, something so loud that it deafened me, and something so soft that it was like an incomprehensible whisper.

"Who cares about this," Portman growled. "I'd give them a better idea."

It probably seemed to him that he was watching how they would beat or shoot his mother, if he had ever known her. I don't know. Harry Cohen, too, let Portman's words float by as if Portman had not said anything the whole evening. This bothered Portman. It was a funeral rite, the singing that was born here and was not performed elsewhere, nor would be in the future. It probably both comforted and disturbed him that there would never be witnesses anywhere to this funereal singing that sounded like a martial song and like a suicide challenge, the invisible gauntlet thrown into the face of the enemy, because those who sang were in mortal danger. I don't know, I really don't know. For me it was something that I had lacked so I could perceive my life as less incomplete, and it was something that I do not understand to this day, somewhat like perceiving an echo that exists only in the memory, or a shadow that fell long ago, or a cry that memory has blurred.

"I don't want to hear them sing so I won't have to hear them cry," Portman added.

The women's voices were still mixing with the snowflakes, with the fire and the darkness. The singing came out of the wild night into which the women had brought it to subdue this night when people were being killed, like every night, every day, Friday, Saturday, and Sunday, on high holidays and on every working day, before the stars came out and when they were already fading in the sky. It came out of the night where words meant less than wind, snow, slivers or clumps of ash before they disintegrate, out of the night where innocent people were being killed in one part of the world, while in another part makeshift barracks were being slapped together for more and more prisoners, until all of Europe was German.

Not very far from us was a whorehouse to which prominent Jewish prisoners, henchmen and informers, and Jewish collaborators were occasionally admitted. We hated and despised them and at the same time feared them almost more than we feared the Germans. A little farther away there was another brothel for the soldiers, and yet a few steps further, mothers and wives of the soldiers and clerks of the Totenkopf SS garrison (entrusted by the Nazis with the cars of the prisons and prison camps) were reading fairy tales to their children about Hansel and Gretel and about the brave Siegfried. In one block, German doctors cut pieces of skin from healthy prisoners for grafting on frostbitten German soldiers from the Eastern front, or frozen airmen fished out of the English Channel. In another block two Jewish women pianists played a concert of Beethoven sonatas before being sent up the chimney. The prisoners and the jailers slept under the same stars.

The voices of the women came from the openness and closeness of the night, flowing together like a refuge created from nothing, like a shelter where you can hide for a moment without fleeing, where you can rest and gather strength or save the remainder of your strength. Their voices became a battleground upon which danger, for a few seconds, did not seem so dangerous. For just a moment, pain changed to painlessness and indifference to solidarity.

Over Auschwitz-Birkenau, below the low-lying snow clouds, in the thick smoke of the chimneys of the five crematoria, the singing of unknown women continued. Their singing came from the darkness, for a few seconds, almost a minute, two minutes or three, from the ever renewing sea of life and death, from the darkness and out of the wind, from the lips of women and from the depth or shallowness of the universe.

I was afraid that the women would stop. Then everything would be quiet. And then the only sound in the night would be that of ashes, snow, and wind.

"They're probably feeding the dogs in the kennels now," Portman mused.

Harry Cohen plucked hair from his nostrils and his ears, so that he would look all right in the morning lineup.

The singing of the women sounded like a river overflowing its banks. It colored the night the way the flames colored it or the red morning sky for those who are still alive to see it and for those who can only imagine it before they perish. It marked the night with a forgotten strength, forgotten tenderness, forgotten defiance, and forgotten understanding.

"How can anyone in Germany today think that a thousand years or ten thousand years from now this will be forgotten?" Harry Cohen asked.

"I can't stand the realization that all that remains of them is song and ashes," Portman said. He did not say that he divided people into those who'd already gone through it and those who were still waiting. He divided women into those whose families had perished long ago and those who had lost them to the trucks only this afternoon or at dusk.

"It's like catching sparrows in a cage," Portman added. "Everyone knows what to expect for that." His voice was full of anger and death. "I wish they'd shove it," he added.

Harry Cohen held a piece of bread and a bloody pork sausage under his ragged jacket. Was he about to eat it? He had to be hungry. But he did not eat. Was he waiting for the women to stop, just as I was, in the same way as before, when we waited for them to begin?

"They've lost their minds," Portman said. "They're more stubborn than I thought. They're more persistent than salt."

I could imagine the way the women looked. Some were swollen from hunger and irregularity, while others for the same reasons lost weight. Some sold themselves for a piece of bread, while others sold a piece of bread for a bowl of water so that they could wash. Sometimes I saw them humiliated because they had to strip naked before the SS soldiers or the Ukrainian guards, or doubly humiliated because their heads had been shaved, destroying their femininity. They had already lost their capacity for bearing children. But now they were singing.

"What's the good of it?" Portman asked. "Why don't the smarter ones make them shut up?"

"With what? With ashes?" Harry Cohen asked.

"It makes my teeth hurt," Portman said. He was afraid that he would get another toothache and that without his teeth he would never be the same.

Did he huddle up so that he would not hear them? Did he pretend that the song no longer interested him? Was he interested only in his own breath? I no longer understood Portman and probably neither did Harry Cohen. The more we exposed our lives, the less we understood each other.

The women sang lullabies that they had brought from their homes, songs about love and joy, about the freshness of children. Their voices brought back a world which no longer existed.

"Yesterday Schiese-Dietz picked out the redheads," said Portman. "He probably knew why."

Portman feared who the Rottenführer would pick out in the morning. Sometimes he picked out people who had prominent chins, or those who had small receding chins—sometimes people with white teeth and sometimes just the opposite. He'd picked out people with small heads and small brains, or with big heads in which he expected big brains even though the selected one would stare at him with the eyes of an idiot. Portman did not know when Schiese-Dietz would start selecting people with big ears or untrimmed fingernails, with thin or flat lips, with thick or sparse eyebrows, cross-eyed ones and people with long fingers.

But Portman sometimes dreamed about being selected for the Canada commando special unit, who cleared the arriving transports and had lots of food and drink, at least for a day, for themselves and their friends—bacon and bread and strawberry or plum jam, and thermoses with coffee, cocoa or hot tea, and vodka or schnapps or French cognac. He dreamed about this in spite of the fact that it also meant clearing the ramp and the railroad cars of dead infants and small choked children, carrying a bunch of them at once like bananas or shot hares or killed or choked hens. Portman dreamed around the clock of a full belly, but there was no chance of his making it into the Canada commando.

"It's my father's birthday today and the anniversary of his death," Harry Cohen said. "I'm as old today as he was when he died. He died in his bed. He was reading a book, closed his eyes and whispered that the end had come."

"Congratulations," replied Portman. "Do you think this has some special meaning here? That it's perhaps some kind of prophecy and you have a right not to let me sleep?"

Suddenly I realized that I was holding on to Harry Cohen's elbows and wrists and that Portman was pressed between us. Our lice crawled from one collar to another. I took their warmth and they took mine.

The unknown women had sparked an image of what could be because it had been once before. From time to time, the wailing wind interrupted the singing. Then the voices became clearer again, although the wailing wind did not die down. They floated through the network of electrified barbed wire, somewhere into infinity. I no longer waited for what Portman would say to mask his envy of the living and the dead. He was sometimes afraid of the dark so he had to talk to hide his even greater fear of the Germans. The singing roused something in him that he thought was dead. It was somewhere on the limits, the nakedness of everything that was still living. It was the heart of his existence, on the thin border of his moral and physical strengths, like the moments of selection, dependent upon decisions made by someone else, but also upon his own decision on how to accept it. Only he and his consciousness, no traitor to himself.

"I'll give them another minute," said Portman. Suddenly it sounded as if he did not have enough air in his lungs. Maybe he had tuberculosis already. "Count to sixty."

The women were singing a popular German song: *"In den Sternen steht es alles geschrieben, du sollst küssen, du sollst lieben . . ."** Probably those who went into the chimney were German or Austrian Jewish women. It was an old coffeehouse hit, but in this moment it carried some immediate, close, pure and direct sincerity and courage, some surprising truth about who is who, no matter where or when. There was in that song now everything that was still unselfish or honorable, even when it was weak and abased.

*"Everything Is Written in the Stars: You'll Kiss, You'll Love . . ."

"Well, really, who cares? When it takes so long it's no good," Portman added. He was being sarcastic about the thought that someone here was singing about kissing and passion or the desire from which children are born.

I held on to my wish that the women would not stop. Did Portman want them to stop because he feared for them? Besides the fact that in the end he turned his anger upon himself?

"Do you want to cry for them?" Portman asked. "This is idiotic solidarity and does no one any good. It just makes the Germans madder. The women have paid for it a couple of times already, as far as I know. Wouldn't it be better to let the Germans sleep at night at least? Do they want to test whether it's true that the devil never sleeps at night? Do they think they're in America—where they can sing whenever and whatever they please?"

Portman turned his face to the planks of the bunk. "They'll knock their teeth out if not worse. I've seen a lot of toothless ones here. And then later I didn't see them anymore, precisely because they had lost their teeth."

I held on to the last bits of the melody, which was already becoming blurred against the night. My brain, along with hunger and the smell of human bodies in the barracks, was floating off into the numb wakefulness that precedes sleep. I suddenly realized that I needed to have Ervin Portman beside me, even with his anger and superstitions and fear, and I needed Harry Cohen, just as they both needed me. Even when the people—both the living and the dead—get on each other's nerves because they have nothing to offer each other and just vegetate, even on the way to the showers. What makes a man forgive others is not born of weakness or of strength, but from the fact that human life is irreplaceable and that nothing else matters.

Portman was bothered by my lice and by Harry Cohen's fleas. Portman picked them out from the ends of his short hair before it would be cut again for blankets and army coats. He pinched them with his nails and threw them down from the bunk in a curve resembling the flight of crows. When someone below grumbled, as though he'd swallowed them, Portman stopped. I felt Portman's hot breath, just as Harry Cohen felt mine.

"When I was a kid, my mother used to believe that if two people started singing the same song it brought them luck," Harry Cohen said.

"I never sing or whistle when I play cards," Portman said. "That's sure to make you lose. I don't rush headlong into hell if I can remain at least a little while at hell's entrance. I prefer to look at smoke instead of looking straight into the fire. That's one thing they've taught me: that it's better to have your hand in the water than to have it in the fire."

"Hell isn't down below, in the center of the earth, in the crevices of mountains, rocks, and passes. It's up above, on the surface of the earth, like scales on the body of a fish. It's in every man, under the stars, under the snow, under the enormous firmament," Harry Cohen said. "It's in every uniform, in every whip, bullet, or dog that is set against people. I think that hell exists fivefold—in each of the five crematoria. Or in the eight-meter pits where they burn people when it's not snowing or raining."

"They're still singing. They're more persistent than salt," I said. "They're braver than geese or wild swans. I wish I had their marrow in my bones."

"A lot of good it would do you. You'd really end up in a fine mess then," Portman said. "Don't count on me."

The uniqueness and worthlessness of human life—like two sisters—floated through the night, over the camp and over all of the camps of Germany and the occupied territories. This is what in five, six years the German war had done. This is what made one doubt man and his existence, victory and defeat, good and evil. What caused the first to be last and the last to be first. The wind distorted the singing of the women. It drove the snow against the wall of the barracks and filled its enormous arms with ashes which it carried away farther or closer, spreading them on all sides—ashes that will remain on the face of the planet like a birthmark, even if every last Jewish man, Jewish woman, or Jewish child should perish.

The ashes silenced the echo of the first shot. It lasted only a few seconds. Definitely less than half a minute. The singing mixed with the shooting, then there was only shooting and its echo. It definitely was nothing unexpected; Portman was right about that. It sounded like the Rottenführer's switch for camels. Last time, Harry Cohen remembered that a camel will never forget if somebody wrongs it. But, Harry Cohen added, they never forget the good that someone does for them, either. I was sure that he was now in his mind with that Aryan girl in Prague, Maruschka. Maybe she could read his mind, but she could never know what he knew.

In the air, among the snowflakes, a kind of echo remained from the shooting of the machine gun. It was all ordinary, just like the snowflakes. Like the smoke from the chimneys. Like mud.

"They could have figured it out," Portman said. "Unless they've become blind, like moles in the winter."

"Maybe they did figure it out," Harry Cohen replied. "But I don't believe it, just as I don't believe that moles go blind in the winter."

"I'll go to sing with them, once they get the idea of throwing rocks at the Rottenführer and at the Scharführer at the same time," Portman said. And then he added, "No one will give a damn."

Someone went to the bucket. Nothing had changed in the least in the daily routine of the barracks. From the sounds of the steps and movements you could tell who had diarrhea, who had dysentery or even typhus, who had tuberculosis, pneumonia, or only asthma—who would not make it till morning or midnight and who would manage to infect his bedfellows before he died.

I bid farewell to the singing of the women and looked forward to tomorrow's, just as I had looked forward to my grandmother's bedtime fairy tales. Where does a man find strength when all that he has is weakness?

"Schiese-Dietz believes that Jewish children are born hairy, like animals," Portman said. "He also selects the hairy girls first, even before those who are skin and bones."

I knew that I would be able to hear the women singing again, and yet it occurred to me that they would be singing when I wouldn't be there—nor Portman, nor Harry Cohen. I thought about life and death in Auschwitz-Birkenau. What is the purpose of human existence? Why is man plagued by feelings of futility and worthlessness? How do you find out how to do the right thing when it comes to it? It was again a moment of unanswerable questions. But there was some change in the air, even if there were no answers. How many people had they shot? What would the women appeal to tomorrow with their singing? I was lost in the echo of their chorus. I felt the snow, the ashes, and the silence around me. I felt the urge to go outside, for which the guard would immediately shoot me before I got to the barbed wire. I wanted to touch with my lips a sliver of ash or snowflake. I listened. There was no sound. The women had sung themselves to sleep. Others cried themselves to sleep. And still others had been shot into an eternal sleep.

"At last," said Portman.

It did not sound like relief or satisfaction, though sometimes Portman thought that women had nine lives, like cats. Perhaps he hoped that it was true. Everything that had to do with cats Portman believed to be ill-fated, an echo or foreboding of misfortune, even when it was born only in his head.

"Couldn't they have gone to sleep long ago?" Portman asked. "Tomorrow they'll be dropping like flies, even those who were not hit." And then he continued, "There's no sense in wasting your time or your strength. It's a waste of every drop of blood when you spill it on your own and for nothing."

I tried to call back the echo of the women's voices, the image of birth, of something that for me was always connected with women, something which I knew little about and probably never understood. I felt the familiar shame, but for the first time, maybe knowing why.

"Do you have the shakes or what?" Portman addressed me. And finally, "They'll drive me nuts with their singing."

Portman curled up to sleep. I wished the remaining women would fall asleep and not be cold. Harry Cohen began to chew his bread and blood sausage. He was, maybe, concerned with the number of possibilities everything and everybody had, has, or should have, or doesn't have at all. He wished to be somewhere else, someone else, where people do not live a miserable life, where they don't have to ask every other second why they were born, about things for which there were no solutions.

Portman's fleas and lice were feasting on me. The smoke rose slowly toward the sky. Black snow was falling. I have never forgotten the black snow with ashes in Auschwitz-Birkenau, which that night I saw for the first time in my life before I fell asleep.

"Good night," Portman said in a conciliatory tone. He repeated to himself the tattooed number on his right forearm. He probably no longer envied the dead.

"Good night," Harry Cohen said. It was one of the possibilities.

"Good night," I managed sleepily.

We were swallowed by the silence into which Samson once disappeared and in which every night and every man seems to be the last. Infinity engulfed us.

20. Adolf Rudnicki

Adolf Rudnicki was born in Warsaw in 1912. Although he began publishing novels in the early 1930s, he found his major theme only after the outbreak of World War II in 1939, when he applied his talents as a psychological novelist to commemorating the fate of Polish Jewry. He fought with the Polish army in its early campaigns, was taken prisoner by the Germans, escaped, and made his way to the Soviet-occupied zone of Poland, where he settled in Lwow. After the Germans occupied the city following their invasion of the Soviet Union, he returned to Warsaw. Although Jewish, he lived on the so-called Aryan side with forged identity papers. Even though he was not involved in the Warsaw ghetto revolt in 1943, he participated in the uprising in the city of Warsaw in 1944.

After the war, he moved to Lodz, where he resumed the project that he had begun during the occupation: creating a fictional monument to the fate of Polish Jewry. He used short stories to explore the complex psychological tensions that ensued for victims and survivors who faced as their legacy a doom of mass murder totally unrelated to the moral reality that normally shapes human behavior. In "The Crystal Stream," from Rudnicki's collection of stories, *Ascent to Heaven,* the spontaneous value of love and friendship has been so tainted by Nazi persecution that, at least in the immediate postwar period, one is left wondering whether and how it will ever be restored.

Rudnicki died in Poland in 1992.

The Crystal Stream

When Abel went out, the day after his arrival in Warsaw, he was still wearing his army uniform, though the war had ended some months before. It was obvious that until recently he had been a prisoner-of-war. He was tall, lean, with black eyes, a head perhaps a little too round, a face which passed quickly from moodiness to sorrow. It was this note of sorrow in his features that attracted the attention.

In Nowy Swiat Street he saw the same devastation as he had seen yesterday, on his arrival from Lodz. Only a short distance separates Lodz from Warsaw. Today the two cities are more closely linked together than in the past. Every day thousands of people travel between them on thousands of affairs; the traffic is lively and continuous. So one would think that an inhabitant of Lodz would be able to imagine what Warsaw looks like, if only from hearsay. But no. A city so completely destroyed as Warsaw, defeats imagination.

From the moment the train had steamed into the outskirts of Warsaw and the first burnt-out houses had risen before him, he had sat speechless. What he saw was impossible to describe in words. After undamaged Lodz, the mind could not take in this prospect. Time after time he told himself: it's bound to end at the next corner. But it did not end at the next, or at any following corner; there was no end to the ruins—house after house, street after street, district after district—which had been struck by a cosmic anger, the same anger of God that thunders from the mountain tops. The empty shells, the empty windows, the mounds of foul earth inside the shells, breathed horror and filth, the dreary dreariness of corpses.

Yesterday he had spent several hours looking at what remained of the city, recalling a similar scene from his childhood after the first world war, in the silent ruins he had rediscovered a long familiar aspect of history. But here the novelty, the incredible novelty was that every one of the houses was reduced to a shell.

Burnt-out houses stood all along the Nowy Swiat. And yet, they stood. The ruddy hue of certain interiors recalled the ruddiness of broken earthenware pots. Where there were missing houses, a gap of two or three in succession, it was at least possible to bring them to recollection. A former resident returning to this district would still be able, after a moment's thought, to say, 'my house stood here'.

But when he crossed the Krasinski Square and reached what formerly had been the Jewish district he looked to left and right of him, before and behind him, and although he had seen other parts of this 'most devastated city in the world', he refused to believe his own eyes. He had expected to see destruction, but on the same scale as in the other districts; he had expected to see some traces which would enable him to re-create what had formerly existed here. There were no traces whatever. There were no buildings, not even gutted shells or partially destroyed; there were no buildings at all. No walls—no chimneys—which cling so tenaciously to life—no outlines of streets, no sidewalks, no tramlines, no roads or squares; on no floor had anything survived, to give the eye a moment's rest. Here there was not one of the elements created by organized human effort, nothing to establish that this spot had been inhabited by man. Over an area which the eye could encompass only with difficulty, where formerly the greatest concentration of Jews in Europe had been housed, there was nothing but rubble and broken brick, with here and there yellow and grey sheet-iron, like untanned ox-hide. Here the city had been rubbed out, removed from the surface of the earth like a tent from a meadow. In other districts there were dead bodies; here there was not even a dead body. In this place the capital had been crushed to powder, not one stone was left on another. And though beneath these fields of rubble rested more dead than in a hundred cemeteries, there was nothing to suggest a cemetery. The city was deleted, and only its frail, uncertain and delusive outline loomed in the onlooker's memory. Faced with this nothingness, Abel no longer felt any urge to seek out the house in which he had lived—an urge as idiotic as it was profound. Before this place had been levelled in death, it had been levelled in suffering. There was no justification for weeping over any particular individual. He sat down on a heap of bricks. All round him was a dead silence, the only note of life was the murmur of the waters flowing through the sewers. Some distance away, several people passed in single file between hillocks of rubble. Over the city reduced to chaos hung a sky cloudy and cold.

As though in a dream, he heard footsteps. As he unwillingly raised his head, he saw a woman coming towards him. His stupefaction was followed by a realization which he could not accept. He recognized and doubted, doubted and recognized. Yet he did not stir. He stood up only when the woman, as moved as he, halted in front of him. And although he felt that he was flinging himself into her arms, and crying, he stood unmoving, unspeaking. Only his pupils dilated and contracted.

She was the first to speak; just as in the old days, she hardly sounded the 'b' in his name:

'Abel!'

His voice died in his throat, as it had during heavy air raids. For some time he could not even say:

'Amelia!'

2

History as we knew it in the days of *blitzkriegs,** the history of recent times, this ultra-modern history in all the fields of struggle, on land, sea, and in the air, was completely different in the prisoner-of-war camps. In ordinary life, when a man changed his place of residence he seemed to be changing the age in which he lived. In some cities one had only to move from one street to another to go back several centuries in history. But the prisoners of war, from being people incessantly occupied, as is the modern fashion, became people cursed with an excess of leisure. Only then could they ponder on many things for which they had never had time before.

Only when shut away in the camp—in the neat, small German town which had been transformed into Oflag 3 E, and holding about 6,000 men—did Abel come to realize Amelia's distinctive qualities, which in the days of freedom had ricocheted at him off other people's appreciation; he himself had never paid much attention to them. He had met her first in 1937, just about the time when he had set up as an architect. She was a thoughtful, thin, and very young girl then. As time passed she grew more and more beautiful, and one day Abel realized that her beauty was of a very distinctive kind. She was tall, dark, with a perfectly proportioned figure, a face that was wholly mild, like the faces of Madonnas, whose unprovocative femininity seems to be a contradiction in terms. Set in its frame of rhythmic curves, her face quivered with light and colour like the interior of a church. Any background against which her profile happened to be set—a whitewashed wall, a scrap of sky—became an arabesque marvellous in itself, like the work of an abstract painter. As they gazed at Amelia, people were speechless, as at the sight of a mountain flooded with moonlight, or of a great fire. She always reminded you of someone. Or rather, it was not that she reminded you of anyone, she awakened the age-old yearning for beauty that is innate in us all.

So long as they were together—a period of two years, one year as husband and wife—Abel was not very conscious of her beauty: as yet he had no need of it. Amelia's home—her father was a chest- and lung specialist—was warm and welcoming. During the early days of their acquaintance Abel loved the family more than the girl; he himself came from a home which never knew laughter, his father hated his mother, and had not spoken to her for years. Amelia conquered Abel's heart above all by her quick response to her mother's good and bad moods. The simple fact that her mother had eaten well, and had gone to see her hairdresser—which indicated that she was regaining her interest in life—or, on the other hand, the fact that she did not wish to see anyone, was an experience in itself. And it was this that originally had most effect on Abel.

Once in love with the family, he deliberately ascribed to Amelia all the

*Lightning wars, describing the swift German invasion strategy in World War II, using the overwhelming striking force of air power and motorized units.

qualities he thought worthy of love. Because he craved for love, he persuaded himself and Amelia that he loved her. So he said many things which were not strictly true. Even as he said them he was aware of some inadequacy, of his own reservations, and he hated himself, for he was not really a hypocrite. He craved for love, but for a long time he was conscious of a gap in his soul.

The years in the camp closed that gap. His fear that one day Amelia would realize his duplicity proved groundless. In the camp he fell head over heels in love with his wife. He discovered her both as a human being and as a woman, and much in her that before had annoyed him now became something to adore. The fact that, unlike him, she was undemonstrative and inscrutable in her feelings; that she was capable of strength of character just when he lacked it; that she could bear the most unjust charges with dignity and not seek an explanation, as long as he remained angry—an art few possess; that she could be more experienced, wiser, older, though she was so much younger; that she could treat him as a sick man—in love, one party should always have the privileges of the sick. Many other fine qualities, attitudes, details, which had not been obvious to him during their everyday life together, now came to life for him, arousing tenderness and delight. Everything connected with Amelia acquired its own savour, the savour of certain early mornings, the savour of sweetness and pain. His heart ached a little at such times, as a lover's always does.

The last phase of their life together grew especially dear to him. In 1939, people were no more afraid of war than a child is of a revolver. But in Amelia's eyes Abel saw a fear of war. She would not listen when people were talking about it, she would leave the table rather than listen. After Colonel Beck's speech on the Danzig Corridor issue, in May 1939, she burst into tears.

One day Abel repeated to her what some important person had said: Poles were not Czechs, they would shoot. When he awoke in the murderous stillness of next morning's dawn, he saw that she was not asleep. She was gazing at him intently. After that morning, whenever he woke up he would find her gaze fixed on him. When he tried to comfort her with the remark that the Germans had no food, that their tanks were made of tin and that England would not allow them to do anything, she did not reply.

In the later days he felt sure she had a premonition of the approaching disaster. But no. Although she was living through a difficult year, she was not unhappy. The first year of her marriage was also the first of her life as a woman, she loved her husband and she had her own home. During their last nights together she received him as a communicant receives the Host, with eyes so pure that he could not look into them.

When he was called up, she went to see him off at the station, through a city as thronged as if it were the eve of a holiday. And then he realized that her voice had taken on a new note; it seemed to flow from some unknown source. She spoke emphatically, underpinning her words, so to speak, evidently afraid she would not be able to get them out. As he stood at the carriage window and took a last look at her strong, beautiful figure, now huddled as though it had suffered an inner collapse, he felt for the first time afraid of the war. And he understood the old truth that pain affects us most through the people we love.

She was inscrutable in her feelings, she was 'impregnable', and this circumstance had always intensified his own reaction; but during those last months before the war he was able to see that the fire he had laid had caught. And then he reproached himself with receiving more than he deserved.

Like the majority of prisoners of war, he lived for letters, for those official forms so well known all over Europe—the *Kriegsgefangenenpost**—of which half was allotted to the reply. Amelia amazed him. He had never expected such ardour. He remembered her as a being of perfect balance, and doubted the existence of such depths of tenderness in her as she now revealed. He doubted because he was naturally sceptical. He doubted, too, because, like the majority of men, he thought that great beauty is satisfied with itself, as if beauty were not exactly the same as comfort is to the man accustomed to comfort: the outer skin, only the outer skin.

Sometimes her replies were delayed. Then he was completely carried away by his suffering, as is so often the case with people who feel deeply. When they did arrive, he did not even need to read them. Lack of news from Amelia drove him apart from the others, made him long for solitude. The arrival of a letter, especially after an interruption in their regularity, made him seek company.

The Germans had such contempt for other human beings that they did not care what the prisoners wrote about themselves or what was written to them. The censors hardly ever blue-pencilled letters, and all the men were aware of events at home. In August 1942, the camp underground paper had an article on the slaughters in the Jewish district of Warsaw. And Abel's letters from Amelia ceased.

It was a summer night painful in its beauty. In two small, connecting rooms, nine men were lying on palliasses, but not one was asleep. Abel was sitting at the window. From time to time one of the others rose and went to him with words of comfort. Nobody had heard a word from him for two days. Nobody had seen him eat. He sat at the window day and night, his ears attuned to the inaudible cry of the murdered city. The voice of the distant suffering floated over his head as though he were a wireless receiver. He could not suppress the watchfulness and attention which is fundamental to life. As of old during his worst attacks of jealousy, so now he lacked the courage to look at Amelia's photograph, it caused him so much pain. For him a crime against Amelia summed up all the crimes against four hundred thousand human beings. During those days his face darkened. 'He's fretting himself to death,' the others said.

Then he received two letters from someone with the unfamiliar name

*Prisoner-of-war mail.

of Anna Zuch. It was Amelia's new name. She was safe. She was living outside the ghetto, in Zoliborz suburb. As he read the new address Abel had the feeling that she had changed her country. After the two letters there was a further interruption, and he reproached himself for having been as imprudent in replying as she had been in writing. But soon she wrote again. Her note included the words: 'I have been through something during these last few months.' As he read, Abel caught his breath. The meaning was clear, for it was universal.

At a later period her replies contained fewer confidences and expressions of tenderness than before. At a time of such suffering what significance—one would think—could a few tender adjectives have? None the less their absence, and the absence or rarer use of certain verbs, some change in her vocabulary, plunged him into a deep anguish of a different kind. He was nothing, less than nothing, he lived with the other Jewish officers in a separate block 14 E known as the *Judenblock.* Their office records were kept separately, they expected the end every day. Everybody in the camp was quite sure that block 14 E would suffer a different fate from the others. And at home in Warsaw Amelia, hiding in the so-called Aryan side, lived continually under sentence of death, and any one might be her executioner. And yet . . . and yet, in his thoughts, desires, and demands he never ceased to be first and foremost a man, and to think of her as a woman.

He wrote to her, saying that as he could give her nothing, he had no right to demand anything in return, and he set her completely free. To which he received a reply so passionate that he could not have hoped for anything better. But then her letters suddenly stopped. When she wrote again, despite his expectations she did not give a changed name or address, so evidently there had been no outward obstacles to her writing. But he noted that she had written with three different kinds of pencil. It was obvious that she was no longer writing at one sitting, and that in fact she no longer liked writing him letters, but had to force herself to write them.

The Catholics, who possess a long-standing knowledge of man, say that the human soul is the real arena of struggle and crisis. Abel had so often felt numb at the thought that Amelia might leave him—might perish, or might leave him, it could not be otherwise—that when her letters grew infrequent, and conversations with other prisoners forced him into the thought that it must already have happened, he bore the shock with comparative calm. Human strength is limited; suffering, like fever, has its upper limit. Tragedies arising out of infidelity were everyday experience in the camp. The women outside entered into new relations, bore children to other men, floated on the varied waves of living existence. The prisoners who were the first to suffer from their wives' 'betrayal' were plunged into the depths of despair; those whose turn came later suffered less, for they had already suffered in sympathy and apprehension. That was the general law, and Abel was no exception.

In October 1944, officers who had taken part in the Warsaw rising arrived at the camp. The majority of them were in civilian dress, with white and red armlets. They inspected the camp with curiosity, and their first question was whether they would be retrained, and whether there would be a joint German-British attack on Russia within the next six weeks. Those who had urged them into the steps leading to their imprisonment had assured them that these questions were settled. Unlike the earlier prisoners, the insurgents were solely interested in politics. They were broken up and scattered among all the blocks.

Among these new arrivals was one of Abel's student friends, Tadeusz Mazurek. He knew a good deal about Amelia, and he thought it wrong to let people keep their illusions. Yes, there was someone else in Amelia's life. But Abel was not surprised by that news. He was surprised by himself. Before the insurgents' arrival he had been apathetic. He thought he had completely exhausted his stock of emotions; but now hell opened again before him. He was astonished at the strength of his suffering, he thought he had lost it all. He had believed himself to be reconciled to the loss of Amelia. Like the others, he had told himself: it had to be. Now it transpired that he did not know himself. In the night he started up from his paliass with horror and defeat in his soul. He choked, as though the room were stifling. His position seemed bitter, unbearably bitter.

As Amelia was no longer his, one would have thought that he had lost all capacity for joy. But no. He was even capable of such delirious joy as is comparable only to that of a dog rushing about a yard on the first warm day after the winter. Warsaw knew that joy on May 8, 1945. Abel knew it the day the Russian forces overran Camp 3 E.

3

As she stood before him she repeated his name, obviously deeply moved. There was a warm light in her eyes. She had received his letter and card announcing his release, and had been tremendously pleased. She had sat down a dozen times to write a reply, and she really could not say just why she didn't. In fact, she had intended to visit him at Lodz, but always something had cropped up to hinder her. All the same, now he was in Warsaw she did not understand why he had not called on her, instead of relying on such a remote chance as this fortuitous meeting. 'You should just have dropped in.'

He gazed at her, succumbing to a tumult of joy, in which forces were liberated which had been suppressed for years; he was enveloped in its gracious, its blessed, invisible ecstasy. He walked along beside her, feeling no desire for talk, which flows only when emotion grows cold. The uttered pain is a pain subdued. At its highest it is a cry, or silence. As he listened to her he was void of all desire. She was at his side.

When he regained what painters call conscious vision, for a moment he saw two faces, as though on a twice-exposed negative: the face he remembered was contained within this face he now looked at. He recognized certain intangible changes. He reflected that it was stronger than formerly. Of recent months he had often said to himself: I am behaving like a fool: by allowing myself to dream of Amelia I am killing myself. There is nothing like meeting her face to face to cure me, by the very act of living she must reveal her weaknesses and make mistakes from which she is now shielded by my yearning, by my very need for beauty. Only meeting her will cure me. I must let life kill the idea. For a lonely man there is no disillusionment to equal that of life itself: having forgotten what it is like, he will certainly find it too crude for him. But his first glance had not brought him disillusionment, not, at least, in her beauty. As of old, she was perfectly proportioned, solid, and tall. Her skin emanated warmth, well being, and serenity; it invited caresses, like a piece of carving, like a child's bare belly. After the holocaust of a nation such women were rarely to be seen in the streets of Warsaw.

She spoke first:

'In the early days we used to wonder what we would say to the people who had spent the war abroad, and who would come back, of course, when it was over. And we used to worry over it. How could we give them some idea of the sort of life we had lived, how could we make them realize that at bottom it was inconceivable to those who had known only the pre-war life? We all agreed that we would have to show them the children the Germans had photographed for publication in their press as "the breed of the Bolshevik paradise". Not everybody on the Polish side was friendly. Some said the ghetto walls—behind which hunger and typhus were killing the inhabitants who were packed together like people in a crowded train—ought to be surrounded with a ring of machineguns, so that if the war came to a sudden end the infection should not be spread. But when a ragged child, overgrown with filth, knocked at some house on the Polish side, and, not knowing, or barely knowing Polish, thought it better to stand mute and humble, it was not sent away without bread.'

Abel remembered that when news of the slaughter reached Block 14 E all the prisoners reacted strongly, but each of them to a different aspect of the tragedy. One said: 'Driving old people into a truckful of unslaked lime!' Another: 'The beautiful girls that are going under! How could any man kill a young girl?' And another: 'My God, they're murdering little children.'

'We couldn't explain to the children what a tree or a river was, for there were no rivers or trees. When a children's entertainment was given in a cinema the crowd wept. Korczak[1] told my father: "I am an old man. I thought there was nothing else I wanted in life. But there is. I want to live to see the opening of the ghetto gates . . ." As you see, our gates have been opened,' she said, pointing to the waste of rubble.

Over this soil of the most terrible of all human suffering, Abel walked, clinging to Amelia's arm, as though it were a rose-garden. He drank in the sweetness of her presence, which dispelled all the horror as a memory. She was almost as tall as he, and was wearing a coat of black lambswool. The plate-shaped little hat had slipped down over one eye; she regarded a hat only as an ornament, like a cornice at the top of a house; he remembered that all her hats had fallen over her eyes, they had varied only in the shape of the crown.

The memory of this detail was like a new draught of sweetness to him, and he lightly brushed her arm with his lips again and again. Her eye was lost in the shadow cast by the brim, he saw only her nose with its slightly tilted nostrils and the provoking curve of her lips, which he badly wanted to kiss. For a long time he could not bring himself to do it. When at last he leaned across and brushed her lips with his as they walked, her smile was suddenly extinguished. That hurt him. They walked across fields of broken brick topped here and there with rusting iron. Iron rods grew out of the rubble like some new kind of bush.

It is difficult to trace the course of living conversation, it sprouts—like some men's beards—in all directions. After a moment or two Amelia turned once more to recalling the past. Every word she uttered quivered with suffering. That in turn gave Abel encouragement to speak.

When he ended, her face was drawn. The warm lights had gone. Abel was reminded of some friends he had met after six years' separation, only to discover that after fifteen minutes he had nothing more to talk about with them.

'Love,' she echoed him in a tone of astonishment, and apparent dislike. For a moment he had the feeling that he had done her some wrong. 'Love—that's a word I haven't heard for a long time. A word quite forgotten. A fiction . . .'

He felt as humiliated as a man who has declared his love to a woman, only to have her say: 'That's nothing, it'll pass . . .' His mind rested on the word 'fiction'. He knew that fiction through and through . . . Set free by the Red Army, he had returned with many others to Poland. Those were wonderful days. He had stopped at Lodz. He had liked that city. During those days he had liked everything. He had taken the first job offered to him, and had found pleasure in setting to work early. He had revived. The last flames of war had been flames of regeneration not only for him.

Everybody had sincerely desired a change, everybody had been yearning for light and joy after the gloom of the occupation. But this zest had not lasted long, the depth of that night had penetrated to the bone. When the great waters subsided Abel realized that he was still eaten up with longing, just as he had been in the camp. Freedom had not stilled that feeling. As a lover listening to a broadcast concert is conscious of the breath of the beloved who is in the hall, so Abel was conscious of Amelia in the voice of his country.

His heart stopped beating when he caught sight of a dress, a coat, a hat, similar to those he remembered from past years, though in all probability they no longer existed. A facial likeness, or a similarity of movement or gesture, no longer caught his fancy, as they had when he had noticed them in fellow prisoners. Now they quickened only pain. His heart was never silent. Worst of all, he could not stop thinking and remembering.

In past days his mother, a frail, little woman, had wept whenever someone mentioned jasmine to her. After six years of war he could not control his voice when he had to utter the word 'love'. When he saw a couple nestling against each other he could not help trembling. He could not bear to hear that someone else was happy. His heart felt like an unhealed wound, and in it was a sucking pain that never ceased. His longing turned to a nagging anxiety. He woke up again and again at night, feeling he was disembowelled of all except his pain. He was one mass of pain, and it alarmed him all the more because he did not know how to defend himself against it.

Unable to forget, he remembered. Remembering, he suffered. His body remembered, his mind remembered, and all memory was pain. He would ask himself what it could mean, but could find no explanation. He surrendered to the rapture from which he could not free himself. It was the earth to his feet, the light to his eyes, the reason for his existence. To him and within him it was the main, avid, self-sufficient reality . . . this—as she called it—fiction.

'After all, you see what is left of our life . . .' said Amelia. 'This is not the result of a half-hour raid by the Royal Air Force. At night we did not go to bed, and in the morning those left alive did not see what we now see. The death agony of the ghetto, the death agony of our community went on over years. Before the Jewish nation was sent up in smoke it was changed into mud and dung. Don't believe it when they say people know how to die. Twice during this war this city has seen that people do not know how to die. People are created to live, not to die. What people say about dying is an exaggeration, or a bitter jest, something that has as much in common with dying as the metaphorical death which young poets talk about so much.

'Millions went to their death still trading and loving, still intriguing and hating, with not one earthly feeling the less. They went to their death immersed in life up to their ears. With the taste of their own blood on their lips, they still did not feel the breath of God. To the very end they clung incomprehensibly to life. Even their death was a hymn to life. Even in the shadow of the gallows they were finicky about the food they ate.'

In the old days she had talked only with reluctance and difficulty. She had never been one of those garrulous creatures. At first Abel had called her a beautiful silent child, for she had been so astonishingly taciturn. She used his forename for the first time only some months after their marriage. She was one of those women who find it difficult to become intimate with a man. But now Abel recognized that her voice, like her face, had gained in strength. It had changed.

'Death dwelt among us and changed much in our ideas of life,' she went on. 'When a man went away he went into the darkness, and usually did not return. We did not dare to let our feelings accumulate, they had to be spent at once. Women gave themselves to men just as though they were doing them a justice, as though they were righting a wrong. And then it appeared that there are a hundred kinds of love, and every one is good. And every one is acceptable.'

Abel also knew the hundred kinds of love. During his first few weeks of freedom—those most happy and gracious weeks—he had believed that he would find a niche in life for himself even without Amelia. At first he had thought that the disaster was not so great after all, and that the Jewish intelligentsia, of whom a few had been saved, would want to build up their life again in a land which, truly, was like one vast grave, yet even that would fade in the memory. At the end of the war those few survivors who had been saved from the flood that had overwhelmed a community of three million people had emerged from underground, from the forests, from the camps. A few had returned to their former addresses. They had been received as people who had experienced such suffering should be received. Here and there sincere tears were shed by both sides.

However, of all human feelings sympathy has the most brittle feet. As they returned after dying a hundred deaths a day for six years, many of these people, in whose presence the world should hang its head with shame, were struck down at the very doors of their homes by bullets which were not German.

Along the roads creatures who called themselves soldiers lurked in wait for them, and with a well-tried method, with *Genickschuss,* a shot in the back, murdered them. When they lay down to sleep in the villages, in small and even large towns, they were almost as uncertain of the morrow as in the time of German rule. These murders dispelled the last feeling of reluctance that troubles every man pondering on the difficult question of leaving his own country.

There was nothing else to be done, the death of a Jew had long since ceased to move anyone, so they began to flee in large numbers. And one day Abel realized that he was living in a wilderness. He met acquaintances who had closed the book of their lives, but none was opening any new book. He could see no attempts at renewal: on the contrary, they did not want to renew their life in this land. They wanted only one thing: flight. And they fled.

Abel sometimes thought that if he were living in a community whose mind was set on regeneration it would be easier for him to forget. But all around him was a wilderness; he was living in a wilderness, surrounded by stumps that were withering away. There was no life, for there were no people. Besides, he was no longer young, he was close on forty, and woe to those who in their fortieth year must begin their emotional life anew, must seek; what had he to offer to any women he came to know only now? In comparison with what he felt for Amelia all else was thin, empty. Dazzled by Amelia's brightness, he thought all others dim. All other women bored him, and he bored all other women. For so it is, that to some we are as the sea, yet to others we are as a miserable, trickling stream. There were a hundred other kinds of love—he knew that—but not one of them was for him.

'The only joy we had in those years,' Amelia said, 'a joy we are already ceasing to understand, though so little time has passed, was to kill a German. That was the only thing that brought us happiness. Nothing else. No one set himself any other task. It would have been imprudent, petty. Maybe there is a time which is favourable to love, but those years were not . . .'

Abel no longer pressed his lips to her coat. Her words parted him from her just as effectively as the camp had, in the past. He felt that she was right, and he felt, too, his own pettiness. And shame. Not for the first time, he thought how badly he fitted in to what was happening in the world.

Nations had perished, cities had disintegrated into dust, the gains of centuries had been lost in a single night. And what was he interested in? What really occupied his mind?

4

The sky was the colour of soapsuds. In one spot the gate of the military prison, and elsewhere sections of the brick wall of the Pawiak prison, had escaped destruction; in the distance a church could be distinguished; only fragments of the prisons and the church had been spared. There was no means whatever of determining the spot on which their home had once stood—a handful of dust amid this expanse of rubble slag-heaps. A bare beech tree stood in the cinnamon-coloured gap of the Pawiak wall. Fastened to its trunk was a black board with a white cross; beneath it two young men were standing, bareheaded. One was talking, the other listening; they were both gazing down at a grave at the foot of the beech. Abel and Amelia walked past them, reached the high road which once had cut right through the Jewish district, then turned back. Their feeling of exasperation persisted.

As she referred again to the fiction, to life in which nothing matters, she laughed an unpleasant laugh—the laugh of a woman false to her husband, the laugh of women delighted with the clever trick they have played. Abel caught a sudden coarseness in her tone, and recalled that it had always repelled him. He took another sidelong glance at her lips, and realized how sensuous they were. In his sudden hatred he seized her by the hands and shook her, then raised his hand to strike her. But he realized what he was doing, and was panic-stricken, afraid that out of all his years of longing Amelia would remember only that moment when his face was distorted with fury. He tried to apologize, but could only stammer. He said words which were meaningless even to himself, words which afterwards he could not even remember.

As he caught her frightened gaze he pulled himself together. For a long time he could not free himself from the effects of his outburst; he felt exhausted, alien to himself. Amelia looked at him intently; his face melted, his lips quivered. She had powers of observation at times when he was far from any such ability. They halted. She said in a quiet, slow voice:

'Abel, you have been entirely concentrated on your own life, I have not. I am no longer the woman you left behind. I am different. How can I hide it from you? Look about you; think of what happened here . . . Love! For many months the blood tingled in my veins, at night my body swelled like the earth in spring, like a river in spate; it ached as though I had been thrashed. My eyes were as restless as those of young girls who feel they are walking about the world with no skin on them. The secret of man is that he always wants more than he can achieve. Love! There is a time in our life when a single being swims in our crystal stream. And then we set ourselves the task, the difficult task, of keeping that stream in all its purity; but the heart delights in difficult tasks . . .'

Now she was speaking tenderly and warmly, she put her hand on his arm. Suddenly he observed that all the strength had gone out of her face. Her cheeks were as delicate as those of convalescence, her lips were puckered, her inscrutable eyes were tired.

'Abel, listen; I will tell you how I escaped. I was taken out of the ghetto buried under the load of a municipal dust cart. It was the only way. It took them half an hour to bring me round, I was almost suffocated. My last thought as I fainted was that I would be tipped out on the rubbish heap down by the river. I wasn't, and I didn't die on a rubbish heap. But after that, of course everything lost all meaning. That which previously I had regarded as most precious now seemed nonsense, a tragic sneer, a fiction. Every day we died, and every day we were born again, but always we were ashamed of the sawdust in our minds. We ceased to understand the world of yesterday. We hated it, and everything that had been bad was now good, and everything we had avoided was now permissible. Once I had escaped to the Polish side of the city I no longer set myself tasks, I did not struggle, I did not go about with cyanide in my bag. Before that, I used to ask myself: "do you want to be like Marysia Werner?" No; I did not want to be like Marysia Werner. But then I did become like her, and did not even give it a thought. The crystal stream had dried up . . .'

She was silent. After a moment or two she went on:

'How strange it is that in face of such destruction, such an end of the world, the heart still retains all these nuances . . . If anyone were to ask me if I regret all my self-denials, all that period when my body, like a dog on a leash, whimpered for the fiction of a crystal stream, I would answer: "No." For in face of this end of the world which we now see, what deserves to be called a fiction, and what does not deserve to be called it?'

They sat down on a fragment of tiled stove, first Abel, and then, after some hesitation, Amelia. Borne on a new wave of hope, he said:

'I deluded myself that you would come. Every time I went out I left a note saying where I was to be found. I regretted every wasted moment. You did not come. I wrote; you did not answer. Later someone told me you were happy and I at once believed it. One never does doubt something which may cause one pain. That news explained everything. You didn't write, you didn't come, because you already had all you wanted. That's just the difference between the happy and the unhappy: the unhappy are never at home. Of course I could have come to you. But what for? She is happy, I told myself, so leave her alone. And I left you alone, though every day, every hour meant so much to me!

'My brother wrote to me from Casablanca; I was glad to have his invitation. I came to Warsaw yesterday in order to arrange my departure. I went and hung about your house, but I lacked the courage to go in. My first thought was: How can I go abroad? All that distance apart I shall surely wither away . . . I came here a ruin, I came with set face, incapable of a smile, with ashen lips, with a frozen heart; but now I can see what happens to me as soon as you are beside me! Without you, here or anywhere on earth, life holds nothing for me. If you leave me, you will be flinging me into my grave . . .'

He fixed his eyes on her. There was no entreaty in them, they only testified to the truth.

'Abel, in this life repetitions are never successful,' she said. 'They are dangerous.' As he insisted, she added: 'Abel my dear, what can I bring you back? This body, this body which is like a conquered land . . .' She did not finish. She drooped her head, and sat thus for some moments. When she raised her head again, Abel saw a face swollen, quivering with emotion. Her upper lip was trembling absurdly.

'*He* has gone to visit his graves. I told myself: I too will go to visit mine, and I came here . . . It's too late for us now, Abel. I often used to think you'd return, and we'd come together, you'd forget like thousands of others. But what one does only for the moment often lasts, and lasts for a lifetime, and it cannot be broken . . .' She added in a voice over which she had lost all control. 'Now it isn't even a question of *him* any longer. I have a child, six months old. And that is quite different . . .'

She said no more. And Abel, too, said nothing. He no longer asked for anything. His silence was profound, the silence that falls when all contact is broken between one and another, when one knows that the decision has already been made.

He realized that he had lost her for ever. This was the end. He was surrounded by a dead sea. And there was no hope, no light. He had dreamed of Amelia as only a man long-hungering can dream; perhaps children too can dream like that. This body was no longer for him. If he had not been too ashamed, he would have asked for it as though asking for alms, but only as alms could he have received it. This body, which after him had borne with so many, and would bear with many yet, was not for him, not exclusively for him, who desired it more than could anyone else in the world.

Because she had loved him, and now had no strength, or else did not desire his love any more, so now he could never draw anything for himself from that body, except shame and degradation. For love which has once been sublime, but then has lost its exaltation, is only degradation. Just because he had been borne along in the crystal stream of her life she could not belong to him now, though she might belong to all others.

He buried his face in his hands. He knew that in a moment she, his life's last joy, would be departing from him. Once more he would experience nights and days of despair oozing through the chinks of life, the old incessant sucking pain in his heart would return. He realized that fate had defeated him, and his breast ached with the pain. He felt her bend over him, he saw tears in her eyes. He felt their warmth as they mingled, hers with his, on his face. Then she left him. He did not call after her. Stunned and still, he gazed after her retreating figure, and at times a groan burst from his throat. He gazed after her until she disappeared, disappeared without once turning round. How long he sat there he did not know. When at last he began to feel cold, he rose and dragged himself away.

It was November 1—All Souls' Eve. Warsaw was celebrating its first post-war holy-day of the dead. The city was full of graves, it was one enormous grave; every few paces flowers were lying on the pavement; little funeral flags peeped out among the flowers, and candles were burning. Small boys and girls were keeping guard over the graves. They stood as motionless as soldiers. Other boys, walking in a long open file, halted before the guards and tried to make them laugh by staring into their faces. The streets were almost empty. In one street Boy Scouts were decorating a grave. The young scouts were thoughtful, simply thoughtful; their thoughtfulness conveyed no memory of the living blood of their dead. For them all that the graves implied was already history. A little farther on, at a curve in the wall, a candle was burning; several vases were standing on the sidewalk, near them a kneeling woman was absorbed in prayer, as motionless as a chair. With wide, resigned eyes she gazed into the invisible faces of her dear ones. Abel halted beside her. Joined with her in resignation, he said farewell to his city, his country, and his dead hopes.

1. A well-known Jewish social worker, who refused to leave the children of his orphanage, but went with them to death.

21. Alexander Kluge

Alexander Kluge was born in Halberstadt, Germany, in 1932. He took a law degree in 1956, and has been a practicing attorney as well as a teacher and writer. In addition, he has created scripts for some highly original and controversial films, the most famous of which are *Abschied von gestern* (*Yesterday Girl* [literally, *Farewell to Yesterday*], 1965–1966) and *Die Patriotin* (*The Patriot,* 1979). He has also taught film production in the city of Ulm. His documentary chronicle of the German defeat at the battle of Stalingrad appeared as *The Battle* (*Schlachtbeschreibung,* 1964). "Lieutenant Boulanger" is taken from a volume of short stories called *Attendance List for a Funeral* (*Lebenslaufe* [literally, *Curricula Vitae*], 1962). Kluge calls these stories "case histories, some invented, some not invented." "Lieutenant Boulanger" is probably a little of each. Professor Hirt, a character mentioned in this story, is based on a historical figure who did, in fact, carry out experiments at the University of Strasbourg that involved measuring the skulls of Jews in an attempt to prove that Jews are biologically inferior because they have smaller brains than Aryan Germans.

In this and other stories, Kluge tried to develop a totally objective narrative style as a protest and defense against the Nazi bureaucracy's corruption of literary language. Kluge's ambition has been to use his art to mediate between Germany's Nazi past and the efforts of his own post–Nazi generation to come to terms with it. Concealed behind the arid stylistic detachment describing Lieutenant Boulanger's grisly war assignment is a horror so unimaginable that the wary reader can reach it only by piercing the ironic façade shielding it from view. As for the unwary or indifferent reader (who would include many of Kluge's countrymen), a kind of inner statute of limitations allows Boulanger's crimes to fade from perception and then from memory. The unconfronted atrocity sinks into obscurity, and the unforgettable ends up forgotten—or, what for Kluge is perhaps worse, disregarded.

Lieutenant Boulanger

I

In February 1942 the head of the Department of Anatomy at the Reich University of Strassburg, Professor A. Hirt, sent the following communication to one of the leading men in the Reich Government:

Re: Securing of craniums of Jewish-Bolshevist commissars for purposes of scientific investigation at the Reich University of Strassburg

Cranium collections are on hand representing almost every race and people. The Jews are the only group for which science has an insufficient number of craniums at its disposal for the obtaining of conclusive results. The war in Eastern Europe now provides an opportunity of remedying this deficiency. Jewish-Bolshevist commissars, the embodiment of a repulsive but typical subhuman species, make it possible for us to lay hands on tangible scientific proof by securing their craniums.

The most efficient method of acquiring and preserving this cranium material is in the form of instructions to the Army henceforth to immediately hand over all Jewish-Bolshevist commissars alive to the military police. The military police will in turn receive special instructions to report to a specified location the number and whereabouts of these prisoners as they are delivered and to see that they are kept under close guard pending the arrival of a specially authorized officer. The officer charged with the securing of this material (a junior medical officer, or a medical student, attached to the Army or possibly the military police, and supplied with car and driver) will take a prearranged series of photographs and anthropological measurements and to the best of his ability will establish origin, date of birth, and other personal data. When the death of the Jew has subsequently been brought about, in a way which will not damage the head, the officer will separate the head from the trunk and, after immersing it in preserving fluid in a metal container (with a close-fitting lid) specially provided for the purpose, will forward it to its destination. The photographs, measurements, and other data pertinent to the head and eventually to the skull will permit the laboratory to embark on comparative anatomical research, as well as study of the racial origin, pathological features of the cranium formation, brain formation and size, and many additional aspects.

By virtue of its functions and objectives the new Reich University of Strassburg would appear to be the most suitable place for the preservation and study of the cranium material thus acquired.

A. Hirt

The Personnel Department of the Army offered Lieutenant Rudolf B., of Flörsheim (on the Main), the command of this special mission. Boulanger had been a medical student. In effect, the acceptance of this special mission meant a short cut to promotion. The prospect of a transfer to research was held out. Boulanger seized the opportunity.

II

In 1942 Rudolf B. was thirty-four years old. He was of medium height. His complexion was olive, his eyelids were hairless. He may have had Romans or (eighteenth-century) Frenchmen among his ancestors. He had volunteered for the engineering corps, was prepared to shine, to seize advantages, to arrive at speedy solutions. For years he found no opportunity of conquering, he did not pass his state medical examination, he had no technical qualifications. All he had were good intentions, and with this he waited for his chance, which came in 1942.

Good intentions

If executing a task consists of proceeding straight ahead along a prescribed path without flagging or allowing oneself to be impeded, it might be said that B. fulfilled such a task in the highest degree. Vigor and intelligence operated—if, like Seneca, one is to regard a human being as a marching army—in this case on the very center of the front. In practice, however, none of B.'s activities proceeded along a straightforward course. Problems of ambiguity arose, and his determination to succeed was not enough to solve these. In cases of this kind there is no such thing as good intentions since good intentions are beside the point. B. decided such ambiguous cases on the basis of the greatest effectiveness.

What he would actually rather have been

Since boyhood B. had wanted to be a hydraulic engineer. However, there was no school of hydraulic engineering near Flörsheim (on the Main). After his graduation, therefore, Boulanger decided to study medicine in Frankfurt (on the Main). Difficulty in passing the state examination put an end to his studies. B. was drafted into the Army.

Advantages of his new post

On taking over his special mission Boulanger became eligible for a front-line allowance of Reichsmark 2.65 per day. Lieutenant Boulanger was responsible for the allotting of duties, more particularly hours of duty. Later on he sometimes took advantage of this in order to make a brief excursion to some place of interest in Russia. Furthermore, there was the possibility of being invited to various staff headquarters where one could make lifelong friendships. Thirdly, there was the chance of procuring extra blankets, supplies, and rations at various commissariats, a relatively easy matter for a mobile outfit such as B.'s special detachment. Finally there was no overlooking the fact that the connection with the academic world, even if only as an agent, had advantages of a prestige nature.

Superiority of the academic world

A connection with the academic world means lifelong security. The academic world is free. The members of the academic world rank immediately behind Party members, at social functions they come after the S.S. but before the German East Africans.

Chain of command

In disciplinary matters B. in his new post came under the command of the Army Personnel Department as represented by the Army Group chiefs, who were in turn represented by the senior Army Corps general commanding the odd-numbered Army Corps. B. had no dealings with any of the senior Army judges. In fact the only superior who could give him orders was the academic world. In a formal sense B. also came under the jurisdiction of the divisional commander of whatever sector of the front he happened to be in. If necessary B. could absent himself from their jurisdiction by moving to a different divisional area. But no conflicts arose.

Relationship to an officer's honor

At times during the years following 1942 B. found his butcher's duties—which other officers compared to a hangman's job—abhorrent. Some officers entrusted with this task might have taken to drink. While working on his very first case B. had to overcome mental inhibitions. In this particular instance B. deserted his post just as the head was being severed, thus incurring the risk of his assistants making mistakes. B.'s thought: You must not take to drink now.

On the other hand, there are no genuine inhibitions involved in killing someone else if one can see clearly enough that it is not one's own death that is taking place. B.'s sense of insecurity during much of the procedure was based mainly on the disapproval of his activities by various officers with whom he was friendly. But this disapproval was not valid. When B. later rejected one or another of the commissars delivered to him by the

Army, these same officers, who criticized his special mission, were in command of the firing squad executing the commissar (which usually resulted in mutilating the head).

Relations with women
Excellent.

Prison life
For many men the conquest of Eastern Europe meant a removal of barriers after years of conforming to a narrow way of life. Hence the occupation of the East should by rights have brought with it rape and pillage, or at least an adequate number of brothels. Instead, the occupation was carried out according to regulations which might just as well have been those of a prison administration. In this respect the great liberation of Eastern Europe in 1941 and the following years coincided for B. with the greatest disappointment of his life.

Initial visit to Professor Hirt
Luncheon at Professor Hirt's consisted of four courses: chicken bisque, fish, saddle of venison, macédoine of fruits. The professor's nieces had been invited for coffee. It was late afternoon by the time Boulanger took leave. He could almost have fallen in love with one of the nieces. Next morning he boarded his eastbound train.

Activities started in the region of Orel. Not everyone handed over to Boulanger was a commissar. It turned out that the number of commissars was being exaggerated and the classifying of commissars as "Jewish-Bolshevist" was arbitrary. Most of them were simply partisan leaders who had been arbitrarily classified as commissars. While the Army showed great reluctance to deliver officers taken prisoner at the front, it was most liberal in handing over partisans, although experience showed that commissars were more likely to be found among the officers. However, the Army officers did take back partisans when required (and then shot them themselves), exchanging them for captured front commissars whom they handed over to the military police.

The main problem consisted of the proper classification of the prisoners as commissars and of the use of the additional criterion "Jewish-Bolshevist." Not every Jewish-Russian officer came into this category, nor did every Bolshevik. Boulanger tried to obtain data on the question: with what rank in the Party hierarchy does an officer become a commissar? From several staff officers whom he questioned he received inconclusive data that only served to convince him even further of the arbitrariness of the selection methods. The care which he devoted to the execution of his special mission, and the extra research that went into the clarifying of its basic principles, was on a par with the care which otherwise in the Army would have only been expended on the awarding of the highest decorations. At some headquarters this earned B.'s unit the title of "Decorations

Detachment." Even a negatively evaluated mission—in fact, especially such a mission—must be accomplished with all the energy and intelligence at one's command. In so doing, one must compensate for what has not been thoroughly thought out or is subject to criticism in the original definition of one's duties. Despite his efforts Boulanger knew he could not avoid making any number of errors, and that these were sometimes to the prisoners' advantage and sometimes to their disadvantage. The errors in favor of Russians (e.g., those who were not classified as Jews although they were Jews) could be compensated for by safety margins, a procedure which was not permissible in the case of the erroneous inclusion of innocent persons (e.g., Jews who were not commissars, commissars who were not Jews), since in such cases a safety margin, while it reduced the probability of error, at the same time jeopardized the effectiveness of the method. In actual practice the only way to determine racial origin was by such primitive factors as appearance, family and first names, or perhaps skull measurements. The strong possibility of error made it seem advisable to discontinue the mission altogether. By the same token it followed that, if the mission were continued, the errors must be tolerated and allowed for in the calculations. So Boulanger was beset by endless doubts, but he felt that these doubts must not be permitted to hamper the careful and conscientious execution of his mission. It was therefore important to him to convey his ideas and good intentions to an area where they could not harm the performance of his duties: during this period he read philosophical works, as he was toying with the idea of later extending his (previously abandoned) studies to this subject.

III

The task of bringing precision into the hopeless end of the captured commissars completely occupied B. during the summer and winter of 1942 in the areas of the Central and Southern Army Groups (later of the two Southern Army Groups). In February 1943 his sphere of activity was limited to the Central Army Group. At that time Boulanger no longer restricted himself to a superficial anthropological examination and to obtaining of personal data before having the prisoners killed; he now had informal talks with the commissars in order to retain an impression not only of their external appearance but of their minds. The body of thought thus committed to paper immediately before the writers' death served, B. believed, as additional research material; this sense of quality enhanced B.'s respect for the mission entrusted to him. It is probably always true that to a certain extent one identifies oneself with the enemy one kills (prior thereto one has just seen him alive). Thus Boulanger identified himself, as it were, with the intellectual achievements of his object of study. He did not know what his attitude should be when he was conscious of such feelings. Perhaps B. could, while conforming to his orders, have adapted his method to his feelings at the time by continually extending the respite accorded the prisoners to make their written statements; but since his feelings had not the degree of clarity evinced by his former methods, he kept to his former methods.

In the summer of 1943 a further difficulty arose, the problem of increasing numbers. First the Central Army Group retreated from the Orel bulge. The prisoner-of-war camps situated behind the front had to be hurriedly evacuated. This in turn necessitated an accelerated screening in these camps on the part of Boulanger's detachment. At times the decision as to whether or not a case in point was a Jewish-Bolshevist commissar had to be delegated. This meant that the depots sometimes received heads and descriptions where obvious mistakes had been made. Any change in the procedure would presumably have increased this confusion.

During the late summer of 1943 the Russians took advantage of the gap forming between the southern and central German fronts (a gap which the High Command was bending every effort to close) by attacking at an entirely different point. On the day of the offensive a crisis developed in the Second Army. The base units and the prisoner-of-war camps were caught up in the retreat. The 33rd Panzer Division brought up as a reinforcement found itself involved in the offensive while it was still aboard the trains. Desperate staffs tried to stem the retreat by maintaining their old headquarters. During these days Boulanger's detachment, which was taking along six Russian officers, found itself near the junction of Schlichta, not far from Smolensk. Close to a wooded area Boulanger's assistants were attacked by partisans. Boulanger himself was wounded and escaped with the vehicles in the van of the convoy. It was later assumed by the Russians that the leader of the "Decorations Detachment" was among those captured by the partisans.

The experience was a shock to Boulanger. Surrounded by victorious German troops, by staffs which raised no objections, and in the service of scientific research, there is no sense of guilt; suddenly the situation changes: like a draft springing up, a sense of guilt arises from which one must protect oneself, just as in early spring one has to be careful of open doors, for these only bring colds or even pneumonia.

IV

Boulanger spent the last days of 1943 and part of 1944 at the Wiesbaden Area Military Hospital. The collapse of the Central Army Group in August 1944 dashed B.'s hopes of promotion. A transfer to research was also out of the question. A transfer to civilian research, which would have been possible on Professor A. Hirt's recommendation, was not feasible because of B.'s complete recovery. Military research at that particular time was concentrating on special problems to which Boulanger's Eastern European experience had nothing to contribute. Instead he was transferred after his recovery to the "Central Administration of Austrian Prisoner-of-War Camps."

One more opportunity for total dedication presented itself to Boulanger before the end of the war, in Vienna. In January 1945 the doomed city, like a shrine, attracted troops and officers within its walls—if one can speak here of walls to be defended rather than, perhaps more properly, palaces, canals, and hills. On January 14 General Rendulic took over the defense of the city and had a thousand soldiers strung up. On January 18 (twenty-four hours later an air raid destroyed the opera house) the young Party Area Chief's augmented staff, which included Boulanger, attended a performance of *Lohengrin.* The night of January 21/22, after two days of quiet, the Russians embarked on a new major assault on Vienna. Within a few hours the opportunity of meeting death here was exhausted. Russian panzers were lined up on the northern bank of the canal; practically speaking the city was in Russian hands. All that remained of the last great chance was the biological factor pure and simple. B. managed to reach the American lines. There he gave himself up.

V

How do people like B. live today?

In the summer of 1961 reporters tracked Boulanger down as a packer in a paper mill near Cologne. His offenses were subject to statutory limitations. A correspondent from *L'Humanité* asked for an interview. B. was ushered into the board room and answered questions. When asked about his present beliefs he replied that he was a Marxist. What was he doing? There was nothing one could do.

Contagion from the enemy

He said he had been infected, as it were, by his enemies, since naturally he had talked to some of the prisoners. Did that mean he thought the imperialists in the German Federal Republic were in particular danger of becoming Communists? Of course not. One would have to be in closer contact with the enemy. Chop off his head? In a sense, yes. That was carrying the Christian spirit too far, said the reporter from *L'Humanité.*

Renewed encounter with the academic world

The war over, there was a chance of resuming his studies as soon as the universities reopened. He was given credit for four prewar semesters.

Expiation and surrogate

Question: I see by your forehead that you joined a students' dueling club?

Answer: That was another compromise.

Question: But how did you land in jail?

Answer: After a row with the principal of the Institute at Marburg in 1953, I tried to start up in business on my own, in the textile branch. This fresh start ended in disaster, that's to say I had trouble making payments and so they put me in jail.

Question: With criminals?

Answer: They don't make any distinction.

Question: How many years?

Answer: Three. The prison padre regarded it as an expiation for my deeds during the war.

Question: You mean he forgave you because they were commissars?

Answer: In his eyes the prison sentence made everything all right. He thought I might go and look after lepers in Ethiopia.

Question: What kind of a job would that be?

Answer: You get leprosy too, but in return you can save the lives of a few lepers.

Question: Not a very fair deal, to exchange some sixty to a hundred well-trained key men for five or six lepers saved in Ethiopia. For that you would have to believe that all men are equal.

Answer: I agree absolutely. Of course it's not a sensible exchange.

Question: Then, if I understand you correctly, you are now carrying on the existences of the men you murdered?

Answer: No.

Question: What are you doing in a positive sense?

Answer: I've already told you: one's not allowed to do anything. Just have a conviction.

Question: That's something!

Answer: But the change in me is not meant either as expiation or surrogate.

How the "decorations" method worked

We usually arrived in the evening, because in the morning we were busy somewhere else with dissecting, embalming, etc., and sometimes we covered considerable distances during the day—our day began at 5 A.M. and was never over before midnight; toward evening we would reach a depot, arriving in our jeeps with trailers carrying the instruments and other material. Sometimes there were headquarters near by, and we would be invited for a bite to eat or a drink, but often they would ignore us, especially during our first visit to the boundary-sector divisions of the Northern Army Group, which was not actually part of our territory. It was almost as if the climate had something to do with it: you might say that, going from north to south, there was a greater acceptance of our mission, i.e., in the south more, in the north less. The reasons had nothing to do with climate: in the Northern Army Group, which had been almost stationary since the winter of 1941, the peace-time influence among the top brass had remained more constant, in other words the atmosphere was conservative; it was different in the southern and central areas, where fresh recruits were continually injecting new ideas. There we were not regarded with the same scepticism. I must say, though, that even among the conservative forces there was no actual disapproval of our work, they just made it clear that they were dubious about our methods. They would have preferred us to shoot the commissars according to military regulations instead of killing them by injection, which we did so as to keep the heads intact. Execution by shooting was the very thing that would have destroyed the heads, and it was the heads that were vital to us.

You are wandering from the subject. What was your procedure? First we looked at the takings of that particular day. Most of those who were supposed to be commissars were not. In one case it would be a noncom, in another an anti-aircraft officer. Anyone who was too outspoken in camp and harangued his men gave the impression of behaving like a commissar. From time to time you would hear the expression "Freemason" commissar. Although it was probably contrary to my orders and meant that later, when I had to carry out the rest of the job, I had the added burden of personal feelings, I frequently had talks with the prisoners, since in my experience this was the best method of selection. The level of education or training gave the best indication of which ones might be commissars. Where there was a certain level of intellectual superiority there was every probability of the man being a commissar.

I don't want to bore you with the finer points of our method. When the selection had been made we took the Jewish-Bolshevist commissars to be measured, this was seldom done the same night, often it wasn't until the next morning, because after the lengthy selection procedure we would be dead beat by the time we returned to our quarters. To save on guard personnel we often took the commissars back to our quarters with us, and this laid us open to the accusation of abusing them homosexually. No such case is known to me—one thing is certain, not even a love affair of this kind could have saved any of the commissars at this stage of proceedings. The local troops were responsible for the bodies that we left behind.

Caution

In future he would refuse to be put in charge of such work or any work like it. He would be extremely careful. And if others did it and he looked on? Caution should be used there too; probably he would do nothing and await developments.

Activity and inactivity

The "Decorations Detachment" was his attempt at activity in his life, and the counterpart to it was now inactivity, so that it was even possible to say: I am a Marxist; but there is nothing one can do.

Road closed to the east

Question: Why don't you go to Eastern Germany?

Answer: They disregard statutory limitations there.

What happened to Professor A. Hirt?

That's a good question. In 1944 I got a postcard from Professor Hirt, when I was in Vienna. In Strassburg the Allies came across the remains of his skull collection—there had not been time to destroy it completely. The Foreign Office asked him for an official reaction to the accusation raised in Swiss newspapers; on April 6, 1945, Hirt promised a prompt reply. Later he disappeared.

Did he have any after-effects from his 1943 wound?

Pain in his right elbow, a periodically recurring inflammation of the ligaments: when that happened he went to a chiropractor for treatment.

No coffee

During the interview a human relationship had developed between B. and the reporter from *L'Humanité.* When the interview was over they would have liked to have a cup of coffee together. This turned out to be impossible. At this hour coffee was not being served in the cafeteria so as not to give the staff an excuse to leave their jobs. And in the cafeteria no one was allowed to sit down. So B. and the interviewer parted without having had a cup of coffee.

22. Jakov Lind

Jakov Lind was born in Vienna in 1927. After the Germans annexed Austria in 1938, he was sent with other Jewish children on a transport to Holland. He survived there and later in Germany itself by using false papers. An account of his bizarre wartime adventures while hiding and later while posing as a non-Jew is given in the autobiographical volume *Counting My Steps* (1969). Lind has also written novels, plays, and short stories. He now writes in English. Lind lives in England, but travels frequently to the island of Majorca and to New York City.

"Resurrection" is taken from *Soul of Wood* (1962). Its irony and undertone of black humor are typical of Lind's efforts to invent a bleak comic style for such unruly material. The story of a Polish Jew named Goldschmied who as a Christian convert prays in Dutch and Church Latin with a Yiddish accent reflects Lind's refusal to transform the gloomy events of the Holocaust into tragic knowledge. Instead, he puts horror to ludicrous uses, as if to suggest that only a grisly form of comedy could open the eyes of humanity to what it has endured and become as a result of the catastrophe of European Jewry. In a mood more sardonic than compassionate, Lind probes the failure of the victims themselves to see clearly the doom that threatened them. In spite of his conversion, the baptized Jew Goldschmied is forced to hide from the Germans in a grave-like enclosure, explaining to an unconverted fellow fugitive in Talmudic detail the reasons for his religious gesture. The events of the narrative dramatize the utter irrelevance of their debate.

But insofar as we, as readers, are caught up in the story's dialogue, *we* become the last naïve victims of Lind's morbid "comedy"; the story's title, in both name and meaning, has been a lure and an illusion, since for Lind neither Jewish nor Christian faith does anything to forestall the Nazis' grim design for the Jews. Weintraub may be right when he punctures Goldschmied's diatribe against his original religious heritage with the charge that Goldschmied's words echo fear rather than conviction. But this layering and stripping of motive and identity are meager psychological feats, especially when we consider how easily Weintraub's frail body coughs them helplessly into eternity. Not everyone will agree with Lind's fictional premises, but within the framework of his story he is determined to expose the futility of dispute when the stake is one's life and not "merely" one's belief.

Resurrection

'Deum Jesum Christum in gloriam eternam est. Nu.' Goldschmied turned over on the other side, put down the prayer book and tried to sleep. He pulled his coat over his head and nearly suffocated, he took it off and the light hurt his eyes. He turned from side to side, but cautiously, so as not to touch either of the walls. His head touched the wall behind him (it was padded) and his toes pressed against the chair between the bed and the fourth wall. He couldn't sleep a wink. They won't let you have your rest, not even in the coffin, you'd expect there'd at least be room to stretch your legs six foot underground. Not a chance. Psiakrew Pieronie! * It was only in Polish that he dared. As a Protestant he wasn't allowed to swear. I hope he's a midget, I'll put him under the bed. How can two people sleep in this place? A hundred guilders a week and he won't let me breathe. Meine goyim. Czort! †

Swiss Alpine Club. Holiday at Arosa. Altitude 6,000 feet. First-class hotels. Reduced prices out of season. It's out of season all right. Who wants to go to Switzerland in October? Too early for skiing, too late for sun-bathing. Now would be the time to go, if I could. The calendar won't mind. He himself had brought the calendar. Every day a stroke. So far he had struck off 184 days, 184 years. Only the pencil had stayed untouched by it all. It hung on its nail, its point as sharp as on the first day.

The motto for October: In golden splendour flows the wine—and the picture: vintner carrying a basket full of grapes, clinking glasses with a young couple. Carriage, vines, women, a team of oxen in the background. Young woman smiles merrily. Husband smacks lips. Vintner holds one hand over his paunch.

I could stand it in Switzerland right now. Not too hot, not too cold.

In November they plough, in December they sing Holy Night, Silent Night beside the Saviour's cradle, in January skiing, in February too, in March they take the cable car up the Matterhorn (does it have to be March?), in April the young lambs playing in the meadow, in May a nightingale singing in the trees, in June they ski and swim.

But how will I live through such troubles?

Nothing about it in the calendar. A book tumbled down from over the

*Damn it to hell!

†My Gentiles. What the devil! (Yiddish and Slavic).

bed. *Introduction to Inorganic Chemistry,* Dr. K. Kluisenhart, Groningen 1902. In 1902 I was still in Cracow. Do I need inorganic chemistry? I'm inorganic enough already. He put the book down and took paper and pen to write van Tuinhout a letter. Dear Mr. van Tuinhout: Nothing doing. I can barely stand it by myself, if there are two of us I'll go mad. I'll give you a hundred guilders, but don't do that to me. Find him another place.

I'm a sociable man, but how can two live in this hole without killing each other? Besides, there's the difference in denomination. Try to understand.

He didn't write the letter, he didn't have time. Van Tuinhout was outside the wall. He gave the prearranged knock. Without having to get up, Goldschmied opened the two hooks.

The trap door was pushed up slightly from outside and van Tuinhout climbed through the opening. Now what does he want? Has he found two more?

As usual van Tuinhout sat down on the bed without a word and for a time said nothing. Van Tuinhout was a pale man, about forty-two, thin hair, short straight nose and small brown eyes. When he spoke, he usually stuck his tongue out as though to give his words a last lick before he let them go; when he was silent, he played with his false teeth.

Meneer Goldschmied, he said finally: not tomorrow, tonight he'll be here. Right after dark.

Thank God, said Goldschmied, I would have had a sleepless night. Van Tuinhout eyed Goldschmied with suspicion. He had taken in roomers for fifteen years, but a Protestant, religious too, by the name of Efraim Goldschmied, origin unknown except that he was a *mof,* a German—that was a new one.

Whatever Goldschmied said in his mixture of Yiddish and Dutch sounded suspicious to van Tuinhout.

How old is he? asked Goldschmied.

Not more than thirty. Maybe twenty.

You mean there's no difference? Nu, we'll see. But remember, you promised. A week at the most, it'll get to be three—then I'll kill myself.

Meneer Goldschmied, it isn't my fault. I have to do what they tell me. It's only a week, then they'll put him somewhere else.

Did you protest at least, van Tuinhout?

Of course. But that's how it is. We can't be finicky. It's getting more dangerous every day. And the Jews have got to be helped.

You're telling me? Of course they've got to be helped, but does that mean putting two grown men in a box? Why, this isn't a room, van Tuinhout, it's a coffin.

It's not a coffin, Mr. Goldschmied, you're always dramatizing, it's a closet. Where one can live, so can two—don't get me wrong—but that's how Mr. Jaap and Mr. Tinus want it. You think I have anything to say about it?

Suppose something terrible happens, Mr. van Tuinhout. You'll be responsible.

Responsible? In the first place the Germans are responsible, in the second place Mr. Jaap and Mr. Tinus. I'm just carrying out orders.

Silence set in. What can I do with this goy? He's an idiot. Goldschmied rubbed his three-day beard. (Every three days he was allowed to use the bathroom in the rear hallway.) It can only end in disaster. That much he knew. A hundred and eighty-four days he had lived through it—and what's to prevent the war from going on for another twenty years? Cholera! He'd never live to see the end. A young fellow in the same hole? Two corpses in one coffin would have more room; besides, corpses wouldn't mind. Van Tuinhout didn't budge. He sat there with his hands in his pocket (it's not that cramped in here), stared straight ahead and seemed frozen. He didn't smoke, he didn't seem to be looking at anything, he just sat there. After four minutes Goldschmied began to feel uncomfortable. He knew exactly what was going on, but now that he had something to complain about, he didn't want to play along. He too waited four minutes, he too put his hands in his pockets and stared at the wall. I can outlast him at this any time. Goldschmied said his twelve Our Fathers. That's more than he can do. How can that man think nothing so long? He gave himself another dozen Our Fathers. Still van Tuinhout didn't budge.

Till the Last Judgment I'll let him sit, Goldschmied decided. He'll get his money anyway, but this time he can beg for it. Van Tuinhout gave a slight cough. Ah, he's starting in. Goldschmied gave a little cough too. (In this hole 'coughing' meant an almost inaudible clearing of the throat, and 'talking' meant a barely intelligible whispering.)

So make up your mind. Spit it out.

Van Tuinhout would rather have hanged himself than remind Goldschmied of his rent. Goldschmied knew the ritual by heart.

It was up to him to start in about the homework. Then came lamentations about Kees, the poor motherless boy, followed by a short speech about the moral degradation of children in wartime, and finally a word of consolation and encouragement. But today he would not start in, Goldschmied had made up his mind to that. The stubborn *mof,* thought van Tuinhout, he knows damn well I won't ask him for money. I can wait. After the third dozen Our Fathers Goldschmied had enough. To hell with him, he'll never be as generous as me.

All right, Goldschmied broke the silence, how's the homework going? Van Tuinhout was overjoyed. He still had the rabbits to feed and the supper to prepare. Perfect. He got the best mark in everything. How do you do it? Why, it's at least forty years since you went to school. I don't understand a word of it. Neither does Kees. Not even the teachers, if you ask me. You must be a genius. Every single answer was right. To tell the truth, the work is much too hard. He's only twelve. Yes, I know he's lazy. Maybe not lazy, but neglected. It's always that way without a mother. I can't keep after him all day long. And he takes advantage. It's lucky we have curfew at eight, or he wouldn't get home until morning. You should see the friends he bums around with all day. A bunch of thugs. Juvenile delinquents the whole lot of them. Every day I expect him to be locked up for theft or murder. He's capable of anything these days. In the street? The kind of people you find in the streets these days. Riffraff, soldiers, and whores. Respectable people don't go out. Don't exaggerate, said Goldschmied in whom this subject (streets, going out) touched a sore point. Look here, Meneer Goldschmied, war breeds criminals—what they see now they imitate later. Wait and see what happens after the war (I should live so long, thought Goldschmied). And said aloud: I should live so long, Mr. van Tuinhout.

What do you mean, Mr. Goldschmied, you think I'm telling fairy tales? What do you know? Sitting night and day in this hole. Have you any idea what's going on outside?

Have I any idea what's going on outside? asked Goldschmied with a slight shake of his head. (Why are the goyim so dumb? After all, I'm a Christian myself, so it can't be the religion: Goldschmied's everlasting puzzle.) That's it, Mr. Goldschmied. You just sit here. Sometimes I envy you. Would you like to change places, van Tuinhout? I didn't mean it that way—but it's hard. Every day new regulations, sometimes you don't know if you're still allowed to use the pavement, because some of the regulations aren't posted. People vanish into thin air for no reason at all. Yes, you can consider yourself fortunate, Meneer Goldschmied, you're out of the rain at least.

So it's raining too?

Van Tuinhout looked at him with suspicion. With Jews you can talk, with Protestants you can talk (he himself was a member of THE BRETHREN OF THE BLESSED VIRGIN,) but with a Christian Jew, a Jewish Christian, you don't know where you are. The Christians are hypocrites, most of all the Protestants, the Jews are too smart. To be on the safe side, he took Goldschmied's question literally.

This morning the weather was good, he said, but it may very well rain tonight. Get on with it, said Goldschmied impatiently, he couldn't stand it any more.

As I was saying, Kees is getting to be more of a gangster every day. Do you know what he did yesterday? He took the ferry across the Ij and found himself a girl, a child, maybe ten years old . . .

Nu? (Goldschmied was growing more and more impatient.) He's only a kid himself—you want him to sleep with an old woman?

Believe it or not, Meneer Goldschmied, he really did sleep with the child, but the police caught him in the act. I'll be surprised if they don't put him in jail.

He is a little young, Goldschmied admitted. At his age I was apprenticed already. Sixteen hours a day. We supplied umbrellas all the way to Budapest. No, for such things I didn't have time.

That's what I've been telling you, you're living here like in a hothouse, so sheltered. You can be glad you haven't any children.

Glad, no. Except maybe right now. Children, that's all I need.

Anyway, Meneer Goldschmied, everything is getting more expensive and the money is worthless.

That was the cue. Goldschmied took out his wallet and gave him the hundred guilders rent he was going to give him anyway. But van Tuinhout had certain principles. And one of them was: You can't ask these poor persecuted people for money. Renting rooms was his profession, hiding people was patriotism. If his protégé wished to contribute something of his own free will, he couldn't refuse. But never in all the world would he have asked.

Bring me the next batch of homework soon, said Goldschmied, or the boy will be left behind.

Just one question. Doesn't the teacher notice that his homework is right and his answers in class wrong?

The teachers these days, Mr. Goldschmied, aren't teachers: they're students who've flunked their exams. Today you could be a professor at the university.

I ask you, Goldschmied shook his head. Is it such an honour to be a professor at a university? My umbrellas are more interesting and it's a better living. But if the war keeps on much longer and the homework keeps coming, I'll be ruined. After the war, I'll need a flood to put me back on my feet.

Meneer Goldschmied, I'd like you to do me a favour.

What? You're asking me a favour?

The gentleman who's coming today, van der Waal his name is, he doesn't know you're paying me one hundred guilders a week. If it's all the same to you, please don't tell him. I have my reasons.

Don't worry, Mr. van Tuinhout. I'll be silent like a tomb. I don't think I'll speak to him anyway. I'll just ignore him.

You promise, Meneer Goldschmied.

I promise. All day I'll look at the wall and pretend he's not there. You can rely on me. You have your reasons and I don't even want to know them. But now you must excuse me, I'm busy.

Between ten and twelve. All right?

A few minutes more or less don't matter, van Tuinhout, and in case you decide to put him somewhere else, it'll be all right with me too. I can manage for money, but the air here is another matter. I could do a good business in air if I had some, it's fantastic.

The knock came at about half past ten. Goldschmied looked up from his book. There he is. Exactly between ten and twelve. You can trust van Tuinhout. Van der Waal, said Goldschmied to himself, that means either Birnbaum or Wollmann. But he was mistaken. The young man who crawled in, Goldschmied guessed him to be nineteen, was called Weintraub.

To err is human. Weintraub had red cheeks, sweaty hands, and short-cropped hair. He had big blue eyes, a fleshy nose (poor boy, the bone is missing, thought Goldschmied) that looked Jewish at the end, but only at the end. He was short and thick-set. Had on a blue sweater and corduroy trousers, introduced himself as van der Waal, and tossed his small suitcase deftly under the bed.

Van Tuinhout showed his face for another two minutes in the opening, darting glances intended to impress it once again upon Goldschmied that the matter of the hundred guilders was a private arrangement between van Tuinhout and Goldschmied.

Goldschmied bent down to van Tuinhout and whispered: 'Don't worry. We practically won't see each other.'

Van Tuinhout handed him two copybooks. These are for next week. I'll bring the other two tomorrow.

Goldschmied took the copybooks and put them too under the bed. The wall was closed and the two sat on the bed.

Goldschmied looked the young man up and down, decided the view was incomplete, and said: Stand up. Weintraub stood up. Goldschmied got up too and stood beside him.

'Good. You get the shorter blanket.'

Maybe the young man was shy, he said nothing and looked the other way. How do you like it here, Goldschmied interrupted. Isn't it cosy?

Weintraub, his first name was Egon, saw the one chair, the bookshelf over the bed, the calendar of the Swiss Alpine Club with the pencil hanging from a nail, and at the foot end of the bed (he couldn't believe his eyes) a cross a foot high with a crucified Jesus on it. He couldn't take his eyes off it.

That, Mr. van der Waal, is mine. So are the calendar and the pencil. But the blankets belong to the landlord. No talking in here except in a whisper, even if it wrecks your voice, and don't breathe too much either. Air is very important, we've got to economize. What you exhale I inhale and vice versa, so I hope your teeth are good.

Still not a word out of Weintraub. He just looked at Goldschmied.

Goldschmied had sagging cheeks, a bald head, an enormous nose, and a chin that receded like a flight of steps. His lips were two thin lines, drawn down at the ends. Weintraub put his age at sixty. Actually he was only fifty-two. His hands were large and broad, he had sunken shoulders and a paunch. Two fingers of his right hand, pointer and middle finger, seemed to be crooked. Reminders of a wound in the First War.

With these fingers, said Goldschmied, I swore allegiance to Franz Josef; God punished me by making them crooked. My name, by the way, is Hubertus Alphons Brederode of Utrecht, but you can call me Efraim Goldschmied, that's what I call myself to show sympathy for the Jews.

Otherwise I'm a Christian, a real Christian, as you probably noticed right away.

(Still no sign of life from Weintraub.) A Christian, see, a goy, not one of us, one of them. Now do you see what I mean?

But baptized?

Thank God, you can talk. I was beginning to think they had cut your tongue out. Yes, baptized. Disgusting, isn't it?

Weintraub shrugged his shoulders. It's a question of taste. But are you hiding as a Jew or as a Christian?

Ha, a khokhem* yet. Both, my young friend. This isn't only the cave of the Maccabees, it's also the catacombs of Amsterdam. I'm hiding double, so to speak. You see, I'm not an ordinary baptized Jew, I'm a convinced and pious Christian. I'd have had tsores† either way.

Either way? Why as a Christian?

Some day I'll tell you the story of my life, but there's no hurry, because I will have the honour of seeing you again. But in a nutshell: I am the deacon of a congregation in the Nederlandsche Gereformeerde Kerk. You know the church on the Overtoom, the Church of Saints Peter and Paul? Well, I, Goldschmied, am the deacon. Yes, the Catholic name is misleading, a leftover from before the Reformation—after the war we'll change it with God's help. And you, Weintraub? You're a Jew, I hope. Because two goyim in here would be too much. And I wouldn't be able to convert you.

Nobody can convert me. I'm not interested in such things. I'm of Polish origin.

Polish? Goldschmied could hardly contain himself. He almost shouted. Polish—don't say another word. Jescze Polska niezginela.‡ He nearly fell on Weintraub's neck.

But, said Weintraub, I was only a baby when I came to Holland.

Doesn't mean a thing. Once a Pole always a Pole. What luck!

What's so lucky, you want to know? It's not so quick to explain. Polish isn't just a nationality. It's not so simple. The Poles are the chosen people the Jews would have liked to be. And why davke§ the Poles? Because the Poles have what the Jews haven't got. Sense and faith.

The Jews have no sense and faith? You're joking.

Pan Weitraub, I ask you, if the Jews had sense would they have gone on being Jews? Not a chance. They would have gone over to the new religion long ago. And because they have no true faith, they are the worst heathen in the world in my opinion. Absolutely.

But, Mr. Goldschmied, Weintraub protested . . .

Don't interrupt me just because I'm right, that's how it is . . . They're always talking about God, but they don't really believe in anything, except money. Sure, they are good at making up ethical laws in God's name, for everything they've got a law, but what's all this got to do with religion? Nothing. The whole Jewish religion is full of practical advice, but the sense of mystery, the feeling for holiness, that they haven't got; just like the Germans and that's not the only reason.

The Germans? What are you talking about?

Goldschmied didn't like to be interrupted. You want to know what I'm

*Sage or wise man. Here, of course, it is used ironically.

†Trouble or problems.

‡"Poland hasn't perished yet." The opening words of the Polish national anthem.

§Here, something like "Why the Poles in particular?"

talking about? Listen and you'll find out. Why do the Germans shout so loud about nation and blood? Because they're not a united nation. It's exactly the same with the Jews; they shout too loud about their Jehovah and His chosen people. There's something fishy about that. So you'll ask what's fishy? Well, I'll tell you: their religion, that's where it begins. That's where everything begins. Between you and me, Weintraub, the Jewish religion is no good. What do I mean, religion? And what do I mean no good? I'll tell you, and Goldschmied whispered mysteriously: Because Jews have no religion and because they stopped being a nation thousands of years ago, that's why they have such a lousy time.

Moralizing, that's what they do. Philosophizing. The Greeks, the Romans and the English, they got somewhere in this world—except as individuals, the Jews never accomplished a thing, not where it counts, and what counts is to find a union between man's need of faith and his individual humanity. The Jews are still what they always were, scattered tribes of merchants and Bedouins with a small group of intellectuals, from a little, insignificant Mediterranean country. Super-chauvinists, all their national feeling is nothing but primitive clannishness—like the Indians. And their racial purity? Racism, my dear Weintraub, was invented by the Jews, not the Germans. Azoi it is.* And only azoi.

Goldschmied leaned back against the wall, exhausted but happy. Come to think of it, thought Goldschmied, a fool is better to talk to than a wall.

Weintraub was a kind of Palestine pioneer—a quarter Zionist, a quarter orthodox Jew (by upbringing), and the other half Dutchman. Until driven underground, that was two years before, he had worked on a farm as a hired hand. He had graduated from secondary school, though very late. Illness had delayed everything in his short life. He had been tubercular since the age of thirteen. Work and fresh air on the farm had done him good—the coughing fits had stopped; and he had been lucky during his two years in hiding, always somewhere in the country—his last hiding place had been raided, someone had denounced him, he had escaped at the last moment. Van Tuinhout's hideout was only temporary; the friends in the underground who were helping him were well aware that a consumptive in a wall was a danger to everybody.

A temporary solution, for a week or two at the most, until they could find him a new hideout with a peasant or gardener.

He had expected it to be small, but not this small; he had expected a bed of his own. How could he share a bed with this old codger when he didn't share a single one of his opinions?

But his ups and downs had made Weintraub philosophical. Well, he said to himself, it's an experience. I only hope he's not homosexual. That I couldn't stand.

Goldschmied was not homosexual—sex seemed never even to occur to him. Sex is not for me, he would say, it's for women and children. What

*So it is.

interests me is my business, making umbrellas, and theology. Everything else is playing around.

They lived through the night, each rolled in his blanket, back to back (twice Weintraub woke up because he thought he was going to suffocate, but somehow he survived) until the grey dawn trickled in through a pipe connected with the chimney. When Weintraub sat up and rubbed the sleep from his eyes, Goldschmied was already sitting on the other side of the bed, his back turned to him, an open book on his knees and muttering something. He swayed his body, fell into a soft sing-song—reminding Weintraub in every way of his father chanting his morning prayers.

Goldschmied prayed in Dutch and Latin—both with the same Yiddish accent, crossed himself three times at the end, kissed the book, and put it back with the others above the bed.

Of course you don't pray. You heathen—now come the exercises, then comes breakfast, such a breakfast you won't get in Krasnopolsky—everything here is home-made—even the scrambled eggs. Stand as thin as possible against the wall—good. And now, one, two, three, four—Goldschmied lay down on the bed, propped up his back and began to bicycle in the air. After five minutes he said: That's for the legs. Now for the arms. He thrust his hands out to both sides a dozen times, each time hitting Weintraub in the stomach. (You hippopotamus, can't you make yourself thinner?) In conclusion a few knee bends.

That's that, said Goldschmied, it's healthier than tennis, and it doesn't make you perspire so much. Now it's your turn. Goldschmied stood on the chair and beat time.

One, two, three, four, one, two, three, four. And so on. That's enough. Save the rest for tomorrow.

At nine van Tuinhout brought in a basin of water.

Mr. van Tuinhout, the young man has to have his own water. We're not married. Goldschmied handed him the urinal to empty. He needs his own bottle too. What you Dutchmen need is a little of our Polish tidiness.

Goldschmied washed from head to foot, showing thin white legs, a blubbery back, and sunken buttocks.

Van Tuinhout came back five minutes later with Goldschmied's breakfast, a large cup of black coffee sweetened with saccharine, two slices of bread and margarine, and a dark-yellow mush on the edge of the plate—fried egg-powder.

He took away the basin and brought it back five minutes later with fresh water. Weintraub, who had decided to spend three days at a public bath after the war, dipped a corner of his towel and rubbed his face with it. Goldschmied looked up: Oh no, my friend, that won't do. After all, we sleep together. You wash yourself properly from top to bottom, or you can move to a hotel.

Weintraub mustered him. The hell with him, he thought. Not even my father had the nerve to tell me to wash and where.

But as a newcomer, he could only give in to the elderly goy.

Breakfast was cleared away, the bed made, and they sat on the bed, Weintraub with his legs crossed, Goldschmied with his elbows on his knees and the book on his chair.

Three weeks later. Weintraub was still there. I predicted it, Mr. van Tuinhout, in six months he'll still be here—with God's help we'll move to the old people's home together. Van Tuinhout had nothing but curses for the situation. The underground had hoodwinked him. As it turned out, Weintraub was penniless. What should I do, Mr. Goldschmied? I can't put him out in the street. He can't pay. What should I do?

I'll make you a proposition, Meneer van Tuinhout. Just forget about the few cents he would have paid you. Put it on my bill. Give me twice as much homework.

Now what does he want? Van Tuinhout sucked his teeth. Does he think he can pull my leg?

I'll speak to my aunt—she has money, I can't ask persecuted people for money. I'll speak to my aunt, that's the best way.

Are you sure?

My aunt is obligated to us. My wife took care of her when she was down with varicose veins. She's got to help. She has more money than you and Weintraub put together.

Weintraub, said Goldschmied, we've known each other now for three weeks—and it looks like we'll be together for ever. To tell you the truth, I've almost got used to it; it's been like a change of air. But now, seriously, if you want to be my friend, you've got to stop coughing like that. That cough will cost us our lives. Coughing is all right in peacetime—you should have done all your coughing before, because now it can cost us our necks. Weintraub flushed. I thought, he said, it had stopped. But now it's started again. I doubt if I have six months to live. Six months, Weintraub, six months is a long time.

Yes, but the end can come any day. I never told you, Mr. Goldschmied, but now I've got to tell you. I have tuberculosis. I can die any day. Goldschmied looked at his new friend sharply. If you're telling me the truth I won't be so hard on you any more. A man marked by death deserves consideration, special consideration. Marked by death? What does that mean in times like these? Weintraub couldn't stand it. What do you think will happen to you if you stick your head out of the door? Marked by death. It sounds so tragic—actually I may live to be seventy. But you, Goldschmied, how long do you expect to live—without tuberculosis?

Goldschmied didn't like the way this conversation was going. He was willing to feel sorry for a poor sick man, but that the candidate for death should predict an early end for him, Goldschmied, that was too much.

Weintraub, said Goldschmied—he wanted to get this thing settled once and for all. I'm not afraid, you see I'm living in the grace of our Lord Jesus Christ. My Jesus loves me, He'll see me through, His mercy is great, His will be done, as it says in our prayer. We'll see.

Now, after three weeks, Weintraub had ceased to live in a dream. The wall had become reality; actually he was very happy to be with this fellow Goldschmied.

He was ashamed of his coughing. Coughing was a sickness and he was ashamed of being sick. At the approach of a coughing fit—Goldschmied had learned to recognize the signs—he wrapped his young friend's head in a blanket, which he removed ten minutes later.

Every day there were three or four fits and the previous night had been especially bad. Something's got to be done about you, Weintraub. Maybe I should keep watch at night.

Weintraub didn't know what to say. Yesterday he had spat blood again. He felt the cough tearing his lungs to pieces, he was simply spitting them out. That cough is deplorable, disgraceful. What could be done?

Van Tuinhout is bound to turn up with good news any day. He's got to get out of this wall—if he doesn't, he'll die and everybody will be in danger. What day is it? he asked.

Goldschmied scrutinized Weintraub. He took the calendar (peasants ploughing a field. The Alpine Club's motto: He who sows will reap). Your twenty-fourth day, Weintraub. It's my two hundred and eighth. You don't catch up with me. This is the last winter. Next year you'll be in Jerusalem and I in my church. One more winter, Pan Weintraub. What's one winter? I'm too old for skiing anyway. I'll stay here.

Knocking. Goldschmied pushed the hooks aside. Van Tuinhout appeared in the opening; a stranger was with him. One after another, they crawled in.

The stranger was large and broad-shouldered, with protuberant cheekbones and a wide chin. He wore glasses and a cap.

He looked like a repair man from the telephone company. This is Verhulst, van Tuinhout introduced him. He knows a peasant in Frisia who'll put van der Waal up.

But it won't be cheap, understand, said Verhulst. Goldschmied understood. How much is not cheap?

Fifteen hundred guilders. We're not getting anything from the underground.

Fifteen hundred guilders. That's a lot of money. Van der Waal hasn't got any. The underground is broke too. So what will we do?

Yes, what will we do? Hasn't he somebody he can borrow from? asked Verhulst.

Have you somebody, van der Waal? Goldschmied looked at him sternly. My parents are gone, said Weintraub. I have relatives, but where they are I don't know. I'll make you a proposition. I'll give you a pledge. He looked through his suitcase and brought out a small tin wrapped in paper. The three looked on eagerly. A pocket watch came to light.

It had a modern dial and was chrome-plated.

It's worth three guilders, said Weintraub and looked from one to the other. He was ashamed of his childish treasure. But it's worth a million to me. After the war I'll give you fifteen hundred guilders for it. It means a great deal to me.

Verhulst looked at him under his glasses. Goldschmied looked away. Van Tuinhout played with his false teeth. A short silence.

Weintraub put the watch back in its box, wrapped it in the same paper and replaced the rubber band around it. He put it in his suitcase and shoved the suitcase under the bed.

Goldschmied was first to speak. In this wallet—he took the wallet from his jacket—there's still a hundred guilders. The last. Until my committee sends me some more money, that's all. He handed van Tuinhout the wallet. Van Tuinhout turned it over twice, thought of opening it to have a look, because he couldn't believe his ears, and decided to let well enough alone. He put the wallet down beside Goldschmied. Verhulst gave him a glance.

Van Tuinhout stood up, followed by Verhulst, they opened the trap, and Verhulst climbed out first. For a few seconds van Tuinhout's head remained in the opening. He gave Goldschmied a look of reproach and astonishment. My aunt won't do it, Meneer Goldschmied. She can't right now.

Goldschmied reached under the bed and gave van Tuinhout two copybooks: Here is the homework, Meneer van Tuinhout. The last. The trap closed. They were alone.

Weintraub, it's hopeless. You can see that. No money no life. I can't keep myself any longer and you're done for too. Weintraub—Goldschmied looked at him out of eyes in which this world was already extinguished—Weintraub, my friend, I think it's all over.

Weintraub's voice had a nervous flutter and seemed to come from far away: I won't survive it, neither here nor in Poland, Mr. Goldschmied, but you, no children and baptized, all you have to do is get yourself sterilized, and you're free.

Goldschmied's whole body swayed and he spoke louder than usual: Jesus suffered more, and that's why He understands. He's got to help, because no one else will. He, the Anointed One, is the only God. How do we know? Do you know the Talmud?

Why so serious, Goldschmied? You forgotten how to laugh? What kind of laughter, young man, did I ever have? Goldschmied continued:

He and He alone is the Anointed One; it is written in your holy Talmud, but one has to know how to read it:

When a man stands up and the others remain seated, does it mean that those who remain seated, as I am seated here, are inferior to the one who stands up? Or does the one who stands up wish to dissociate himself from those who are seated? People stand up for various reasons. For instance, to mention only three: a man stands up because he has something to say and wishes to be seen; or he stands up because he wants to see something that he can't see when he is seated (for instance, if I want to see what is written on the Cross—as it happens I know it by heart—I have to stand up), or he stands up simply because he doesn't want to sit down any more.

In the first case—he wants to speak and be seen, in other words, he wishes to exalt his spirit, but to exalt one's spirit means to come closer to the Holy One, may He be praised. This standing up is therefore a good work.

In the second case, however—when a man stands up because he wants to see what he can't see sitting down—it means that his soul thirsts for wisdom, for wisdom does not come down to a man who is seated.

Therefore this standing up is also good.

And now to the third case—if a man stands up because he doesn't wish to be seated any longer, he is likewise doing a good work, for the heart in which dwells the love of the Almighty, holy is His name, is filled with joy and jubilation and wishes to be seated no longer. To sit, is it not to mourn? Therefore it is good to stand up: but what does this mean?

It means that the spirit, the soul, and the heart lift themselves out of their abasement, and standing up is to sitting as life is to death. When Rabbi Gershon ben Yehuda asked his student Rabbi Naphtali: Why do some stand up while others remain seated? he was really asking: How is it that some rise up from the dead and others do not? What does this mean? It means above all one thing: Some can rise up from the dead and others cannot. So you see, Rabbi Gershon admits (would he otherwise have asked such a question?) that there is such a thing as standing up, or resurrection, from the dead. But who can rise from the dead before he is judged? Who doesn't have to wait until the Prophet Eliyahu announces the Messiah? Who? Only someone who doesn't have to wait for the Messiah. But if someone can stand up from the dead without waiting for the Messiah, can he be an ordinary man? Not in the least. Can he be an extraordinary man? No, because an extraordinary man is still a man. Therefore he must be what no one else can be, namely, the Messiah Himself. Therefore He who has stood up from the dead is the Anointed One. His name is Jesus Christ. Who else?

As a baptized Jew without children, Mr. Goldschmied, you'd only have to be sterilized and you'll be a Messiah yourself. You'll be able to stand up as much as you please—even in the tram, in the train, anywhere. And when you go to the cinema, you can take standing room.

Young man, Goldschmied gave him a friendly tug on the ear, you are making fun of me. But sterilization is no joke.

Take it from me, Mr. Goldschmied, if they'd let me. This very minute. But they won't let me. They need me the way I am, half dead. But you? Goldschmied, who had grown fond of his young Polish friend, looked at him with a fatherly tenderness. They don't exactly need you, and aren't you being a little frivolous, van der Waal? Tuberculosis isn't enough for you, you want to be sterilized too?

Anything, Mr. Goldschmied, anything is better than to die before your time. Even if they left me nothing but a mouth and a lung, believe me . . .

Goldschmied wagged his head: Yes, I admit, in your case breathing is the most important thing in life, and maybe if I had your . . . maybe if I, myself, well, you know what I mean—maybe I'd talk the same as you.

But as it is? Am I a mad dog? Weintraub, who had come to love Goldschmied like his own uncle, was dismayed. The moment Verhulst disappeared through the trap, he saw himself getting out of the train at Westerbork.* Westerbork, stopover on the way to the end. This has been going on for two years and twenty-four days. The Germans aren't to blame, or the Nazis, or the Verhulsts; it's this disease that's come down in my family. He died of TB, they'll say, nobody has him on his conscience, they'll say. Nobody will have me on his conscience, said Weintraub aloud, he died of TB, they'll say. A lump rose in Weintraub's throat.

But Weintraub, Goldschmied laid a hand on his shoulder, what do you care what they're going to say? Who dies for his obituary? Do you really think this world still needs more examples of murdered innocents? There's no shortage. No one will miss you except a few friends and relatives. Sad, but that's how it is.

Although Weintraub had his eyes on Goldschmied's lips as he was saying these words, his thoughts were far away: It's all an accident, pure chance that there was no other place that week; chance that I had to fall in with this van Tuinhout, who has to make his little deals. The fifteen hundred guilders wouldn't have done me any good either, or would they? The J is the meat hook. Everybody has to carry his own—if you've got it, they gas you right away, if you haven't they kick you and torture you until you admit you've got the hook. And then they hang you up on it.

It's not the J that matters, even without it you can be sentenced to death, it's the admission that counts. Admit you're a Jew. If it comes to that, what will Goldschmied do with his Jesus and his Talmud? He can live. He has only to say the word. He's not a Jew. He doesn't want children. He doesn't bother with sex, or not very much. Has he a martyr complex? Why does he want to die when he can live? If they find him here, that's an admission in itself. They'll smash the Cross over his head. As deacon of a Christian congregation, he had no need to hide.

He looked at Goldschmied, who was passing a finger over the mountain ranges on his calendar, and tried to read his thoughts.

The word sterilization had but one effect on Goldschmied, to throw him into utter confusion.

The possibility of saving his life was more than his nerves could stand. How could he explain this to Weintraub? But he had to explain (or Weintraub would die with mistaken ideas and false hopes). The essential difference is between killing and being killed. Murderers after their deed need human mercy—but the murdered need divine mercy in advance. Goldschmied also knew it was all over, not with life, that would be no problem, but with hiding. Two hundred and eight is a cabalistic magic number, if you could only discover its meaning. Goldschmied knew the Talmud, Rashi's commentaries, and of course his Old Testament (how he

*Transit camp in Holland from which Jews were deported to death camps in central and eastern Europe.

had time left for his umbrellas was a mystery to his closest friends); when he wanted to start on the Cabala,* it was impossible to find either teachers or books.

How easy it is to miscount, Weintraub. The years were too short. Two hundred and eight is a mysterious number. Why just two hundred and eight? Is two hundred and twenty better, or two hundred and fifty? The highest number is the best, but is there such a thing as the highest number? There is only infinity. But I've taken out my insurance on that. Is there any better life-insurance, with lower premiums, than Christ? If there were, Weintraub, wouldn't I have taken it out? A Jew who takes up Christianity has lost nothing and gained everything. For good Christians such a Jew is a Christian, but for anti-Semites I'm still a Jew. So I turn anti-Semite; that way I can go on seeing myself as a Jew (between you and me, I was an anti-Semite before and as a Jewish anti-Semite I couldn't stand myself). So now you know why I turned Christian. It makes everything so simple. With one exception: the regulation about the childless baptized. On one rotten condition they let me live—as a Jew, no conditions, they just kill me. They let me choose something I wouldn't wish on a dog. You have no choice. You don't have to turn into a dog; you can die like a normal, healthy human being.

That's why I don't want the day to come, because tomorrow I'll have to make up my mind and I can't choose. Because if a normal, healthy human being lets himself be killed when he has a choice—is that normal and healthy? And I'll tell you what's sick about the Jews: their religion. As Christians or Mohammedans they could have trampled on the world and established the Jewish justice they're always raving about. But no, they didn't want to. They didn't have the imagination or the power; to succeed they'd have had to become Christians. But they didn't feel like it. Instead of martyring, they let themselves be martyred; looking on is impossible. For Jews. Now I'll tell you the truth. I didn't hide because I'm a Jew, I hid to avoid choosing.

Then you can stand and look on, Mr. Goldschmied: Weintraub was furious at the Talmudic complications with which Goldschmied tried to talk himself out of his fear. Why won't he admit it? The Nazis are to blame that a man like Goldschmied has to think such thoughts. By hair-splitting he had turned their guilt into a guilt of his own.

Jewish conceit, Mr. Goldschmied; you won't even let the other fellow keep his guilt. How can anybody know where he stands if the victims take the guilt for themselves? No wonder they all climb into the trains of their own free will; they think it serves them right, and not that they're wronged.

Maybe we all of us suffer injustice, but does it really matter, Weintraub? If tomorrow I decide to be sterilized, I can look on as they finish you off.

The Nazis would have fired their ovens in any case, believe me, if not

*System of Jewish mystical thought.

with Jews, then with Poles, Russians, gipsies. If they had let the Jews, God forbid, look on, or even worse, if they had let them help make the fire, not one, Weintraub, but the majority would have gone over to the Nazis. When it comes to anti-Semitism and organization, the Nazis could learn plenty from some Jews. But how could such a thing have been justified in the eyes of God?

Mr. Goldschmied, you talk like that because you're scared stiff.

I talk the way I do because tomorrow you'll lose your life and I my sanity. To tell you the truth, I've considered it from time to time—but for the last three weeks, since you came, my last chance is gone too.

Do you love me as much as that? asked Weintraub bewildered.

Like my own flesh.

Just admit you're a Jew, Goldschmied, and we'll go together.

What's that, Weintraub? You know I am a pious Christian. My mazel! *

Frankly, Mr. Goldschmied, I have no sympathy for you. I'd rather be a live onlooker than a dead victim. You talk and talk. Religion, holiness, the Jews' mission. All a lot of phrases, slogans. Choice, dog, guilt. I don't give a shit about all that. In a few days they'll strangle me and burn me like a leper, and that's the end of Sholem Weintraub. They'll give me a number on a mass grave, coloured with gold dust, and I'll never, never be alive again. Resurrection is nothing but Talmudic hair-splitting, mystery, smoke and sulphur, hocus-pocus, theological speculation. There is no second time, not before and not after the Messiah, and He doesn't exist anyway. I want to live, Mr. Goldschmied, I want to live and breathe and I don't care how—like a dog or a frog or a bedbug, it's all the same to me. I want to live and breathe, to live.

Weintraub's face turned dark-red and his glands swelled. Goldschmied reached for the blanket and threw it over Weintraub's head. But Weintraub shook it off. His eyes glittered, sweat stood out on his forehead, and his hands trembled as he shouted: Live, breathe, I want to live, live. Goldschmied flung himself on Weintraub, and tried to put his hand over his mouth, but Weintraub flailed like a wild beast and went on shouting. Live, live, I don't want to die like a dog. Cut off my balls and my cock with it, cut off my hands and feet, but let me live and breathe!

Weintraub broke into a coughing fit, and he spat and wheezed blood. Goldschmied sat stiff and pale on the chair and watched his young friend Weintraub who was beginning to decompose even before he was dead. Goldschmied's eyes stared into the void. There was a knocking and drumming on all four walls. Shouts were heard and a car stopping. Weintraub flailed about on the bed and seemed to choke with coughing. The drumming grew louder, angrier.

Goldschmied stood up, climbed on the chair, and tore the Cross off the wall.

Shouts and stamping feet were heard, followed by unexpected silence, then boots pounded through the corridor, the trap was pushed up with

*Yiddish word for "luck."

rifle butts, and a voice under a helmet shouted: Come on out, or I'll take the lead out of your ass.

An ambulance, Goldschmied heard himself saying from far off, he wants to live, but he's going to die on us.

Goldschmied crawled out first, he stood with upraised hands, the Cross protruding from his coat pocket, waiting for them to bring out Weintraub.

Two of them reached through the opening and picked Weintraub off the bed like a sack. Goldschmied was unobserved for a moment: running isn't in my line, he decided.

In the living-room stood two more men in uniform, through the window a small crowd and a patrol car could be seen. Van Tuinhout sat there with bowed head, staring into space.

The policeman with the most stripes was in Dutch uniform. He turned to Goldschmied.

You can take your things, of course, or just wait here and I'll get them. The first to come down was Weintraub, looking pale and sick—escorted by a policeman. Then came the Dutchman and his German colleague.

Each carried a small suitcase. I'll take them to the car, gentlemen, your friend seems unwell.

He carried the suitcases to the car. Yes, Meneer, said the Dutchman, it's disgusting work, but what can you do. I'm only doing my duty. I have a wife and three children. One of my sons is just about your age, he said to Weintraub. Just lie down on the bench and if we drive too fast for you, please knock and we'll slow down a bit. There's no hurry. Goldschmied had recovered from his terror and Weintraub too felt newborn in the fresh air, even though it was damp and cold. So you know what it's like to feel sick? he asked the Dutchman.

I know plenty. O.K., he said to the German driver, but not too fast. The Dutchman turned round to Goldschmied: I've been suffering from headaches for years and this work is driving me crazy. I've got a good recipe for headache, Inspector, you should try it some time, said Goldschmied. Sugar water, bring it to a quick boil, mix it with honey and melted butter, and drink it down while it's still hot.

You don't say? And it helps? I'll have to tell my wife about that, she'll make me some up tomorrow. We menfolks are lost when it comes to cooking and such. Am I right? Ha-ha-ha.

Yes, that's a good idea, the German driver put in. I'll have to try it. I'm crazy about sweet things. Chocolate, candy, and all that kind of stuff, that's for me. I used to work in a chocolate factory, that was a few years back, it belonged to a Jew, but not any more. I should have known you then, called one of the policemen in the rear, a ramrod of a man in his forties. I'm crazy about chocolate myself. A nice piece of chocolate, as I always say, is as good as a meal. You can keep your chocolate, said the second policeman, who was standing with his rifle beside van Tuinhout. What I like best is fresh dill pickles and marinated herring. Naw, sweets ain't for me.

Why argue, Goldschmied interrupted. It's all a matter of taste. One likes sweet, another likes sour.

That's the truth, the Dutchman agreed. How's your friend? he asked Goldschmied. I hope he's feeling better.

Weintraub listened to the whole conversation with closed eyes—chocolate, dill pickles, sugar water with honey—they're talking about normal things. In the last three weeks the conversation was all about religion and Jews and guilt. I almost died in that hole. If Goldschmied hadn't got me so riled with his high-flown speeches, maybe we'd still be sitting in that hell.

Suddenly life seemed to him reasonable and simple again, and he was ashamed of having acted like a madman. Now that there was air to breathe, all his fear had left him. The air has done you good, said Goldschmied, glad to see Weintraub looking normal again. Get a good lungful. You never know when there'll be more.

Van Tuinhout, who had so far sat silent and motionless, turned to the Dutch police officer: Who's going to take care of my boy when I'm gone?

The state, I suppose, I don't know exactly how it works—but the Germans always look after the younger generation, you've got to hand it to them.

They were taken to Gestapo headquarters. Don't be afraid, said Goldschmied to the livid Weintraub next day as he was carried from the cell to a waiting ambulance, we'll meet again, I'll take bets on it.

Weintraub didn't have one word to say for himself—his case was clear. After a lengthy cross-examination Goldschmied's case was also settled, and a week after his arrest he too arrived at the transit camp. No sooner had he passed through the gate than he ran into his friend, looking healthy and cheerful. They hugged each other. There were tears of joy in Weintraub's eyes. I can breathe again, Goldschmied, he cried with joy, what do you say to that, I can breathe again.

Well, said Goldschmied, the air here isn't bad (it was a warm autumn day and the children were playing in the sun), it's nice in the outside world. You look new-born.

You look much better too, Goldschmied. Why, you stand up straight as a soldier. I always thought you had a hump.

Yes, Weintraub, if they just leave you alone, if they just let you stand and sit and walk up and down . . . it's like a second life.

I've missed you, Weintraub. When are you leaving?

Weintraub said blandly: My train leaves today. At five o'clock.

Today? When I've just come? Can't you take a later train? What's the hurry?

Today, Mr. Goldschmied. Today at five. We shall meet again.

Still in this world?

Why, naturally, in this world, Mr. Goldschmied, do you really believe those stories about Poland? Now that I'm feeling better, I don't believe them any more. Sick people get such crazy fears.

Goldschmied looked at Weintraub for a long moment, then turned and left him. As he left he said: You're right, Weintraub, we've got to keep our health, with all this fear we might as well be dead. Keep healthy, have a good trip, and make sure you get there all right. We shall meet again.

23. Pierre Gascar

Pierre Gascar is the pseudonym of Pierre Fournier, who says he chose to write under the name Gascar because it reflects his Gascon origins. He was born in Paris in 1916, but was raised by uncles in southwestern France, where he developed an intimacy with nature that is reflected in his writing. In 1937, he was drafted into the French army and spent time on the Maginot Line. After the outbreak of World War II, Gascar was involved in an abortive attempt to defend Norway from German invasion. He accompanied his unit to Scotland, and then returned to active duty in France, where he was eventually taken prisoner by the Germans.

After several unsuccessful attempts to escape, he was sent to a punishment camp at Rawa-Ruska in Ukraine. His experience there, working with a crew of gravediggers in a cemetery for French soldiers who died in captivity, formed the basis for the following novella, "The Season of the Dead." According to Gascar, the cemetery still exists, cared for during the postwar years by Soviet schoolchildren of the village.

Gascar was liberated by Russian troops in 1945. After returning to France, he worked as a reporter and literary critic. During this time, he began publishing short stories. His volume *The Beasts* won the Prix des Critiques in 1953, and in the same year these stories, together with "The Season of the Dead" (the entire collection translated as *Beasts and Men*), won the Prix Goncourt. His early vision was built around the tensions between our human and our animal nature, particularly as the latter was unleashed by the chaos of World War II. His later novels and stories address themes other than the Holocaust.

Gascar has said of himself: "Divided between a private, a secret world and the outer world of my fellow men, in everything I write I accept this dualism and find in it a sort of equilibrium. From this comes the diversity of my work, which gives almost equal weight to reality and to the dream."

A similar dualism energizes "The Season of the Dead," though it is questionable whether in this narrative Gascar achieves the equilibrium he describes. In the novella, "normal" dying shrinks in importance as the assault on the Jews gradually infiltrates the reality of the story and replaces the narrator's more manageable concern with the fate of his comrades. As he accidentally unearths the twisted bodies of previously executed Jews in the cemetery he thought he was preparing for French prisoners-of-war who had died in captivity, he is forced to face the terrifying chaos of mass murder and adjust his carefully nurtured vision of ritualized burial and lovingly tended gravesites to a doom from which even the healthy Jews cannot escape. The process by which his imagination shifts from one orbit to the other illustrates the power of Holocaust fiction to shape our response to a reality that more closely resembles a nightmare than a dream.

The Season of the Dead

Dead though they be, the dead do not immediately become ageless. Theirs is not the only memory involved; they enter into a seasonal cycle, with an unfamiliar rhythm—ternary perhaps, slow in any case, with widely spaced oscillations and pauses; they hang for a while nailed to a great wheel, sinking and rising by turns; they have become, far beyond the horizons of memory, rays of a skeleton sun.

We had reached the first stage. We were opening up the graveyard—in the sense in which one speaks of opening up a trench; in this place, there had only been life before there was death. And this freshness was to persist for a long time, before the teeming dust of the charnel should dim it, before, eventually, when all the earth was trodden down, oblivion should spread with couch-grass and darnel, and the writing on the tombstones should have lost its meaning; and the arable land should regain what we had taken from it.

For a graveyard to become a real graveyard, many dead must be buried there, many years must pass, many feet must tread on it; the dead, in short, must make the ground their own. We were certainly far from that point. Our dead would be war dead, for whom we had to break open a grassy mound. It was all, in short, brimful of newness.

War dead. The formula had lost its heroic sense without becoming obsolete. The war had lately moved away from this spot. These men would die a belated and, as it were, accidental death, in silence and captivity, yielding up their arms for a second time. But could one still use the word "arms"?

From the slope of the mound where the new graveyard lay I could see them walking round and round within the barbed wire enclosure of the camp, looking less like soldiers than like people of every sort and condition brought together by their common look of sleeplessness, their unshaven cheeks and the cynical complicity of gangsters the morning after a raid. Following several abortive escapes through Germany, some thousand French soldiers had just been transferred to the disciplinary camp of Brodno in Volynia. It was a second captivity for them, a new imprisonment that was more bewilderingly outlandish and also more romantic. That word gives us a clue: it was an imprisonment for death.

I had been granted the title of gravedigger in advance of the functions.

When you dig a ditch, it's because you have already found water. Just now there was nothing of that sort. The ditch we were digging was too long to have a tree planted in it, too deep to be one of those individual holes in which, at that time, throughout Europe, men in helmets were burrowing, forming the base of a monolithic monument which was hard to imagine, particularly here. It could only be a grave. Now we strengthened it with props, we covered it with planks. Nobody was dead. The grave was becoming a sort of snare, a trap in which Fate would finally be caught, into which a dead man would eventually creep. He would thus have been forestalled and would glide into the darkness through wide-open doors, while we would shrink back as he passed, hiding our earth-stained hands behind our backs.

The German N.C.O.* had rounded us up in the camp. He needed six men. When he had got that number he took us to the gate and handed us over to an armed sentry. We skirted the wall outside the camp until we came to a small rough road which, a little farther on, led over the side of a hill. At this point a track took us to the verge of the forest. The N.C.O., riding a bicycle, had caught up with us. He went to cut a few switches from an elm tree.

"Who knows German?" he asked without turning round.

"I do."

He called me to him. He was trimming the leaves from the switches with vigorous strokes of his penknife. I disliked the sight: swift-working fingers, pursed lips, and at the end of the supple, swaying branch a ridiculous tuft of leaves dancing as though before an imminent storm. There is no wretchedness like that of flogged men.

"There's to be a graveyard here . . ." he said to me, suddenly handing me the trimmed branches. "These are yours. Follow me and tell your mates to pull off some more branches."

I passed on the order and followed the N.C.O. to a place where the skirt of the wood dipped down into a narrow valley at the end of which lay a round pond like a hand mirror. The German dug the heel of his boot into the grass: "Here." As carefully as a gardener I planted a branch at the spot he showed me. Then he straightened his back and made a half-turn; staring straight in front of him, he walked forward, stopped and dug his heel into the grass, started off again and stopped again. My companions came up with their arms loaded with leafy branches. The task of planting began, and soon the branches stood lined up there in the still morning, marking the footsteps of the man as he doggedly staked his theoretical claim.

When the enclosure was thus demarcated, the German called me. We had to mark the site of the first grave. When this was set out, my companions, in a fit of zeal, immediately began to lift up clods of turf. Then the earth suddenly appeared as it really was; it lay there against the grass like a garment ready to be put on.

*Noncommissioned officer.

"That's enough, you can dig it tomorrow," the German told us. "We must always have one ready. Death comes quickly these days. War's a shocking thing."

He collected us together and the sentry took us back to the camp. As we were going through the gate one of us got from him, after some pleading, a leaf from his notebook with his signature. We rushed to the kitchens where extra rations of soup were sometimes distributed to the men who were working in gangs about the camp.

"Graveyard!" cried the prisoner who held the voucher, waving it. The man with the soup-ladle looked at us uncertainly for a moment, as though trying to remember to what burial ground this irregular privilege could suddenly have been allotted.

"Camp graveyard!" someone else repeated. The man took the can that was held out to him and filled it. Death had spoken; moreover, death's voucher was in order.

From that day, and still more from the following day when the first grave had been dug and shored up, I began to look out for death in the faces of my comrades, in the weight of the hour, the color of the sky, the lines of the landscape. Here, the great spaces of Russia were already suggested; I had never known a sky under which one had such a sense of surrender.

Sometimes the earth, dried by the early spring sunshine, was blown so high by the wind that the horizon was darkened by a brown cloud, a storm cloud which would break up into impalpable dust, and under which the sunflowers glowed so luminously and appeared suddenly at such distances that you felt you were witnessing the brief, noisy revenge of a whole nation of pensive plants, condemned for the rest of their days to the dull quietness of sunshine.

Close to us, the town was shut in with a white wall above which rose a bulbous church spire, some roofs, and the white plume of smoke from a train, rising for a long time in the same spot, with a far-off whistle like a slaughtered factory.

We had reached Brodno one April morning. The melting snows and the rain had washed away so much earth from the unpaved streets that planks and duckboards had been thrown down everywhere to let people cross, haphazard and usually crooked, looking like wreckage left after a flood subsides. The sentries could no longer keep the column in order and we were all running from one plank to another, mingled with women wearing scarves on their heads and boots on their feet, with German soldiers, with men in threadbare caftans; and here and there jostling one of those strange villagers who stood motionless with rigid faces, their feet in the mud, idle as mourners, with white armlets on their arms as though in some plague-stricken city.

It might have been market day, and the animation in the main street of the village might have been merely the good-humored bustle of the population between a couple of showers, such as one sees also on certain snowy mornings, or on the eve of a holiday. . . . In any case, that first day, the star of David, drawn in blue ink on the armlets of those painfully deferential villagers who looked oddly Sundayfied in their dark threadbare town clothes amongst that crowd of peasants, seemed to me a symbol of penitence, somewhat mitigated, however, by its traditional character.

It was not until later on that their destiny was clearly revealed to me. Then, when I saw them in a group away from the crowd, they ceased to be mere landmarks; exposed to solitude as to a fire, that which had been diluted among so many and had passed almost unnoticed acquired sudden solidity. All at once, they became the mourners at a Passion: a procession of tortured victims, a mute delegation about to appeal to God.

A certain number of Jews had been detailed by the Germans to get the camp ready before our arrival. When we entered the gates they were still there, carefully putting the last touches to the fences, finishing the installation of our sordid equipment, and thus implacably imprisoning themselves, by virtue of some premonitory knowledge, within a universe with which they were soon to become wholly familiar.

The camp consisted of cavalry barracks built by the Red Army shortly after the occupation of Eastern Poland at the end of 1939. A huge bare space separated the three large brick-built main buildings from the whitewashed stables which housed the overflow of our column, according to that mode of military occupation that disdains all hierarchy of places—thus identifying itself with the bursting of dams, the blind and inexorable progress of disasters.

Inside every building, whether stable or barracks, wooden platforms, superimposed on one another, had been set up the whole length of the huge rooms, leaving only a narrow passage along the walls and another across the middle of the structure: tiered bunks like shelves in a department store, where the men were to sleep side by side. Our captivity thus disclosed that homicidal trend which (for practical rather than moral reasons) it usually refused to admit: for the Germans, the unit of spatial measurement was "a man's length."

The great typhus season was barely over at Brodno. It had decimated the thousands of Russian prisoners who had occupied the place before us and who had left their marks on the whitewashed walls—the print of abnormally filthy hands, bloodstains and splashed excrement—messages from those immured men jostling one another in the silent winter night, while death and frost exchanged rings: faintly heard calls from far away. Because of the lingering typhus and the risk of propagating lice, we were given no straw.

We were given little of anything that first day. The Germans, except for a few sentries established in their watchtowers, had retired no one knew whither, as though it were understood that at the end of our trying journey we must be granted a day's truce, an unwonted Sunday that found us standing helpless, leaning against the typhus-ridden bunks with our meager bundles at our feet. A louse crawled up one's spine like a drop of sweat running the wrong way, and within one there was that great echoing vault, hunger.

In the afternoon, however, a few pots full of soup were thrust through the kitchen door. A thousand men lined up on the path of planks that led to it. Hardly any of us had a messtin, but a great rubbish dump full of empty food cans supplied our needs. When the cans were all used up we unscrewed the clouded glass globes covering the electric lamps in the building, and made empty flower-pots water tight by plugging the holes with bits of wood. When these uncouth vessels appeared in the queue they were greeted with shouts of envy, provoked rather by their capacity than by their grotesque character. A sort of carnival procession in search of soup took place, and the owner of an empty sardine-can might be seen gauging a piece of hollow brick half-buried in the mud, wondering if those four holes like organ-pipes might perhaps be stopped up at the base, and turning over the problem with his foot while the column moved on a few yards. Fine rain was falling.

"It's millet!" shouted a man coming towards us from the kitchens, clasping his brimming can in both hands.

A cry of joy, in an unknown voice: a fragmentary phrase, as though cut out of its context, which, uttered in the dying afternoon in the heart of the Volynian plain, seemed to have escaped from a speech begun very far away, many years earlier, and to have returned now—just as, in the hour of death, words half-heard long ago, neglected then and despised, recur to one's memory, suddenly whispering out their plaintive revenge, suddenly gleaming with a prodigious sheen because they hold the last drops of life.

It was not until much later that somebody died.

From the time of that first burial I felt certain that death would never move far from our threshold. A dead body is never buried as deep as one thinks; when a grave is dug, each blow of the pickax consolidates the boundaries of the underground world. Though you lie sepulchred in the earth, like a vessel sunk in quicksands, and the waves of darkness beat against you from below, your bones remain like an anchor cast.

That day a group of German soldiers accompanied the convoy. They were armed and helmeted. They fell into step with the handful of Frenchmen—the chaplain, the medical orderly, the *homme de confiance*[1]—who were walking behind the *tarantass* on which the coffin was laid. They moved very fast and they seemed to be upon us in a few minutes, as we stood watching from the graveyard on the side of the hill (had they remembered to bring the cross and the ropes?); they were charging on us, a crowd of them, two by two, clad from head to foot; they were coming for us, making us realize in a flash what a terrible responsibility we had accepted when we dug that hole, what echoes our solitary toil had roused over there.

We had to face them, to lay the ropes down side by side on the spot where the coffin would be placed, to put down the two logs at the bottom of the grave on which it would lie so that we might afterwards haul up the cords, the ends of which would flap against the coffin for a minute like the pattering footsteps of a last animal escaping. The chaplain recited prayers. The medical orderly sounded Taps on a bugle, picked up somewhere or other. We grasped our ropes and, leaning over the grave, began to slacken them. At an order from their N.C.O. the German soldiers, who were presenting arms, raised the barrels of their guns towards the sky and fired a salvo.

There is always somebody there behind the target of silence. A shout or a word uttered too loud or too soon, and you hear a distant bush crying out with a human voice—you run towards a sort of dark animal only to see it clasp a white, human hand to its bleeding side; it's the tragedy of those hunting accidents where the victims, emerging from silence, are the friend or stranger—equally innocent—who happened to be passing by; there's always somebody passing just there, and we are never sufficiently aware of it.

The Germans' salvo re-echoed for a long time. We had lifted our heads again. Lower down, on the little road, some peasants and their women, coming back from the town, who had not witnessed the beginning of the ceremony (at that distance, in any case, they could not have observed its details) began to walk suddenly faster, casting a quick look back at us. Some women drew closer together and took each other's arms, a man stumbled in his haste, and all of them swiftly bowed their heads and refused to look at what was happening in our direction.

They seemed possessed not so much by anxiety as by a kind of shuddering anticipation, making them shun a spectacle which they dreaded as though it were contagious and hurry slightly despite their assumed indifference. They betrayed that tendency to deliberate withdrawal which, at that time,, was making the whole region more deserted than any exodus could have done. Had we run towards them, clasped their hands, gazed into their faces crying "It's all right, we're alive!" they would no doubt still have turned away from us, terrified by fresh suspicions, feeling themselves irremediably compromised. . . . Now they had vanished. The Germans slung their rifles. We stood upright round the grave, like a row of shot puppets.

This incident and others less remarkable gave us a feeling of solidarity. We tried to secure official recognition for our team from the camp authorities by presenting a list of our names to every new sentry—to those that kept guard over the gates, those that supervised the kitchens, those that inspected our block. Every week we made out several lists, in case the camp administration should prove forgetful. We gave notice of our existence to remoter authorities, to prisoners' representatives, shock brigade headquarters, divisional commanders, with the stubborn persistence of minorities ceaselessly tormented by the nightmare of illegality.

We guessed that our proceedings met with secret opposition from the Germans, who were unwilling to give public recognition to this peculiar team, the granting of legal status to which would, for them, have been equivalent to admitting criminal premeditation—and from the prisoners too, since they did not need so realistic a reminder of the gruesome truth.

Between two deaths, it was only owing to the force of habit and the routinist mentality of the guardroom officer that we found a couple of sentries waiting each morning to take us to the graveyard. This, lying on the side of the hill amongst long grass, was in such sharp contrast to the almost African aridity of the camp as to enhance the feeling of separateness and even of exclusion which the failure of our advances to the administration had fostered in us. We belonged to another world, we were a team of ghosts returning every morning to a green peaceful place, we were workers in death's garden, characters in a long preparatory dream through which, from time to time, a man would suddenly break, leaping into his last sleep.

In the graveyard we led that orderly existence depicted in old paintings and, even more, in old tapestries and mosaics. A man sitting beside a clump of anemones, another cutting grass with a scythe; water, and somebody lying flat on his belly drinking, and somebody else with his eyes turned skyward, drawing water in a yellow jug. . . . The water was for me and Cordonat. We had chosen the job of watering the flowers and turf transplanted on to the first grave and amongst the clumps of shrubs that we had arranged within our enclosure.

Its boundaries were imaginary but real enough. We had no need to step outside them to fill our vessels at the pond which, from the graveside, could be seen between the branches, a little lower down; there, the radiance of the sky reflected in the water enfolded us so vividly, lit up both our faces so clearly, that any thought of flight could have been read on them from a distance, before we had made the slightest movement, before—risking everything to win everything—we had set the light quivering, like bells.

The only flight left to us was the flight of our eyes towards the wooded valley at the end of which the pond lay. the leaves and grass and true trunks glowed in the shadow, through which sunbeams filtered and in which, far off, a single leaf, lit by the sun's direct fire, gleamed transparently, an evanescent landmark whose mysterious significance faded quickly as a cloud appeared.

Flowers grew at the very brink of the pond: violets, buttercups, dwarf forget-me-nots, reviving memories of old herbals; only the ladybug's carapace and the red umbrella of the toadstool were lacking to link up the springtime of the world with one's own childhood. When we had filled our bottles, Cordonat and I would linger there gazing at our surroundings, moved by our memories, and in an impulse of greedy sentimentality guessing at the beechnut under the beech leaf, the young acorn under the oak leaf, the mushroom under the toadstool and the snail under the moss.

Sometimes the sun hid. But we could not stir, for we had fallen down out of our dream to such a depth that our task—watering a few clumps of wood-sorrel in a remote corner of Volynia—appeared absurd to the point of unreality, like some purgatorial penance where the victims, expiating their own guilt or original sin, are forced to draw unending pails of water from a bottomless well, in a green landscape, tending death like a dwarf tree—just as were were doing here.

Actually, I did not know whether Cordonat's dream followed the same lines as my own. I had lately grown very fond of Cordonat, but he was so deeply consumed by nostalgia that maybe I only loved the shadow of the man. He was ten years older than I, married, with two children; home consists of what you miss most. This vineyard worker from Languedoc showed his Catalan ancestry in his lean, tanned face, with the look of an old torero relegated to the rear rank of a *cuadrilla,** his delicate acquiline nose and wrinkled forehead with white hair over the temples which predestined him for the loneliness of captivity.

It so happened that, with the exception of myself, all the men of our graveyard team, who belonged to the most recent call-up, were natives of the South of France, and all showed a tendency to nostalgic melancholy which was highly appropriate not only to their new duties but to our peculiar isolation on the fringe of camp life. This distinction enhanced a characteristic which was common to all the prisoners of Brodno, who, by their repeated attempts to escape from Germany, had in effect escaped from their own kind. At a time when under cover of captivity countless acts of treachery were taking place, they had set up on the Ukrainian border, in a corner of Europe where the rules of war were easily forgotten, a defiant Resistance movement, a group of "desert rats" whose most seditious song was the Marseillaise.

Homesickness creates it own mirages, which can supersede many a landscape. But that amidst which we were living now was becoming so cruelly vivid that it pierced through all illusory images; it underlay my companions' dreams like a sharp-pointed harrow. This became clear only by slow degrees.

When Cordonat and I were sitting by the pond, we would look up and see peasants and their wives on their way back from the town, passing along the path through the trees and bushes at the end of our valley. We would stand up to see them better and immediately they would hurry on and vanish from sight, imperceptibly accentuating the furtiveness of their way of walking, stooping a little and averting their eyes as though they were eager to avoid the sight of something unlucky or, more precisely, something compromising.

As we stood at the foot of this hillock, somewhat apart from the other prisoners, we must have seemed to be in one of those irregular situations which were not uncommon here, like cases of some infectious disease. Were we escaped prisoners, obdurate rebels? Were we in quarantine, or about to be shot? In any case we were obviously trying to make them our accomplices, determined to betray them into a word or a look and

*Team accompanying a bullfighter, here signifying a demotion.

thus involve them in that contamination that always ended with a shower of bullets and blood splashed against a wall. And the forest in which, only a minute before, spring flowers had awakened childhood memories, now emerged as though from some Hercynian flexure, darker and denser, more mysterious and more ominous, because of the fear and hunger of men. Fear can blast reality.

But it was when we left the skirt of the forest and reached the plain where the town lay that this devastating power of fear seemed actually to color the whole landscape. The white road, the far-off white house fronts, the lack of shadows, all this was deprived of radiance by the subdued quality of the light; but it exuded a kind of stupor. At first, you noticed nothing.

But when we drew near we would suddenly catch sight of a man or a woman standing motionless between two houses or two hedges, and turning towards us in an attitude of submission, like people who have been warned to prepare for any danger. The man or woman would stare at us as we passed with eyes that revealed neither curiosity nor envy nor dread: a gaze that was not dreamy, but enigmatically watchful. A few men, also dressed in threadbare town clothes, were filling up the holes in the pavement. They did not raise their heads as we went by; they kept on with their work, but performed only secondary, inessential tasks, like factory hands waiting for the bell to release them from work and staying at their posts only because they have to.

In every case we were aware that, as we approached (or more precisely as our sentries approached), some final inner process of preparation was taking place (but maybe it had long since been completed?) and that one of the Germans had only to say "Come on!," load his gun or raise the butt to strike, for everything to take its inevitable, unaltering course. The tension of waiting was extreme.

They had long ago passed the stage when your pulse beats faster, spots dance before your eyes and sweat breaks out on your back; they had not left fear behind, but they had been married to it for so long that it had lost its original power. Fear shared their lives, and when we walked past with out sentries beside us it was Fear, that tireless companion, that began in a burst of lunatic lucidity, to count the pebbles dropping into the hole in the pavement, the trees along the road, or the days dividing that instant from some past event or other—the fête at Tarnopol, or Easter 1933, or the day little Chaim passed his exam; some other spring day, some dateless day, some distant day that seemed to collect and hold all the happiness in life.

Sometimes, in the depth of their night, fear would flare up and wake them, like the suddenly remembered passion that throws husband and wife into each other's arms; then they would embrace their fear, foreseeing the coming of their death like the birth of a child, and their thoughts would set out in the next room the oblong covered cradle in which it would be laid. Morning would bring back their long lonely wait, tête-à-tête with fear. They tied round their arms the strip of white material with the star of David drawn on it. Often the armlet slipped down below the elbow, and hung round the forearm slack and rumpled and soiled like an old dressing that has grown loose and needs renewing. The wound is unhealed, but dry. But why should I speak of wounds? Hunger, cold, humiliation and fear leave corpses without stigmata. One morning we saw a man lying dead by the roadside on the way to the graveyard. There was no face; it was hidden in the grass. There was no distinguishing mark, save the armlet with the star of David. There was no blood. There is practically no blood in the whole of this tale of death.

One Monday morning two new sentries came up to join us at the camp gates. They belonged to a nondescript battalion in shabby uniforms which had been sent from somewhere in Poland by way of relief, and had arrived at Brodno a few days earlier; one of those nomadic divisions to which only inglorious duties are assigned, and whose soldiers get killed only in defeats.

Our two new sentries were a perfect example of the contrasts, exaggerated to the point of grotesqueness, which are always to be found among any group of belatedly conscripted men, since neither regulation dress nor *esprit de corps* nor conviction can replace the uniformity of youth. One of the two soldiers was long and thin, with a high-colored face; the other, short and squat, was pale.

Each morning, when we went into the enclosure, we would quickly dismiss and, one after the other, go and stand before the graves giving a military salute; it was the only solemn moment of the day. That morning, as I was walking after my comrades to pay my homage to the dead, I heard a click behind me: the taller of the two soldiers was loading his gun. We had been running rather quickly towards our dead, because it was Monday and we felt lively. He had been afraid we were trying to escape or mutiny. But we were merely hurrying towards the graves like workmen to the factory cloakroom, discarding discipline and, at the same time, hanging up our jackets on the crosses.

There were only three graves at this time. We made up for this by working on the flowerbeds, those other plots of consecrated, cultivated ground. Cordonat and I were already grasping our bottles, eager to resume our reveries beside the pond, at the bottom of which could be seen a rifle and a hand grenade thrown there by a fugitive Russian soldier. This filthy panoply, sunk deep in the mud, mingled with our reflected images when we bent over the water, as though we had not been haunted enough for the past two years by the memory of our discarded arms.

I went up to the short German soldier (the other had alarmed me by that performance with his rifle a little while before) and explained to him that it was our custom to go and draw water from the pond. He nodded, smiling but silent; then, before I went off, he said to me in French: *"Je suis curé."* *

* "I am a pastor."

The word, uttered with an accent that was in itself slightly ridiculous, had a kind of popular simplicity that, far from conferring any grandeur on it, seemed to relegate it to the vocabulary of anticlericalism, made it sound like the admission of a comic anomaly. Such a statement, made by this embarrassed little man wearing a dreaded uniform in the depth of that nameless country, suggested the depressing exhibitionism of hermaphrodites, the sudden surprise of their disclosure. I could find nothing to reply.

"Protestant," he went on in his own language. "In France, you're mostly Catholics."

So he was a pastor. The rights of reason were restored—so were those of the field-gray uniform, since I found it easier to associate the Protestant religion with the military profession. I acquiesced: we were, apparently, Catholics. Generously, the Germans granted us this valid historical qualification; we might graze on this reprieve. The little pastor, however, did not try to stop us from going to draw water for our flowers; he encouraged us to do so and walked along with us down the path leading to the pond.

Without giving me time to answer, he chattered in his own language, of which Cordonat understood barely a word. I had never before heard German spoken so volubly; it poured forth like a long-repressed confession, like a flood breaking the old barriers of prejudice and rationalism, and the clear waters rushed freely over me, carrying the harsh syllables like loose pebbles. He was a pastor at Marburg. His family came from the Rhineland and one of his ancestors was French. In the Rhineland they grew vines; the country was beautiful there. Marburg lay farther east; and there they still remembered Schiller and Goethe and Lessing (nowadays people seldom talk about Lessing). The pastor's wife was an invalid. On summer evenings he would go into the town cafés with his elder daughter. They were often taken for man and wife.

He went on talking. It was a sunny June morning. The Ukrainian wheat was springing up. By the side of the sandy roads, you could sometimes see sunflower blossoms thrown away by travelers after eating the seeds. And we carried on our Franco-Rhenish colloquy, squatting at the foot of the hillock, beside the pond, in the shade of the trees, while the peasant women, suspecting some fresh conflict, some subtle and wordy form of bullying, hurriedly passed by higher up, with a rustle of leaves under their bare feet.

In order to keep our hands occupied and to justify our long halt by the pond in the eyes of the other sentry, Cordonat and I had begun to scour our mess-tins. We always carried them about with us. I had fixed a small wire handle to mine so as to hook it to my belt. This habit was partly due to the constant hope of some windfall, some unexpected distribution of food, but also no doubt it expressed a sort of fetishistic attachment to the object that symbolized our age's exclusive concern with the search for food. These mess-tins, which had only been handed out to us on our second day in camp, were like little zinc bowls; we went about with barbers' basins hanging from our waists. As part of this instinctive cult, we felt bound to polish them scrupulously. Cordonat and I were particular about this, to the point of mania. It was largely because of the fine sand on the edge of the pond and also because we were waiting—waiting for better days to come.

"Ydiom! Ydiom!" ("Come on!") we heard a peasant woman on the hill above us calling to one of her companions, who must have been lingering in the exposed zone.

The little pastor kept on talking.

"What d'you think of all this?" I asked him.

"It's terrible," he said. "Yes, what's happening here is terrible."

That simple word assumed the value of a confession, of a dangerous secret shared. It was enough in those days (a look, a gesture, a change of expression would have been enough) to lift the hostile mask and reveal the pact beneath. However, I dared not venture further and I began pleading our cause, describing our destitution, making no major charge but only such obvious complaints as could give no serious offense to the Germans. The petty sufferings we endured acted as a convenient salve for one's conscience; I realized this as I spoke and I resented it. Even this graveyard, so sparsely populated and so lavishly decorated, had begun to look like "a nice place for a picnic," with its green turf overlying the great banqueting-halls of death.

"Terrible, terrible," the little pastor kept saying. It was the word he had used earlier. But it had lost its original beauty.

Back in camp, after the midday meal of soup, I waited impatiently for the two sentries to come and take us to the graveyard. I looked forward to seeing Ernst, the pastor (he had told me his name), with a feeling of mingled sympathy and curiosity that was practically friendship; it only needed to be called so.

I was not surprised when, on reaching the graveyard, Ernst took me into the forest, explaining to his mate that we were going to look for violet plants. It was high time, the last violets were fading. The other German appeared quite satisfied with this explanation. Since the morning, he had been so good-humored that I felt inclined to think that that sudden business with the rifle had been an automatic gesture; a gun never lets your hands stay idle. He had a long shrewd face.

"I'm a Socialist," he told my friends while I walked away with Ernst.

Cordonat watched me go; was he envying me or disowning me? He sat there beside the pond like some lonely mythological figure.

But now the forest was opening up in front of me: that forest which hitherto I had known only in imagination, which had existed for me by virtue not of its copious foliage or its stalwart tree-trunks but of its contrasting gloom, the powerful way it shouldered the horizon and above all its secret contribution to the darkness that weighed me down. We walked on amidst serried plants; he carried his rifle in its sling, looking less like an armed soldier than like a tired huntsman, glad to have picked up a companion along the homeward road. But already the forest and its dangerous shadows had begun to suggest that the journey home would be an endless one, that our companionship was forever; once more, in the midst of a primeval forest, we were shackled together like those countless lonely damned couples—the prisoner and his guard, the body and its conscience, the hound and its prey, the wound and the knife, oneself and one's shadow.

"Well, when you're not too hungry in camp, what d'you like doing? You must have some sort of leisure. Oh, I know you don't like that word. But I don't know any other way to express the situation where I'd hope to find your real self—for hunger isn't really you, Peter. Well, what else is there?"

He pushed back from his hip the butt of his rifle, which kept swinging between us.

"I walk round the camp beside the barbed wire, or else I sleep, and when I find a book I read it. . . ."

"And sometimes you write, too. . . ."

I looked at Ernst suspiciously.

"A few days ago," he went on, "I was on guard in one of the watch-towers. I noticed you. You stopped to write something in a little note book. This morning I recognized you at once."

This disclosure irritated me.

"You see, I can't even be alone!" I cried. "Isn't that inhuman?"

"But you were alone. You were alone because at that moment I didn't know you, and you didn't know me. Really, nothing was happening at all."

He was beginning to sound a little too self-confident. It was a tone that did not seem natural to him; there was a sort of strained excitement about it. I did not pursue the matter; without answering I bent down to pick up my violet plants. Ernst stood in silence for a while then, returning to the words I had spoken a few minutes earlier, as though his mind had dwelt on them in spite of what had followed:

"Books," he repeated with a schoolmasterish air of satisfaction and longing. "They were my great refuge too, when I spent a few months in a concentration camp two years ago. I'd been appointed librarian, thank Heaven. . . . Look here," he went on in a livelier tone, as though what had been said previously was of no importance, "put down your violet plants. Come and I'll show you something."

He was giving me line enough, as fishermen say; but I was well and truly caught this time. I stood up. He had started off ahead of me along a path that led to the right. I caught up with him: "I suppose the camp was on account of your political ideas?"

"Political . . . well, that's a word we don't much care for. Rather on account of my moral views. However, they did let me out of that camp. It was just a warning. I need hardly tell you that I haven't changed. . . . You see that wall?"

I saw the wall. It was decrepit, overgrown with moss and briars, its stones falling apart, and it enclosed a space in which the forest seemed to go on, to judge by the treetops that appeared above it. We walked some way round the outside and came to a gap in the wall. Within the enclosure gray stones stood among trees which were slighter than those of the surrounding forest, contrary to my previous impression. Here and there, pale grass was growing; it had begun to take possession of the ground again. The upright stones marked Jewish graves, a hundred years old no doubt. Eastern religions lay their dead at the foot of slabs of slate, stumbling blocks for the encroaching wilderness. Time had jostled many and overthrown a few of these old battlements of death, monoliths on which Heaven had written its reckoning in the only tongue it has ever spoken.

Ernst knew a little Hebrew. His small plump hand was soon moving over the stone, beginning on the right as in fortunetelling by cards. He read out some long-distant date, lazily coiled up now with a caterpillar of moss lying in the concavity of the figures, some name, with the knowledgeable curiosity of an accountant who has discovered old statements, bills yellow with age that have been settled once and for all. His religion was based on the belief that death belongs to the past and, when he looked at a grave, whether worn down by time or black with leaf-mold, he would say to himself with visible gladness that "all that's over and done with." Meanwhile I was overwhelmed by the symbolism of these graves; on most of the stones there was carved a breaking branch—you could see the sharp points at the break and the two fragments about to separate forming an angle, a gaping angle like an elbow. It was like the sudden rending that takes place high up in the tree of life, its imminence revealed by the flight of a bird, of that other soul which has hitherto deafened us by its ceaseless twittering, whereas our real life lay in the roots. On the stone, the branch was endlessly breaking, it would never break; when death has come, has one finished dying? Ernst raised his head and straightened his back.

"I've been rereading your classics," he said to me with a smile.

I did not understand.

"I mean that this ancient, traditional burial ground, close by your own fresh and improvised one, is rather like the upper shelf in a library. . . ." He laid his hand on a carved stone. "The preceding words in the great book . . ."

"Who can tell? Perhaps the moment of death is never over," I replied, thinking of the symbol of the broken branch. "Perhaps we are doomed to a perpetual leavetaking from that which was life and which lies in the depth of night, as eternal as the patient stars. Perhaps there is no more identity in death than in life. Each man dies in his own corner, each man stays dead in his own corner, alone and friendless. Every death invents death anew."

"But I believe in Heaven and in the communities of Heaven, where no echo of life is heard," replied Ernst joyfully. "In the peace of the Lord."

"I cannot and will not believe that those who are murdered here cease their cries the moment after. . . ."

"Their cries have been heard before," said Ernst. "What do you mean? Do you need to hear dead men's cries? Isn't it enough that God remembers them, that we remember them? I will tell you something: they are sleeping peacefully in the light, all in the same light."

"That's too easy an answer!" I cried. "That's just to make us feel at peace."

"Don't torture yourself," Ernst replied. "In any case, neither you nor I is to blame."

We were not to blame. I had taken up my violets again from the place where I had left them a short while before and, with this badge of innocence, my flowers refuting what Ernst's rifle might suggest, we made our way to the graveyard through the silent forest.

"Tomorrow we'll go for another walk," Ernst said. "We might take the others too. We'll find some pretext."

But next day we were deprived of any pretext. Two men had died in the camp. Graves had to be dug; the dead had to be buried.

Ernst directed our labors skilfully and with pensive dignity. It was he who taught us that in this part of Europe they lined the inside of graves with fir-branches. Not to lag behind in the matter of symbolism we decided to bury the dead facing towards France. As France happened to be on the farther side of the forest against which the first row of graves had been dug, we were obliged to lay the dead men the wrong way round, with their feet under the crosses.

These two deaths occurring simultaneously aroused our anxiety. Sanitary conditions within the camp had worsened; underfed and weakened by their sufferings during the escape, and the subsequent journeys from camp to camp, the men gathered in daily increasing numbers in a huge sickroom. A few French doctors had been sent to Brodno by way of reprisal; lacking any sort of medicaments, they went from one straw mattress to the next making useless diagnoses, reduced to the passive role of witnesses in this overcrowded world whose rhythm was the gallop of feverish pulses and where delirious ravings mingled in a crazy arabesque, while men sat coughing their lungs out.

We buried four men in the same week. June was nearly over; it was already summer. By now, the white roads of the invasion spread like a network over the Russian land, far east of Brodno, right up to the Don, then to the Volga, to the Kuban, milestoned with poisoned wells. The woods were full of hurriedly filled graves, and the smoke rose up straight and still from the countryside, while the front page of German illustrated papers showed bareheaded soldiers, with their sleeves rolled up, munching apples as they set off to conquer the world. Hope dried up suddenly, like a well.

The Germans' victorious summer, as it rolled eastward, left us stranded on that floor formed by the hardened sediments of their violence, their extortions, their acts of murder, which already disfigured the whole of Europe. The drift of war away from us had only removed the unusualness of these things; the things themselves remained, only instead of seeming improvised they had assumed a workmanlike character; ruins were now handmade, homes became prisons, murder was premeditated.

The Jews of Brodno had practically stopped working inside the camp, where everything was now in order. Those who were road menders spent their time vainly searching the roads for other holes to fill; the sawmill workers, whom we used to see on our way to the graveyard, kept on moving the same planks to and fro, with gestures that had become ominously slow. More and more frequently you could see men and women, wearers of the white armlet, standing motionless between houses or against a hedge, driven there by the somnambulism of fear.

Inside the camp the same fatal idleness impelled the French to line up against the barbed wire, with their empty knapsacks slung across their back and clogs on their bare feet. To Jews and Frenchmen alike (to the former particularly) going and staying were equally intolerable fates, and they would advance timidly towards the edge of the road or of the barbed wire, take one step back and move a little to one side, as though seeking some state intermediary between departure and immobility. They would stand on the verge of imagined flight, and in their thoughts would dig illusory tunnels through time.

The tunnels we were digging might well have served them as models. Only ours had no outlet.

"Indeed, yes, Peter, graves have outlets," Ernst told me.

He had just brought me the stories of Klemens von Brentano. Death having left us a brief respite, we had gone off for a walk in the forest. A little way behind us, my comrades were gathered round Otto, the other sentry.

"What's he saying?" they shouted to me. I turned round. Otto repeated the words he had just uttered.

"He says that at home he was the best marksman in the district," I translated.

"Yes, we got that, he won some competitions. But he said something about birds. . . ."

"He can kill a bird on a branch fifty meters off with one shot. He's ready to prove it to us presently."

Otto was smiling, his neck wrinkling. His boast sounded like a public tribute when repeated as an aside by somebody else. I went back to Ernst.

"He's a nuisance with his stories of good marksmanship! Shots would make too much noise in the forest and in these parts they mean only one thing. I was going to take you to see the girls who work where the new road's being made. If he shoots we shall find them all terrified. . . ."

The project had a frivolous ring but Ernst forestalled my questions:

"They are young Jewish girls, unfortunate creatures. I've made friends with one of them. Every evening I take her some bread—a little bread; it's my own bread—at least, part of my own bread," he added in some confusion, recalling the daily agony of sharing it, his hunger and his weakness.

"Why are we going to see them?" I asked him. "We can't do anything for them. You know that."

"You are French, and the point is this: these are girls from the cultured classes of Brodno. They are better dressed than the peasants, they don't speak the same language, they had relations in various other countries, and now they are isolated, as though by some terrible curse. Nothing can save them from their isolation. Even if tomorrow the peasants round here were to be persecuted, the girls would find no support amongst them, no sense of kinship, no co-operation. Believe me, they've always been exiles in the East, even before the Germans settled here. And perhaps only you and I share what they've got, what makes them different. So don't run away."

"How could I run away?" I pointed to his gun. Ernst reddened with anger and shook his head. Behind us, big Otto had seen my gesture: "Don't talk to the pastor about his gun," he called out to me. "He thinks it's a fishing rod."

I turned round. "Please note that I'm just as much a pacifist as he is," Otto said. "Only I know how to shoot!"

He burst out laughing. Cordonat was near him and beckoned to me. "Ask him to choose a biggish target," he said as I came up, "a rabbit, for instance, or a jay or a crow. Let's at least have something to get our teeth into!"

I passed on the request to Otto.

"But then it wouldn't be a demonstration," he cried. "It's got to be quite a tiny animal."

I tried to explain to him that he could still keep to the rules of the game if he stood farther away from his living target. The argument seemed likely to go on indefinitely.

Ernst walked on ahead of us, indifferent to our conversation, with his back a little bent, like a recluse, and in addition that pathetic look that small men have when they are unhappy. He was going forward through the forest without keeping to the paths, and the forest was growing thicker. Already, we felt cut off from the outside world here, just because the shadows were a little deeper and the ground was carpeted with dead leaves, moss, myrtles and nameless plants. For us, as for millions of others, war meant "the fear of roads." These enslaved men looked at you sometimes with eyes like horses'.

There was still the sky, the sky between the branches when you raised your head, an unyielding sky, still heavy with threats. Daylight is up there. I must be dreaming. It was as if when you pushed open the shutters after a night full of bad dreams the influx of light proved powerless to dispel the terrifying visions of the darkness from your eyes. And yet everything is there, quite real. You need a second or third awakening, the maneuvering of a whole set of sluice-gates, before the morning light yields what you expect of it—not so much truth as justice.

After a few minutes I caught up with Ernst. I was afraid of disappointing him by showing so little interest in the visit he had suggested to me.

Perhaps he had given it up; the thought caused me no remorse, for he was a man with too clear a conscience, too easily moved to compassion. But even if I had interfered with the execution of his plans, I did not want him to think me indifferent or insensitive.

"I'm hungry," I said as I drew near him, so as to avoid further explanations.

"So am I. But we shall soon find these girls, and they'll give us some coffee—they make it out of roasted grain. That'll help to appease our hunger. Afterwards we'll go and fell the trees. And then Otto will have plenty of time to fire his shots."

The trees that we were to fell were intended to build a fence round the graveyard. Ernst had discovered this excuse to justify our walks. We were now on a path that led down to the verge of the forest. The new road that was being made skirted it at this point and the workmen had set up a few huts for their gear and for the canteen. Eight Jewish girls worked here under German orders on various tasks, and two of them kept the canteen.

These were the girls we saw first when Ernst, telling his mate we were going to get some coffee, led me towards the hut. The two girls had come out over the threshold. They were wearing faded summer dresses whose original colors suggested Western fashions and cheap mass production. That was enough to introduce into this woodland setting an urban note which would have struck one as strange even without the added impression of bewilderment and weariness conveyed by the pale faces and wild eyes of the two girls. In spite of their youth, their features were devoid of charm. Beautiful faces were rare in this war; those faces which were daily taken from one were like commonplace relatives full of modest virtues, known intimately and loved and now gone into the night, unforgettable faces with their freckles and their tear-stained eyes.

When they recognized Ernst the two girls nodded gently. They stayed close together on the doorstep. Ernst spoke to them, calling them by their names. We were Frenchmen, he told them, prisoners too, hungry and unhappy. "Nothing was happening." We looked at each other, helplessly. What conversation could we hold? Everything had been said before we opened our mouths. It was not to one another that we must listen but to the far-off heavens, to which someday perhaps would rise the noise of our deliverance.

"We can't give you coffee," said one of the girls to Ernst. "The *Meister* hasn't given us any grain today."

Ernst had poked his head inside the hut to peer round. "Who's that?" he asked in a whisper. I looked too. In a corner of the room a man in black was leaning over a basin, with his back towards us. He was dipping a rag in water and from time to time raising it to his face. One could guess from the stiff hunching of his shoulders and the timidity betrayed by his slow awkward gestures that he felt our eyes upon him.

"It's a man from the sawmill," quickly replied one of the girls. "He hurt his face at work."

"But he doesn't work up here," said Ernst. "If the *Meister* finds him here there'll be a row."

"He'll go away when you've gone," said the smaller of the two girls, speaking for the first time and with an ill-disguised nervousness. She was dirty, with untidy red hair hanging down each side of her face. "The other sentinel mustn't see him," she went on in a low voice. "Please, Herr Pastor, be kind and take your Frenchman farther off."

"Where is Lidia?" asked Ernst.

"She's working on the dump trucks," said the other girl. "She's being punished because she broke the cord of the siren this morning."

"It's not fair!" went on the smaller girl, with bitterness. "The cord was rotten. When she tried to stop the siren, at seven o'clock, the cord broke off in her hand. The siren went on wailing long after everybody was at work. All the steam was being wasted. The *Meister* called it insubordination. It's not fair. . . . Yesterday they hanged four more men at Tarnopol."

"The way that siren went on wailing," said her friend. "There was something sinister about it—it was like a warning. But what was the use of warning us? How could we move? What could we do?" she added anxiously.

Ernst did not answer. He stood with downcast eyes.

"Perhaps I'll come back tonight," he said after a while. "Tell Lidia. God be with you."

We joined the group of prisoners who were waiting for us, guarded by Otto, and plunged once more into the forest.

"Somebody must have hit him," muttered Ernst as I walked at his side. "I'm speaking of that man we saw in the hut," he added, with a look at me. "He didn't want us to see the blood."

"Why not?"

"Because they feel that to let their blood be seen is not only a confession of weakness, of impotence, but moreover it marks them out as belonging to the scattered herd of bloodstained victims who are being ruthlessly hunted down. While they are whole, they'll carry on as long as their luck holds; when they're wounded, they go out to meet their own death. The order of things that has been established here is all-embracing, Peter. If a civilian—if one of these civilians is found with blood on him, the authorities think the worst. Where did he get it, who gave him leave to move about? Why is he branded with that mark, and why is he not with the herd of the dead? There's something suspicious about it. That man was well aware of this. He was hiding. He'd 'stolen' a beating. . . ."

"Do you think those girls are really in danger?"

While I was asking Ernst this question I realized that I had no wish to learn from him whether their danger was great or small. For the last few minutes I had experienced that slight nausea which, at the time, was more effective than any outward sign in warning me of imminent peril. Before Ernst could answer me a shot rang out behind us. Otto lowered his rifle and Cordonat ran towards a bush: "It's a jay!" he called out to me.

But I was observing the amazement created all around us, in the lonely forest, by the sudden report. I thought I caught sight of a gray figure disappearing swiftly between distant tree trunks. Then everything was as it had been, unmoving. I did not want to talk any more, and Ernst seemed not to want to either. That rifleshot had been like a blow struck with a clenched fist on a table, a call to order, silencing all chatter, even the private chatter of the heart. Otto was looking at us with a smile. He was looking at Ernst and myself. The bullet had passed just over our heads. He stopped smiling just as I was about to speak, to try and break out of that clear-cut circle which henceforward would enclose the three of us and within which we were now bound to one another by the dangerous silence that had followed our commonplace words.

The dead used to be brought to us in the morning, like mail that comes with habitual irregularity and provides no surprises; their belongings had to be classified, bills of lading for ships that have long since put out to sea; a cross had to be provided, with a name and a date. Now the coffins no longer showed those once ever-present wooden faces like those of eyeless suits of armor. Now a dead man in his coffin was no longer a human being wrapped in a door.

All we knew of the dead was their weight. However, this varied enough to arouse in us occasionally a sort of suspicion that somehow seemed to rarefy this inert merchandise. Unconsciously one was led to think in terms of a soul. A dead body that felt exceptionally heavy or, on the other hand, too light, reassumed some semblance of personality, smuggled in, as it were, in that unexpected gap between the weight of an "average" corpse and that of this particular corpse.

Things had begun to take their course. We might have been tempted to open the coffin, to examine the inscrutable face, to question the dead man's friends and search out his past history. But it was too late; the coffin was being lowered on its ropes, it lay at the bottom of the grave on the two supports we had placed there; the body was laid down, laid on its andirons and already more than half consumed with oblivion, loaded with ashes.

By now our burial ground comprised seventeen graves. The flowers had grown and we were preparing to open a new section. It would be another row of graves set below the first, which lay alongside the edge of the forest. We were thus tackling the second third of the graveyard, for its limits had been strictly set from the beginning. Hardly two months had passed since our arrival at Brodno and we were already beginning to wonder if the graveyard would last out as long as we did, or if it would be full before we left the camp, so that someday we might find ourselves confronted with an overflow of dead bodies which would have to be disposed of in a hasty, slapdash, sacrilegious way.

We foresaw that times would have changed by then; we might have a snowy winter, for instance, full of urgent tasks, and the fortunes of war would be drooping like a bent head. The limits of the burial ground, in a word, were those of our future, of our hope; all summer was contained within them. This summer had begun radiantly.

Fed by marshlands, fanned by the great wind blowing off the plain, the forest was aglow with its thousands of tree trunks—beeches and birches for the most part—and its millions of leaves; carpeted with monkshood and borage, it projected into the middle of the wide wilderness, somewhat like a mirage no doubt, but above all like a narrow concession made to the surrounding landscape, to the past, and, in a more practical sense, to the dead.

We had gone through it now in all directions. As soon as a funeral was over Ernst and Otto would take us off to the hamlets that lay on the other side of the forest. There we would buy food from the peasants in exchange for linen and military garments from our Red Cross parcels; the barter took place swiftly and in silence, as though between thieves. We brought home a few fowls or rabbits, which we hurriedly slaughtered in the forest, using our knees and cursing one another, while the sentries guarded the paths.

We would wipe our hands with leaves. Then the dead creature had to be slipped inside one's trousers, between one's legs, held up by strings tied round the head and feet and wound round one's waist. The volume of the body, the feel of fur or feather, the temporary invasion of the lice they harbored, the smell of blood and a lingering warmth made us feel as though we were saddled with some ludicrous female sexual appendage, as though we had given birth to something hairy and shapeless that obliged us to walk clumsily, with straddled legs.

This stratagem enabled us to get back into the camp safely; I had no hesitation about resorting to it under the eyes of Ernst, since the very vulgarity of my movements, my grotesque gait, freed me from the mental complicity that bound me to him. It was a sort of revenge against despair. Sometimes, on the way back, we would catch a distant glimpse of the men and women at work on the new road. Ernst pointed out Lidia to me; she was wearing a light-colored dress and pushing a dump truck full of earth. Rain was threatening: "She's going to get soaked to the skin," Ernst muttered.

"Why Lidia?" I asked.

"Yes, why Lidia . . ." he echoed, in torment, his head downcast.

"I meant why are you particularly interested in her?" It was hard to keep up this tone with a barely dead fowl stuck between your thighs.

"One has to make a choice, Peter," Ernst murmured. "One can't suffer tortures on every side at once. Mind you, that doesn't mean . . ."

Otto was behind us, though far enough not to catch our words. I felt sure that if we went on talking like this he would soon shoot at a jay, a magpie, the first bird he saw. There's always one ready to fly off from an empty branch over your head, a little way in front of you or to one side. At every word, like involuntary beaters, we "put up" a covey of pretexts, birds with strange plumage and taunting cries. I turned round; Otto was smiling. Since I had begun carrying slaughtered animals between my legs his smile had grown broader; I was a fellow. Suddenly his smile froze: "Halt!" he called to me, pointing into the depths of the forest. Two figures were rapidly disappearing between the trees.

"Partisans," he said, hurriedly smiling again.

The word "partisans," as he spoke it, seemed to indicate some species of big game, rare and practically invulnerable, some solitary stag or boar shaking the unmysterious undergrowth with its startled gallop. Yet my heart had leaped. One morning, when we reached the burial ground, we caught sight of black smoke drifting over a hamlet in the plain. Otto explained to me that some partisans, having been given a poor reception there in the night, had burned it by way of reprisal. After that the open German cars that drove through Brodno assumed a warlike aspect, despite their gleaming nickel and brightly polished bodies. For beside the officers they were carrying sat two helmeted soldiers armed with Tommy guns.

We did not see them for long; the partisans seemed to have disappeared. In the villages, the Germans had distributed arms and formed militias. A watchman blew a bugle at the first alarm; it was a sort of long wail in the depth of the night which, generally without cause, made those who had taken refuge in treachery stir in their uneasy sleep. During the day Ukrainian policemen passed along the little path below the graveyard. They wore a black uniform with pink, yellow or white braid and badges of the same colors in checks, circles or triangles, like the gaudy signals of some obscure code hung out on the semaphores of terror.

The Germans acknowledged the stiff salutes of these liveried men with a nonchalant air. The Jews shrank back when they drew near: the Ukrainian policemen had the wild cruelty of certain sheep dogs, and above all their eyes and their voices recalled traditional pogroms. A few days earlier they had killed three prisoners who had escaped from the camp, shooting them down in the wood.

When they passed near us Cordonat would taunt them under his breath in the patois of Languedoc. These muttered insults showed the total disconnectedness of everything; what mazes of recent history one would have to explore to account for this absurd conflict between a Ukrainian peasant dressed in a stage uniform and a *vigneron** from the South of France who, before the war, used to vote anti-Communist and who was now talking, in his patois, about joining the Volynian partisans.

Otto made fun of the Ukrainian policemen out loud when they had gone past us, and embarked on a conversation with Cordonat in which gestures to a large extent filled the gaps in their vocabulary; after a lengthy exchange of naïvely pacifist opinions, they had ended by discussing hunting, poaching and mushrooms.

Ernst appeared to be growing somewhat mistrustful of his companion.

*Wine grower.

In order to be able to speak freely to me he evolved the plan of taking me almost every day to the sawmill to get our woodcutting tools sharpened. I had never expressed any wish for these tête-à-tête conversations. They only lasted, actually, during the time we took to walk from the burial ground to the sawmill. This sawmill was worked on behalf of the Germans by a Jewish employer and his men. I met there the man with the wounded face of whom we had caught sight a few days earlier in the girls' hut.

I had asked the boss of the sawmill for a glass of water, as he stood talking to us in front of the workshop door. He turned round and asked someone inside, whom I could not see, to bring me one. A few minutes later Isaac Lebovitch came up to me carrying a glass of water. It was a glass of fine quality, patterned with a double ring, the remnant of a set no doubt, and its fragility and bourgeois origin made me uneasy. In this token of hospitality I recognized an object which might have been a family heirloom of my own.

Isaac Lebovitch (he was soon to tell me his name) seemed to be about thirty years old. He had a long face with a lean beak of a nose. His dark curly hair, already sparse, grew low on his forehead. He stood beside us while I was emptying my glass.

"It may be that our hour has struck," the master of the sawmill was saying to Ernst. "It may be that the end of our race is in sight. There are things written up there," he added, pointing to the sky which was empty of birds, empty of hope.

I noticed a vine climbing round the door of the sawmill. In that region where vines had never been cultivated it looked strange, like a Biblical symbol.

". . . And yet," the sawmill boss was saying, "I myself fought during the last war in the Austro-Hungarian army. I was an N.C.O. and I won a medal. That means I was on your side, doesn't it?"

I felt embarrassed as I listened to his words. He was quite an old man, no sort of rebel against order or established authority, ready to accept a strict social hierarchy and even a certain degree of victimization to which his religion exposed him. . . . But not this, not what was happening now! These things were on the scale of a cosmogony. Or worse: they took you into a universe which perhaps had always existed behind the solid rampart of the dead, and of which the metaphors of traditional rhetoric only gave you superficial glimpses: where the bread was, literally, snatched from one's mouth, where one could not keep body and soul together, where one really was bled white and died like a dog.

Like novice sorcerers inexpert in the magic of words, we now beheld the essential realities of hell, escaping from the dry husks of their formulae, come crowding towards us and over us: the black death of the plague, the bread of affliction, the pride of a louse. . . . Seeing that Ernst was listening, with a look of deep distress, to the old man's words, Lebovitch plucked up courage to speak to me. He first addressed me in German, asking me whereabouts in France I lived, what sort of job I had, whether my relatives over there suffered as much from the German occupation as the people here did; then he suddenly spoke a sentence in English.

It was quite an ordinary sentence like "Life isn't good here." There was really no need to wrap it up in the secrecy of another language, since it was no more compromising than the words that had preceded it; and no doubt he had only had recourse to these English words for the sake of their foreignness. A language does not always remain intact; when it has been forced to express monstrous orders, bitter curses and the mutterings of murderers, it retains for a long time those insidious distortions, those sheer slopes of speech from the top of which one looks down dizzily. In those days the German language was like a landscape full of ravines, from the depths of which rose tragic echoes.

"No, life isn't good here," I answered in German. "But out there in the forest . . ."

"I don't go there any more. The other day . . ." Lebovitch showed me the dry wound on his forehead. "Besides, you saw . . ."

"Probably you didn't go far enough, you didn't venture into the depths. . . ."

"I had just gone to see Lidia," said Lebovitch, surprised by my remark. "I have no reason to go farther. You're liable to meet the partisans. . . ."

"That's just what I meant."

He stayed silent for a moment. We had moved a little farther off and had turned our backs on Ernst and the old man. Lebovitch stepped still farther to one side and, realizing that he wanted me to come away from the other two, I went up beside him.

"The partisans," he said with an anxious air. "I'll tell you this: we don't know much about them. If they saw me coming up to them they might shoot me down. And how could I live in that forest? I've got no strength left," he gasped, striking his thin chest with his fist. "And then there's so much violence, so much bloodshed every day, and all those farms set on fire. . . . After all, we're managing to hold out here. I've held out up till now. Perhaps we've been through the hardest part now. Listen, I may perhaps be dead tomorrow but I think it's better for me to save my strength. Don't you think that's best for you too?"

It would have been cruel to tell him that the dangers that threatened us seemed less terrible than his own, and I left him to his patience.

"That girl you call Lidia, who works on the new road," I remarked, "isn't she the one my sentry knows?"

"He ought to stop trying to see her. She told me so. It's likely to do her a lot of harm. I think she's had herself sent to the dump trucks on purpose because of him, so that he can't try to see her during the daytime. What's he want with her, anyway? He knows that relations of that sort are forbidden. He'll only get her hanged, and himself after her."

"He wants to help her."

Lebovitch grasped me by the arm, after making sure that the other two could not see him: "Let me tell you this, Frenchman: the Germans can help nobody, d'you understand, nobody. They couldn't if they wanted to, they couldn't any longer. Imagine a hedgehog struggling to stop being a hedgehog; you wouldn't want to go near him then!"

Ernst called me. The tools were ready. We had to get back to the graveyard. As I left Lebovitch I was careful not to tell him that we would soon be back. It was no doubt bad for him and his friends for us to be seen at the sawmill too often. We must each keep to his own solitude; fraternization had become conspiracy.

As we walked back, I spoke to Ernst about Lebovitch. "He's a friend of your Lidia's," I told him. "He's afraid that your interest in her may compromise her in the eyes of the German authorities."

"He's wrong," replied Ernst in an offhand manner. "You know I'm a pastor, and although I'm not a chaplain, the Commandant, who comes from my home town (only yesterday he was asking me for news of my family), gives me tacit permission to make some approaches to these people in a priestly capacity; to behave with a little more humanity, in other words. . . ."

"Humanity," I echoed.

"I know what you're thinking, Peter," murmured Ernst. "Well, even if your thoughts correspond to the reality, what about it? Won't you ever understand?" he cried, appealing, far beyond me, to some unknown body of critics. "We are all lost. There's nothing for us to fall back on; there never will be. Even the earth has begun to fail us. In such conditions, who can forbid me to love in whatever way I can? Who can forbid me? This is the last form of priesthood open to me, Peter—the last power I've got. It's inadequate and clumsy, it needs to be exercised upon a living object, a single object. . . ."

His lips went on moving. I said nothing. I would not have known what to say. And then Otto was already watching us come, from the top of the burial ground; the evening sun was behind his back and we could not see his face. He was standing motionless with his rifle on his shoulder. Behind him the graves were casting their shadows to one side, like beasts of burden relieved of their packs.

From that time on, discovering into what abysses I might be dragged if I followed Ernst, I fell back on the position of safety provided by the graveyard. Here was the only innocent place. Here we seemed to find a sort of immunity. When an officer came to inspect us we each bent as low as possible over a grave, assiduously weeding it as though pressed for time, without raising our heads, and the visitor refrained from speaking to us, wondering (probably for the first time in his life) if we were acting thus in response to some urgent appeal from the dead, such as he himself might perhaps have heard (it suddenly came back to him) in the middle of the night, in the days when he still felt remorse.

"The weeds are the white hair of the dead," Cordonat would say, and his words savored of that senile cult whose hold on us grew in proportion to the dangers that threatened us.

These dangers now assumed the shapes and sunburned faces of S.S. men and military policemen, a few detachments of whom had recently arrived in the region of Brodno. But it was above all the growing silence of that summer, the pallor of the sun at certain hours, the oppressive heat that secretly frightened us. Our religion, which had never actually been a cult of the dead, was becoming a cult of the grave. As we dug and then filled up our pits, we appeased some haunting dream of underground.

With our twenty-two dead, we had already opened up and explored a real labyrinth. We were familiar with its passages, its detours, its angles. It was a sort of deep-down landscape. We knew just where a tangle of hanging roots clutched clods of earth, where you could catch the smell of a distant spring, where you passed over a slab of granite. In the course of our work of excavation we had grown used to the coolness of this universe of the dead and we found our way about it mentally with the help of these particularities of structure rather than with the help of the names of the dead men who had drifted there accidentally, like foreign bodies.

This longing for the depths, unsatisfied by our task of weeding on the surface of the graves, impelled Cordonat and myself to try and open up a trench which would run a few yards into the forest and drain away the rain water that poured down on to the graves and scored deep furrows in their unstable earth. Since we now no longer left the burial ground to walk in the forest, which was patrolled by the soldiers who had recently come to Brodno, I was looking for an occupation.

Digging this trench had another very different result; it was through this narrow channel that I happened on something that I had been anxiously anticipating for many months. It began under my feet, like a forest fire. Right at the start, we had got on fairly fast with our trench and were now digging in the sandy soil of the forest, among the live roots of the trees. Below us our comrades were lying beside the graves, plucking the weeds from off the dead with one hand. Otto and Ernst were sitting in the shade of the trees, Ernst reading *Louis Lambert,** which had just been sent to me and which I had lent him. As we worked, Cordonat was telling me about his *landes.*† When we stopped to change tools, for we used pick and spade in turn, we divided up a little tobacco. The war seemed endless but here, at this precise moment, under this white silent sky, it had a flavor of patience, a flavor of sand; it bore the same relation to life as a fine sand to a coarser sand.

It was not with the pickax but with the sharp edge of the spade that I cut open the arm of the corpse. It was lying flush with the side of the trench and as I was leveling the walls I struck right into the flesh. It was pink, like certain roots, like a thick root covered with black cloth instead of bark. My blow had ripped off a bit of the sleeve. I started back, spellbound with horror. Cordonat came up and then called the sentries. Ev-

*Novel written in 1832 by Honoré de Balzac.

†Describing sandy soil in the Bordeaux region of France. While digging, Cordonat is probably reminded of his native landscape.

erybody was soon gathered round the unknown corpse. A little earth had crumbled away and his elbow and wounded arm were now projecting into the void; he was literally emerging from a wall.

"Cover it up with earth," said Otto.

"We'll make a cross of branches," murmured Ernst.

They went back into the graveyard with our comrades; the problem of burying this corpse was beyond them. It is easy, it is even tempting to throw earth on a dead body. Often, after our burials, we managed, using boards and spades, to push into the hole at one go most of the heaped-up earth that lay at its brink. Here, one would have had to cover over the arm that projected from the wall of earth, to enclose it in an overhanging recess. We therefore decided to fill up our trench and start it again lower down, skirting round the corpse. I did not know what name to give it. But all the indications (the color of the clothes, their "civilian" appearance) suggested that the body was that of a Jew who had been killed there before our arrival or during one night, or maybe on a Sunday, very hurriedly no doubt, in a hush like that of a suicide.

I felt slightly sick. It was very hot. We were working with fierce concentration now, in silence. A few hours later, when we were digging our trench at some distance from the corpse, Cordonat, who was wielding his pickax ahead of me, suddenly started back. A sickly, intolerable smell arose; he had just uncovered a second body. This one was lying at the bottom of the trench, slightly askew and concealed by a thin layer of earth, so that Cordonat had trodden on it before—surprised by the elasticity of the soil—he exposed its clothing and upper part of a moldering face.

I was overwhelmed by the somber horror of it and the truth it revealed. This was death—these liquefying muscles, this half-eaten eye, those teeth like a dead sheep's; death, no longer decked with grasses, no longer ensconced in the coolness of a vault, no longer lying sepulchred in stone, but sprawling in a bog full of bones, wrapped in a drowned man's clothes, with its hair caught in the earth.

And it was as though, looking beyond the idealized dead with whom I had hitherto populated my labyrinths, my underground retreats, I had discovered the state of insane desolation to which we are reduced when life is done. Death had become "a dead thing," no more; just as some being once endowed with great dignity and feminine mystery may, after a slow degeneration, surrender to the grossest drunkenness and fall asleep on the bare ground, wrapped in rags; here, the rags were flesh. Death was this: a dead mole, a mass of putrefaction sleeping, its scalp covered with hair or maybe with fur: wreckage stranded in the cul-de-sac of an unfinished tunnel: surrender at the end of a blind alley.

Cordonat discovered three more bodies. We had struck the middle of a charnel, a heap of corpses lying side by side in all directions, in the middle of the wood; a sort of subterranean bivouac which even now, when he had exposed it to the light, lost none of its clandestine character. We shouted, but in vain; this time nobody came to us. We turned over the earth till we were exhausted in an effort to cover up the bodies. We were practicing our craft of gravediggers in sudden isolation. And now it had assumed a wildly excessive character; we were grave-diggers possessed by feverish delirium. Night was falling. We had ceased to care who these men were, who had killed them or when; they were irregular troops on the fringe of the army of the dead, they were "partisans" of another sort. We should never have finished burying them.

Their very position close by our own graveyard cruelly emphasized its prudent orderliness. Our dead, meekly laid out in rows, suddenly seemed to exude servility and treachery, wearing their coffins like a wooden livery.

The appalling stench of these accidentally exhumed corpses persisted for a long time in the forest, and spread over our graveyard. It was as though our own dead had awakened for a moment, had turned over in their graves, like wild animals hazily glimpsed in the sultry torpor that precedes a storm. There could be no doubt that something was going to happen. The smell warned one that the tide was about to turn.

The thundery heat and the horror of my discoveries made me feverish. Back in camp, I lay prostrate for several days on the wooden bed. Myriads of fleas had invaded the barrack-rooms and were frenziedly attacking us, while in the shadow of my clothes I traced the searing passage of my lice. Towards evening Cordonat brought me a little water. This was so scarce that at the slightest shower all the men would rush outside clutching vessels, bareheaded, like ecstatic beggars, and when the rain stopped they sprawled on the ground, still jostling one another, round the spitting gutters. Then my fever dropped. When I went back to the burial ground the first trains had begun to pass.

A railway line ran over the plain that we overlooked from our mound. It was only a few hundred yards away. Until then we had paid little attention to it, for the traffic was slight or nonexistent. During my absence it had increased without my noticing it; these trains sounded no whistles. If they had, I should have heard them from my bed in the camp; I should have questioned my comrades and they could have enlightened me, for from certain windows in the building you could see a section of the line, beyond the station which was hidden by houses.

The first trains had gone past behind us, full of stifled cries and shouts, like those trains that pass all lit up, crammed with human destinies and snatched out of the night with a howl while, framed in the window of a little house near the railway, a man in his shirt sleeves stands talking under a lamp, with his back turned, and then walks off to the other end of the room. And now, from the graveyard, I could see them coming, panting in the heat of the day, interminable convoys that had started a long time ago, long freight trains trickling slowly through the summer marshaling-yards, collecting men on leave and refugees like a herd of lowing cattle.

I could hear their rumbling long before they appeared past the tip of the forest, and then when they were in sight (sometimes almost before) I could hear another sound, superimposed and as elusive as a singing in one's ears, a buzzing in one's head, or the murmur in a sea shell: the sound of people calling and weeping.

The trains consisted of some twenty freight cars sandwiched between two passenger coaches, one next the engine and one at the rear. At the windows of these two coaches (they were old ones, green, with bulging bodywork) stood uniformed Germans smoking cigars. All the rest of the train was an inferno.

The cries seemed transparent against the silence, like flames in the blaze of summer. What they were shouting, these men and women and children heaped together in the closed vans, I could not tell. The cries were wordless. The human voice, hovering over the infinite expanse of suffering like a bird over the infinite sea, rose or fell, ran through the whole gamut of the wind before it faded into the distance, leaving behind it that same serene sky, that store of blue that bewildered birds and dying men can never exhaust. On the side of each van* a narrow panel was open near the roof, framing four or five faces pressed close together, with other halves and quarters of faces visible between them and at the edges, the clusters of eyes expressing terror.

But it was more than terror, it was a sort of death-agony of fear; the time for beating their breasts was over and now they watched the interminable unrolling of that luminous landscape which they were seeing for the last time, where there was a man standing free and motionless in the middle of a field, and trees, and a harvester, and the impartial summer sun, while your child was suffocating, pressed between your legs in the overcrowded van and weeping with thirst and fright. Here and there a child was hoisted up to the narrow opening. When its head projected the German guards in the first carriage would fire shots. You had to stand still there in front of the opening and bear silent witness to what was going on in the dense darkness of the van: women fainting, old men unable to lie down, newborn babies turning blue, crazed mothers howling—while you watched the symbols of peace slowly filing past.

The trains followed one another at short intervals. Empty trains came back. Beside the narrow openings the deported victims had hung vessels in the hope of collecting water—messtins, blue enamel mugs—like pathetic domestic talismans which a mocking Fate kept jingling hollowly as the train disappeared in the dusty distance. They had no thought of displaying sacred draperies or waving oriflammes† at this window; death was yet another journey, and they set out armed with water bottles.

Empty trains came back. I recognized them by the sound of the engine's panting.

"Now then, get busy, boys," Otto would tell us.

We had been given a third sentry. Brodno was crammed with troops and they had to be made use of. Ernst was some distance away from us, pale, his lips tight.

* Boxcar.

† Brightly colored banner, sometimes symbolizing devotion or courage.

"Do you know why they're being taken off and where they're being taken to?" I asked the sentry, whom I did not know.

"Delousing," he answered calmly. "Got to make an end of this Jewish vermin. It's quickly done. It happens some thirty miles away. I'm told it's with electricity or gas. Oh, they don't suffer anything. In one second they're in Heaven."

This man, as I learned later, was an accountant from Dresden. He might just as well have been a blacksmith from Brunswick, a cobbler from Rostock, a peasant from Malchin, a professor from Ingolstadt, a postman from Cuxhaven or a navvy* from Bayreuth; he would have used the same language. And he did use the same language under all these different aspects, shifting from one to another like an agile actor impelled by Evil, altering his voice to suit each of these thousands of masks, imbuing it with the atmosphere of profound calm appropriate to Ingolstadt, Malchin, Bayreuth and countless other equally humane cities, and repeating, "Oh, they don't suffer, they don't suffer!"

Trains came down from the far depths of Volynia and the Ukraine, loaded with death agonies, with tears and lamentations. At one stop, farther up, the German guards had tossed dead children onto the roofs of the vans; nothing had to be left by the way, for each train was like the tooth of a rake. High up in the wall of the van, a little to the left in the narrow opening, there was a face; it seemed not living, but painted—painted white, with yellow hair, with a mouth that moved feebly and eyes that did not move at all: the face of a woman whose dead child was lying above her head; and beside the opening the little blue enamel mug, useless henceforward, shaken by every jolt of the train. Death can never appease this pain; this stream of black grief will flow for ever.

Towards the end of the morning I succeeded in drawing near to Ernst, who had moved away from the other two soldiers.

"It had started three months ago, at Brest Litovsk, when I was there," he told me in a low, tense voice. "But it wasn't on this scale. In a few days, tomorrow maybe, they'll begin on the people of this place. Do you think she ought to go away? To take refuge in the woods?"

He was looking at me bewilderedly, seeming more like a priest than ever with his smooth, babyish face, his indirect glance.

"It's probably the only chance they've got left," I muttered, bending towards the turf on the grave, since the new sentry was slowly coming towards us.

"But you've seen her: she's not strong, she'll never stand up to such an ordeal!" cried Ernst, without noticing that the soldier was now standing quite close to him.

The soldier looked at Ernst in some surprise and then went off, humming. I stood silent.

"Won't you speak to me, Peter?" asked Ernst humbly. "You're saying to yourself: He thinks only of her. . . ."

*Unskilled laborer.

"I mistrust the other sentry," I answered. "You're exposing yourself unnecessarily. If you're willing to run the risk, throw away your gun and go off into the forest with her."

"I've thought of that," said Ernst, hanging his head. "I've thought of that. And then nothing gets done. You're horrified and you stay where you are. In this war, every man looks after himself. But we ought to realize that there can't be any true life afterwards for us, who have endured these sights. For me, there'll be no more life, Peter, do you hear, no more life. . . ."

His two companions called him and he went off abruptly. I lifted my head. Down below, there were only empty trains passing along the line. Towards evening all traffic ceased. Ernst did not speak another word to me as we went back to the camp. And I never saw him again.

Next day, when our gang turned up at the guardhouse, the Germans sent us back to our barracks. Soon rumors reached us: the Jews of Brodno were going to be taken away in their turn. Towards noon, looking out of the windows of our building, we saw the first procession of doomed victims appear on the little road leading to the station. Many of them—nursing what hopes, trusting in what promises?—had brought bundles and suitcases. The hastily knotted bundles frequently let drop underclothes or scraps of cloth that nobody had time to pick up, since the soldiers were continually hurrying on the procession, with curses on their lips and rifle butts raised. Other victims, arriving later, thus found themselves confronted with a scene of dispossession whose causes were as yet unknown to them, with the signs of ominous disorder.

The same signs were visible within their own group, where old men, children and adults were mingled; clearly there had been no attempt to sort them out, as had always happened hitherto before the removal of groups of workers or some other utilitarian deportation. The German soldiers from time to time struck at the sides of the column, but as we could only see them from a distance their gestures seemed to be slowed down: silent, clumsy blows, aimed low, more like stealthy misdeeds than like acts of violence. I had turned away from the window.

"Here are more of them!" somebody cried behind me.

Should I see Lidia, her friends, Lebovitch or the old man from the sawmill in this group, or the man who mended the road, or the fair woman who often stood waiting between two houses? Their packs made their silhouettes misshapen. Some of the women hugged them against their stomachs like bundles of washing. The dust was rising and I could not make them out clearly.

Somebody said: "It's their children they're carrying. . . ."

Somebody said: "One of them has fallen. The guards are hitting him. . . . Now he's up; he's starting off again. . . ." Somebody said: "Oh, look at that woman running to catch up with the group!"

The untiring commentaries, despite transient notes of pity, disclosed a sort of detachment, for passionate feeling will not allow you to see things through to the end, whereas these men followed the whole business with the mournful eagerness of witnesses. Evening drew on. I lay stretched out on the wooden bunk. At my side Cordonat was smoking in silence. Later on, flickering lights and distant rumblings rent the night. It was a stifling night, tense with anguish, and you could not tell whether those distant flares came from thunderclouds or from armed men on the march, carrying torches.

The arrests went on all next day. We were told by sentries that the cottage doors had been smashed in with axes and the inhabitants cleared out. Towards noon stifled cries were heard from the direction of the station. The victims, who had spent all night packed together in the vans standing in the sidings, were clamoring for water. By evening not a single train was left in the station. One man was walking along the track, bending down from time to time to check the rails. He went off into the distance, till he was almost invisible. Beyond, the pure sky grew deeper.

It was not until three days later that we set out once more on the road to the burial ground. Two new sentries accompanied us. The road was empty, and most of the houses shut up. In one of them some Poles were setting up a canteen. A few men were working in front of the sawmill. I recognized none of them.

At least we still had our dead, that faithful flock, each of whom we could call by his name without raising up a murdered man's face staring wide-eyed in the darkness. Our dead were already beginning to get used to their earth. After each of our absences they had "put on green," as we said when the grass had once more invaded the graves. Our cult of the dead consisted in wiping away the shadow of a meadow, day by day. And once more the sentries surrendered.

They had come to the graveyard armed with mistrust. But here, amongst the dead, we got the better of them. In front of our tidy graves, so tirelessly tended, their antagonism dropped: our accounts were in order. The liquidations (in the business sense of the word) which the Germans were carrying on all round us took place only spasmodically, as though at an auction, in an atmosphere of anger and excess which, once the payment had been exacted, increased the insatiable credit of the murderers with a debt of bitter resentment; they found it hard to forgive their victims. Our dead, on the other hand, had needed no dunning. Although the Germans were never paid fast enough, these had not worn out their patience like the others; nobody was conscious of having forced them to die.

Then, too, they seemed to be lying at attention under their three feet of earth in properly dressed lines, whereas the others, who had been shot point-blank, hid their heads in the crook of their arms; they had to be pulled by the hair, in fact it was an endless business bringing them to heel.

When the charm of the graveyard had worked, I plucked up courage to ask one of the two soldiers if he knew what had become of Ernst.

"Oh, the little pastor!" he replied. "He got punished. He's been sent to a disciplinary company. He knew a Jewess. They even say he burst out crying in the Commandant's office. Oh, don't talk to me about such people! Anyhow, I'm a Bavarian myself, so you see I'm of the same religion as you. I'm a Catholic, yes, but one's country comes first! Now get along and see to your graves."

I had not had time to inquire about the fate of Otto. In any case I was not deeply concerned about it. I felt calmer; the punishment inflicted on Ernst seemed a light one. Moreover, it cemented our friendship more firmly. He stood beside me now in the rebels' camp, like those guests at a party whom you've expected for a long time, wondering whether they'll come, and who suddenly appear, all made up and grotesquely disguised, with their familiar kindly, serious eyes looking at you from under their hats, when the music has already started and the dark wine is being poured out all round you.

All this was happening in 1942. Our friendship needed some discipline, imposed from without, naturally. I was thinking of Lidia, too. Her relations with Ernst seemed to me solid ground, but there were stars, too. They still wheel round in my dreams. I do not know what became of them on earth, but in the map of heaven, and in the map of my heart, I know where to look for the distant glimmer of that unattainable love.

We never went back to the new road. Our kingdom grew narrower. No more walks in the forest; we went down to the pond only with sentries on either side of us, and we were forbidden to linger there; we were more than ever confined to the burial ground. Otto and Ernst had already passed into a previous existence, threatened with oblivion, when we got news of the former: he sent us a corpse.

This was a fellow that I knew fairly well, a sullen-faced lad from Lyons obsessed by the urge to escape. Seeking an opportunity, he had asked me a few days earlier to let him take the place of a member of our gang who was too sick to leave the camp. He hoped to be able to make an easy getaway from the graveyard. I managed to get him accepted by our sentries. Underneath his uniform he was wearing some sort of escape outfit, and round his ankles there dangled long white laces belonging presumably to the linen trousers formerly issued to the French army. These were only too obvious. Silent, his jaws stiff with anxiety or determination, he thus attracted the guards' attention immediately. They never took their eyes off him and, that evening, he came back to the camp with us.

A few days later he got himself engaged in a gang of "road commandos" (prisoners employed on road-mending at some distance from the town) guarded by several sentries, among whom was Otto. Otto, being indirectly involved in Ernst's punishment, had been posted in charge of this detachment, which performed the duties of a gang of convicts and was rated as such. One evening, as the group was about to pass through the town on its way back to the camp, the young Lyonnais broke ranks abruptly and began to run off into the fields. Otto promptly shouldered his gun and fired. The fugitive dropped; the bullet had gone in through his back and pierced his heart.

It was our first violent death. On the day of the funeral the Germans sent a wreath of fir-branches tied with a red ribbon. The wooden cross bore the inscription FALLEN instead of DIED. Our graveyard had been sorely in need of this heroic note, as I suddenly realized; it was as though it had received the Military Cross.

"You shall soon have some stones, too," a sentry told me. "I know that in the camp they've been asking for a stonemason. It's the Jews' tombstones you're going to get; their graveyard's full of them, down there in the forest. We shall use them for the roads too. You people can't complain, your graveyard's much finer than the one where our own men are buried. . . ."

It was quite true. With its always green turf, its flowerbeds, its carefully sanded paths edged with small black fir trees which we had transplanted, with its rustic fence of birch boughs, against the dark background of the forest edge, our graveyard seemed an "idyllic" place, as the Germans put it. On Sunday the soldiers from the garrison used to come and photograph it. The more we adorned it the greater grew its fame, and it aroused a wave of curiosity like that which carries crowds to gaze at certin baroque works of art or at others which, devoid of any art, are yet prodigies of patience and skill: houses built of bottle ends, ships made of matches, walking sticks carved to fantastic excess—monstrous triumphs of persistence and time.

On the fringe of the war, on the fringe of the massacres, on the fringe of Europe, sheltering behind our prodigious burial-ground, we seemed like hollow-eyed gardeners, sitters in the sun, fanatical weeders, busily working over the dead as over some piece of embroidery.

But the thought of those stones horrified us. We did not want to rob the Jews of their gravestones; that savored of sacrilege, and also of an incipient complicity with the Germans; and anyhow it "wasn't playing fair." We had made our graveyard out of earth and grass and to bring in marble would have been cheating. I went to lay the problem before the prisoners' French representative, the *homme de confiance* as we called him. He promised me to protest to the Germans.

A few days later, in one corner of the camp, I saw a man sitting on the ground sawing and planing a tombstone on which was carved a breaking branch.

He seemed in some doubt. Should each of the slabs of marble or granite intended for the graves (merely for epitaphs, not for funerary flagstones) represent an open book, a cushion (he could quite well picture a cushion slightly hollowed out by the weight of an absent head) or a coat of arms? He asked for my opinion. I told him angrily that I wasn't interested in his stones, and that they should never cross the threshold of the graveyard. However, he went on cutting them, full of delight at getting back to his trade, heaping them one on the other when they were ready, although the Germans were now busy with something else and never came to ask for them. In any case, he could not have handed them over as they were, for the essential part of the inscription was missing.

Though my ill-will discouraged him at first, he soon tried to get from one of us the names of the dead and the dates which he needed. He met with the same refusal. Besides ourselves, only the Germans possessed these essential facts, but no prisoner had access to their offices. Thenceforward, he began a patient investigation, going from one building to the next, questioning the men on the deaths that had taken place there. I soon got wind of this. It made me angry. The stonemason's obstinacy, although of a purely professional character, had begun to look like an intrusion into our field. For the last few months we had managed to keep our tasks secret, and now I felt this secrecy threatened by the publicity that he was causing by his noisy investigation of our past.

One day I saw him turn up at the graveyard, following a funeral. The man we were burying that day had come most opportunely in the nick of time, after the lad from Lyons, who, having introduced an appropriately heroic note into the place, had since seemed to be inaugurating a sequence of violent deaths. For here the latest corpse set the tone; it was in front of the latest corpse that we made our military salutes each morning; it was his bare, sparsely sown grave that attracted the attention (albeit vacuous) of visitors. It seemed important therefore that the tone thus set should not be, for too long, that of violent death: after all, a habit is easily acquired. And so I should have been quite glad to welcome this newcomer who restored order to things, had I not perceived, among the handful of Frenchmen walking behind the cart, the stonemason, sporting a swordbelt.

He had probably managed to slip into the procession by posing as one of the dead man's friends; he may even have been one really. . . . But we were already convinced of one thing: he had come to the graveyard to take measurements and pick up names. We decided to keep an eye on him, and we buried the corpse in a state of nervous tension although, in the depths of our innocent hearts, we had longed for it. We did not take our eyes off the man during the whole ceremony which, with its dying bugle-call and the report of arms fired into the sky, was like that which marks the close of a war. When the burial was over the Germans and the Frenchmen who had formed the procession hung about. As the fame of our graveyard increased, funerals had tended to become, for those who were able to attend them, a sort of summer excursion, a trip into the country from which you might well picture yourself returning with armfuls of flowers picked on the spot. The stonemason had moved towards the first row of graves. I followed him. He was already taking a notebook out of his pocket.

"What are you doing there?" I asked him, my voice distorted with anger.

"I've been told to make tombstones!" he said, very loud. "You shan't prevent me! They don't belong to you, after all!" he added, indicating the graves of the dead.

"More than to you, anyhow!" I answered. "We wouldn't go and put stolen stones on their graves. . . ."

"Stolen?" He shook his head. "They're given to us and it's not our business to ask where they come from. . . . Well, I know, of course!" he went on, seeing that I was about to answer. "And so what? Would you rather see them laid on the roads?"

"Yes, I'd rather see them laid on the roads. I suppose you get soup and bread from the Germans for doing this job?"

"Oh, don't you talk about that!" cried the stonemason. He brought his angry face close to mine. "Everybody in the camp knows what you've wangled with your famous graveyard!" He stopped suddenly.

"Well then, tell me! Tell me!" I cried.

"Listen to me," went on the stonemason in a quieter voice. "Can't we make peace?"

Everything seemed conducive to this. The heat of the hour had led Frenchmen and Germans, in separate groups, to sit down beside the graves, where the forest trees cast their shade. Only our words disturbed the silence.

"Don't expect to get the names of the dead, whatever happens," I answered.

"You can keep them," said the stonemason, sitting down. He was a thin-faced man a little older than myself, with short grayish hair. "We're an obstinate pair. After all, I understand your feelings; you don't want it to be said that those stones had names taken off them in order to carve these on. Well, I'd never have had the courage to take them off myself if they hadn't been written in Hebrew. But in Hebrew they mean nothing! I'm not even sure that they were names and dates. And then the stones were there in the camp, without anybody lying beneath them. Put yourself in my place! I'm bored, I need to keep my hand in for after the war, and I'm presented with tools and stones! But it's all right; I give up the names," he added, pulling out his tobacco pouch and handing it to me.

I rolled a cigarette and he did the same.

"I give up the names," he went on, puffing at his cigarette, "but all the same, something really ought to be done with those stones—they're not all in good condition, you know, some of them are molding away. Besides, the dead people or their relations, if there are any left, wouldn't see any harm in our making use of these stones now that the Germans have pulled them up. They'd surely like that better than to see them crushed and scattered on the roads. . . . After all, we're on the same side as they are, aren't we? . . . So this is what I thought. Let's not talk of carving names on them, but let's make them into little ornaments, cornerstones for instance. We must think of our own dead too. I needn't comment on what you've done for them; it's quite unbelievable. But after all, there's nothing permanent about it. Suppose they take us off into Germany next month; after a single winter there'll be no sign of your graves. The rains and the melting snows will have washed all the earth away, and grass and briars will have grown over it. But with stones . . . Oh, I'm not suggesting making monuments," he cried, raising his hand to forestall objections. "I assure you, very little is needed. And I'm speaking from experience—I'm in the trade. A little pyramid at each corner of the grave for instance, or a ball on a little pedestal if you prefer, although that's much harder to make, or else carved corners joined together with chains, only unluckily we haven't any chains . . . well, you get the idea, something not very high, firmly fastened into the ground, preferably with cement, and above all something decently made. . . ."

But I was no longer listening to him. For the last few minutes I had been listening to the rumble of a train and now it was growing louder. The train was about to emerge round the tip of the wood. I could tell, without waiting for it to roll past before my eyes, what sort of freight it carried. Its slow, jolting sound warned me of the other sounds that would follow although for the moment a contrary wind delayed them. I should soon hear the weeping, the cries of despair. The silence, no doubt, was due to the wind; but perhaps, too, those who were being transported, knowing what fate awaited them, had deliberately refrained from sending out their lamentations into that empty, sun-baked plain, in which the great migrations of death had never yet awakened any lasting echo.

And so it all began again. Every day one or two convoys crossed the plain, and then were no more to be seen; and when night fell a train would rumble, too slowly, through the silence. New processions appeared on the little road that led to the station. They were smaller and more infrequent than the previous ones and seemed to be made up of belated recruits, of survivors from some ancient and now almost forgotten disaster, of beggars or vagrants rounded up in the middle of their wasted summer. Nothing rolled out of the bundles this time; nobody seemed to be in a hurry now, and the soldiers who struck at the sides of the column did so with the lazy indifference of cowherds.

The massacre was drawing to a close, but it lingered interminably like the raw gleam of a lurid sunset on walls, between patches of shadow. We said to ourselves: "Surely this must be the end of these torments." The plain seemed to have nothing left to offer death save its quota of vagrants. We were wrong. On the contrary, those whom the Germans were now dispatching to be slaughtered had been taken from the ranks of a scattered resistance movement which, up till now, without our knowledge, had been fighting to the last ditch for the right to live.

If they seemed wearier than those who had gone before them along the same road that led between darkly gleaming slagheaps and through engine-sheds to Calvary, it was because they had suffered a twofold defeat. They had been surrounded in the forest, they had been arrested at night on the roads or among the brambles in the ravines, where for so long they had been wandering round and round in dazed despair. Now, in the evenings near the graveyard, we often caught sight of nonchalant armed soldiers making their way in extended line into the forest, and bending down from time to time to pick a strawberry. Brodno was encircled with a military cordon, and the whole region was being combed step by step.

This state of siege, of which we shortly felt the oppressive effects, brought back a certain animation around the camp and around our graveyard. The inhabitants—Poles, Ukrainians and Ruthenians—realizing that for the moment the Germans had no designs on them, suddenly felt the need to move around in all directions within the circle that hemmed them in. They came to look at the graveyard, the fame of which had reached them. They stood still at some distance from it, motionless, communicating with one another by gestures.

One evening two girls came forward as far as the verge of the forest, close to the spot where we had discovered the charnel.

One of them was plain and awkward; the other was slender and seemed younger. The brightness of her blue-green eyes disturbed me. They shouted to the soldiers that they were Polish. The setting, the gathering dusk and my own troubled heart made the younger girl's smile seem like that of a vision. To the German who, with one foot on a grave, asked them their names and ages, she called out "Maria!" and there was still joy in her voice. I had gone up to the sentry; the girl looked at me and waved to me as she went off. Next day as we made our way to the graveyard I caught sight of her at the door of the new canteen and we made signs of greeting to each other.

From that time on I clung to her image. The first rains of autumn had begun; the graveyard had ceased to be a garden and was once more a burial ground. The earth sank under one's feet; it was deep again, and heavy, like a morass. Autumn promised to be a season rich in deaths. More trains came through from the further end of the plain, with white faces framed in the narrow windows, trains full of condemned creatures who, this time, uttered no cries of thirst but stood motionless, clutching their despair between their hands like a twisted handkerchief. My mind was fixed on her image.

I turned to it again when new processions appeared on the road leading to the station or when, towards evening, groups of soldiers made their way into the forest with more speed than usual. Three or four times a day, whether we were going to the graveyard or coming away from it, Maria, standing at the door of the Polish canteen, would watch me thoughtfully and smile at me. My friends nudged one another but let fall no word. From their dealings with the dead they had learned to respect mysteries. And this was undoubtedly a mystery—my devotion to the image of this girl about whom I knew nothing to suggest that she was worthy of it surprised me more than it surprised them. I pushed back my hair from my forehead—it was damp from the drizzling rain—and stood upright; there were still many things to which I should have to bear witness later, many sufferings to be shared, many hopes to be nurtured, many steps to be taken which would add up to something someday. But as soon as I had reached the graveyard and was standing under the branches at the edge of the forest, where raindrops rustled, the image recurred.

One morning I was sunk in this sort of reverie when Cordonat called me. He led me to the end of the last row of graves where, covered with planks and with an old tarpaulin to keep off the rain water, the spare grave lay empty, awaiting its corpse. He had just been inspecting it and had found there, besides more subtle traces which, as an expert poacher, he had picked out, a cigarette end of unfamiliar origin. It was rolled in a scrap of paper from a child's exercise book and made of coarse unripe tobacco, presumably taken from one of the plants which, in those penurious times, the peasants used to grow outside their houses.

"It wasn't there last night," Cordonat told me. "For several days I'd noticed that somebody had been taking up and putting back the planks and the tarpaulin during the night. So I began to keep an eye on the grave. The other proofs are more tricky, you might not believe them. But this one! There's no possible doubt about it: a man comes to sleep in our grave at night."

"Well, what then?" I asked.

"Well, so much the better," he cried. "It must be a hunted man. For once, let this graveyard be some use to a living man!" (Cordonat had never been really enthusiastic about the graveyard.) "Only we ought to help him. This evening, for instance, we might leave him some provisions."

I did not doubt Cordonat's charitable intentions, but I also suspected that he was anxious to secure a further proof by this method. We had lately been receiving a little food from France and we were able, without too great a sacrifice, to deposit in the grave for the benefit of the stranger who inhabited it a handful of sugar, a piece of chocolate or a few army biscuits deducted from our store of provisions, which were meticulously arranged and counted, dry and crumbly as rats' provender. When evening came, before leaving the graveyard we slipped a parcel between the planks and then replaced them as before. Next day the parcel had disappeared. Cordonat found at the bottom of the grave scrap of paper on which a message of thanks was penciled, in English. The two words had been written with a trembling hand, no doubt by the first light that filtered through the parted planks as dawn brought back panic.

That evening our gift was made up merely of a few cigarettes. I added a message: "Who are you?" Although the answer consisted only of two initials, it was clear and it did not surprise me. It was written on a piece of packing paper: I.L. An arrow invited me to turn over the page: "You know me Peter = [the arithmetical sign *equals*] I know you. Keep quiet, both of you, keep quiet. Thank You." I had recognized Lebovitch. But how could he know that it was I who had put the cigarettes there and that only one other knew the secret?

"During the day he must stay hidden by the edge of the forest and watch us," said Cordonat.

It would have been madly imprudent, and the Germans would have discovered him long ago; Cordonat admitted it. These communications through the trap door of a grave, these notes with their anguished laconicism, the condition of "semi-survival" in which Lebovitch existed and the second sight with which he seemed to be endowed—all these things concurred to give me the impression that our continual contact with death was beginning to open for us a sort of wicket gate into its domain. I almost forgot Maria. Sometimes, in the evening, she would walk a little way along the road below the graveyard just as we had slipped a few provisions into the grave and were replacing the planks, the tarpaulin and the stones that held it down with furtive care, as though we were laying snares. Cordonat rediscovered the pleasures of his poaching days. I would stand up and wave to Maria. The sentries were amused by my performance and it distracted their attention from the mysterious tasks which we were performing over the empty grave.

"How do you live?" I wrote to Lebovitch, leaving him a few sheets of blank paper. His answers grew longer but also more obscure. He lived with difficulty. During the day he remained hidden, no doubt, in the high branches of a tree, for he wrote: "I am very high up. Do not look for me. A glance might betray me. I see them come to and fro. Please tell me what is happening about the dogs [these last words were underlined]. How soon will autumn be here? Have I the right to try and escape from God's will? Anyhow all this is unendurable and I shall not hold out much longer! If only they would let me speak! I should be exempted. Yes, they should let me speak! Have you heard tell in the village of anyone being exempted? Keep quiet! Thank you."

During the days that followed I had difficulty in preventing Cordonat from staring up at the tree tops on the forest border; he would have attracted the guards' attention. Instinctively, as one accustomed to roaming the woods and starting animals from their lair, he found it an exciting game to hunt for Lebovitch's aerial shelter. One morning I surprised him sitting down with his back turned to the forest and staring into a pocket mirror concealed in his hand which he was slowly turning in all directions.

"I tell you he's not found a perch in any of these trees," he told me, putting back his little mirror into his pocket. "Even if he'd been well camouflaged I'd have discovered him. There's no foliage thick enough to conceal a man. He'd have been obliged to surround himself with other branches, cut off from other parts of the tree; and believe me, I know from experience that the color of leaves changes as soon as they're cut. Ask him about it, once and for all. . . ."

I wrote a note to Lebovitch to this effect, accompanying it with a handful of army biscuits.

"I can't tell you where I am," replied Lebovitch. "You haven't told me anything about the dogs. And the exemptions? Do they ever exempt anybody? Yesterday three more trains went past. During the night there were luminous things drawn on the carriages! You can have *no idea of it* [the last words were underlined]. This morning I vomited because of all the raw mushrooms I'd eaten. God rises early just now. So does the wind! Couldn't they have pity on me? Tell me if it's humanly possible?"

"We shan't learn anything more," said Cordonat when I had read him this letter.

The incoherence of these notes depressed him. We were too close to the world beyond death not to be aware of its dank breath when speech became so sparing and sibylline, when a human being's presence proved so elusive, while these brief messages expressed a tortured silence pierced by a thousand exclamation marks, like nails. We continued to offer food to this Egyptian tomb, which a couple of days later had to remove farther off; death provided Lebovitch with a new neighbor.

Cordonat offered to dig the spare grave into which, that same evening, Lebovitch would creep to rest, and at the bottom of the grave he arranged a little pile of earth for the sleeper to lay his head on. I helped him in this task which, as we soon admitted to one another, filled us with a strange uneasiness; we had the feeling that we were preparing to bury an unseen friend. The present that we left in the grave that evening was more generous than usual. I avoided putting any message with it, however. The tone of the answers distressed me.

Lebovitch broke silence only on the second morning (he must have written his notes during the day, up in his tree or inside some unknown retreat). He had been deeply touched by Cordonat's thoughtfulness in providing the earthen pillow. Perhaps, also, by the nearness of a newly dead Frenchman, whose burial he must have watched, since he never took his eyes off the graveyard. He wrote: "I know that one day there will be no more morning. Last night I went on knocking for an hour against the earth, on the side where all the others are lying. I say an hour, but my watch has stopped. I wanted to go on knocking all night. As long as I'm knocking I'm alive. Even here, where I am now, at this moment, I'm knocking. And I keep saying: have mercy, have mercy! They've killed them all, Peter, killed them all! What is loneliness?"

I could no longer keep up this dialogue, and I could hardly bear the abstract presence that now filled my narrow universe. I could no longer look with confidence at the forest trees or into the hollow grave, nor gaze out over the plain where one of those trains was always dawdling. Like those tireless birds that drop to the ground like stones and as soon as they have touched it dart back to perch on one of a hundred quivering branches, then dizzily gravitate to the ground once more and once more rebound upward, as though in avid quest not of earthly or aerial prey but of pure trajectories, of secretly deliberate flights, of prophetic tangents, or as though irrevocably doomed to this endless to-and-fro, Lebovitch moved between the grave and the treetops every day; he was only fit for shooting down.

Yet I would have liked to save him. His talk of exemptions, although I had at first put it down to insanity (and I was beginning to find out how rich and full was insanity's account compared with the meager bankbooks of reason), had made an impression on my mind. I felt I must sound the Germans, or Maria, who doubtless knew what was happening in the village now.

The Germans told me that the fate of the Jews of Brodno (and elsewhere too) was old history now. "Let's talk of Maria instead," they said to me. "You're keen on her, aren't you, you rascal?"

I endured their mockery. As it happened I was anxious to speak to Maria. If she passed along the road tomorrow, might I not beckon to her to come? I'd only want a couple of minutes with her. I felt infinitely sorry for myself. Evening with its swift black clouds was falling over Volynia and its crowd of dead, over the distant fires of war, and I was standing there, eager to strike my pitiful bargain—two minutes of that time!—I was standing there with Lebovitch at my back, weighing me down, his hard hands against my shoulders. The Germans made fun of my anxiety. "Maybe," they said, sending me back to my graves. I slipped a note into the pit: "In two days I'll know for sure about the exemptions."

Next evening Maria appeared on the road with her friend. "You can call her," cried the sentries, laughing. I called her. She saw me. I beckoned to her to meet me at the graveyard gate. But she shook her head with a smile. She walked away. I had turned white with vexation to which, in the depths of my heart, I gave a bitter name. On the new message that I found in the grave Lebovitch wrote: "I'm knocking harder than ever, Peter. It's the only thing to do: knock, knock, knock!" He said nothing more about exemptions.

A few trains full of condemned victims still passed through the plain. When the sound of them had faded away I still seemed to hear behind me the dull persistent rhythm of blows hammered against the earth, against a tree trunk, mingled with the throbbing of my temples, the tapping of summer's last woodpecker, a far-off woodcutter's blows and the rumble of a passing cart, in a soothing confusion.

Summer drew to a close. In the darkened countryside all life was slowed down; even the great convoys of death became more infrequent—those harvests, too, had been gathered in. But the dawn of a new season was less like the morning after a bad dream or the lucid astonishment of life than the final draining away of all blood, the last stage of a slow hemorrhage behind which a few tears of lymph trickle, like mourners at life's funeral. Autumn brought a prospect of exhausted silence, of a world pruned of living sounds, of the reign of total death. What had I left to delay this consummation, when every gust of wind in the branches of the trees, every leaf blown away, every corner of the naked sky, reminded me of its imminence?

It was from that moment that the life of Lebovitch, his dwindled, precarious existence, became for me the last remaining symbol of a denial of death—of that death which was so visibly being consummated all around me. I renewed our dialogue. I sent him urgent messages: "Where were the rest? What had happened to him? and indeed, who was he?" I urged him to tell me about his own past and that of all the others.

For nothing makes you feel so impoverished as the death of strangers; dying, they testify to death without yielding anything of their lives that might compensate for the enhanced importance of darkness. Thus what did he know of Lidia's fate? He must tell me; the survival of all my hopes depended on it.

These questions remained unanswered. Lidia sank in her turn, with all the others, like them consecrated to death, behind those distant horizons of memory where, even after we have forgotten everything, there lingers a pale light, an endless comforting twilight, a thin streak of radiance which will perhaps serve us for eyes when our eyes are closed in death. Lebovitch soon caught up with Lidia on that dark slope where never, not in all eternity, should I be able to reach them.

One morning on arriving at the graveyard we found that the planks covering the empty grave had been thrown to one side. At the bottom of the grave there lay a black jacket without an armlet. One of its pockets was full of acorns. I knew then that Lebovitch would never come back. Had he been surprised in his sleep or, in a fit of madness, had he suddenly rushed out into the forest to meet his murderers? I raised my head. Clouds were rising towards the west. The wind had got up. My companions were taking away the dead leaves that fell on the graves. In the flowerbeds the summer flowers had turned black. Soon no more convoys passed along the railway over the plain, which in the mornings was drowned in mist. There were no more soldiers to be seen patrolling the woods, where the trees were growing bare. Autumn was really there now.

Maria chose this time to visit the burial-ground. I had been waiting for her there for a long time, secretly convinced that she would come. One evening she came along the ill-paved road, her thin summer dress clinging to her in the wind. She walked slowly, her face uplifted, quietly resolute, and her fair hair was fluttering over her brow. A violent fit of trembling possessed me. I turned towards the single sentry who now guarded us. He was a prematurely old man, full of melancholy resignation. He knew about my romance, and nodded his head.

I darted towards Maria like a dog let off the leash. Seeing me come, she hurried forward without taking her eyes off me and, passing the gate of the graveyard, quickly made for the edge of the forest. The sky had grown dark and the wind was blowing stronger. I only caught up with her under the trees. I seized hold of her arm, and she drew me on involuntarily while I spoke to her in breathless tones. I did not know what I was saying; I was in a sort of ecstasy. Suddenly I drew her to me and pressed her closely. My face groped feverishly for the hollow of her shoulder. For so long I had been waiting for this moment of blindness, of oblivion, this ultimate salvation! It was the only refuge within which to break the heavy, clipped wings that thought had set growing on one's temples, the only place where the mind, like a heavy-furred moth dazzled by the great light of death, could for an instant assuage its longing to return to the warm, original darkness of its chrysalis. . . . Frightened by the desperate wildness of my movement, Maria sharply withdrew from my arms, kissed me on the lips and fled. For a moment I tried to follow her. Then I leaned against a tree. Within me and about me a great silence had fallen. After a moment I wiped away my tears and went back to my dead.

1. One of the French prisoners chosen by the rest to represent their interests in dealing with the Germans.

IV

Drama

Although the Holocaust seems to offer ripe material for tragic drama, that promise has not led to visible fruition. The traditions of tragedy require a protagonist in reasonably equal combat with an inner or outer antagonist, but these conditions mock the reality that Holocaust victims faced. Whether we consider classical, Renaissance, or modern theater, the tragic figure is always somehow partly the agent of his fate—less so, perhaps, in our own unheroic times. Even as passive a figure as Arthur Miller's Willy Loman helps to create the crisis that leads to his unhappy end. If, as Nietzsche believed, we gain human stature from our willingness to risk our lives through *choice,* few Holocaust victims were allowed by their oppressors to be candidates for this role. Who, indeed, could be expected to maintain dignity and avoid moral shame by marching undaunted toward death in the gas chamber? Annihilation of the Nazi sort was not a test of character.

This is not to say that courageous, and perhaps even legendary, figures have not emerged from the Holocaust experience. But to celebrate them in drama as tragic heroes or heroines is to misread the circumstances that drove them to their doom. Although the image of Janusz Korczak walking bravely at the head of the orphaned children under his care to the place of deportation in the Warsaw ghetto continues to inspire later generations, including writers—a play by Erwin Sylvanus is called *Dr. Korczak and the Children*—Korczak did not *choose* gassing in Treblinka as a way of shaping his tragic destiny. His German murderers planned to kill him and his charges all along but bided their time until they were ready, indifferent to (and unaffected by) any resolve Korczak may have had.

Similarly, Mordechai Anielewicz was driven by external forces and a desperate situation to help organize resistance in the Warsaw ghetto. Although no one can fail to admire his valor, the awful fate of the fighters (as well as the remaining ghetto residents) leaves little room to declare his exploits a moral victory. The tragic spirit is crushed not by a weakness of the protagonist's will, but by the Nazis' contempt for their victims, and especially their calculated malice in denying those victims the dignity of significant choice. No one *volunteered* for starvation, deportation, or extermination for the sake of a higher principle; but in the absence of such unions between choice and fate, dramatists are left to invent them (and thus falsify history) or devise other means to capture this event on the stage.

They have chosen both routes, with mixed success. The Holocaust is a threat to art as well as to moral vision because the details of its human ordeal are so cruel that they are difficult to portray and nearly impossible to imagine. Normally, a play's audience grows familiar with its premises as the performance proceeds, until a moment of identification with its central tensions arrives, and play and audience become one. Because of the nature of its subject, Holocaust drama follows a reverse pattern of *defamiliarization,* until that climactic moment when the audience realizes it can *never* identify with the experience of the characters on the stage. Literary precedent primes us to expect people matured or redeemed by their suffering; historical truth requires us to confront victims battered by the anguish of their physical trial. Playwrights are not immune to the appeal of either; but in trying to balance the two, they often capsize on the rocks of their own compromise.

Two well-known examples of Holocaust drama are Rolf Hochhuth's *The Deputy* (1963) and Peter Weiss's *The Investigation* (1965). Hochhuth's work condemns the papacy during World War II for its timidity in refusing to denounce publicly and explicitly the Nazi murder of the Jews. The setting alternates between Italy and Germany, ending in Auschwitz, though Hochhuth in a note dismisses documentary naturalism as a stylistic principle for representing a deathcamp: "No matter how closely we adhere to historical facts, the speech, scene, and events on the stage will be altogether surrealistic." This is a legitimate point of view, but when he shifts his attention from locale to character, Hochhuth creates such conventional figures that he forfeits any surrealistic benefit gained by the vague movements of the doomed toward the gas chambers in the closing moments of the play. The final dialogue between the disenchanted doctor and the sympathetic priest turns into a familiar metaphysical dispute between cynical nihilism and Christian compassion—an example of the commonplace disguised as the profound. There is little surrealism behind the verbal premises defining the play's structure: good, evil, conscience, spirit, truth, and love. They do not carry us very far into the defamiliarized zone of Auschwitz.

Weiss's tactic is to veer away from metaphysics, stressing instead what we might call the materiality of the deathcamp. *The Investigation* replaces dialogue with monologue. Its substance derives almost entirely from testimony offered at the Frankfurt trial of more than twenty Auschwitz guards in the early 1960s. Testimony, at least, is rooted in the historic details of an event. By carefully grouping and adapting statements from the victims and the accused, Weiss creates from the futile courtroom strife a fresh vision of the clash in Auschwitz between moral space and destructive place. Although naked statements may lead the imagination into the silence of despair, the form Weiss imposes on them achieves the opposite effects. Guards and officers in the end indict themselves by their relentless denials of complicity, whereas the language of the victims ultimately sheds a terrifying light on the ordeal they succumbed to—or so wretchedly survived.

That ordeal betrays the limitations imposed on the dramatist by Holocaust reality. While victims furtively maneuvered their environment in order to remain alive, their oppressors curtly manipulated them at will, deciding the terms of their lives and the moment and manner of their deaths. Deception and self-deception meet at the heart of the human drama, though combining defiance and insight into the heroic stance is hardly a proper plan for such leaner fare. Dramatic roles thrive on the inner tensions that goad characters to understand and control the external forces that nourish or threaten them. But the foes ruling Holocaust existence at its harshest level—roundups, deportations, illness, exhaustion, starvation, torture, mass murder—were unrelated to character, and the conflicts they bred cannot be shown by conventional means. In many ways, the playwright most responsive to this sense of characters stripped of their usual guise and the denuded milieu that surrounds them is Samuel Beckett. Gogo and Didi's desperate antics in *Waiting for Godot,* as they try to face if not shape their elusive destiny, echoes the futile prospects of Holocaust victims once they were caught in the jaws of the Nazi death machine.

Beckett's use of dialogue and monologue reflects the shrunken value of language to ease friction in a Holocaust-dominated world. Talk is a charade, not a challenge to his moral resources, when man's fate is the gas chamber. This may be why Beckett pares down some of his briefest minidramas to gesture alone. Perhaps it is asking much of dramatists to have them reimagine the reality and the theatrical conventions that have nurtured their art for generations, but failure to do so leads to versions of the Holocaust that have little to do with the truth of that brutal event. The pallid, sentimental staging of *The Diary of a Young Girl,* based on Anne Frank's experiences, is a classic example.

Another perplexing issue for the Holocaust dramatist is how to portray the personalized evil behind the Nazi mentality and its verbal disguises ("special handling" and "final solution"). Whatever drives an Iago, an Edmund, or a Macbeth—jealousy, envy, the will to power, or even sheer malice—we witness their acts and are aware of the choices that breed them. William Styron's valiant but, I think, unsuccessful effort in *Sophie's Choice* to humanize the man behind the infamous commandant of Auschwitz, Rudolf Höss, reflects the difficulty that any writer—whether novelist or dramatist—must face when conjuring up a figure whose motives are often no more unusual than "orders from above." In spite of Styron's specific and intimate facsimile of the Auschwitz commandant, in the end we are no closer to understanding how or why Höss supervised the murder of more than 1 million Jews than we were before he made the transition from historical person to character in a novel. This is not necessarily a failure of art, but an exposé of the limits of traditional fictional means for gaining entry into a universe of evil unlike any we have known before.

As readers of Holocaust drama (and I suppose this holds true for the playwright, as well), we need to clear our minds of certain theatrical expectations. Just as the historical victims in the camps and ghettos found themselves vexed by threats to their very existence that they could not identify with distinct antagonists, so we, as the audience, must be prepared to meet a wasting away of individual lives without apparent human agency. Since protagonists of evil do not exist in these works, they cannot reveal their intent with the candor of a Shakespearean villain; the sources of mass murder during the Holocaust cannot be traced to, and thus cannot be represented by, private voices or visions driven by the will to destruction. "Inconsequential dying" is neither a happy idea nor a welcome expression, but in fact death seems to have ensued severed from any victim's risk; it resulted as much from the *indifference* as from the hatred of the killer. This is hardly the stuff of dramatic conflict.

What, then, is the playwright to do? Victims and killers certainly existed, but do not fit into traditional parts. They did not pit their wit or will against each other in a contest for triumph or survival; the daily rhythm of mass murder continued regardless of individual protest or resistance. When the "enemy" or antagonist is a fanatic ideology rather than any of its particular exponents, a play cannot draw on the strength of its major roles for its impact. Charlie Chaplin's early comic rendition of Hitler in *The Great Dictator* seems less humorous and suitable when filtered through our knowledge of subsequent events. But would any other portrayal of Hitler on the stage be more compelling? One sad consequence of the Holocaust is that it exiled everything it touched (including ourselves and the victims who survived, as well as those who did not) *beyond* the margins of tragedy, leaving us with the charge of finding new and more modest forms to express it.

The single play included in this section reflects the problem rather than its solution, and the same could probably be said for all Holocaust writing. Joshua Sobol's *Ghetto* has been received with mixed enthusiasm, but few critics have suggested how his material might have been better used. Choosing a familiar strategy in Holocaust literature, Sobol blends fact with fiction. What theme could be more dramatic than the efforts by Jews in the Vilna ghetto to avert their doom, which a modern audience, with the benefit of hindsight, can interpret more literally as "postponing their execution." What could be more "surreal"—following Hochhuth's admonition—than the painful antics of Sobol's characters as actors in quest of their own survival. Little that they mimic flatters their dignity or augments our image of their stature. But this does not reflect a flaw of the playwright. The milieu from which they emerge determines the frontiers of their conduct. If this fails to exalt their nature or increase their appeal as creatures of the dramatic imagination, the blame lies with the waning prospects for their lives in a ghetto marked for destruction. This forces them to seek a reprieve from, rather than a victory over, their fate.

The question that victims had to ask themselves was not "How can I best be myself in this hostile place?" but "What role must I play in order to survive?" The transformation of self into role has always been a *formal* challenge to the dramatist. Sobol's play raises the question of whether art is a distraction from or an entry into, reality. The actor who plays a character impersonating an actor in order to entertain or to escape from an intolerable life spurs the audience to redesign its sense of identity. The presence of a wooden dummy that is the most fearless voice in the play introduces the issue of whether all dialogue in that hopeless situation may be nothing more than a kind of ventriloquism. Years after the event, former victims who recount their ordeal often fail to recognize the disguised self whose features they dredge from memory while insisting that it does not resemble the self they know today. The ventriloquist and his dummy in Sobol's play are potent reminders of how the urge to survive splintered what used to be called the integrated self.

The role of the German commandant is especially troublesome, since his affable manner frequently contradicts his official task of managing the murder of Vilna Jewry. How else might he behave? Clearly, everything he says or does is a pose or ruse, a performance in which his desperate victims hope to glimpse some sympathy. What stance *should* they assume vis-à-vis their German persecutors to mute the fury that was consuming their fellow Jews? If the behavior of Sobol's characters seems unsatisfactory today, given our knowledge of the villainy of the German authorities with whom they dealt, this only highlights the poverty of available choice. Any other form of dialogue would be equally futile, as the commandant's final violence brutally confirms.

A crucial feature of the Holocaust experience that drama perhaps can better convey than straight testimony is the struggle between what we might call compressed and expanded time. Every Jew in the ghettos and the camps, but especially those who were given somewhat greater flexibility because of their positions of leadership or responsibility, had to distinguish between the short-term and the long-term effects of their actions. What should be your instant concern when the ultimate fate is the same for all—selection, and death? Compressed time concerns events of the day and the moment; expanded time involves the issue of life's last goal—mortality—and the quality of existence on the journey toward it.

In everyday society, as Freud suggests, most of us postpone the challenge of expanded time by assigning it to others while believing in our own immortality. But residents of the Vilna ghetto could not so easily court this illusion, since the execution pits at Ponar were a constant threat. The vaudeville atmosphere of much of Sobol's *Ghetto* (like the cynical façade of Brecht and Weill's *The Threepenny Opera*) betrays a community where consoling distinctions between near and distant time have collapsed. All the gestures and deeds by members of that community, including their bid to create a musical theater despite the feeling of doom, reflect for those of us who enjoy the luxury of expanded time an evasion rather than an expression of the period's inner truth. Moral life festers when one is forced to adopt the pretense of planning for the future, whereas in reality the only options are to save some by condemning others or to see others condemned without trying to do anything to prevent it. Moral life disintegrates when the contrast between choosing and, as one of Sobol's characters puts it, "letting it happen" disappears. Compressed time then cancels expanded time, and one is left to drift and hope.

Like so much of Holocaust literature (and testimony), Sobol's *Ghetto* is staged in the memory. The drama of creation has passed; this is a theater of re-creation, whose theme forbids it from lapsing into mere recreation. We live in a time contaminated by recall, and as disguises fall away in the closing moments of the play, revealing an elderly, maimed survivor staring into the darkness, we are left dazed in a present whose future the Holocaust has immutably scarred. The energetic burlesque before our eyes cannot void this pitiless truth: despite their fervent efforts to promote culture and sustain dignity, most residents of the Vilna ghetto, including Gens and his aides, ended up dead—victims of their Nazi murderers. Our foreknowledge of their doom frames the play and engages us in a way that must have drawn ancient Greek audiences to the fate of Oedipus. But there the resemblance blurs: hubris was a common trait; mass murder was a unique ruin that we cannot share. Whatever impact Holocaust drama achieves must sprout from that necessary estrangement.

24. Joshua Sobol

Joshua Sobol was born in Palestine in 1939. He was active in the Ha-Shomer Ha-Tza'ir (Young Guard) youth movement, and lived on a kibbutz from 1957 to 1965. He tried his hand at fiction, studied philosophy at the Sorbonne, and later taught at Tel Aviv University and a teacher's seminary and drama school. Eventually, the theater became his profession: from 1985 to 1988, he was artistic director of the Haifa Municipal Theater.

Sobol's first play to draw critical attention, *The Wars of the Jews* (1981), is set in the first century during the Jewish rebellion against Rome and the subsequent Jewish civil war. The title of the play is taken from the Jewish historian Josephus's account of these events. The characters are the ghosts of these historic Jews. The Zealots among them, who prefer martyrdom to negotiation and thus bring disaster on the community, are presented in an unsympathetic light. Sobol views ardent nationalism and the violence with which it is associated as a threat to contemporary Israeli society. His plays invite audiences to link the issues dividing Jews in the past with those splintering his countrymen in the present.

His next play, *Soul of a Jew* (1982), was performed at festivals in Scotland, Germany, and the United States, as well as in Israel. It dramatizes the last night of the Austrian writer Otto Weininger, a Jewish antisemite who lived in the late nineteenth century and committed suicide at the age of twenty-three. Weininger gained notoriety when Adolf Hitler was said to have spoken of him as his "favorite Jew." Sobol describes *Soul of a Jew* as a dramatization of "what it means to be a victim of hatred, the ways one interiorizes it and becomes one's own enemy." Weininger's failed struggle to find an identity that might ensure survival, involving surrealistic visits from Sigmund Freud and August Strindberg, leads to self-destruction. The play illustrates Sobol's free-ranging imagination and his willingness to incorporate expressionistic techniques into his drama. It also reflects his desire to use the theater as a springboard to help his country forge with integrity an unthreatened and unthreatening identity.

Ghetto (1983), the first panel of a dramatic triptych, was performed in England, Germany, and the United States, and won the German Critics Award for Best Foreign Play in 1985. It was also chosen by England as Play of the Year in 1989. It has been translated into English, German,

French, and Italian. The work has roused enthusiasm among its supporters and animosity from its critics. Some have interpreted the portrayal of the difficult situation in the Vilna ghetto under Nazi occupation as analogous to the Israeli–Palestinian controversy. The choices thrust on individuals, communities, and peoples threatened with extinction are germane to Vilna's Jews and Israeli society, though the moral space available to postwar Israel must be distinguished from the claustrophobic atmosphere of the Vilna ghetto. Freedom to negotiate today is a real option, but, as we now know, it was only an illusion for the Jews in the Vilna ghetto; the Nazis had determined their fate from the beginning.

One of the most noteworthy features of the play is its *theatricality*. The action focuses on an actual dramatic group that performed in the Vilna ghetto, even while other Jews were being sent to the suburb of Ponar to be shot. The audience is invited to experience simultaneously history as performance and performance as history: the characters enact on a stage human conflicts that both we and they know exist in the shadow of mass murder. Perhaps this is one reason why some ghetto residents objected to what they called "theater in a graveyard," while others celebrated the moral courage of actors engaged in a form of cultural resistance. *Ghetto* leaves unresolved the ultimate value of such activity, as it does the question of which kind of individual and communal behavior most favored survival in such a restricted milieu.

The play eschews easy judgments. It raises in dramatic form some of the concerns voiced more bluntly in the selections from ghetto diaries and chronicles that are included in this anthology. The characters in the play, many of whom are based on historical figures from the Vilna ghetto, choose diverse strategies to increase the chances for survival—clinging to hope in the midst of despair. Jacob Gens, chief of the Jewish police and later the German-sanctioned head of the ghetto, believed that by cooperating with the Germans in the selections he might save the few by sacrificing the many. In this, he shared the view of Mordechai Chaim Rumkowski, Elder of the Jews in the Lodz ghetto. Hermann Kruk,* the librarian and a member of the socialist Bund, was firmly opposed to collaboration of any sort, preferring death through armed revolt. Other personalities, like the entrepreneur Weiskopf and Dessler of the Jewish Order Police, are opportunists who stop at nothing to ensure their survival. All exist subject to the stern control of the SS officer Kittel, another historical character, who was sent to Vilna to supervise the ghetto's final liquidation.

Few Holocaust writers have had the courage or resources to attempt to portray such agents of evil in literature because they defy our efforts to imagine their motives or to gauge the effects of their deeds on their moral natures. Some critics deem Kittel's presence on the stage, mingling with the ghetto's Jews, an insult to the memory of the victims. But in his role as congenial villain, Kittel prods the audience to pierce the façade of his

*Kruk's detailed and invaluable *Diary of the Vilna Ghetto* is currently being translated by Barbara Harshav.

theatrical persona to see whether an appalling moral vacuum or a complex apostle of destruction hovers behind it. How to portray in art those who staged the Nazi ideology in history remains a dilemma. Although the challenge may exceed Sobol's achievement, his effort remains one of the most imaginative available to us.

A drama such as *Ghetto,* which contains music, song, and dance as well as dialogue, does not translate easily to the printed page. Even in the theater, its success depends on the resourcefulness of the cast and the director; some performances have been greeted with enthusiasm and others with dismay. One critic has complained that "stylization and realism get into each other's hair and rebuke rather than reinforce each other," but it is not clear whether this is a comment on the text or on the manner of mounting the production. Holocaust art provokes dispute more often than it gains consensus, and perhaps this is an affirmation of its enduring importance. Whether it represents, like the play *Ghetto* itself, a form of cultural resistance to genocide or a more modest resistance to cultural genocide, it reflects the vital summons from an awesome and awful moment in history to find creative forms to contain, if not order, chaos.

Adam (1989), the second play in Sobol's triptych, examines the partisan resistance movement. The last play in the cycle, *Underground,* which premiered at the Yale Repertory Theater in 1991, portrays the efforts of doctors in the Vilna ghetto to treat a serious typhus epidemic while concealing it from the Germans.

Ghetto

Playwright's Note

On various festive occasions our tradition demands that we *remember*. On the first night of Passover, which is the celebration of the freedom of a nation born of slavery, every adult participating in the feast is invited to tell the story of the Exodus in as much detail as possible. In other words, everyone is encouraged to reimagine the past, to revive it for his children by telling it as if he had lived it himself. History should be a constant and permanently living presence, the fruit of creative and imaginative memory. Maybe because it is the only possible way to assume it and yet to go on living, to survive.

Characters

SRULIK, puppeteer, singer
DUMMY
KITTEL, SS officer, saxophone player (he also plays DR. PAUL)
GENS, head of the ghetto
WEISKOPF, director of a tailoring factory
CHAHA, female singer
KRUK, director of the library
The Acting Troupe, who also play:
HASID, a palm reader
DR. WEINER, a young doctor
DR. GOTTLIEB, an old doctor
JUDGE
RABBI
WOMAN
GRODZENSKI, a young smuggler
JANKEL, a young smuggler
GEIWISCH, a young smuggler
ELIA, a young smuggler
HEIKIN, a klezmer clarinetist
DESSLER, chief of the ghetto police
LEWAS, officer of the Jewish police
JEWISH PROSTITUTES, JEWISH POLICE, GESTAPO OFFICIALS.

Time/Place

The play takes place in the mind of Srulik as he recollects the Wilna ghetto during 1941–43.

Act One

The living room of an apartment in Tel Aviv. 1984. A bourgeois home, typically orderly and clean. A one-armed MAN *sits in an easy chair, wearing a bathrobe.* HE *speaks with the difficulty and confusion of old age. His name is* SRULIK.

SRULIK The last performance? I don't want to talk about it. No, I don't remember anymore. It was a long time ago . . . . Well, some things I remember, but what's the point? It was on the evening before Kittel murdered Gens. Gens was the head of the ghetto. Exactly ten days before the liquidation of the ghetto. That was the last performance. A full house! Of course. It was like that at every performance. The audience came right up to the last days of the ghetto. People who were loaded up and sent away in trains the next morning, still put on their best clothes the evening before and came to our theatre performance.

But as for what happened on the stage—that's all gone now. I'm the only one who could remember it, and I'm . . . (HE *shrugs, as if to say "hopeless"*)

We had arranged a competition for plays about life in the ghetto. We got lots of plays. Everyone wrote: Katrielka Broide, Liebele Rosenthal, Hirshke Glick, Israel Diamantman. Everyone wrote for us. Wonderful plays. Songs too! (HE *sings*) "Frozen toes and frozen fingers frozen to the bone!" All hand-written, all gone now. Life in the ghetto. I mean, we lived in a world—a mad world—where people disappeared and their clothing stayed behind.

(*Suddenly struck by his own words*) Yes! The clothing! *Di Yogenesh in Fas.* I still have a few songs from that satirical revue. *Di Yogenesh in Fas.* That's Yiddish for "chasing around in a barrel" because the theatre was so small. But do you get the pun? "Di-yog-en-ez"/ Diogenes? Diogenes, the Cynic philosopher? He lived in a barrel and so did we. He roamed the world carrying a lantern in broad daylight looking for justice and truth. And in our little barrel? The ghetto? Did we have justice? Truth? We had a chase. A manhunt. Chasing around in a barrel. Of course, I'm no Diogenes . . .

But for a time I thought I saw . . . really saw what was happening. And we had a number in that show about clothes. "Finish it quickly!" No. "Disappear." Hmm. "Finish up quickly! Something, something disappear!" I get confused . . . "Finish it up! Today you're here, tomorrow . . . disappear . . ."

A great rumbling and banging is beginning to swell behind him.

I don't want to talk about it!

With a stunning crash, the walls of his apartment collapse, the apartment vanishes, and we're faced with a bare stage. At the rear wall is a tremendous, chaotic mountain of clothing.

The explosion of the apartment leaves us in a dark, undefined space. Then we hear the clatter of keys, the shooting of a bolt, and a man enters. HE *is dressed in underwear, but* HE *takes from the pile of clothing a German officer's uniform and puts it on. As* HE *does so, the stage fills with* OTHERS, *dressed in rags and sorting through more rags. The German officer,* KITTEL, *finds a machine gun and a long black case, and walks among the crowd shining a flashlight from place to place.*

KITTEL (*Observing the mess*) Chaos!

WEISKOPF, *a mousy little man dressed in rags, walks up behind him.* KITTEL *ignores him.*

Light! Get some light on in here! Separate 'em! Sort 'em! Keep moving!

WEISKOPF *throws a switch. Industrial light goes on.*

SRULIK (*To the audience, as the scene materializes around him.* HE *is young now, and has both arms*) That's Kittel. I was there too. We had to sort the clothes that came in.

KITTEL More light!

HE *moves to the group of ragged* PEOPLE *pawing disconsolately through the clothes, indicating different areas for different piles of clothes.*

Men's. Children's. Wet. Dry. We've got truckloads outside. Move!

The PEOPLE *separate into* TWO GROUPS. ONE *sorts, the* OTHER *brings in more piles from outside.* KITTEL *stands in the middle of all this, indifferent.* CHAJA, *a young woman, appears and approaches.* SHE *shivers with cold and pulls the sheet* SHE*'s wearing around herself tighter. Her hair is tangled and unkempt. Her feet are bare and filthy.* SHE *approaches* KITTEL *and stands, watching the* WORKERS *and* UNPACKERS. KITTEL *shines his light in her face.*

CHAJA May I . . . please . . . a pair of shoes.

KITTEL Come.

HE *points to the spot in front of him.* SHE *moves to him, frightened.*

You need what?

CHAJA A pair of shoes.

KITTEL If you knew where these shoes had been, you'd stay barefoot.

A pause.

Please. Help yourself. (*To* WORKERS) Again! Pile 'em again!

CHAJA *hesitates. Then* SHE *goes to a pile of shoes and begins to try some on.* KITTEL *spots her and hits her with his flashlight beam. The* WORKERS *stop and watch.* CHAJA *finds a pair, puts them on and starts to run.*

KITTEL Halt! (*Pointing to the spot in front of him*) Come!

CHAJA *returns.* HE *points to her sheet.*

Take it off.

SHE *complies. Beneath,* SHE *wears only a torn-up slip.*

You need a dress.

CHAJA *shakes her head.* KITTEL *points to the mound.*

Take a dress!

SHE *goes; takes a dress.*

Put it on!

SHE *does.*

(*To* WORKERS) What are you staring at? This isn't a show

A coat!

SHE *goes to the coat pile, picks one up quickly.*

Hat!

SHE *moves to the hat pile.* SHE *takes a beret, but doesn't put it on. Again* KITTEL *points to the spot in front of him.*

Come!

Defeated, SHE *goes to him.* HE *shines the light in her face.*

Hold your head up! Fix your hair! Put on the hat!

SHE *complies in something like panic.*

Not bad. With a little effort, you people can really be first-class. Turn.

SHE *turns, the reluctant model.* HE *looks at her belly quizzically.*

What's that? (HE *points the tip of his gun to the slight swelling in her belly*) What is that? A baby? Do you have any idea what happens to Jews who get pregnant?

SHE *is silent.*

Do you?! Come! Wait!

CHAJA *takes another step toward him.* HE *puts a hand on her belly.*

What is this?

CHAJA *takes a small bag from under her dress.* KITTEL *holds out his hand.*

Let's have it.

CHAJA *hands it to him.*

A pound of beans! (KITTEL *turns over the bag. Beans fall out and cascade across the stage*)

Black market? Who sold you this? I want names!

CHAJA *is silent.*

No names? So. You stole a pound of beans. (HE *looks piercingly at her for a moment*)

In a whirlwind KITTEL *grabs* CHAJA *and pulls her up to the pile of clothing.* HE *turns and crosses downstage, cocking his gun as he moves.*

Hands up!

SRULIK, *who has been carrying clothes to the pile, runs to him. In his hand* HE *holds a life-sized dummy, who is completely beside himself. The* DUMMY *screams at* KITTEL *while* SRULIK *tries to silence him.*

DUMMY Stop! Halt! *Arrêtez!* Whoa! For God's sake, hold your fire!

SRULIK (*To the* DUMMY) Cut it out, Ignatz, he's got a gun.

DUMMY Oh, terrific! She's about to be blown to bits and you're pissing in your pants.

SRULIK (*To the* DUMMY, *a well-worn routine*) Well at least I know how!

DUMMY So much for chivalry! Whatever happened to the days of yore, when damsels in distress were rescued by knights in shining armor?

SRULIK So get a knight. I'm a Jew.

DUMMY You don't have to brag about it.

SRULIK Who's bragging? I'm trying to stay alive.

KITTEL Hold it! Who are you?

DUMMY He stole the beans.

KITTEL Is that right?

SRULIK Would you take evidence from a dummy? He's pathological!

DUMMY Pathological?! I'm not even Jewish!

KITTEL That's enough!

SRULIK You heard the gentleman. That's enough.

DUMMY He was speaking to you!

SRULIK No, you.

DUMMY No, you.

SRULIK (*To* KITTEL) You see, I'm helpless. He's driving me crazy. I can't get rid of him.

DUMMY He's quite right Herr Kittel, but he's got it backwards. I can't get away from him.

SRULIK Yes, you can!

KITTEL *Shut . . . up!* Or I'll tear your head off. (*To* SRULIK) Did you give her the beans?

SRULIK If I had beans they'd be in *my* stomach. Besides, this woman's not a thief. She's an artist. A great artist.

DUMMY Yes, she's an artist. And besides . . . (*Sotto voce*) He *loves* her.

SRULIK Before the war, she was a star. Now she hasn't worked in months. She'll starve. I turn to you only because you are an artist too.

DUMMY Ass-kisser! Kittel *hates* ass-kissers!

KITTEL (*A satisfied laugh*) How true, how true. (*Suddenly serious,* HE *roars*) You! Get a scale.

All work ceases. PEOPLE *gather round.*

Juden! Achtung! You all have exactly one minute to find every bean she stole. A pound of beans and not an ounce less. Go!

A massive bean hunt; a scale is produced. EVERYONE *crawls around the stage, retrieving beans.* KITTEL *looks at his watch.*

STOP!

ALL *freeze.*

. . . On the scale!

EVERYONE *puts the remaining beans on the scale.* CHAJA *checks the scale.*

CHAJA Eleven ounces.

KITTEL Five ounces short. Well, well . . . you now have a choice. Careful here. Think. This? (HE *points to the gun in his hand*) Or this? (HE *points to the long black case* HE*'s been carrying*)

CHAJA (SHE *points to the case*) That.

KITTEL Ah, hah. Yes, indeed. (HE *takes out an object from the case.* HE *whirls on the crowd holding a saxophone as* HE *would hold a machine gun.* HE *laughs.* HE *puts it to his mouth and plays a few bars of Beethoven's Ninth Symphony*) Do you know that one?

CHAJA *nods.*

Well, by all means, let's try it.

KITTEL *plays.* CHAJA *opens her mouth but no sound comes out.* HE *stops.*

According to this Jew, you're a singer. A great artist. If he lied, then tomorrow morning, you're both off to Ponar* where you will be stripped, marched to the pits, and shot. I hope that's sufficient inducement. Now sing!

SHE *opens her mouth wider, but still nothing comes out.* SHE *points to her throat.*

CHAJA My throat is dry . . .

KITTEL Well, why didn't you say so?

HE *gives her his flask.* SHE *takes a gulp and gives the bottle back.*

*Wooded area near Vilna, where tens of thousands of Jews were shot and buried in mass graves.

CHAJA Thank you. If I could sing one of our songs instead . . .

KITTEL S'il vous plait, Madame!

CHAJA (*Sings "Shtiler, Shtiler" [Be Still]*):

Be still, be silent, be a shadow, softly draw each breath
The day of wrath is here, my child, we're in the house
of death
Your papa's vanished like the wind and I with anguish burn
And pray that, like the wand'ring wind, he one day will return
The past's a fading fantasy, the present inky black
Our single road leads to Ponar with no road leading back
The world has turned its face away, we're banished, battered
and reviled
But still, be still, we may yet live, my child

I saw a woman on the street whose manner was bizarre
Some people spoke in whispers of her children and Ponar
This is the winter of our souls, the blackest hour of night
But nature's timeless wheel must turn and bring a new day's light
Then there will come another time before your eyes grow old
For us a warmer season when our foes will feel the cold
We'll greet your papa at the door, we'll be a fam'ly as before
And you will sing out loud forevermore—forevermore.

KITTEL My God you really are an artist. Jesus, look at this. (HE *rubs his eye with a finger*) Real tears. Well. That experience, in my opinion, is worth an ounce of beans. I mean . . . tears are easy. What about the other four ounces? You can't find work at your trade, so how do you propose to pay? You're all artists, yes?

DUMMY Yes sir. All artists. At your service, sir!

KITTEL Well. Interesting. Prove it to me. Prove your art is worth four ounces of beans. You'll get your opportunity. But watch out. I'm no fool. Don't play me for one.

HE *turns and goes.* EVERYONE *exits after him except* CHAJA, SRULIK *and, of course, the* DUMMY.

CHAJA (*To* SRULIK) How can I thank you?

DUMMY How do you think?

SRULIK Don't be silly. You don't need to thank me.

DUMMY Hypocrite!

SRULIK I'm just happy you're still alive.

DUMMY Give the man a shovel!

CHAJA But you risked your life for me.

SRULIK Well, my life. What's it worth, really?

DUMMY In beans?

SRULIK After all, when you get right down to it, who am I? A puppeteer from the Meidim Theatre. An actor, not much of a man.

CHAJA I think you're a very brave man.

DUMMY Brave?! Are you kidding? Did you see him trying to muzzle me? Chajale! I'm the one who saved you.

CHAJA (*Stroking the puppet's head*) You're cute.

DUMMY Oh, my. Mmm . . . that is *heaven*. Do you know how long it's been, since somebody stroked me like that. Would you like a backrub? I could—

SRULIK Enough! You ought to be ashamed of yourself.

DUMMY (*Confidentially, to* CHAJA) He's insufferably jealous. (*To* SRULIK) Now go. Sit over there.

SRULIK Where?

DUMMY There.

SRULIK Here?

DUMMY *Further!* And don't bother us! (*To* CHAJA) You do love me, don't you.

CHAJA How could anyone not love you, *mazik?**

DUMMY And I love you, too! From the moment I saw you, my heart leapt. And I . . . you must be hungry. And here I am making idle chit-chat. (*To* SRULIK) Get her some food!

SRULIK Food?! Where from? I'm starving.

DUMMY Hah! What's in your pocket, then?

SRULIK My pocket? Lint. (HE *turns his pocket inside out. It's empty*) Not even lint.

DUMMY The other pocket, *gonif.*†

SRULIK Oh the *other* pocket. Well . . . (HE *finds a carrot*) I must have forgotten . . . I, uh . . . please, take it. (HE *gives her the carrot*)

CHAJA And what will you eat?

DUMMY Don't worry about him! When it comes to carrots he's a bottomless pit.

SRULIK *takes a second carrot from his pocket.*

SRULIK *Bon appétit.*

THEY *eat in silence a moment. Then* SRULIK *speaks, getting up his courage.*

Do you . . . have a place to sleep?

CHAJA Under the stairs.

DUMMY Well, come to our place. We have an enormous blanket.

SRULIK What a suggestion!

DUMMY You think it's any fun snuggling up to you at night? You're all bones, and it's freezing at our place. Chajale, I only meant we could, you know, bundle up together. For warmth. Is that a sin?

*Little devil. Term used affectionately for a mischievous child.
†Thief.

CHAJA No, it's no sin, you little rascal, not in this world. I'm cold at night, too. All right, let's go!

CHAJA, SRULIK *and* DUMMY *sing "Hot Zich Mir Di Shich Zerissn" (Dance, Dance, Dance)*.

CHAJA

Frozen toes and frozen fingers
I'm frozen to the bone

DUMMY

But what's the good of freezing
When you're freezing all alone

So dance, dance, dance
A little dance with me
Let's do a sultry tango
It'll warm us up, you'll see

CHAJA

Not a penny, not a crust
Nor twig do I possess

DUMMY

But with you beside me
I remember happiness

So kiss, kiss, kiss
Share a little kiss with me
And maybe we can reinvent
What kissing used to be

SRULIK

Colored papers, pink and yellow
Yellow saves your life*
I've got my yellow papers
But I haven't got a wife

DUMMY

So marry, marry, marry me
Love's not beyond recall

SRULIK

And maybe we will live
To see tomorrow after all.

CHAJA *exits*. SRULIK *looks around the stage like a professional appraiser*. GENS *appears and calls out to him*.

GENS So? What do you think?
SRULIK (*Dubiously*): Well, it's a place . . .
GENS Did you count the seats?

*At various periods in the Vilna ghetto, the color of one's identity card could temporarily mean the difference between life and death.

SRULIK I counted.
GENS Six hundred seats!
SRULIK I counted.
GENS And have you had a look around at the stage?
SRULIK It's a stage.
GENS This place, I give to you. Take it and create the Ghetto Theatre here. Not just plays: discussions, lectures . . . concerts. Culture!
SRULIK Culture. Well, yeah, why not?
GENS Clean the place up. I need an inventory—what's here, what you need—a list.
SRULIK When do you need it?
GENS Yesterday.

A HASID *enters.*

HASID (*Calling toward* GENS) Your honor, your honor! Mr. Police Commissioner, sir!
GENS What do you want?
HASID I want to read your palm. Give me your hand, I'll give you the future. Your hand please!
GENS What is this? A sideshow? Get a job.
HASID (*Suddenly trancelike*) This summer will change your life forever.
GENS Wait, wait, wait. How do you know that?
HASID (*Businesslike again*) I do ears, too. But it is in the palm that all of the details lie. Your palm, sir.
GENS Oh, for God's sake. (*Giving over his palm*) Make it quick, I'm busy.
HASID Three changes into eight. Can you see that?
GENS What if I can?
HASID Well, it's obvious. Three represents the third letter, which is Gimel. Eight is the eighth letter, which is Chet. You do see that?

GENS *is silent, impatient.*

Well, Gimel is G, like Germany, and Chat "Ch" which is like Cherut—that means independence and freedom. You will deal with the Germans with independence, and lead us to freedom. There will be a great revolution. But before that, you will take command of the ghetto. You will lead us to freedom.
GENS Great. You have dates for this event?
HASID (*Staring at* GENS*'s palm intently*) In three more time periods.
GENS Three more time periods? . . . What the hell does that mean?
HASID Could be three weeks, three months . . . even three years.

GENS *laughs. The* HASID *sticks out his hand for payment.*

Three marks, please.

GENS What?!

HASID Three marks.

GENS (*Paying*) Go find some decent work. You peddle the future around the ghetto, you'll starve.

HASID Another thing I noticed in the future is an acting company . . .

GENS You were listening!

HASID And, I'm an actor. (*The* HASID *pulls off his wig and beard*) You want comedy, tragedy? "Has a Jew not eyes? Has a Jew not hands? If you prick us do we not bleed?" . . . (*Sings*) "So dance, dance, dance, do a little dance with me."

GENS Alright, alright. Go dance over there.

GENS *pushes the* HASID *upstage and looks to* SRULIK.

You'll find a place for him, yes? Now, we were saying . . . will the place be all right?

CHAJA The place is fine.

SRULIK But it's the time, Jakob. Three weeks ago fifty thousand Jews were murdered on this spot. The blood's not dry yet and you want us to create theatre here? It's not the time.

GENS (*Goes to a door at the side of the stage*) Not the time, you say. (GENS *opens the door and calls out*) Let 'em in! Now!

MEN *and* WOMEN *are shoved into the room: the acting company.* THEY *look around, dazed, frightened, blinking.* SRULIK *stares at them dumbfounded, as if seeing ghosts.*

SRULIK Lionek, my God. No, no it's all right. You're safe. It's me. Umma . . . I was sure you were dead.

CHAJA (*Spots* HEIKIN *and runs to embrace him*) Heiken, my klezmer. You're alive.

HEIKIN *embraces her, in a daze.*

SRULIK Where . . . Jakob . . . where did you find them all?

GENS Living in garrets, working in forced labor, hiding in root cellars, Srulik. These were the ones I could still save.

CHAJA *has been running from person to person as the room fills with* ACTORS, *showing off* HEIKIN, *who still looks about blindly.* GENS *hands him his clarinet.* HEIKIN *stares at it in disbelief, but can't bring it to his lips.*

CHAJA Heikin, please. For me.

HEIKIN *stares at her balefully.* SHE *begins to sing "Ich Benk A Heym" (I Long For Home) for him.*

CHAJA *and the* ACTING COMPANY (*singing*):

When you are young,
Yes, young and strong,
You long to try your wings.
You leave your home,
Your childhood nest,
Forsaking childhood things.
But when old age draws near,
Scenes from your past appear
And, oh! What feelings rise,
From those forgotten ties,
From childhood scenes
When seen through older eyes.

I long to see my home once more.
Is it the way it was before?
The weathered porch, the slanting stairs,
Four walls, a table and some chairs,
My poor old home.

I long to know them once again,
The things I took for granted then
The songs I sang, the dreams I dreamed
Are more enduring than they seemed.
I miss my home.

I hear a breeze
As gentle as a sigh,
Recalling a mother's lullaby.

HEIKIN *begins to play, and the* ENSEMBLE *sings.*

ALL

Though not of brick nor made of stone,
A stronger home I've never known.
As strong as steel, as light as air,
Built from my mother's loving care.
I long for home!

I long to know them once again,
The things I took for granted then.
The songs I sang, the dreams I dreamed
Are more enduring than they seemed.
I miss my home.

I hear a breeze
As gentle as a sigh,
Recalling a mother's lullaby.

CHAJA

Though not of brick nor made of stone.
A stronger home I've never known.
As strong as steel, as light as air,
Build from my mother's loving care.
I long for home!

The ENSEMBLE, *momentarily uplifted, is lifeless again.*

SRULIK Jakob, this is not going to work.

GENS Tell me, Srulik, what these men and women have in common.

SRULIK They're artists!

GENS Artists? Pathetic! What they have in common is no work permits. Also no food rations. They're next in line for Ponar—that's what they have in common. You want that on your conscience? (*Gradually working himself up*) The right time for theatre, the wrong time for theatre. You intellectuals amaze me. When all this is over I can hear you telling your grandchildren: "Even though police chief Gens tried to force us to perform on the site of the massacre, I refused." The pride! The moral conviction! Well, look at them, Srulik! Look in their eyes! If you make them into an acting company on this spot, I can get them yellow work permits—and bread. And butter—and potatoes. And soap! A half ration of soap for every one of your "artists."

ALL (*Ad lib in whispers*) A half ration!

GENS That's right, for each of you. (HE *marches back and forth like an officer dressing down his troops*) But there's more. . . . As a troop of Jewish actors, you can make a difference! Look around! Our self-esteem is in the sewer. People walk the streets staring at their shoes—if they have shoes to stare at. You can change all that—give them back their self-respect. That's what I want: show them they have a culture—a language, a powerful inheritance: an inner life. Nu? Begin the rehearsal! In three weeks, I want to see a play! A performance!

SRULIK With all due respect, Mr. Gens, what are we supposed to rehearse? What kind of a play?

GENS Do I ask you how to run the ghetto? How should I know what kind of play? Something good, something funny, something cultural. Whatever will make us feel like men again. You know what I mean—you're artists!

SRULIK *and the* ACTORS *exit. As* THEY *go,* GENS *and* WEISKOPF *remain. Across the stage, in the library,* HERMAN KRUK *appears.*

WEISKOPF Mr. Gens! Mr. Gens!

GENS You, too. You, too! Go rehearse.

WEISKOPF Mr. Gens, I'm not an actor.

GENS So who the hell are you?

KRUK speaks, as if dictating, far from the actual scene. GENS and WEISKOPF remain motionless while KRUK dictates.

KRUK That's Weiskopf. A few weeks ago he was a nobody, a *vonce** with a tailor shop. Who could know what the war would make of this man? In a few weeks—the king of the ghetto. A man to keep your eye on . . .

WEISKOPF If you could spare me just a minute or two, Mr. Gens . . . you won't regret it.

GENS I hope not.

WEISKOPF Do you know how many first-class tailors we have in the ghetto?

GENS Tailors?

WEISKOPF And sewing machines? How about sewing machines?

GENS Sewing machines . . . ?

WEISKOPF Take a look. (*Giving* GENS *a small notebook*)

GENS What is this?

WEISKOPF I made a few inquiries. Tailors . . . seamstresses . . . machines . . .

GENS A list. With names, addresses . . .

WEISKOPF I went from room to room.

GENS But what for?

WEISKOPF Have you seen the trains, traveling from the Russian front to Germany?

GENS I don't understand.

WEISKOPF Do you know what their cargo is?

GENS How should I know?

WEISKOPF Uniforms, Torn, bloody German uniforms. And why are they sent from the Russian front all the way to Germany?

GENS I imagine . . . to be put back in shape.

WEISKOPF You can appreciate the crucial point. We are next door to the Russian front here in the ghetto. Hundreds of tailors, seamstresses, sewing machines. . . . It's *meschugge,*† no?

GENS And you think the Germans . . .

WEISKOPF Approval on the spot. We'll erect a giant uniform repair factory. Instead of sending the stuff thousands of miles, burning coal, losing time, stopping up the tracks, they can deliver the uniforms here. Good for them, good for us.

GENS And every tailor, every seamstress, will be indispensable. . . . How many can you use?

WEISKOPF We could begin with, say, a hundred. Later we'll raise it to a hundred and fifty. I've picked out a spot.

* Bedbug.

† Crazy.

GENS A hundred and fifty families, kept alive indefinitely. Come to my office tomorrow morning.

WEISKOPF Tomorrow? What happened to today? If we work through the night, tomorrow morning you'll lay out the whole program for the Germans—precise figures, just the way they like it—and we're in business.

GENS (*Regards* WEISKOPF *for a moment*) I like people like you. What's your name?

WEISKOPF Weiskopf.

GENS Come into my office, Mr. Weiskopf—Mr. Factory Director.

THEY *exit. Across the stage,* KRUK *dictates, leafing through some documents and letters.*

KRUK An interesting chapter in the life of the ghetto. An interesting case. Weiskopf. (HE *picks up a letter from his desk and reads*) "January 17, 1942. On Sunday, January 18, you are cordially invited to the premiere performance of the new Ghetto Theatre. On the program: dramatic scenes, songs, music . . ." (HE *turns to his invisible secretary*) Put this in capital letters: YOU DON'T PERFORM THEATRE IN A GRAVEYARD.

A rush from all sides of the stage. MEMBERS OF THE RESISTANCE *scurry about, leafleting and postering the stage with* KRUK'*s slogan: YOU DON'T PERFORM THEATRE IN A GRAVEYARD.* THEY *exit, leaving* KRUK *where* HE *was.* GENS *enters the library, carrying a poster* HE *has ripped from the wall.*

GENS Herman Kruk? (HE *holds up the sign*) I know who did this. I know everything that happens in the ghetto. Everything. Why this slogan?

KRUK Why this invitation?

GENS Every important person in the ghetto was invited. You run the library.

KRUK And did you really think I'd come to your vaudeville show?

GENS What have you got against theatre?

KRUK The invitation is an insult.

GENS Insult?

KRUK A personal insult. (*Pause*) In any other ghetto, perhaps. There might be some entertainment, even some fun. One should always try to live in the presence of art—if possible. But here? In the middle of the tragedy of Wilna? In the shadow of Ponar? Of the seventy-six thousand Wilna Jews, how many are left? Fifteen thousand.

GENS Sixteen thousand.

KRUK A theatre? Now? It's shameless!

GENS All right. Let's talk about shame. On Saturday, September 6, 1941, the Germans herded us into the ghetto. Hell on earth. People were driven through the rain; wagons stranded in muck. And what did you see? Books. Books slipping from people's hands, into the mud. While our people stumbled through the streets as if they were drugged, Herman Kruk made a book collection. Next morning, you opened this library. Well, for that I salute you. Relax, Herschel, I don't expect anything in return. Look, let's face it, you're a socialist, a Bundist. I am a Zionist, a revisionist. Worlds separate us. Jakob Gens didn't fish any books out of the slime. I admit it. *I pulled human beings out of the slime.* I found them clothing and food, and I forged them into an orchestra and a theatre troupe. I've given them back their jobs, and I'll get them work permits. Heikin, the klezmer? I took him out of forced labor. I took a pickax out of his hand and gave him back his clarinet. Is that a crime? (*A pause.* HE *fishes the invitation to the theatre opening off* KRUK*'s desk*) This is no invitation. It's an order. You and your friends, Bundists, socialists, leaders of the workers' union, all of you, will be at the theatre.

KRUK (*Regards* GENS *for a moment, struck by his ferocity*) Why is this so important to you?

GENS Solidarity. Everyone in the ghetto must agree that we are one people. A great, brave people, a creative force through history. I'm going to unite the ghetto. Everyone, everyone—without exceptions. It's not a matter of choice anymore.

KRUK I don't think you'd miss us. After all, the Jewish police, the foremen and brigadiers of the forced labor groups—you'll have a distinguished crowd. Plus whatever guests you've invited from the outside—German staff officers and their wives. I gather our famous chanteuse is feverishly looking for a few *Deutsche lieder* to add to the program, in case the Germans—heaven forbid—should want a few songs from the Fatherland.

GENS Look, we can continue this little discussion the day after tomorrow. In the meantime, I expect to see you at the theatre.

KRUK Enjoy it. Play up to the Germans any way you want to. The workers' union has decided to boycott. None of us will join this concert of crows.

GENS In that case, Mr. Kruk, your workers' union is hereby dissolved.

KRUK (*Outraged*) What?!

GENS Forbidden.

KRUK It's the only Jewish organization in the ghetto that was democratically chosen!

GENS Aye-aye-aye—*Vusszugstute,** Herschel.

KRUK What are we, Jakob? Nothing more than your own personal king-

*"That's what you say."

dom? And this theatre is your Versailles? We refuse to take part in this revisionist farce!

GENS You think you can play party politics with me? Listen to me, Kruk: One more poster like this, you go straight to Ponar.

KRUK (HE *dictates again*) The boss of the ghetto has ordered the dissolution of the only democratically elected group we have: the workers' union. It seems the workers have been playing partisan politics. Just to make sure he is clearly understood, he has also threatened to send Kruk and his comrades to Ponar.

GENS History will judge which one of us helped the Jews more in this catastrophe, Kruk. Write *that* down in your diary. (HE *leaves*)

KRUK I write, I write. What else is there to do? (*Dictating again*) We are living through such a dismal time that people can no longer recognize the true nature of the world around them. They don't want to see. They can't see. And the sad truth is, as long as we are made helpless by fascism, it is our duty to write. That's why my notebook will see all; it will hear all; it will be the mirror and conscience of this catastrophe.

During KRUK*'s monologue, the* ACTORS *and* MUSICIANS *become visible on stage behind him.* CHAJA *sings "Wei Zu Di Teg" (Crazy Times). It is a rehearsal.*

CHAJA (*Singing*)

Oy, little ones . . .
What times we live in!
Life grows harder day and night . . .
Cruel, crazy times . . .

Times of hydroplanes and Dreadnoughts!
Times of iron, steel and lead!
Ships that sail beneath the water!
Trains that rumble overhead!

Oy, it's all topsy-turvy!
Nothing is the way it was before.
Once the world was simple;
Now I don't understand it anymore!

The future's a puzzle,
The present's a maze!
Confusion and madness
Disfigure our days!

The future's a puzzle,
The present's a maze!
Confusion and madness
Disfigure our days!

ALL

Oy, it's all topsy-turvy!
Nothing is the way it was before.
Once the world was simple;
Now I don't understand it anymore.

*Oy, little ones . . .
What times we live in!
Life grows harder day and night.
Cruel, crazy times . . . (*Repeat three times**)

Oy, it's all topsy-turvy!
Nothing is the way it was before.
Once the world was simple;
Now I don't understand it anymore!

Oy, it's all topsy-turvy!
Nothing is the way it was before.
Once the world was simple;
Now I don't understand it anymore!

At the end of the song, WEISKOPF *swings onto the stage with elan.*

WEISKOPF What is it with the moaning and groaning, people?

SRULIK Weiskopf, this is a rehearsal.

WEISKOPF Life is a rehearsal. Hard times? Nu? When do the Jews ever have it easy? Suffering makes us strong people. Take me for example. Who has better grounds to moan and groan than Weiskopf? Before the war I was a Schneider* with nine tailor shops; nothing much but it was mine. Then the Germans came and squeezed us into the ghetto. No more shop, no more nothing. Kaputt, my whole life, into the toilet. Could I have cried? And how I could have cried! But no! I said to myself: Why are you called Weiskopf? Weiser Kopf. Wise head. All brains. (HE *taps himself on the head*) So. I took my little Jewish brains and said to myself, Weiskopf, you have lost your store. Are you kaputt? Sure! And if you lose your head what? Double kaputt you'll be. But you have your brain, and as long as they don't remove it with a scalpel, you'll stay alive! So you've got a brain, why not use it? I looked around me: A ghetto—walls, walls, walls, walls. Closed up tight. But always: always always always there is a way out. And who found it? (*Triumphant*) *Weiskopf!* So before the war what am I? Just a nebbish with a shop. And today what do you see before you? The director of a tailoring factory—the largest in the district! A hundred and fifty Jews work for me. And who are my partners?

SRULIK The Germans!

*Tailor.

WEISKOPF The Germans give *me* contracts. It's a great big business and all it does is grow. Each morning I wake up to a bigger income, and does that make me stingy? Do I hoard it? The opposite, my friends: I'm very generous. If someone comes to me with a good cause, I'm open handed; five thousand rubles I give, minimum. I'm free with my money and everyone should know it. I'm not ashamed of what I do. My way is the only way! I set an example for the community! I am an ordinary Jew, but we Jews, we're gifted. If you would follow my example, instead of kreching and moaning, we would have a productive ghetto, and they would find us indispensable, the Germans. And once we're indispensable, we stay alive.

KITTEL *emerges from the pile of clothing.* HE *has a large case in each hand.* HE *puts them down and applauds* WEISKOPF*'s speech.*

KITTEL Bravo, Weiskopf, bravo! I like you. And when I like someone, he stays alive. (HE *goes to a* GIRL *and slaps her*) Doesn't anyone say *shalom aleichem* around here?

EVERYONE *stops and greets him.*

I'll let it go this time because I took you by surprise. I didn't come through the gate. So no one could send out the warning—"Kittel is in the ghetto." (HE *laughs*) Be careful, though. Kittel doesn't use the gate. Kittel slithers into the ghetto like a snake. You dig a tunnel to hide yourself, Kittel reaches out to grab you. You take an idle stroll down the street, Kittel strikes from an attic window. Don't hide in the basement. Kittel is already there. (TO SRULIK, *suddenly*) What is in the cases? The wrong answer will cost you more than you can pay.

SRULIK (*Pointing to one case*) The gun.

KITTEL (*Opening the case*) The gun. Very good. (HE *gets it out and loads it. Then to* CHAJA) And in the other one.

CHAJA The saxophone.

KITTEL Well, let's see. (HE *opens the case*) The saxophone. The gun and the sax! Haaa! (*Suddenly cutting and threatening*) Why do I love you Weiskopf? Why?

WEISKOPF I'm productive.

KITTEL *laughs.*

And I make a contribution to the war economy.

KITTEL And who made you productive?

WEISKOPF You did.

KITTEL I . . . (HE *smiles, then becomes enraged*) . . . can't stand ass-kissers. (*To* SRULIK) Why can't I?

SRULIK Because you're an artist.

KITTEL I'm an artist. And what do artists invariably love?

SRULIK The true, the good, the beautiful.

KITTEL Did you hear that, Weiskopf? The true, the good, the beautiful. I didn't make you productive, *you* made you productive. All I did was create the climate for you, so that a previously hidden tendency in your Jewish character could emerge full-blown. I mean, this mad energy you Jews possess, Christ almighty. I look at the Lithuanians trudging through the streets; Jesus, what scum they are. We should be wiping *them* out, not you. What a mistake. No wit, no spirit—they're the walking dead. Then I slither into the ghetto, and it's another world. Raw energy is spilling out into the street, it's a sight. It's beautiful. You probably didn't even notice; I mean, those who live in paradise take it for granted. Yes? But for me . . . the cafes, the shops, the sense of people doing business . . . the spirit! When you run out of real food you chop up some beets and call it caviar. Sauerkraut juice becomes champagne—so it's champagne and caviar every night. Combine that Jewish spirit with German soul, and something great will be born. Did you ever dream you'd come so far Weiskopf? Tell the truth now!

WEISKOPF Never.

KITTEL So. The good, the beautiful and the true. All right, we've covered the good and the beautiful, now for the truth, without which there is no real art. A question, Weiskopf. The true answer lies at the essence of your Jewishness, and it is a question of truth. Let me remind you, Weiskopf . . . (HE *gestures to the two cases at his feet*) The gun, the saxophone. (*Picking up the gun*) Now, Weiskopf, what is the difference between partial liquidation and total liquidation? That is the question.

WEISKOPF *looks at him, thinks. Tension builds.*

WEISKOPF If you kill fifty thousand Jews but not me, that's partial liquidation. If you kill me, that's total liquidation.

KITTEL Bravo! Very nice, Weiskopf, absolutely on the mark. The wit! The sting! No one but a Jew could think of such an answer. Well, forget about the gun, this is hardly the time for the gun. It's time for the saxophone. I'm telling you, when the ghetto is liquidated, I'm going to have a piano set up right by the gate, and while you march to the trains, I'll play Schumann. "Scenes from Childhood," I think. Or "Carneval"? (HE *muses*) Well. I'll tell you what I'm doing here. Ah! The orchestra?

The MUSICIANS *step forward, warily.*

Gentlemen! To your instruments!

The MUSICIANS *move hesitantly to their instruments.*

The reason I came . . . is Gershwin. Can you imagine? All of a sudden I had this hankering for Gershwin. Funny, isn't it? These pigs at the Ministry of Culture in Berlin have banned him. "Death to jazz," they shout. So, when the simple urge for a bit of Gershwin begins to get to me, where am I supposed to go for satisfaction? Ah, the ghetto. Your Mr. Gens told me you had a jazz band, and that's why I'm here. (HE *looks over the* BAND, *missing someone*) Where's your vocalist? There's a singer here, she owes me two ounces of beans . . .

CHAJA *steps forward out of the group and stands face to face with him.*

Mademoiselle. Well, let's see if you can pay your debt. (HE *leaps onto the improvised podium, and directs with his saxophone*) "Swanee"! Do you know it? You must! Everybody—"Swanee"! Alright, I'll make it simpler for you. Sing it in Yiddish.

SRULIK We'll have to . . .

KITTEL Now!

The BAND *launches into "Swanee."* CHAJA *gives it all* SHE*'s got, fighting for her life.* KITTEL *moves between the* ACTORS, *sax in hand, and orders them to dance.* THEY *improvise a jazz number to "Swanee."*

CHAJA (*Sings "Swanee"*)
Ich bin avek fun dir a lange zait
Ich benk noch dir, ich gai ash oiss
Ch'ob a gefil, du libst mir fil
Swanee, du rufst mir oiss.

*Swanee, wi ich lib dir
Wi ich lib dir
Main taire Swanee,
Ich shenk di ganze velt
Zu zen noch ein moll D-I-X-I-E
—Ven now my mamee warten oif mir
Davwen far mir
Dort bai der Swanee
Di fremde weln mir nisht zen shoin mer
Ven ich kum zu dem Swanee mer . . . (*Repeat**)

*Swanee
Swanee
Ich kum zurick zu Swanee
Mammy
Mammy
Ich will zurick a heim! (*Repeat**)

KITTEL Ladies and gentlemen, I thank you! What a unique artistic experience. Not perfect, of course. The choreography was a bit . . . how can I say it . . . ragged? And a touch . . . heavy-handed. In jazz, the body must be light, utterly relaxed. (HE *dances a bit to demonstrate, singing his own accompaniment.* HE *stops; turns to* CHAJA) And the vocal? Not bad at all. I'd say you have a bright future in front of you. In a few years you could sink your teeth into—what—*Porgy and Bess?* Or even *Carmen.* As for today's performance, let's call it an ounce of beans on the Kittel scale. A magnificent score. (*Turning to* WEISKOPF) Look at them, Weiskopf, they can't step out on stage in these rags! You may donate the costumes from your treasure-house, yes? And make it lavish, Weiskopf. Extravagant.

WEISKOPF Only the finest.

KITTEL Good man, Weiskopf. Don't forget: I'll be there.

KITTEL *starts to leave without his gun and saxophone.* WEISKOPF *runs and picks them up.*

WEISKOPF Mr. Kittel . . .?

KITTEL (*To* EVERYONE) And the rest of you—remember: Kittel can turn up anywhere—anytime, anyplace. Lift up a rock, Kittel is coiled underneath. Kittel, the snake.

HE *laughs, and disappears, without the cases, into the orchestra pit. Two* GESTAPO OFFICIALS *run in and grab the cases from* WEISKOPF *and exit.*

WEISKOPF Okay, okay, people, let's get going. Come and get it. Take what you need, whatever you want. *Meine* is *deine.** Any cut, any size, any amount. The one thing we have in abundance is clothing. Everything sorted, everything stacked. Help yourselves, try it on for size. (WEISKOPF *grabs a dress from the pile and hands it to* CHAJA) Here Chajele! Try it on. Slips, suits, cloaks and coats. The finest workmanship, the latest styles. From Lodz. (HE *holds up a little girl's dress by mistake*) Children's wear?

A momentary shock-wave, as the ACTORS *realize what this clothing is.* WEISKOPF *tries to cover, dropping the dress quickly.*

*"What's mine is yours."

Well, it's not a children's theater. What do you need in fabrics? You want wool, you want corduroy? What do you want, rags? Of course, you need rags, for a play. Rags I also have no shortage of. And professional stuff? Whatever your play calls for: police, judges, doctors—I got cloaks for Hasidim, caftans for rabbis in genuine velvet. And for ladies and gentlemen of leisure, the finest tweed suits from Manchester, and haute couture from Paris. Wait, wait! What about uniforms? The Polish cavalry I got in heavy supply. Uniforms of brave heroes, some from Warsaw, some from Danzig, men who galloped on horses with fixed bayonets and met the tanks of the German army. Can you imagine what these uniforms looked like when they turned up in my shop?! Like from a sausage grinder! The blood, the bullet holes—and now? Take a look I beg you. Like new. Their past life has been erased with an invisible weave. Step into them and step right out on stage. You don't want Polish uniforms, how about German? Believe me they don't look so much better than the Poles when they get here from the front. If these uniforms could talk, my God, the tales they'd tell! I'll tell you, you should see the action in our laundry when these uniforms come in, *that* you could make a play from. Real drama! Germans and Poles chasing each other in the wash kettles, fires blazing in the ovens, the muck and filth that comes boiling over and runs across the floors and out into the sewer. An entire world fogged over with chlorine, grease and steam. What a battle! And then you should see the repair shop! An enormous tailoring station, a hundred and fifty sewing machines rattling away around the clock—RAT–TAT–TAT–TAT–TAT—it sounds like a railroad station. And all for you! The theatre company! Ladies and gentlemen, help yourself, whatever you don't see, just ask. There is no shortage of clothing here!

The ACTORS *have dressed.* TWO *are dressed as* DOCTORS GOTTLIEB *and* WEINER, ONE *in a rabbi's caftan, and* ONE *in a judge's robes. A* WOMAN *has dressed herself in rags, and immediately turns screaming to* WEISKOPF.

WOMAN Mr. Weiskopf, Mr. Weiskopf! They've arrested my husband.

WEISKOPF (*Still lost in thought*) Don't worry, please. It'll all work itself out.

WOMAN Work itself out? They caught him with five pounds of flour, and locked him up in Lukischki.*

WEISKOPF I said, be calm. It will work out.

WOMAN But he's a diabetic. Without insulin he won't last a week. Everyone says, you're the only one—

WEISKOPF Please, I'm talking now. I was right in the middle of—

*Jail in Vilna.

WOMAN I have to come up with twenty thousand rubles *today.* If I don't pay them—

WEISKOPF I beg your pardon. Do you know who you're talking to?

WOMAN Weiskopf! But there's so little time! From Lukischki they send them straight to Ponar, and—

WEISKOPF My good woman. Why is my name Weiskopf? You can go home now. *It will all work out.* I'll speak to my German before the day is out. Weiskopf will never give up. (HE *exits regally*)

Another ACTOR, *who will later play* DOCTOR GOTTLIEB, *immediately waltzes on behind him, doing a dead-on imitation.*

GOTTLIEB I'll speak to my German! Weiskopf will never give up! Weiskopf will never shut up, Weiskopf will never wake up!

WOMAN (SHE *looks around, confronts the* JUDGE *and the* RABBI, *still in character*) We should kiss Weiskopf's shoes! He gives carloads of food to the poor. He gets people out of jail. . . . We should kiss Weiskopf's shoes I say!

SHE *hands* WEISKOPF*'s shoe to the* RABBI. HE *passes it to the* JUDGE.

JUDGE Well, there's no legal precedent for shoe-kissing, but in these extraordinary circumstances, I'll make an exception.

RABBI Yes, an exception should be made. This man Weiskopf may not be a scholar, but in view of his amazing generosity and . . . ah . . .

ALL Power . . .

RABBI Power! He should immediately be elevated to a honored place in our council.

JUDGE Next case! (*Slams shoe*)

Music. GOTTLIEB *and the* WOMAN *move over to the others.* GOTTLIEB *quickly dons his doctor's coat and falls asleep. The* WOMAN *becomes a nurse. What's left on stage is a set-up for vaudeville scene: five characters dressed in exaggerated stereotype.*

WEINER Rabbi. Your Honor. Dr. Gottlieb. I'm glad you're here. The situation is desperate.

RABBI So what else is new?

JUDGE Who is this guy?

RABBI Dr. Weiner.

JUDGE Oh! Dr. Weiner. Well, introduce yourself.

WEINER I'm Dr. Weiner—in charge of the ghetto hospital, and here is my problem. I've got too many diabetics and not enough insulin. In three months all the insulin in the ghetto will be gone, and all our diabetics will die. Now, some could live long, full lives, others are already old and dying. I have to give out the insulin. You're the moral pillars of the ghetto. You tell me: do we have the moral right to . . . ah . . . choose . . .

JUDGE *and* RABBI Gesundheit!

WEINER Do we have the right to select?

ALL What?

WEINER Select . . .

ALL WHAT?

WEINER Select.

ALL Oh! Select. . . . WHAT?!

WEINER Cut off insulin to the old and dying, and allow the fittest to survive? Or do we condemn them all to death?

A long silence. It grows oppressive, then embarrassing.

JUDGE (*Finally*) That's a good question. All right, the case is clear. You want to know whether you can condemn specific people to death. As a judge I can give you a judicial answer: Sure. But not just anybody. Only in cases where the crime carries the death penalty. So I ask you: What is the nature of the crime of these old diabetics? The answer is: They're accused of being seriously ill. Proof? Plenty—blood sugar count, lab tests, occasional coma.

ALL Guilty, guilty, guilty.

JUDGE But, I make a concerted search of every law book in my library, and nowhere do I find serious illness listed as a capital crime. I'm sorry, Doctor.

RABBI There is a passage in the Talmud which relates to this theme. Maybe. It's hard to tell, but let's give it a try. An enemy attacks a city and demands, let's say for the sake of argument, twenty hostages. By the death of the twenty, the city will be saved. Now, the Talmud asks: Should one deliver the hostages or not? And the Talmud answers: Maybe. If the enemy presents a list of names, then those people should be delivered so that the city may be saved. But, if the enemy does not give a list of names, then not one man may be delivered to the enemy. Better that the whole nation should be destroyed than that anyone should have the power to decide who lives and who dies. That leaves us with a big question: Who is the enemy here, and who makes the list?

WEINER The list? Here's the list. (HE *produces the list*)

EVERYONE *dives for cover.* HE *chases them around the room with the list.*

Here. Each patient and his medical history. Name, age, legal status. Occupation, contributions to the community. Look!

JUDGE Get away from me with that list!

RABBI I don't want to see! I don't want to know!

WEINER Listen to it! Who gets the insulin? Here's a seventy-eight-year-old widow, no children, critically ill. And here a thirty-six-year-old father of three with a law degree—should I name names?

JUDGE *and* RABBI No names! Never name names.

RABBI What's the point of your list! God alone gives life, He alone may take it back. Can you tell me who will live through the night . . . if any of us? Human beings don't know. Human beings have no right to know! Only God.

KITTEL (*Offstage*) Gens! Gens! Gens! (HE *pops up from the corner of the stage and calls out*) Gens! Oh, there you are. You've got to help me Gens. I've got a problem of logic here—you're the only man who can untangle it. A man and woman get married, right? They have a child. Have they added to the race or not?

GENS Not ultimately, no.

KITTEL Good. Very good, Gens. So far, excellent. Now, they have a second child. Have they added to the race yet?

GENS No. Two parents, two children . . .

KITTEL Right. Of course. Now: three children.

GENS Well, three children . . . that, ah . . . that would be an increase.

KITTEL Three children is an increase! Exactly as I thought. Thank you, Gens. The Führer, you realize, has ordered an immediate stop to the natural propagation and increase of the Jewish race. Which means that the third child . . .

GENS Is excess?

KITTEL Precisely the word. Excess. I knew I could count on you, Gens, to solve the problem with that exquisite logic I've come to expect. Right! Now: selection of the third child—let's get to it, Gens. One mother, one father, two children. The third child, out. Move it, Gens.

HE *tosses* GENS *a cane with a bent knob on top.* GENS *stands on a platform and conducts the selection.*

GENS Father, mother, child, child. Father, mother, child, child. Move it! Double time! Father, mother, child, child . . .

Music is heard in the distance. As GENS *works,* KRUK *comes downstage and dictates.*

KRUK Eight rows ahead of me were five people: husband, wife, and three children. Now what? Gens counted them like this:

KRUK *and* GENS Father, mother, child, child, child.

KRUK The youngest boy, a twelve-year-old, he smacks with the cane and knocks him back off to the side. The rest of the family he shoves into the group that lives. The family stands open-mouthed among the chosen survivors, wailing: "Our child is lost, taken from us by a Jew; Jakob Gens has murdered our child!" Rage sweeps through the crowd. They surge forward, whispering, "Gens, the Jewish Jew-killer. The traitor." Only abject terror keeps them in check. Meanwhile, another family moves past his cane: a mother, a father, a child. Gens counts: "One father, one mother, one child. . . ." He stops, turns to the father, and shouts: "You idiot! Where the hell is your twelve-year-old?" The father begins to stutter uncontrollably, denying the existence of a second child, but Jakob Gens, the Jewish Jew-killer, won't hear of it. He smacks the man with the cane, shrieks at him, creates a complete uproar of protest. In the midst of this chaos, Gens grabs the stray twelve-year-old, the third son of the other family, and pushes him at the bewildered father: "Schmuck! Here's your son, for God's sake keep track of him!" The child goes with the new family, and the family stands among the survivors. And Jakob Gens, the Jewish Jew-killer, saves another child.

Lights down on KRUK. *The white curtain goes up again, reintroducing the* DOCTORS, *the* RABBI *and the* JUDGE, *in the pose where we left them. But something has changed. Their vaudeville costumes have been replaced by realistic clothes, and their jovial style has vanished as well.* THEY*'re sober, serious.*

RABBI Look Dr. Weiner. It just can't be. We are neither legally nor morally authorized to decide who lives and who dies. It's God's decision. God and God alone.

WEINER With all due respect, Rabbi, what world are you talking about? Surely not the world *we* live in—the world of the ghetto. God has deserted us here. Men decide everything here. Everything! And what you shrug and call the will of God, is the will of a group of evil men.

RABBI That's blasphemy.

WEINER What keeps us going? Hope. Hope that—what—that the Red Army will march in here and liberate the ghetto while we're still alive to appreciate it. Hope! That's all we have. (*Producing insulin ampule*) And this ampule here, for the men and women who live by it, contains the little word, "perhaps." It's hope in a bottle. In a little bottle of insulin is more hope for these men and women than all your principles could give them. For the others, the ones we call "hopeless cases," what's the point? If that day of liberation comes, what will it mean to them? It's too late for them. Heaven has nothing to say about these things anymore.

RABBI Blashpemy!

WEINER Look, Dr. Gottlieb. You're older than me, you have far more experience. Say something!

GOTTLIEB I'm sorry. I'm walking out of this meeting as a protest. I won't stand for discrimination against the ill. No matter how ill.

WEINER Are you suggesting I should become a robot? A medicine dispensing machine that sees nothing, feels nothing—is that your idea of ethics?

GOTTLIEB You're on your own. I wash my hands of it; you're a monster of medical science. Selection among patients? It's Nazi medicine!

WEINER I am a monster?! You condemn them *all* to death. You walk away from your responsibility to humanity and let them die? And you call me a Nazi?!

GOTTLIEB *looks from the* JUDGE *to the* RABBI *and back again. Without a word, the three of them turn and depart, leaving* WEINER *alone.*

What are you walking out on? Me, or your conscience?

GENS *steps into the hospital basement.* HE *has a bottle in his hand, and is drunk.* HE *mumbles past* WEINER.

GENS Father, mother, child, child. . . . Father, mother . . . (HE *waves at* WEINER *and collapses on the pile of clothes*) How come you stay here, Doctor? (*Pause*) Why not just . . . take off? Go join the partisans for God's sake. You even look Polish. You've got the forged documents, I *know* that. (*Self-mocking*) "Gens knows everything that's going on in the ghetto!" Sure, sure, sure. On top of everything else, you *sound* like a Pole. One hundred percent Warsaw Pole. What's keeping you, for God's sake; there's no future in the ghetto, y'know. So?

WEINER *remains silent.*

Go on. Join the partisans. Get out!

WEINER I'm afraid.

GENS (HE *laughs, mirthlessly*) What, you're staying here because it's so safe?

WEINER I'm afraid of deciding.

GENS But you have the courage to stick around here.

WEINER Here in the ghetto, I never decide anything. I go from day to day, not responsible, not responsible. I let it happen. Day after day you let it happen. In the ghetto there is nothing more beautiful than the philosophy of passivity.

GENS There's no future here, you know that very well.

WEINER What about you, Mr. Gens? You could leave anytime. What are *you* doing here?

GENS (*With the insolence of a drunk*) My place is here. With my people. I stand with the Jews in the ghetto. I won't flee into the forest.

WEINER Flee? You know the partisans aren't running away. Whoever goes into the forest goes to fight. They're heroes.

GENS (*Sits bolt upright. His drunkeness vanishes*) Hear me! There are many forms of resistance, my good Doctor. You want gun running and sabotage? Go to the forest. Don't bring it here in the ghetto.

WEINER *tries to object, but to no avail.* GENS *is gaining power as* HE *speaks.*

Don't play dumb, Dr. Weiner. Are you blind? Don't you understand what the Germans have in mind? Blowing our bodies to bits? That's easy. They can have any one of us they want. No, Dr. Weiner. They're after our souls. They're trying to get inside—reach down our throats, to the essence that's inside. Our souls. And that must never happen. This is the ultimate test of Jewish history: they'll lose the war, of course, that's a matter of time. But they could lose the war and still conquer the Jewish spirit, still infect us with their deadly sickness. Do you see? Can you understand? That's what the Resistance can never prevent out there. We have to protect our spirit, our essence in here, in the ghetto. And to do that, we have to save those who are strong. Physically, spiritually strong. Selection! We have no choice. The sick, the weak, the hopeless ones: let them go. They're a sacrifice. So much for insulin. (*Pause*) What will our children, our grandchildren think? Will they be able to justify our actions in their minds? Can they possibly understand the world we had to survive in? Well, it's not my problem. I stay here. I've got to save what there is to save. (HE *slugs from the bottle, and goes off mumbling*) There's no future in the ghetto. No future. That's why I stay.

SRULIK *crosses downstage. Blackout.*

END OF ACT ONE

Act Two

Four young gutter rats climb through a hole in the ghetto fence, maneuvering a coffin with them. THEY *are* LIEB GRODZENSKI *and his henchmen,* JANKEL, GEIWISCH *and* ELIA. GENS *shines a flashlight on them.* THEY *freeze in its beam.* GENS *recognizes one of them.*

GENS Halt! Good God. Leibele Grodzenski?
GRODZENSKI Evening, chief.
GENS (*Pointing to the coffin*) What the hell is that?
GRODZENSKI They buried a man in the cemetery outside, so we're bringing back the coffin.

GENS Through a hole in the wall? Has the gate disappeared, what?

GRODZENSKI (*Thinking fast*) It's a short cut.

GENS (*Cuffing* GRODZENSKI *good-naturedly*) What the hell is in that coffin Grodzenski? You running guns for the Resistance?

GRODZENSKI Who, us? Transport illegal firearms into the ghetto? You gotta be kidding, chief . . .

GENS All right, all right, so whaddaya got? Salami? Coffee? What? Sugar?

GRODZENSKI A ghost, chief. The ghost of a dead man, that's all.

GENS (*Trying to lift one end of the coffin*) This ghost has rocks in his pockets.

GRODZENSKI Well, there are ghosts and ghosts, right, chief . . .

GENS I can't have it, Lieb. You can't do this in public and expect to get away with it. You be at my office tomorrow morning at nine A.M. I want a five thousand ruble contribution to the juvenile delinquents home. In cash. Is that clear? Call it a tax.

GRODZENSKI Three thousand rubles, sure chief, sure . . .

That's all GENS *can take.* HE *collars* GRODZENSKI.

GENS You don't negotiate with me, friend! You're under arrest.

GRODZENSKI Wait, chief, wait. Tomorrow morning, nine A.M. Five thousand rubles. For juvenile delinquents. No problem, chief. You have the word of Leib Grodzenski.

GENS Too late. You spend the night in a cell while your friends dig up a ransom. You'll have some time to think it over, Momser. Next time you'll think twice before you tangle with Jakob Gens.

GENS *pulls him out. The* OTHERS *remain, look on in disbelief.* JANKEL *and* GEIWISCH *are talkative.* ELIA *lurks in the shadows.*

JANKEL Nu? What, do we follow him? You want to go spring him, what?

GEIWISCH That's a nice idea. Your five thousand rubles or my five thousand rubles?

JANKEL Oh, yeah. That's a problem. Well, you know what? Maybe we should make—a plan.

GEIWISCH (*Looks at him incredulous, then defeated*) We'll wait here for our contact to come from Weiskopf. When he collects the goods, he'll give us the dough, and we'll spring Leib from the can.

The TWO *of them settle down on the coffin and roll cigarettes.*

JANKEL That was a good plan.

GEIWISCH It wasn't that complicated.

ELIA, JANKEL *and* GEIWISCH (*Singing "Isrulik" [They Call Me Izzy]*)

Come buy my fine tobacco

Or buttons for your shirt.
A lower price you'll never have to pay!
Thank heaven for the ghetto:
Where life is cheap as dirt!
A penny lets me live another day!

They call me Izzy,
A kid right from the ghetto.
Always busy,
I hustle all day long.
In my pockets,
Less than nothing;
My only assets:
A whistle and a song.

A coat without a collar,
Galoshes but no shoes;
There's room inside my pants for two or three.
And if you think that's funny,
Well, mister, I got news:
I'll teach you not to laugh at guys like me!

They call me Izzy,
A kid right from the ghetto.
Always busy,
I hustle all day long.
In my pockets,
Less than nothing;
My only assets:
A whistle and a song!

I wasn't born an orphan
Or raised by hearts of stone.
My parents loved me just as yours loved you.
But they were taken from me.
Since then I'm on my own
And like the wandering wind, I'm lonely too.

They call me Izzy,
A kid right from the ghetto.
Always busy,
I hustle all day long.
In my pockets,
Less than nothing;
My only assets:
A whistle and a song!

They call me Izzy,
A kid right from the ghetto.
Always busy,

A smile from ear to ear.
Still it happens,
All too often
When no one's looking,

Sudden stop. Slowly.

I wipe away a tear.

JANKEL Hey! Shh. Someone's coming.

GEIWISCH From Weiskopf. Our contact.

The HASID *enters.*

HASID Good evening, boys, how are you?

GEIWISCH You come from Weiskopf?

HASID I can read your palm.

GEIWISCH Oh, for God's sake! Get the hell out of here!

HASID (*Moving to* ELIA, *the third gangster*) Ah hah! What's this? The coming week will bring a fundamental change in your life.

GEIWISCH I said beat it!

ELIA Hold on a minute. How do you know that? You haven't even looked at my palm.

HASID I do ears, too, but it is in the palm that all of the details lie.

ELIA *wrestles with his better judgment for a moment, but* HE*'s hooked.*

ELIA Okay, okay! Read my palm.

The HASID *looks at his palm.*

HASID Very interesting. Very interesting. Your palm is made up of eight and three. Chet and Gimel are at war, yes? In your hand, Chet conquers Gimel. Chet is like Cheruth—freedom, independence. Gimel is like Germany. Therefore, in your hand, freedom and independence will win out over the Germans. In three more time periods.

ELIA In what? What is that—"three more time periods"?

HASID Could be three weeks, three months, three years . . . (*Extending his hand*) Thirty rubles please.

ELIA What about three seconds?

HASID Well, if it's three seconds that will be forty rubles!

ELIA Here, take it.

HE *sticks a knife into the* HASID*'s gut. The* HASID *gasps.* ELIA *drags him a few steps, pulls out the knife, and the* HASID *collapses in a pool of blood.* ELIA *rifles his pockets.*

GEIWISCH My God, Elia you're crazy! What the hell did you . . .

ELIA *stands up with a fistful of money.*

ELIA Crazy, eh? A thousand, two, six, *ten thousand rubles!* Crazy is right.
GEIWISCH Stash it in the coffin! Quick!

THEY *open the coffin lid. A* FIGURE *wrapped in a shroud sits bolt upright. The* THREE GANGSTERS *gasp. Then the shrouded* FIGURE *stands and climbs out. The* THREE *scream and scatter. The* DEAD MAN, *now alone on stage, starts to unwrap his shroud. It's* KITTEL. *At that moment,* KRUK *enters the library and starts to dictate.*

KRUK This is the second murder and robbery in the ghetto. There's no doubt now that the perpetrators are members of the ghetto underworld. According to my sources, the crimes are tied to the flourishing black market here. The ghetto elite acquire whatever their taste dictates, and the underworld keeps the goods flowing to those who can pay.

KITTEL *throws the shroud into the coffin and removes a thick book.* HE *puts on a pair of horn-rimmed glasses, sticks the book under his arm and is thus transformed into a new character*—DR. PAUL. HE *enters* KRUK*'s library.*

DR. PAUL Do I have the honor of addressing Mr. Herman Kruk?
KRUK Indeed. And to whom, may I ask—
DR. PAUL My name is Dr. Paul. I'm from the Rosenberg Foundation. For the investigation of Judaism without Jews? It's a pleasure. You have heard of us.
KRUK I've heard the name . . .
DR. PAUL The foundation, Mr. Kruk, works with scholars and experts in all areas of Jewish culture, sending them into specially chosen ghettos to conduct our research. Our goal is to document the intellectual, spiritual and religious components of your culture, and separate the chaff from the wheat, so to speak. We're after the essence of Judaism—in terms of artifacts. Then, when we've collected certain cultural objects we send them to the Central Institute in Frankfurt.
KRUK I see.
DR. PAUL It's a difficult task, and it must be completed soon, before the transmitters of your rich heritage, ah . . . cease to exist. I have been given the great honor of being sent to Wilna, and I hope that you and I can develop a close working relationship, as befits a couple of scholars embarking on a noble, arduous task. I've heard much about you—your mind, your abilities. No doubt you've

heard nothing whatever about me. Allow me to give you my most recent work.

HE *gives* KRUK *the book* HE*'s been carrying.* KRUK *leafs through it.*

KRUK Investigations of the Talmud . . .

DR. PAUL Precisely. The Jerusalem Talmud. That's my particular area of expertise. I'm ashamed to admit that I still haven't mastered Aramaic. But I've begun. I've even done some work on the Babylonian Talmud.

KRUK And how, may I ask, did you find me?

DR. PAUL I used the method of Rabbi Jochanaan Ben Sakkai. But in reverse.

KRUK I beg your pardon?

DR. PAUL I was speaking metaphorically. The Rabbi escaped the occupied city of Jerusalem, as you no doubt know, on the eve of the temple's destruction. He was carried out in a coffin by four of his students. And I have entered your ghetto—the Jerusalem of Lithuania, to employ another metaphor—by the same method. Wouldn't you like to sit down?

KRUK Thank you.

DR. PAUL As your great poet Bialek said: "As man *schtejt—redt* man, as man *sitzt* . . ."

KRUK *"Redt sech . . ."**

DR. PAUL *laughs.* KRUK *joins him uneasily, then stops.*

How does it happen . . .

DR. PAUL How can a goy like me converse so easily in Yiddish? Do you think the question offends me? Or did you take me for a Jew? Even in Jerusalem they thought I was Jewish. Well. The Arabs did. I almost got killed in the pogroms of '36. The kids—Arab street gangs—pummeled me. I'm only here today because the Jewish fighters from the Haganah† saved me. Good men, nice fellows. (HE *chuckles as if at the memory of them, then his laugh turns threatening*) Were you there? In Jerusalem?

KRUK No.

DR. PAUL What a shame.

KRUK I've never even been to Palestine.

DR. PAUL Shame on you.

KRUK I wouldn't say that. I'm no Zionist.

DR. PAUL Communist?

*"When you're standing, you speak your mind, but when you're sitting, you listen (to someone else)."

†Jewish fighting organization in Palestine that preceded the Israeli Defense Forces.

KRUK (*After a moment to consider*) I *was* a communist.

DR. PAUL You were a founding member of the Polish wing of the International Communist Party, as a matter of fact.

KRUK So. You know all about me.

DR. PAUL Does it embarrass you, this . . . episode . . . in your political life.

KRUK Embarrass me? (*A pause*) Not at all. I mean, during the October revolution all of us were drunk with enthusiasm. I really thought—I really *knew*—that the revolution meant the end of all injustice, the end of persecution—even of Jews.

DR. PAUL But you left the party. *Before* Stalin's excesses made it fashionable.

KRUK Long before. Stalin had nothing to do with it. I left because the Jews in the party were vilifying their own heritage. It was outrageous.

DR. PAUL And that bothered you. Even though you're not a religious man.

KRUK My atheism has no more to do with it than Stalin did. It's just incomprehensible to me that Jews would spit on their own beliefs. The self-loathing, at the time I just couldn't understand—

DR. PAUL . . . But now you do. Understand, I mean.

KRUK That's right. Thanks to the Germans.

DR. PAUL (*Looks at him a moment*) Now *I* don't understand.

KRUK It's the circumstances that allow us to see ourselves so clearly. I mean, when Jews like that Gestapo-agent Dessler, and his henchman Lewas routinely storm around the ghetto beating up Jews to show the Germans how friendly they can be, what am I supposed to think? When the Jewish police invite German officers to the Jewish council building to have a party, get drunk together, sing songs together, invite in a truckload of Jewish whores for the night, I suddenly understand just how deeply Jewish self-hatred is rooted.

DR. PAUL (*Regards him with apparent sympathy*) Yes. I see. But you're a socialist, right? A Bundist, as you people call it? A believer in the Diaspora. Jews wandering the world, as opposed to a Jewish state.

KRUK That's right. Does that seem odd to you?

DR. PAUL After everything that has happened to you, you still believe that socialism in the Diaspora would allow for the survival of Jewish culture?

KRUK Absolutely. Maybe not in my lifetime, but someday.

DR. PAUL You remind me of a Hasidic legend. A king had a fight with his son, and threw him out of the castle. Then he thought better of it, and sent a messenger out to find him. He said to the messenger: "Go seek my son, and ask him what he'd wish for if he had three wishes." The messenger found the son living in filth, clothed in rags, and put the question to him. And the son said: "Three wishes? Bread, clothing, and a place to sleep." The messenger reported back to the king, and the king said: "My son has forgotten that he's a prince. If he'd remembered who he was, he would have only had one wish—to come back to the castle. With that, all his other wishes would also have come true. He would have had food, clothing and far more than shelter. My son is really lost forever." So you see? You dream of native cultural riches for a people wandering the world? Why fight for a few privileges among people who don't want you when you could go back to the castle? In Palestine you could have it all. You wouldn't have to walk around in rags *hoping* that one day—"not in my lifetime"—good would triumph over evil.

KRUK Are you a lobbyist for the Zionists, Dr. Paul? Allow me a question. What did you do in Jerusalem?

DR. PAUL Allow me a question, Mr. Kruk: I know you deplore the way Gens uses people like Dessler and Lewas to control the ghetto. Strongarm tactics and—

KRUK Gens does what he can under the circumstances you created.

DR. PAUL Are you defending these characters? Believe me, I don't like them any better than you do. All they want to do is mimic us, look like us . . . but what they look like is a horrifying caricature—something you might see in a funhouse mirror. I, for one, don't enjoy staring at such things. This slavishness will get them nowhere. You, on the other hand, are different. You go your own way. You have good instincts, Mr. Kruk. (*Confidentially*) Listen: I know you were in the underground in '20 and '21. During the anti-Semitic riots you behaved heroically, I know all that. Suppose I told you Gens's days are numbered? He and his henchmen are no good to us anymore, and we'd like to put you in their place. You could handpick the people you'd work with. Carte blanche.

KRUK Are you joking? I wouldn't take a job offer from you!

DR. PAUL Not even if you crown Gens by your refusal?

KRUK *Nebuch!** Heaven protect me from a king who depends on *your* favor. And as for the Hasidic legend, you've got it all wrong. The son understood the situation perfectly. He wished for what he needed: Men aren't at home because of any particular soil—they're at home with their heritage, with their traditions. *That's* the loss they must guard against. Without culture, they lose their identity.

DR. PAUL Very well. If you side with the Diaspora, you leave the future of Zionism to the Genses of this world. *They* don't turn down offers of power. Not from us, not from anyone.

KRUK What people do in the ghetto, Dr. Paul, may have nothing to do with their behavior in Palestine. Palestine's another world.

DR. PAUL Permit me to disagree. I've been there, you haven't. I'm also familiar with the leaders of the same Zionist movements that Gens

*Roughly, "Lord help me!"

and Dessler subscribe to. I know whereof I speak. (*Pause*) I admire *your* brand of Judaism, Mr. Kruk. *Your* Judaism might stand a chance of creating a balance between our two peoples. I regret that you won't assert it, just at the moment when you might gain some political influence. You might have corrected a terrible historical injustice.

KRUK What balance? What historical injustice? What are you talking about?

DR. PAUL Well, never mind. We'll discuss it another time. (HE *stands up, hands* KRUK *a list*) In the meantime, I need these books wrapped. I have to send them to Frankfurt. Artifacts, as I said. In addition, I have a less tempting job offer, one that you'll hardly find tainted, though. I need a report on the sect of Karaites in Wilna. In Berlin they want a scientific opinion as to whether the Karaites belong to the Jewish race of not.

KRUK Of course they're Jews, but . . .

DR. PAUL You're not above making such a report?

KRUK *stares at him. In the background, the noise of a crowd.* DR. PAUL *and* KRUK *stand.*

DR. PAUL (*Points to the noise*) You see? Those are the *other* Jews. They understand power. Those are the faces in the funhouse mirror. You think I enjoy looking at them? (HE *laughs a bone-chilling laugh and disappears*)

GENS, *the* JUDGE, DR. GOTTLIEB, DESSLER *and* LEWAS *enter. The latter two are dressed as ghetto police.* THEY *lead the three condemned gangsters,* ELIA, JANKEL *and* GEIWISCH, *who are tied together. A wooden frame is shoved onstage by* TWO ACTORS *dressed as butchers. On the upper crossbar are three meathooks, each dangling a hangman's noose. Three stools are placed under the hooks.*

GENS (*Calls out*) Your Honor!

JUDGE The Jewish court of the Wilna ghetto, in its session of June 4, 1942, has reached a verdict in the case of Jankel Polikanski and the brothers Itzig and Elia Geiwisch. The accused are found guilty of the murder of actor and palm reader Joseph Gerstein on the night of June 3. Having been found guilty of this murder, the three are sentenced to death by hanging.

GENS (*Trying to remain sympathetic*) Your Honor, members of the Jewish Council, police officers, ladies and gentlemen. Of the seventy-six thousand Jews who once populated Wilna, sixteen thousand, thank God, are still alive. It is the duty of these remaining Jews to be upright, hard-working and honest. For those who do otherwise, we have no comfort. We must investigate and prosecute all criminal cases within the ghetto, and carry out the sentences with our own hand. We have no choice . . .

KITTEL *enters and stands near* GENS. THEY *greet each other with a look. Suddenly* GENS *is a transformed man, aggressive and blunt.*

The execution of the three convicted slayers—Jews who murdered Jews—will be carried out in the courtyard of the old slaughterhouse, at number nine, Butcher's Street. The sentence will be carried out by the Jewish police, whose duty is to protect life, law and order in the ghetto. (*To the* POLICE) Gentlemen: your duty!

DESSLER *and* LEWAS *lead the condemned* MEN *to the gallows.* GENS *raises his cane, and gives the signal by sharply dropping it.* KRUK *dictates.*

KRUK The rope around the neck of Jankel Polikanski broke, and Jankel fell to the ground. He was still alive. Gens, citing the oldest traditions of judgment, wanted to pardon him. Kittel, looking like a Roman Caesar, turned thumbs down . . . and Jankel was hanged again.

KITTEL (*Moves to center stage, raises his arm to speak, and pulls a sealed notice from his pocket.* HE *ceremonially breaks the seal, unfolds it, and reads*) On this solemn and impressive demonstration of orderly self-rule by the Jews of the Wilna ghetto, which has been carried out flawlessly in every respect, I hereby declare: Whereas the Wilna ghetto leadership is about to embark on an important new task; and Whereas the present Judenrat is an unwieldy and slow-moving body, now therefore, the Jewish Council is hereby dissolved. In its place I name Jakob Gens as the autonomous and sole leader of the Wilna Ghetto.

Applause.

He will be assisted by Mr. Dessler, as police chief and Mr. Lewas as chief guard of the gate.

Applause.

GENS (*Raising his hand, like* KITTEL *did*) Thank you. In honor of this change in ghetto leadership, I invite the police chief and other public officials to a celebration. And I make a solemn promise to all of you: The new ghetto leadership will do everything, everything in its power to promote well-being and security in the community. I thank you.

KITTEL I accept your invitation to the celebration if—if—you can guarantee good music, and a first-rate show. It will be a great honor to see your unforgettable chanteuse again. I still have a small bill to settle with her, perhaps we can do some business at your celebration. (HE *puts his proclamation away, turns to go, then turns back for a moment*) One other thing. In honor of this event, I will waive one of our strictest regulations. For the celebration—and only for the celebration—you may once again bring flowers into the ghetto.

HE *goes. A jazz* BAND *appears and strikes up a cheerful number. It's not a real band, but the ragged* ACTORS *who we met earlier with their instruments, spruced up a bit for the occasion.* WEISKOPF *sweeps on stage, very much the successful businessman in authority. During his next speech, his instructions are followed, causing a riot of activity.* WEISKOPF *himself runs to and fro, the happy despot in the midst of his prosperity. At the end, the stage is utterly transformed.*

WEISKOPF Flowers! For God's sake more flowers! It's a once in a lifetime event, I want to see a riot of colors. Petals, blooms—everywhere. (*To a pair of* ACTORS *carrying a buffet*) Look, the cold buffet can go over there, and the roast chicken over here. (*To some* COOKS *setting up the food*) No, no! The gravy next to the chicken, what is it with you people? You've never been to a dinner party before? Shlemiel. All right now. (HE *turns, looking for something*) Wait a minute, where's the *cholent?* What the hell happened to the *cholent?* What, are you gonna serve it for desert? Get it out here!

More COOKS *emerge with a huge stewpot.*

Okay. Now the bar, the bar . . . the bar can go there. No, wait. Better idea. Divide the bottles up among the people and they can drink all they want.

WAITERS *begin to place bottles on tables.*

Open 'em up! Open 'em up! I want it lavish—leftovers we can donate. We'll show these pigs we know how to celebrate. What happened to the kvass? (*To an idle* WAITER) Hey, you! Shmuck! Bring the kvass. (*To another*) And you! Didn't I say open all the bottles? Don't talk to me about waste, it's my money, right? Is it your money? No, it's my money, I'll spend it my way. Besides, it's business. I mean what the hell, *shmeikel* these pigs today, tomorrow the orders start rolling in. I'll bring in a hundred times what I'm spending today. Today's nothing. Especially the way it looks now—who the hell is responsible for this?! All right, all right, the orchestra goes here and the stage . . . (HE *turns to the makeshift platform where the entertainment will take place*) My God, the stage! Who the hell arranged this place? What is this, poverty theatre? This is supposed to be a "Follies," not a tragedy—who thought this up? Oh, what the hell, strew the damn place with flowers, who'll notice? (*Shouting off*) Another truckload of flowers! (*Under his breath*) I'll kill that Srulik, where does he find these set decorations?

To the STAGEHANDS *as* THEY *work with the flowers.*

I want it gorgeous. Gorgeous and plush, like a bower in heaven. Their eyes should pop out of their heads when the lights come up. Their eyes should pop out and they should never find them! May the plagues of the Pharaohs and the trials of Job befall them! (*Going a bit mad now*) And the food—gorgeous too! And with a smell—they should eat like no tomorrow. Seconds, thirds, stuffed from head to heel! Plug 'em up at both ends and may the worm of Titus dance a tango in their brains. *We'll pleasure them to death!* Rice, meat, roast chicken and gravy! Cake! And kvass, kvass, kvass! Build up the pressure, tangle their guts till their asses explode! Strangle 'em on their entrails and feed 'em to the dogs! (*Looking around, in a sweat*) Magnificent! Spectacular! Brilliant! Orchestra ready? *Play!*

The stage is unrecognizably lavish. The ORCHESTRA *strikes up. The guests enter:* GENS, DESSLER, MUSACHKAT, KITTEL, *two* GESTAPO OFFICIALS, *three* JEWISH PROSTITUTES, SRULIK *and his* DUMMY, *and* CHAJA, *in a spectacular evening gown.* WEISKOPF *receives the guests, offers drinks and hors d'oeuvres. While* THEY *eat and mingle,* CHAJA *sings "Friling" (Springtime). During the song, the orgy begins. Dancing, drinking, eating. The* PROSTITUTES *move through the* CROWD *offering their favors freely.*

CHAJA (*Singing*)
I wander through alleyways, lost and distracted
Until I arrive at the wall.
The weather is warm and the breezes are gentle
But I don't feel April at all.
I stand there and listen to laughter and street sounds
That come bubbling in from outside.
Your face is before me wherever I go
And it's almost a year since I cried.

Springtime
Where is my loved one?
Why do the birds sing up there in the trees?

Springtime
There's music in the flowers
But till I join you I won't hear the melodies.

I pass our old house ev'ry morning at six
On the way to my long daily grind.
The doors are all padlocked, the windows are shuttered
Like me, it is deaf, dumb and blind.
Each night I am drawn to the same shady corner
We'd meet there each day, way back when.
But why do I go there? I guess I can't help it . . .
Tonight I will go there again.

Springtime
Where is my loved one?
Why do the birds sing up there in the trees?
Springtime
There's music in the flowers
But till I join you I won't hear the melodies.

I peer in dark waters, the face that's reflected
Is someone I no longer know.
Wherever you are, be it earth or in heaven
Or hell, that's the place I will go.
I wander through alleyways, lost and distracted
My odyssey's practic'ly through
And spring will be warm in the April of Aprils
When I'm reunited with you.

Springtime
Where is my loved one?
Why do the birds sing up there in the trees?
* Springtime
There's music in the flowers
And when I join you I will hear the melodies. (*Repeat**)

When the song is done the AUDIENCE *applauds.* KITTEL *raises his hand and there is silence.* HE *moves to* CHAJA.

KITTEL Close your eyes.

SHE *does.* KITTEL *pulls a long string of pearls from his pocket and places it around her neck. Amazed admiration from the* CROWD.

Now open them.

CHAJA (*Discovering them*) Oh!

KITTEL Unfortunately, they're only pearls. But if you knew where I'd gotten them . . .

CHAJA *tries to take them off, but* KITTEL *stops her.*

Now, now. She who begins with shoes ends up with pearls. But you still owe me three ounces of beans. Well, two and a half, counting that last song.

DUMMY Careful Chajele, your price is dropping.

KITTEL (*Leaves her and moves to* SRULIK *and the* DUMMY) And how's our little wooden friend? Still taking chances?

DUMMY Taking chances? Please. It's just ordinary, run-of-the-mill chutzpah.

KITTEL Ha! Chutzpah, you say. A great Jewish tradition. Let's hear a little chutzpah, if you dare.

DUMMY All right, but . . . my, my, Herr Kittel, you don't look too well.

KITTEL It's nothing. A little headache.

DUMMY A headache. Well, by all means, I'd prescribe head baths. A miracle cure guaranteed to eliminate all pain.

KITTEL (*Not understanding*) Head baths?

DUMMY Stick your head in the water three times and pull it out twice.

KITTEL *laughs and* EVERYONE *joins him. Suddenly* HE *stops.* EVERYONE *stops. Tense silence. Then* KITTEL *laughs loudly again.*

DUMMY Do you know why a German laughs twice when he hears a joke?

KITTEL No, why?

DUMMY Once when he hears it, and once when he gets it!

KITTEL (*Laughs once, goes silent. Then laughs again, goes silent again, ominously*) So. That's chutzpah. Very good. I bet you wouldn't dare to take it one step further.

DUMMY How much do you bet?

SRULIK (*To the* DUMMY) What?! Cut it out.

DUMMY You cut it out. We could get rich here. (*To* KITTEL) All right. You can put up fifty thousand rubles, and I'll wager my life.

KITTEL (*Reaches in his pockets and comes up with some paper money*) I'm a little short . . .

WEISKOPF (*Butts in, forcing a wad of bills on* KITTEL) Please, Mr. Kittel, be my guest.

KITTEL Do you have a pen, I'll give you an IOU.

WEISKOPF *pats his pockets for a pen.*

DUMMY What does he need an IOU for? Germans always give back, it's in their character. You took Krakov—you gave it back. You took Stalingrad, you gave it back. Germans give *everything* back—you'll pay Weiskopf back every penny, of course!

The proceedings come to a dead, stunned halt. KITTEL *stares at the* DUMMY. *A horrified silence.*

KITTEL (*Handing the money to* SRULIK) Here. You win. This time.
SRULIK No, please . . . he didn't mean . . .
DUMMY Shut up and take it. A thief who steals from a thief is no thief.
KITTEL (*Sharp, angry to* SRULIK) *Es reicht.**

Another dead, mortifying silence. WEISKOPF *breaks in with a bottle of cognac.*

WEISKOPF Some cognac, Mr. Kittel, please. You've never tried a finer. Exquisite French cognac . . . only the best for you.

HE *forces a glass on* KITTEL. *Both* MEN *drain their glasses in a single swallow.*

KITTEL Aaahhh! Paree! Paree! That reminds me of Paris. Paris . . . Paris. Very nice, yes.

SRULIK *quickly orders the* ORCHESTRA *to strike up some French music.* CHAJA *begins to sing a chanson. The party has now begun to deteriorate into a debauch, crushed flowers and clothing litter the stage, and many empty liquor bottles.*

WEISKOPF (*To* KITTEL) I'm in a position to make you a fantastic offer.
CHAJA (*Sings "Parlez-moi d'Amour"*)

Parlez-moi d'amour
Redites-moi des choses tendres.
Votre beau discours
Mon coeur n'est pas las de l'entendre.
Pourvu que toujours
Vous répétiez ces mots supremes:
Je vous aime.†

Parlez-moi d'amour
Redites-moi des choses tendres . . .

* "That'll do!" or "That's enough!"
† Speak to me of love
Tell me once more those tender things.
My heart will never grow weary
Of hearing your beautiful speech
Provided you always
Repeat those supreme words:
I love you.

KITTEL is now dancing suggestively with CHAJA as SHE sings. The COMPANY interrupts her with "Mir Lebn Eibek," carrying her away from him.

ACTING COMPANY (*Singing*)

Mir lebn eibek, ess brent a velt.
Mir lebn eibek, on a groshn gelt.
Un oif zepukenish alé sonim
Voss viln unz farshvarnz unzer ponim—
Mir lebn eibek, mir zeinen do!
Mir lebn eibek, in yeder sho.
Mir viln lebn un derlebn
Shlechte zeitn ariberlebn.
Mir lebn eibek, mir zeinen do! *

At the end of the song WEISKOPF calls for silence.

WEISKOPF Ladies and gentlemen! I have good news. Mr. Kittel and I have just closed the most tremendous deal in the history of laundry! We will take in four hundred railroad cars of uniforms in need of repair. Work for everybody! But there's more: I've just received word—I'm afraid I can't reveal my source—that I am to have a meeting with Göring himself! I will travel to Berlin to work out a five-year contract between the German army and the factory. We will build a new place—a gigantic plant with all new machinery and equipment for making uniforms, fatigues, combat boots, dachrician—everything for the modern soldier. The success story continues! So . . . to your health! L' chayim.

ALL L' chayim!

KITTEL Prosit!

ALL Prosit!

EVERYONE drinks. The orgy gets further out of control. KITTEL sees GENS standing alone, observing.

KITTEL Gens! Gens!

*We live forever, a world is burning.
We live forever, without a penny.
And to spite our enemies
Who want to blacken our days—
We live forever, we are here.
We live forever, always.
We want to live and see the future
Surviving the bad times.
We live forever, we are here!

GENS *goes dutifully to* KITTEL. KITTEL *lays a hand on* GENS*'s shoulder, and walks with him.*

What's your problem, friend? You're not enjoying this party.

GENS I'm having a wonderful time.

KITTEL Please! Dessler is having a wonderful time. Muschkat is having a wonderful time. Lewas is having a wonderful time. You are standing like a stone. You never have any fun. You know how to throw a party, but you have no idea how to enjoy one. And do you know why? Because you're an asshole, Gens. You want to *use* a party. You want to make sure we're fraternizing in a useful, productive way. You want favors. You want to prove something. I just want you to have a wonderful time. How does the song go? You know the song: "I want to be happy, but I can't be happy . . ."

GENS ". . . till I make you happy too." I'll do my best.

KITTEL It's not hard! Look, I'll help you.

KITTEL *raises his hand. Immediate silence.*

Ladies and gentlemen! I have good news. You'll all be very proud. I've decided to expand the empire of our friend Gens. I hereby annex the Oschmany ghetto. From this moment on—with the aid of the Jewish police and the honorable Mr. Dessler, the Jews of Oschmany are your subjects.

Applause. DESSLER *stands and bows.* KITTEL *raises his hand. Immediate silence.*

There is one . . . small . . . thing. As you may know, there are four thousand Jews living in Oschmany. Unfortunately, that's two thousand more than we need. So, this evening, a battalion of the Jewish police, under the direction of Dessler . . .

DESSLER Yes, sir!

KITTEL We will conduct a selection process. Of course, we could send our own people or Lithuanians, but our presence in the ghetto always upsets you people so much. No need for unnecessary panic. Your people speak Yiddish, the population will stay calm, and the job can be done smoothly. (*Issuing an order*) Police officers. Attention!

The JEWISH POLICE *rise, untangling themselves from the* WHORES. THEY *are stripped to their underwear.*

Ah. How convenient. You see, everything works out for the best. To celebrate the Oschmany plan, you will get new uniforms. Russian officers' uniforms, complete with caps and coats, leftover from the Czar's army. We found them in the Wilna warehouse. Bring on the uniforms!

A GERMAN SOLDIER *brings fresh uniforms. The* JEWISH POLICE *dress.* ONE *of the* WHORES *distracts* KITTEL, *who collapses on top of her for a quickie. As* HE *paws her,* GENS *stands over them.*

GENS Excuse me. Uh . . . pardon me, Mr. Kittel. Out of four thousand people, more than half have got to be productive.

KITTEL (*Distracted, but listening*) You think so?

GENS Absolutely. According to our experience, no more than one thousand are really unproductive.

KITTEL Is that so? Really? All right, make it one thousand.

GENS Well, wait, wait. Let's suppose that in the selection process, we discover that there are only eight hundred who are really unproductive.

KITTEL Take it easy, Gens. Productivity is a vague concept, I don't need to tell you that. A Jew can prove anything he wants to, yes? Is an eighty-year-old man in a wheelchair productive? *You'd* connect a generator to the wheel chair and claim he's making electricity when he rolls off to take a shit!

GENS All right, all right. Look, that brings up age, and age is very clear. What would you say to a selection of all Jews eighty and older.

KITTEL (*Looking* GENS *straight in the eye*) Seventy.

GENS Seventy.

KITTEL But a minimum of seven hundred.

GENS No fewer than five hundred, no more than seven hundred.

KITTEL Listen, Gens, what's a hundred head more or less between you and me? Call it six hundred and we'll shake on it.

GENS Six hundred. Deal.

KITTEL *turns to the* POLICE, *who are now in uniform and armed with truncheons.*

KITTEL Nice. Very handsome. We'll send along eight Lithuanians from the Ypatinga militia. You hand over the old people, they'll do the rest. Dessler!

DESSLER Yes, sir!

KITTEL The troops are at your command!

DESSLER Yes, sir! Thank you, sir!

KITTEL Would you care to address the troops, Dessler?

DESSLER Gentlemen! MOVE! We have been given an order, and we will execute it to the letter! Every detail. Any questions? All right then: Left face! Division, march! Left-right, left-right . . .

THEY *march off in front of* DESSLER, *when* HE *is almost off,* GENS *shouts after him.*

GENS Dessler! Get back here!

DESSLER *returns.*

You're about to do a filthy job. There's no choice, I understand that. You needn't jump at it like a famished dog in front of these butchers.

DESSLER We're doing it, aren't we? You think it matters *how* we do it? You think your broken heart buys you anything? We're none of us going to heaven, Gens, and you gave the order. You stay here and drink—I go out into the field and do the job. Don't you preach morality to me!

DESSLER *wheels and departs.* GENS *spits after him.*

GENS Scum! (HE *upends a bottle of cognac*)

KRUK *appears and dictates.*

KRUK Four hundred and ten old and sick Jews were selected and penned together in the square in Oschmany. An old Jew began to sing "El Moleh Rachamin," and everyone started to cry. Some of the Jewish police, who had rounded up the elderly, broke into uncontrollable wailing. The 410 were driven six miles out of town, and the eight Lithuanians went to work, liquidating the throng while the seven Jewish police from Wilna stood by. The action was overseen by Dessler, Nathan Ring and Mosche Lewas—all Jews. The three of them were armed with pistols. During the entire process, selection and extermination, the seven Jews and eight Lithuanians consumed one hundred bottles of vodka and schnapps. The Jewish Council donated a baked lamb for the event. When it was all over the astounded citizens of the Oschmany Ghetto lined the streets in dumb incomprehension and stared at the departing Jewish police. One of the policemen, Isaak Auerbuch, was seized by hysteria during the selection process and had to be given emergency medical treatment. Another, Dressin, began to sing:

"We came to warm your heart,
Good night, we now depart."

GENS (*Drunk, lurches forward and speaks to* KRUK*'s voice*) Kruk—no, no, no, listen—Kruk is an honest man. A courageous man. He tells the truth, and not many people want to hear it. He's all right. Fearless. Many of you think just the way he does, I know it. I know it. You think I'm a traitor, you wonder what the hell I really want out of all you upright, innocent people. After all, it's me you're talking about, isn't it? I'm the one who has your hiding places blown up, right? Gens! Well, Gens has his own way of hiding Jews from the butchers. I wheel and deal, right? I'm a monster, right?! Well, for me only one thing counts. Not Jewish honor. Jewish life! Jewish lives. If the Germans want a thousand Jews from me, they *get* 'em! Because if we don't do it their way, they'll march in here and take a thousand Tuesday, and a thousand Thursday. And a thousand Saturday. And ten thousand next week. You people are all saints, I know! You don't dirty your fingers. Gens gets down in the mud and wrestles with the pigs. And if you survive, then you can say: Our conscience is clear. But me? If I live through this I'll walk through life dripping shit, blood on my hands, and I'll turn myself over to the Jewish tribunal and say, "Look at me! Everything I did I did to save as many Jews as I could. To save some, I led others to their deaths with my own hands. And to preserve the consciences of many, I had no choice—I plunged myself into the sewer, and left my conscience behind." A clean conscience for Jakob Gens? I couldn't afford one!

END OF ACT TWO

Act Three

The PEOPLE *slowly start to awaken and clean up the orgy. As* THEY *do this* CHAJA *sings a Yiddish version of "Friling."*

CHAJA (*Singing*)
Ich blondzshé in geto
Fun gessl zu gessl
Un ken nit gefinen kein ort;
Nito iz main liber
Wi trogt men ariber,
Mentshn, o zogt chotsh a vort.
Ess loicht oif main heim izt
Der himl der bloyér—
Voss zshé hob ich izt derfun?
Ich shtei vi a betler
Bai jetvidn toyér
Un betl, a bisselé zun.

Friling, nem zu main troyer,
Un breng main libstn,
Main trayen zu-rik.
Friling, oif dainé fligl bloyé,

O nem main harz mit
Un gib ess op main glik.*

The scene evolves to a May Day celebration in the ghetto. Red paper flowers, red flags and posters decorate the stage. EVERYONE *wears red neckerchiefs or scarves. As the music ends,* KRUK *speaks.*

KRUK After the events in Oschmany, the atmosphere in the ghetto was never the same. The people grew restless, irritable. Normally, ghetto inhabitants sleepwalk through life. They live like flies—from day to day. Things calmed down so they calmed down. What choice did they have? But there is a faction that refuses to be calm. At this moment Zalman Tektin lies in the prison hospital in critical condition. Yesterday he was shot during an attempt to rob a German munitions depot. He's eighteen. (HE *pulls a manuscript from his pocket*) This song was written by a comrade of Zalman's, if he were here he'd want us to do it for him. Let's sing it! (HE *hands it to* CHAJA) Can you do it? Or do you only know tangos?

CHAJA *steps on stage and takes the music and sings "Zog Nit Keinmol" (Never Say You Can't Go On).*

CHAJA (*Singing*)
Never say you can't go on, your day is done.
Yes, a thick and smoky mist enshrouds the sun;
But the sound of marching feet is drawing near
And they're beating out the message: We are here!

*I wander in the ghetto
From alley to alley
But can't find any place.
My beloved is gone:
How does one live through it?
People, tell me at least one word.
The blue sky
Now shines on my home—
What do I gain from that now?
I stand like a beggar
By each and every gateway
And beg for a little sun.

Spring, take away my sorrow,
And bring back my beloved,
My faithful one.
Spring, on your blue wings
O take my heart with you
And give it to my love.
Less literally translated stanzas from this song appear on pages 520–521.

From the land of waving palms and drifting snow,
We are coming braced by pain and steeled by woe.
And where torrents of our blood have
Stained the earth,
There our strength and dedication find rebirth.

Gradually, the ENSEMBLE *joins in, a militant energy driving them.*

Some day soon the sun will bless us with its light
And our foes will be devoured by the night
But until the sun can burn away the mist
Then this song shall be our theme as we persist.

Not a song the robin sings throughout the land
But a song a people sings grenade in hand!
It was written not with ink but with our blood
Mid collapsing walls and storms of flying mud!

So
Never say you can't go on, your day is done,
While a thick and smoky mist enshrouds the sun;
For the sound of marching feet is drawing near
And they're beating out the message:
    We are here!
Yes, the sound of marching feet is drawing near
And they're beating out the message:
    We are here!

GENS *bursts on stage.* EVERYONE *turns to him.*

GENS What the hell is this? What are you thinking? A song like that—it could sink the whole ghetto. Do you know what would happen if the Germans even heard about this?

SRULIK Mr. Gens . . .

GENS I'm talking!

SRULIK But Mr. Gens . . . you're the one who ordered us to create a theatre.

GENS Not this kind of theatre! Not a song that incites riots! We've just calmed the place down for God's sake.

DUMMY Like a graveyard.

GENS People, please, there's nothing to fear now. The ghetto is finally secure.

DUMMY Well . . . compared to Ponar . . .

GENS This is no time to provoke audiences. Theatre, yes—but entertainment. Something to take people out of themselves. We need to show the world we're industrious, hardworking people, not maniacs.

DUMMY Present company excepted.

GENS You want to make fun of something, fine, make fun of your own: take a crack at people who won't look for work, at parasites.

DUMMY (*Pointing at* SRULIK) Here's one. He won't even help our German friends pick out their favorite Jews.

GENS Look, I've got nothing against satire, but watch who you aim it at.

DUMMY Whoever passes by. Parasites, traitors.

GENS Who's a traitor? Me? The Jewish police? Just what the hell are you trying to say? The Jewish police will lead this place to freedom! Not you! It's a delicate balance we've achieved here. Upset that balance, and *you'll* be the traitors.

KRUK That's what you call solidarity Mr. Gens? . . . One people.

GENS Don't you preach morality to me! If there's a Jewish patriot here, it's me! I'm going to bring Hebrew into the ghetto starting right now! Tomorrow morning we begin Hebrew lessons in the schools. Required. No more Yiddish bibles. Hebrew bibles. Hebrew in elementary schools from day one. In kindergarten. And I'm also introducing Palistinography to the schools. Any objections? Me. I object. I object to the utter lack of Jewish national conscience in Wilna. I object! And in the theatre! A gala performance in Hebrew, how about that for an idea! Hebrew lectures, readings, a blue and white evening dedicated to reading the poetry of Bialik! Anyone not in agreement with the new policy of nationalism is hereby barred from all key positions. Alright now. Does anyone object?! I thought not.

KRUK I'm sorry Dr. Paul isn't here. He was right. The Germans have been more successful than they could have dreamed.

GENS What was that?!

KRUK Nationalism breeds nationalism.

GENS I beg your pardon. Are you saying the Germans have done this to me?

KRUK Take it however you like.

GENS This rehearsal is hereby dissolved. Everybody go home. Now!

The GROUP *disperses singing "Zog Nit Keinnmol." Only the* DUMMY *and* SRULIK *remain.*

DUMMY Go home! Get out! Study Hebrew!

SRULIK Please, not now. I can't take it anymore.

DUMMY Pay attention! A historic moment. You'll want to tell your grandchildren you were there for the Hebrewization of the ghetto. At this moment in history, what could be more important than Hebrew. To say nothing of Palistinography. You don't know what Palistinography is, do you?

SRULIK Yes I do. You stick your head into Palestine three times and you take it out twice.

DUMMY But let's get down to brass tacks. How do you tell Chaja you want to sleep with her in Hebrew?

SRULIK How did I get myself into this mess?

DUMMY *"Ani chafetz bach,"* I think. Or maybe *"Yesh li chefetz bach"!** What do you think? Which one? Could be important. Could be the difference between a "yes" and a "no."

SRULIK Don't make me *meshugge!*

THEY *start to leave, the* DUMMY *still chattering.*

DUMMY Chajale! Chajale! *Ani chafetz bach!* It's love! It's love! What do you say, Chajale? You wanna . . . *chafetz bach?* You think so?

THEY*'re offstage now, the* DUMMY*'s voice fading.*

My, my, my. The Hebrewization of the ghetto of all things. What next . . .

The scene shifts to the library. CHAJA *is looking for a book.* SHE *has changed. Her dress is simple, almost masculine. Her hair is pulled back.* KRUK *approaches and looks at her carefully.*

KRUK May I help you?

CHAJA No thanks. Just looking . . .

KRUK Excuse me, but you come here every day and search for hours. Surely I could help you . . . you must be looking for something.

CHAJA (*A little defensive, protecting something*) I like to browse.

KRUK I thought you were wonderful in the revue—"Pesche from Resche."

CHAJA Wonderful?

KRUK There's no need to be ashamed. It's a fine thing to be an actress.

CHAJA Is it? I don't think so.

KRUK No?

CHAJA What good is theatre in our situation? It's trivial . . . even insulting.

KRUK I thought that way once. I was against the theatre company from the beginning.

CHAJA I know. You were right. You are right.

KRUK (*Shakes his head*) No. Every form of cultural activity is essential here in the ghetto. It's the battle plan in our fight to remain human beings. The fascists can kill us at will—it's not even a challenge for them. But they can't achieve their real aim: They can't obliterate our humanity—not as long as we cling to a spiritual life, not as long as we reach for the good and the beautiful. They forbid flowers in the ghetto, we give one another leaves. And suddenly, leaves are the most beautiful flowers in the world. Theatre is essential.

*Both expressions mean "I desire you." The first uses the Hebrew active voice with direct object; the second, an indirect and passive form.

Silence.

You must be looking for a book on theatre. Come.

CHAJA I don't want a book on theatre. I want a book . . . on explosives.

KRUK (*Smiling*) Why didn't you say so? I could have saved you precious time.

HE climbs a ladder and takes a thin book from the top of the shelf. HE climbs back down and hands her the book. SHE looks through it quickly.

CHAJA Is this Russian?

KRUK It's a Soviet army manual. I stole it from the university. It's the only book I've ever stolen.

CHAJA But I don't know the language.

KRUK You must have friends. Show it to your friends. Unless I'm mistaken, one of your friends will know Russian.

CHAJA Thank you. I'll bring it back.

KRUK Please, I don't even know you took it. A thief who steals from a thief is no thief.

CHAJA Thank you. (SHE *heads for the door*)

KRUK (*When* SHE *is almost there*) Wait.

From a tin box HE takes a leafy stem and hands it to her. SHE takes it, and begins to sing "Dremlen Feigl" (Drowsing Birds) to him. As SHE sings the scene shifts.

CHAJA (*Singing*)

Baby birds in summer branches
Drowse in a downy nest
Down below a baby nurses
Softly on a stranger's breast
Lullaby, lullaby, croons the stranger
Rest, sweet little one, rest
Lulu lulu lu

There's a story I must tell you
Though you're much too small
Baby dear, your mama's gone
She won't be coming back at all
And when your daddy tried to save her
These eyes saw him fall
Lulu lulu lu

If you get a little older
This sad story you will know
Carry it with you like a blessing

Everywhere you go
Lullaby, lullaby, klaine schaine
Orphan child, I love you so.
Lulu lulu lu.

As SHE *sings and strolls, lost to this world, a strong flashlight beam catches her.* SHE *freezes, startled. It's* KITTEL.

KITTEL Where have you been? Rehearsal?

CHAJA Yes. Working on a new piece.

HE *reaches for her book.*

KITTEL You sang very well at the party. Is this your new play? (HE *looks it over*) In Russian?

CHAJA That's right. Do you . . . know Russian?

KITTEL Sorry.

CHAJA Pity. It's a good play.

KITTEL What's it called?

CHAJA *Beneath the Bridge.*

KITTEL And you're performing it in Russian?

CHAJA No, no. We'll adapt and improve it.

SHE *reaches for the book.* KITTEL *holds on to it.*

KITTEL You dance and sing your way through the war, eh?

CHAJA When I'm happy I laugh. When I'm sad I sing.

KITTEL (*Laughs*) Very good. (HE *hands her the book*) Perhaps—I hope—you'll wipe out your debt to me with this one.

CHAJA I'll try.

SHE *runs off suddenly.* KITTEL *looks after her.*

KITTEL What an exotic group of people. My God, they're strange.

HE *puts on his black, horn-rimmed glasses and becomes* DR. PAUL. HE *enters the library.* KRUK *appears from behind a bookshelf.* DR. PAUL *waves a manuscript* HE *has in his hand.*

DR. PAUL I've read your study on the Karaites in Lithuania. Brilliant work.

KRUK Thank you.

DR. PAUL You reach the conclusion that there's no connection whatever between the Karaites and the Jews.

KRUK That's what the research shows.

DR. PAUL You argue the case so convincingly that I almost believed you were telling the truth.

KRUK I was.

DR. PAUL (*Chuckling*) Mr. Kruk, there's no doubt whatever that the Karaites are Jews. You know it and I know it. You've constructed a monumental superstructure of falsehoods, half-truths and suppositions, so skillfully built that it looks like "proof." And for no other reason than to save the Karaites from annihilation. The Karaites—a race that despises your own people. Why?

KRUK You commissioned the study, Dr. Paul.

DR. PAUL I didn't commission you to reach this conclusion.

KRUK My conscience is clear.

DR. PAUL Scientific conscience? Or human conscience?

KRUK Is there a difference?

DR. PAUL And the truth? What about the truth?

KRUK All my research rests on a single truth.

DR. PAUL When you say "truth" you mean "lie," and vice versa. True?

KRUK False.

DR. PAUL Talking with you is a sublime experience, Mr. Kruk. A great intellectual pleasure. I think you'll agree that the two of us—just by talking, yes?—the two of us have managed to wipe out the distinction between what is true and what is false. Do you think?

KRUK I don't think you came here for a symposium on truth and falsehood, Dr. Paul.

DR. PAUL Look, you had an opportunity here. All you had to do was write the truth, and you would have had instant revenge on a sect of collaborators who have been betraying you since the war began. But no. You decide to protect them, and why? Don't you have *any* aggressive urge, you people?

A moment of silence.

KRUK You promised to send my report to Berlin.

DR. PAUL You know what your friend Freud has written about the origin of aggression?

KRUK That it derives from a basic impulse toward death.

DR. PAUL The death instinct, yes. So. German aggressiveness proves there's a death instinct in our souls, yes?

KRUK You'd know more about that than I would.

DR. PAUL While the Jews show no aggression at all, which means they have no death instinct, right?

KRUK Could be.

DR. PAUL You don't seem very engaged by this theory. As a Bundist—and an anti-Zionist—I would have thought you'd be very interested.

KRUK I don't think I see the connection.

DR. PAUL I think I can clarify it for you. The Zionist Jews in Palestine are completely different from you. They're an effective military organization, and they don't necessarily wait to be attacked. When I was in Palestine, I watched them make pre-emptive strikes on villages before the enemy had a chance to get organized. They're not like you, Kruk, they're no strangers to aggression. Is that the death instinct that you lack coming out in them, Kruk? Have we succeeded in transplanting it from the German soul into the soul of the new Jew?

KRUK What are you talking about? Zionism existed long before you came to power.

DR. PAUL I'm not just talking about Germany—I'm talking about two thousand years of anti-Semitism—persecution, pogroms. Please understand, Mr. Kruk: nothing in the world is more irritating than your endless capacity for suffering. It drives us wild. When we see the utter lack of killer instinct in you, we taste blood in our mouths—murder erupts in our souls. That's just the way it is—I sometimes think it's a chemical reaction. (*Pause*) The Jew only wants to survive, yes? He swallows degradation, humiliation, inhuman suffering—all just for the privilege of staying alive under appalling conditions. We rip the basic necessities of human survival out from under you and what do you do? Build a theatre. Sing and dance. All right, perhaps the German killer instinct is strong—but we're only carrying out the wishes of every nation in Europe. No other country would dare, that's all.

KRUK No theory will justify your crimes.

DR. PAUL You think not? Then why don't the Allies destroy the death camps? They're fair game, but no one touches them.

KRUK *stares at him, unable to summon a response.*

Well, never mind, time is short and I have another assignment for you. You'll survey all the monastaries in the district and catalog all the books in monastary libraries.

KRUK I'll what? What's the point of that?

DR. PAUL Point? No point. Beyond the fact that it'll keep you alive. If you work for me, you're safe.

KRUK (*Trying to get the real answer*) I don't much care whether I live or die, under the circumstances . . .

DR. PAUL How can you say that? The Eastern Front is collapsing, the Russians will invade Lithuania any day now . . .

KRUK (*Still prying*) It's an attractive vision of events, Dr. Paul. I'm afraid I don't buy it.

DR. PAUL It's true, believe me. They're sending untrained officers to the Eastern Front—men with no combat experience whatever.

A light begins to dawn for KRUK.

Everyone is needed to stop the Russians.

KRUK (*Filling in the pieces*) Then why should I detain you here? You must be itching to join your brothers on the Front.

DR. PAUL We're both intelligent men, Kruk. I'd like an inventory of the monastery libraries. Is that clear?

KRUK Absolutely clear.

DR. PAUL Very good. Goodbye, Mr. Kruk. (HE *departs*)

KRUK (*Dictates*) A strange symbiosis is developing between this German and myself. I don't want to die in Ponar, he doesn't want to die on the Front. So, he's attached himself to me; I carry out a series of pointless tasks under his direction, and both of us remain alive. I spend my days traipsing from monastery to monastery in Wilna, places I'd never set foot in before. As long as it keeps me alive . . . why not?

GENS *and* WEISKOPF *enter.*

GENS Well, this is the spot. Three thousand square feet of warehouse space. And a sewing machine takes what—six square feet?

WEISKOPF That's not the issue here. What do I—

GENS You could put five hundred machines here.

WEISKOPF What do I need with five hundred machines? I'm trying to tell you—I don't need the machines, I don't need the workers. I can do perfectly well with—

GENS They're dumping four hundred carloads of uniforms on you. You know how much work that is? The Germans'll give us another factory, no questions asked. It's a golden opportunity.

WEISKOPF I already got a factory!

GENS What?

WEISKOPF Look, I need, at the most, fifty more operators. They can fit in the old place. With those fifty I can handle the German order, no problem.

GENS No you can't.

WEISKOPF Are you telling me my business? Do you doubt Weiskopf? Look, I worked it out exactly—I know my business. (HE *pulls a large spreadsheet from his pocket and unfolds it*)

GENS What the hell is that?

GENS The numbers. You can graph it right on a graph. Production rates per worker per hour, number of workers. I add fifty workers, I put them on split shifts, two hours additional per worker and the job gets done. It's in black and white.

GENS Let's see.

WEISKOPF Be my guest.

HE *gives* GENS *the spreadsheet.* GENS *looks it over.*

GENS You're a thorough man, Weiskopf. You don't miss a trick.
WEISKOPF Naturally. I know my business.

GENS *tears up the spreadsheet.*

What the hell are you doing? My numbers! I spent all morning—
GENS Piss on your numbers!
WEISKOPF Don't you speak to me that way!
GENS Your numbers! You think I give a shit about saving money for the German army? Are you crazy? Five hundred more workers means five hundred more families saved. Does that make sense to you?
WEISKOPF What am I, a welfare fund now? I run a factory, it's a business. I pay a decent living wage to my employees, and what little is left over I live on.
GENS "What little is left over . . ."? The ghetto's newest millionaire! Believe me I don't care—you earned it, I'm no socialist. But even greed has its limits. Now listen to me Weiskopf: You'll build a second factory on this site, and employ five hundred workers supplied by the Ghetto Work Authority.
WEISKOPF The Ghetto Work Authority! Cripples I'll get!
GENS What are you running now, the Olympics?! You run a lousy factory stitching together lousy uniforms for the goddamn *enemy.* And Jewish cripples are very good at that, if it saves their lives.
WEISKOPF It is not a "lousy" factory. It's a successful business enterprise. I built it with my own hands from nothing. The factory is my life—I *am* the factory. I'm not about to let you blow it to bits with your lousy philanthropy!
GENS Tomorrow you start organizing a new factory!
WEISKOPF No, no, no, no, *NO!* Understood?
GENS Let me put it another way. It's an order!
WEISKOPF An order? You think I take orders from you? Fuck your order. You're not the ultimate authority around here.
GENS Oh really? So who is? Not you, by any chance?
WEISKOPF Kittel.

A pause.

GENS So help me, Weiskopf, if you speak to Kittel about this I'll—
WEISKOPF You'll what?
GENS Weiskopf.

WEISKOPF *stares at* GENS *defiantly.*

You will not speak to Kittel about this . . .

KITTEL *peers out from the bundle of clothing and climbs out, standing directly between the two.* HE *seems surprised to find the two men here.*

KITTEL My goodness, Gens . . . what are you doing here? I've looked everywhere. You planning your next show?

GENS No show. This is now the site of Weiskopf's new factory.

KITTEL Well, that shows admirable initiative. What's wrong with the old factory?

GENS We've outgrown it. The five hundred sewing machines won't fit.

KITTEL Five hundred sewing machines . . . my, my. Growing by leaps and bounds. You planning to drape Europe in newly-minted shrouds?

GENS It's the uniforms. Four hundred carloads . . . we have a contract to fulfill.

KITTEL (*Turning to* WEISKOPF) A contract, yes. You need five hundred workers for this . . . contract?

WEISKOPF Well, maybe not five hundred . . . I mean, perhaps we could . . . uh . . . make some adjustments . . .

GENS Don't make promises you can't keep, Mr. Weiskopf.

KITTEL You seem to have a slight disagreement here.

WEISKOPF Yes, well, there are always different estimates in any situation. That is . . . (HE *begins to cough uncontrollably*)

KITTEL Weiskopf! Are you hiding something?

WEISKOPF Me? Why would I hide?

KITTEL That's good.

WEISKOPF Except—

KITTEL Except what?

Silence.

Is there a disagreement here or not? Answer me!

GENS No!

WEISKOPF (*Simultaneously*) Yes!

KITTEL Well, we cleared that up. (*Turning serious*) Now I want the truth. Disagreement? Weiskopf, you're a reliable businessman. How many more workers do you need? The truth, now.

WEISKOPF I need about—

KITTEL Precisely.

WEISKOPF Fifty.

KITTEL Fifty. So why this place? Gens, are you hiding something?

GENS Nothing. He needs five hundred workers, not fifty.

KITTEL (*To* WEISKOPF) Where are the figures you showed me? Your graphs?

WEISKOPF I gave them to him.

KITTEL (*Turning to* GENS) May I see them please?

GENS I tore them up.

KITTEL You what?

GENS The figures were a joke. Mr. Kittel, the man is deluded, believe me.

KITTEL Is that so, Weiskopf? Did you waste half my morning on a joke?

WEISKOPF Mr. Kittel, if you give me an extra two hours per day per man and if you let me hand-pick fifty choice workers, if I don't have to hire cripples and half-wits, and if—

KITTEL If, if, if . . .

WEISKOPF If no one interferes with my work schedules—

KITTEL Your next if, Weiskopf, is your last.

WEISKOPF Mr. Kittel, I'm only trying to help save you money. I'm sure that if I were to speak to Göring about this—

KITTEL That's enough! Talk about "chutzpah"! You're worse than the dummy at your theatre—at least he made me laugh.

WEISKOPF You promised me a meeting with Göring!

KITTEL Jesus, Weiskopf, can't you take a joke? I mean, people who can't take a joke get on my nerves.

WEISKOPF I swear on my wife's head! Give me fifty workers and I'll finish the job. Believe me—

KITTEL Weiskopf, you're hysterical!

WEISKOPF (*Trying to calm himself*) Please. Let me explain it once more. With fifty workers . . .

KITTEL Gens says you're deluded. He says it's impossible.

WEISKOPF Well, Gens *is* the head of the ghetto . . .

KITTEL You don't say.

WEISKOPF I'm only trying to point out . . . he could have his own agenda, his own motives . . .

KITTEL Are you sweating?

WEISKOPF Not at all, I—

LEWAS *breaks in, carrying a bottle of cognac and a salami.*

LEWAS You okay chief?

GENS Not now!

LEWAS But chief, you told me—

GENS Piss off!

KITTEL What is that, Lewas?

LEWAS We searched his house. Weiskopf's.

KITTEL My, my. In Weiskopf's house. (HE *takes the bottle and looks at the label*) Contraband cognac. And Hungarian salami. Weiskopf!

LEWAS There's more. Rice. Olive oil, half a sack of sugar . . .

WEISKOPF It was left over from the party. I—

KITTEL Weiskopf, really. Nothing warms my heart like the sound of a man apologizing.

WEISKOPF All right then. I'll tell you why he wants five hundred extra workers. I'll tell you the real reason. I'll tell. The real reason—

KITTEL Gens? Free me from this leech.
GENS Lewas! Take care of him.

LEWAS *turns to* WEISKOPF *and slaps him hard.* WEISKOPF *reaches up to protect his face and* LEWAS *lunges at him, pulling the buttons off his pants. His pants drop to the floor.* WEISKOPF, *more embarrassed than hurt, bends down and quickly retrieves his trousers. But* LEWAS *has taken the opportunity to get some brass knuckles on, and when* WEISKOPF *straightens up, holding his pants with both hands,* LEWAS *lets him have it again, this time tearing into his face.* WEISKOPF *topples to the floor with a scream. When* LEWAS *picks him up there is blood dripping from his face.* LEWAS *delivers one more battering punch to the face, and* WEISKOPF *goes down for good.*

Lock him up.

LEWAS *drags out* WEISKOPF. KITTEL, *who has been watching impassively, opens the cognac and takes a slug.*

KITTEL Nicely done, Gens. (HE *gives* GENS *the bottle*) You learn quickly, you people. Call in your gorilla.
GENS Lewas!

LEWAS, *who has deposited the insensible* WEISKOPF *at the back of the stage and left him there, returns.*

LEWAS Yes, chief.
KITTEL Bring me the theatre troupe. All of them. Now.
LEWAS Yes, sir!

LEWAS *goes back to* WEISKOPF, *drags him off stage.*

KITTEL Now let's get down to business. I've been getting reports. People are escaping from the ghetto.
GENS But that's impossible.
KITTEL Horseshit, Gens. It's a fact and you know it. Since the day the bomb exploded beneath the bridge—the big fire—you remember?
GENS Beneath the bridge, yes . . .
KITTEL Since that day, thirty people have disappeared from the work crews.
GENS What's the connection?
KITTEL You tell me.
GENS I don't get it. The Jews don't figure in it—the bridge was in a Lithuanian village. I thought . . .
KITTEL You could blame the Lithuanians. Well, it played out that way, didn't it? Forty Lithuanians were shot for that firebomb; let's not

mourn them, Gens. Let's just hope they were innocent creatures whose souls are in heaven—a place where they'll be safe from the likes of you and me, right Gens? (HE *laughs, then turns serious*) Tell me Gens, what do the words "mutual responsibility" mean to you?

GENS It's a Jewish principle.

KITTEL Right. What does it say in the Old Testament? "One hand washes the other"? Something like that.

GENS All Jews answer for one another . . .

KITTEL Right, close enough. Well, this beautiful Jewish adage is now the law of the ghetto, Gens. If anyone disappears, his family will be shot. If a family disappears, everyone who shared a room with them will be shot. If a roomful of people disappear, everyone who lives in the same house with them will be shot. All workers will be divided into groups of ten. One runs, the other nine die. Is that clear to you, Gens?

GENS Clear. Yes.

KITTEL Oh, and as for Weiskopf's factory . . . no expansion. No new work. Any questions?

GENS *remains silent.*

The weasel was right, of course. His calculations were accurate down to the minute. He explained it to me clearly this morning. What are you going to do about it?

GENS Get him back?

KITTEL What?

GENS Weiskopf. Reappoint him, give him back his factory.

KITTEL Jesus Gens! You people are really baffling. My whole life all I hear is how smart the Jews are, how resourceful. . . . Don't you know anything about the way the world works? There's no second chance in this world, Gens. A man collapses, you bury him. You don't prop him up with a stick. You think I care about who was right and who was wrong in your pitiful squabble with the weasel? Gens, really. Learn a little German philosophy: Among reasonable people the only conversation worth having is the one about *how* to achieve a goal. Who cares about the wisdom of the goals themselves? Who's got time? There are no just goals. *I* justify a goal. My will. When you and Weiskopf square off, the only thing that interests me is: Whose will prevails? Who is stronger? Not even really a challenge for you, was he? He caved in the minute you poked him, like a house of cards. I mean, it wasn't even any fun. He'll be in Ponar by the weekend. You're too damn good at what you do, Gens. Now, let's see the theatre troupe.

A series of massive, discordant notes from the BAND. *From out of the clothing pile there is a sudden, eerie explosion of clothes.* GARMENTS *wave about on the pile and cry in agony.*

SRULIK'S VOICE Welcome to the last performance!

The "Finale" begins.

COSTUMES

We have been living in hell!
We have been living in hell!

THEY *rise off the pile and begin to dance—stylized versions of the theatre troupe. The stage is eerily vacant of human form, but is filled with the swirling, dancing costumes.*

(*Singing*)
Scalding steam!
Choking fumes!
Thrashed with sticks in the
Laundry rooms!

Boiled in suds!
Soaked in lye!
Jabbed with naptha until we die!
Ai! Ai! Ai!

Stitched—and stretched!
Wrung out—and hung out!
And always that smell!
That disinfectant smell
That nothing can dispell!
We have been living in hell!

But when at last
We are free
Our threads will tremble
With ecstasy.
Ai! Ai! Ai! Ai!
Coats and pants
Will find a reason
To sing and dance.

Ai! Ai! Ai! Ai! (*Etcetera*)

SRULIK (*Over the music, to the* DUMMY) What the hell's going on? You promised me these clothes would be neatly folded, sorted and stacked. Now they're all over the place. Can't you control them?

The DUMMY *beats the clothes. As* HE *sings and attacks them,* THEY *quietly go back to the pile and curl up, defeated.*

DUMMY (*Singing*)
Ai! Ai!
You must be steamed and soaked with lye.
Ai! Ai!
You know that's your fate so don't fight it.
SRULIK Shut up and lie down.
ALL No!
SRULIK No? No? Where else do you think you can go?
ALL
We will find a wardrobe,
An armoire.
Quality clothes need room to breathe.
Quality clothes, that's what we are
And quality clothes should have their own armoire.

The HASID*'s head and upper torso burst out of the pile.*

HASID Your Honor! Mr. Police Commisioner, sir. (HE *sings*)
I read sleeves,
I'll tell your future!
Blessed is he who believes!
Chet conquers Gimmel!
Three becomes eight!
The change in your future
Will be great!
Could be three days . . .
Could be three weeks . . .
Could be three months . . .
Something in threes . . .
Three million marks, if you please!

The CLOTHES *begin to agitate again.* THEY *make the pile swarm with life, holding out their palms to be read.*

COSTUMES (*Chanting*)
OUR FUTURE!
OUR FUTURE! OUR FUTURE!
SRULIK Their future?! Jesus, Ignatz, can't these people take a joke? I thought I could depend on you. Well, learn a little German philosophy. If you want a job done well . . .

HE *begins to unpack his saxophone case as the clothes continue to climb back off the pile.* SRULIK *pulls the saxophone from the case and begins to gun down the* COSTUMES *with it. A series of rim-shots from the pit and the* COSTUMES *all lie quietly in and around the pile. There is dead quiet for a moment. An oversize costume of* CHAJA *appears.*

CHAJA'S COSTUME That's enough! That's enough! (*Sings*)
We're done with suds! We're done with lye!
We're done with naptha, you and I!
We're going home . . . it can't be far . . .
Home to our cozy old armoire!

COSTUMES (*Singing*)
We'll flee this damp and soapy spot
Where some are spared and some are not.
From fumes and steam we seek release
So we can go and live in peace.

ALL (*Singing*)
La la la la la la la la (*Etcetera*)

DUMMY (*Singing*)
You're only rags, no more than rags!
How dare you dream of being free!
You're only rags, how stupid can you be?

ONE COSTUME Look!

ANOTHER COSTUME A wardrobe!

CHAJA An armoire! (SHE *sings*)
We're going home
To our armoire . . .
The finest clothes
Is what we are . . .

ALL (*Singing*)
We're going home . . .
It can't be far . . .
Home to our cozy
Old armoire!

We're going home
To our armoire!
The finest clothes
Is what we are!
We'll have a home . . .
A gracious home . . .
A spacious home . . .
We're going home!

We'll have a home . . .
A gracious home . . .
A spacious home . . .
We're going home!

CHAJA (*Singing*)
Ich benk a heym.
ALL (*Singing*)
We're going home!
We're going home!
We're going home!
KITTEL (*Applauds*) Bravo! Bravo!

HE *whispers something to* GENS. GENS *exits.*

All right, everybody out. I said out!

Silence, the armoire stays shut.

ROUSE!

The armoire opens and one by one the PIECES OF CLOTHING *emerge.*

Line up!

The CLOTHES *form a line in front of* KITTEL.

Very good work. A great satire. I'd like to see the actors.
DUMMY What actors? We're only clothing.
KITTEL I said I'd like to see the actors. Now!

The faces of the ACTORS *slowly emerge from the clothing. One by one* THEY *become visible. But* ONE DRESS *remains empty.* KITTEL *is drawn to it.*

You too, my chanteuse. I want to see your face.

Silence. It grows uncomfortable. KITTEL *looks down the neck of the woman's costume.* HE *looks up, stunned.*

It's empty. Unbelievable! I said I want to see the entire troupe! All right, you've put on a great show. Full of Jewish wit and . . . what do you call it . . . *chutzpah?* Okay. The performance ends, I applaud loudly, and what has happened? An actress—an actress already in debt to me for two-and-a-half ounces of beans—is missing. Vanished from under your nose.

Silence.

You know what this means, according to the new rules of the ghetto? Mr. Gens can tell you about "mutual responsibility," an old Jewish concept, recently adopted.

Silence.

So. A Jewish satire. Very funny. (*Suddenly, the nightclub compere**) And now! A German satire: *Up Against the Wall!!*

EVERYONE *turns in terror and faces the back wall.*

Machine-gunner, front and center!

A blood-curdling screech, and GENS *enters, pushing a wheelbarrow. On it is a huge jam-pot labeled "JAM." Next to the pot is a basket loaded with sliced challah.*

Machine gun here! Load! Release safeties! (HE *releases the safety of his own gun*) Actors! About face!

The ACTORS *turn, their faces frozen death-masks of terror. It takes them a moment to even see the jam pot.* THEY *stare, bewildered.* KITTEL *roars with laughter.*

What, you thought I'd have you shot? After that fantastic number? German satire, I said. Look in these hard times, with the Russians threatening to march in here any day, you've given me a moment of transcendent joy that is solely the province of the arts. Your virtuosity has saved your lives, and here is my thanks. Break bread with me: Fresh-baked bread and red currant jam. Please. Help yourself. Here's to your next performance! (KITTEL *takes a slice of bread, dips it in the jam and eats, bliss crossing his face*) If there's one thing I'm a sucker for it's red currant jam. Mmm-mmm. Come on, try some. Please . . . join me?

The ACTORS *move hesitantly to the pot, their minds a scramble. But little by little,* THEY *join in, first slowly, then ravenously.* THEY *pack in close to the pot, jockeying for position. As* THEY *eat, the* DUMMY *sings.*

DUMMY (*Singing*)

Gobble it up! Gobble it up!
Quickly! Quickly!
Today—you're here.
Tomorrow you'll all disappear!
Sooo . . .
Quickly! Quickly! Quickly!
Gobble it up!

*Speaking like a master-of-ceremonies or an announcer in a revue.

As the DUMMY *sings and the* ACTORS *eat,* KITTEL *steps back. In a burst of energy* HE *swings round, raises his gun and fires hundreds of rounds into the crowd. Pandemonium. The* TROUPE *goes down in a hail of bullets and* ALL *fall forward with their heads in the jam pot. Only* SRULIK *and the* DUMMY *survive.* THEY *stagger back away from the pot.* SRULIK *is still in his Kittel costume. As* KITTEL *turns to* SRULIK, HE *confronts a grotesque mirror image of himself.*

DUMMY (*Singing impudently to* KITTEL)
Finish it up! Finish it up!
Quickly! Quickly!
The master race?
You poor deluded man!
Go ahead! Brag!
Tomorrow you know what you'll be?

KITTEL *shoots the* PUPPET. *Blood runs from his mouth. As* HE *falls,* SRULIK*'s Kittel costume falls away with him. Underneath,* SRULIK *wears a bathrobe, as at the opening. One of his arms falls away with the Kittel costume.*

SRULIK A rag!
Finish it up!
Finish it . . .

HE *stares at the audience, an old, one-armed man. The clarinet is heard again, coming closer. As the music finally drifts away . . .*

Darkness.

END OF PLAY

The Songs

Shtiler, Shtiler (Be Still) Lyrics by Schmerke Kaczerginski; music by Alec Volkoviski; English lyrics adapted by Jim Friedman.

Hot Zich Mir Di Shich Zerissn (Dance, Dance, Dance) Anonymous; English lyrics adapted by Jim Friedman.

Ich Benk A Heym (I Long for Home) by Lev Rosenthal; English lyrics adapted by Sheldon Harnick.

Wei Zu Di Teg (Crazy Times) Lyrics by Katrielke Broide; music by Misha Veksler; English lyrics adapted by Sheldon Harnick.

Isrulik (They Call Me Izzy) Lyrics by Lev Rosenthal; music by Misha Veksler; English lyrics adapted by Sheldon Harnick.

Friling (Springtime) by Schmerke Kaczerginski; English lyrics adapted by Jim Friedman; music adapted by Gary William Friedman.

Zog Nit Keinmol (Never Say You Can't Go On) by Hirsh Glick; English lyrics adapted by Sheldon Harnick.

Dremlen Feigl (Drowsing Birds) by Leah Rudwitzky; English lyrics adapted by Jim Friedman.

Finale (Dance of the Clothes) Lyrics by Sheldon Harnick; music by Gary William Friedman.

Postscript

A Theatre in the Wilna Ghetto by Joshua Sobol

In 1942–43, with the horrors of the Holocaust well under way, a theatre was involved in putting on plays in the Wilna ghetto. It had its debut on June 18, 1942, about four months after the Jews of Wilna were deported to the ghetto, and a mere two months after the mass extermination in which over fifty thousand of the seventy thousand Jews were massacred.

The decision to found a theatre at such an agonizing time met with a stormy response in the ghetto. In his diary on January 17, 1942, the ghetto librarian and record-keeper Hermann Kruk, a Bundist, wrote: "In a cemetery there can be no theatre." When the slogan also appeared in the alleyways of the ghetto on the following day the head of the Jewish police, Jakob Gens, warned Kruk: if any such slogans appeared again, the librarian and his cohorts would be sent to Ponar.

Whose decision was it to found a theatre in those excruciating days? At a party marking six months since the theatre was established, its director Israel Segal noted that Gens himself had been behind it. Gens, head of the Jewish police at the time, was to become sole ruler of the ghetto as of July 1942; it was his brainchild, the ghetto theatre.

Despite the strong protest of political circles and intellectuals within the ghetto, the first performance was held as scheduled on January 18, 1942. The evening's artistic success was reflected in the ticket sales as well. Kruk himself noted in his diary that the proceeds were as high as four thousand rubles, all earmarked for charity, in keeping with the slogan: "Let no man go hungry in the ghetto." As news of the opening night success spread through the ghetto, it became clear that the theatre was there to stay. Fear of offending the public in a time of anguish and grief proved to be unfounded. "People cried and laughed—and their spirits were lifted."

Driven by his conviction that normalization and productivization of the ghetto were the key to saving as many people as possible, Gens regarded the theatre not only as a source of livelihood and employment for the actors but also as an invaluable emotional outlet which would boost morale and help to normalize ghetto life. This accounts for his insistence on cultivating the theatre and turning it into a permanent feature of ghetto life. A week after opening night, there was a repeat performance. Thanks to the warm reception the theatre had had at its debut, public objections subsided—despite the presence of German and Lithuanian officers in the audience. Among these, according to Kruk, were the Nazi officer Herring and the commander of the Lithuanian militias charged with the mass exterminations at Ponar. Kruk notes that the two of them left during the intermission.

In its first year, the theatre put on no fewer than 111 performances, selling a total of 34,804 tickets. The hall was usually packed and the shows were often sold out weeks in advance. By the time the ghetto was liquidated on September 20, 1943, the number of performances had doubled. A population of twenty thousand people had bought seventy thousand (!) tickets.

The performances included both light and serious theatre. In its two-year history, the theatre put on four variety revues based on original material written in the ghetto, mainly by Katriel Broide and Leib Rosenthal, whose poems and songs were especially popular.

The revues included programs like *Karena Yahren Un Wie Zu Die Tag* (July 1942), *Men Kann Garnicht Wissn* (October 1942), *Peshe von Reshe* (June 1943), and *Moshe Halt Sich* (August 1943). The last was staged just when deportations to Estonia were at their worst, and continued until the liquidation of the ghetto on September 20, when the songs of the revue accompanied the last of the Wilna Jews being taken to the camps.

The repertoire also included the following five plays: *Grineh Felder* by Peretz Hirschbein, first performed in August 1942; *Der Mentsh Untern Brik* by the Hungarian playwright Otto Hindig, November 1942; *Der Otzar* by David Pinski, March 1943; *Havehudi Hanizchi* by David Pinski, first performed (in Hebrew) in June 1943; *Der Mabul* by Swedish playwright Henning Berger (August 1943), performed in the final weeks of the ghetto.

Rehearsals of *Tevya the Milkman* were well under way when the ghetto was liquidated. On its first anniversary, the theatre held a "Theatre Week in the Ghetto" at its home at 6 Rodnitzki Street. This was a full-fledged festival, including a revival of the first concert in the ghetto, two performances of *Grineh Felder* and one of the *Men Kann Garnicht Wissn* revue, the Yiddish choir, recitals, light music concerts and a jam session of the jazz ensemble, as well as one symphony concert and a performance of the Hebrew choir. In its January 24 write-up on Theatre Week, the *Ghetto News* (printed under the auspices of the Judenrat) proudly noted that its pace of activity would do credit to a cultural center in any European metropolis.

Cultural activity in the Wilna ghetto was not confined to the theatre. Two days after the Wilna Jews were forced into the ghetto, a library was opened at 6 Strashon Street by Hermann Kruk, one of the cultural leaders of the Bund party, who had established hundreds of libraries and cultural centers in the Jewish communities of pre-war Poland. By September 19, with the extermination *Aktion* fully under way, 1,485 readers had registered at the library. Books were being borrowed at a rate of four hundred a day. Even during October 1941 with the *Aktionen* being implemented more intensely than ever, the library continued to supply its readers with books. On Yom Kippur (October 1, 1941) three thousand Jews were deported to Ponar. On the following day, 390 books were borrowed from the library. This continued at a rate of three hundred books a day, so that within a little over a year after it opened on December 13, 1942, the library proudly celebrated the borrowing of the 100,000th book. The readership had reached 4,700 by then, out of a total population of 17,000 in the ghetto.

The thriving cultural activity was apparent in other areas as well. There were competitions in music, the plastic arts, poetry and playwriting. The theatre foyer featured an exhibition of works by painters and sculptors in the ghetto, among them the nine-year-old Shmuel Back, whom Kruk mentions as being unusually gifted. Meanwhile, plans were under way to open a ghetto museum and Gens was working on a blueprint for a printing press and a publishing house which would go on operating after the war as well. This was in the early summer of 1943, only weeks before the liquidation of the ghetto. . . .

This vitality, so intensely reflected in the different areas of creative and intellectual activity, was channeled into other areas as well, such as the health-care system, schools and institutions for delinquents. There were dozens of lectures and symposia and morning assemblies for the workers. In addition to this cultural plethora, there was a strange vitality on the economic and commercial planes as well. In his diary, Kruk recorded the proliferation of café-theatres with every easing of pressure, along with an amazing rise and fall of the "King of the Ghetto" or the "Caliph of the Ghetto," Weiskopf, head of the clothing-repair shop.

Reading through the chronicles of the ghetto and the diaries of inmates—survivors as well as those who perished—and delving into the day-to-day affairs of the Wilna ghetto, one is overawed by the burst of vitality. Without it, there is no accounting for the ability of the defenseless survivors to cling dauntlessly to life, to retain their joy of life in the face of armed tormentors and murderers.

It is to the mystery of this vitality that I owe the play.

V

Poetry

In one sense, reading Holocaust poetry is a frustrating exercise, because much of it has been translated from a foreign tongue. This is also true of Holocaust literature, but the narrative and dramatic modes allow for close approximations to the original that the more concentrated language of verse usually resists. Moreover, the dense cluster of associations emerging from a particular verbal and historical culture cannot be shifted from one poetic vernacular to another without some loss. For example, approaching Whitman's *Leaves of Grass* in translation without the power to imagine the American landscape and its epic sweep, to say nothing of its native idiom exploiting a "barbaric yawp," leaves the reader lurching on a vessel bound for nowhere. Celebrating an "open road" that no one can identify or visualize raises the danger of facing a phantom reality.

Readers acquainted with the details of the Holocaust experience escape at least part of that danger. But problems remain. Translations unavoidably sacrifice sound (and often rhythm) for sense, injuring the original form. No accurate rendition of the recurrent half-line from the final strophe of Paul Celan's celebrated "Death Fugue"—". . . der Tod ist ein Meister aus Deutschland" (. . . death is a master from Germany)—can capture the hissing sibilants and pounding dentals that echo ominously like the fate theme in the closing moments of Bizet's *Carmen*. In addition, the meeting of Scripture and Goethe's *Faust* in the poem's final allusions to the golden hair of Margarete (Gretchen) and the ashen hair of Shulamith echoes ironic references to German culture that are plain to a native audience but must be mined by others. These are but the most obvious examples of an issue that nothing short of learning the original language and mastering its culture can resolve, though the patient and informed student for whom this is not possible can still enter the terrain of Holocaust verse in translation using the poet as a guide.

By reading enough of a poet's work, even in translation, we gain a glimpse of how he or she has struggled with the problem of converting the murder of European Jewry into poetic vision. This means nothing less than finding a form for chaos by including chaos as part of the form. Since single poems cannot illustrate this dilemma in a meaningful way, I have chosen to include many poems by a few authors, all of them major figures who have devoted a portion of their careers to invoking the Holocaust as a theme for their poetry (though some might consider Miklós Radnóti the sole exception). They are linked by the premise of a people's extermination, and by the need to enter into what Paul Celan called a "desperate conversation" with their audience—a conversation that evidently has already occurred within their own imaginations. This bond reduces, if it does not eliminate, the cultural differences separating poets writing in Hebrew, Yiddish, German, and Hungarian; it also solicits readers to forsake their estrangement from the theme and to embrace a community of craftsmen who ply their trade in words.

The ties joining these poets to the Holocaust experience vary. None draws on the stability of a single locale. Jacob Glatstein left Lublin (later to be the vicinity of the Maidanek deathcamp) for the United States just before World War I, returning to Poland only once, in 1934. He launched his career as a poet long before the outbreak of World War II, and turned to the Holocaust as a theme midway through that career. Abraham Sutzkever, after a brief sojourn as a child in Siberia, returned to Vilna and spent the early war years in the Vilna ghetto (where he wrote his first poems), and then fought as a partisan in the nearby forests. Eventually, he emigrated, via Moscow and Paris, to Israel, where he still resides.

Nelly Sachs left Germany for Sweden in May 1940, and remained there essentially until her death. Already past fifty, she turned to the Holocaust as a topic during the war and devoted most of her subsequent poetic activity to that theme. Paul Celan grew up in Czernowitz, worked for a time in a Romanian labor camp during the war, moved to Bucharest after his release, then to Vienna, and finally to Paris. Dan Pagis shared Celan's Romanian background. After enduring three years in a Nazi concentration camp, he emigrated in 1946 to Palestine, where he eventually faced Israel's efforts to merge the Holocaust into a rhythm of destruction and renewal that ran counter to his vision of that event. The private tensions in his Holocaust poems respond to the public views of his adopted land.

Miklós Radnóti, the only explicit murder victim of the Holocaust among the poets included here, was a Jew who converted to Catholicism, possibly to escape an unbearable estrangement from the centers of cultural life in Hungary, which had become increasingly antisemitic. Ironically, his government refused to recognize his conversion; classified as a Jew, he was drafted three times into forced labor. The last time, he did not survive; too weak to continue, he was killed in November 1944 on a death march from Yugoslavia back to Hungary.

Much of the Holocaust poetry of these writers appears in the guise—or the disguise—of autobiography. But readers must be wary. An "I" is not always the poet's self; frequently, it is only a mask, a voice to evoke some truth through eyes that may not be a personal "I" at all. Glatstein and Sachs witnessed from afar what Radnóti did not outlast and Pagis, Celan, and Sutzkever survived. Readers of these poems have a chance to watch unfold points of view as diverse as the encounters with atrocity themselves. Moreover, because metaphor is the heart of verse, finding similitudes for the incomparable becomes not only a challenge for the poet, but often a condition of the internal dynamics of the poem itself. Two lines from a poem by Sutzkever warn us of this plight: "Inside me, a twig of sound sways toward me, as before. / Inside me, rivers of blood are not a metaphor." Analogy suggests a system of order in the poetic universe that words evoke, and this in turn may inspire a feeling of consolation from that hint of verbal and imagistic coherence. Not entirely unjustly have some commentators feared that the aesthetic stylization of the Holocaust experience, especially in the condensed expression of verse, might violate the inner (and outer) *incoherence* of the event, casting it into a mold too pleasing or too formal. It is instructive to follow the poet's imagination as it pursues various solutions to this unavoidable dilemma.

One need hardly stress that traditional themes of lyric verse, such as love, nature, beauty, and even ordinary death, could furnish insufficient inspiration to poets creating verbal tombs for a murdered people. Although it is impossible to duplicate the process of recasting reality that must have afflicted the poetic impulse as the ruins of the Holocaust were emblazoned on history, we can trace an example through Glatstein's poem "I Have Never Been Here Before." As the poet's dazed voice moves from a quiet rendezvous with familiar uncertainties to the spectacle of an unimaginable catastrophe, it introduces a cluster of images that become fragments of an embedded reality. These images clamor for release, causing both poet and poem to stray from the initial thrust. The poem begins:

I always thought
I had been here before.
Each year of my patched-up life
I mended the fabrics
of my decrepit, tattered world.

Glatstein does not pretend that the past provokes no agitated memories, but they are neither unusual nor obscure, and he can control them through the forms of his imagining. Suddenly, however, "these last ragged years" have conjured up a landscape whose terrain is cluttered with singular shapes and images:

shreds of hair—
inventive deaths—
are my days and nights.
My warped destiny
I have lived to see.
The frozen reverie,
burnt fields,
cartography of cemeteries,
stony silence,

emblems of vicious joy—
I don't recall them.
I have never seen them before.
I have never been here before.

Stricken by the view but consoled by his conviction that "Blasted ornaments will bloom again," the poet momentarily falters in the new role demanded of him by these vivid impressions. But they prove too potent for his nostalgia, their protest against anonymity drowning his hope:

And yet, the dead will still cry midnight prayers—
each corpse, a trickling voice.
Like a tiny candle over each grave,
a cry will burn,
each one for itself.
"I am I"—
thousands of slaughtered I's
will cry in the night:
"I am dead, unrecognized,
my blood still unredeemed."

Holocaust poetry acknowledges this cry as a form of mourning to replace redemption. It searches for a vocabulary to depose that tired, indecorous term.

These tributes to the wounded self serve also as dirges to the wounded word. A language that once may have celebrated the transcendent splendors of the universe limps now on splinted limbs, bereft of its certitudes. In her poem "Chorus of Comforters," Sachs captures the impasse facing the poet deprived of theme and language by the murder of European Jewry. It begins:

We are gardeners who have no flowers.
No herb may be transplanted
From yesterday to tomorrow.
The sage has faded in the cradles—
Rosemary lost its scent facing the new dead—
Even wormwood was only bitter yesterday.
The blossoms of comfort are too small
Not enough for the torment of a child's tear.

An unanswered interrogative refrain—"Which of us may comfort?"—runs through the poem like a single thread in search of a garment or, to borrow Sachs's later image, like a seed in need of nurturing. But it also raises the question of the poet's role in recording this grief. The very words of the poem clash in a struggle for domination—"blossoms" with "tear," "scent" with "dead," "yesterday" with "tomorrow"—but the issue remains in suspension; the rupture between content and form frustrate efforts at a satisfying union. The poem ends:

We are gardeners who have no flowers
And stand upon a shining star
And weep.

Unwilling to allow despair to sabotage hope but too honest to let hope supplant despair, Sachs hovers between an ancient tradition of suffering and the modern legacy of atrocity. She picks through the rubble of the Holocaust to rescue separate words that may have survived the disaster, while sadly confessing that their place in the moral architecture of language will never be the same.

Not all the poets included here might identify with Sachs's activity though Celan and Pagis at least would feel like kindred spirits. But they would certainly sympathize with the sources of her inspiration, to which her own words testify: "[T]he dreadful experiences that brought me to the very edge of death and eclipse have been my instructors. If I had not been able to write, I would not have survived. Death was my teacher. How could I have occupied myself with something else? My metaphors are my wounds. Only through that is my work to be understood." The concept of wounded metaphors does much to illuminate how these poets approached the fractured universe of the Holocaust.

How do we speak authentically of victims who were denied the last vestige of inner freedom—a sense of connection between the manner of one's living and the mode of one's death? Perhaps no other poem better helps us to imagine this challenge than Pagis's "Written in Pencil in the Sealed Railway-Car." The poet has concentrated into a title and six lines—a mere nineteen words in the original Hebrew—the dilemma of finding a poetic language to speak truly about the destruction of European Jewry:

here in this carload
i am eve
with abel my son
if you see my other son
cain son of man
tell him that i

The "you" of the poem is a plural pronoun, so it is not a private message but a public appeal, drawing in the entire reading audience. Both this plural appeal and the poem's abrupt ending force us to concede that *we* are made the agents of its testimony, the transmitters of whatever meaning it implies. We need to decide what silenced the potential victim, what bequest she might have left behind, what language—if any—is sufficient to the dimensions of the catastrophe that consumed her.

But Pagis packs much more into his tiny poem. Eve, the archetypal mother, has changed the locale of Jewish experience from the spacious Garden of Eden to a sealed boxcar on its way to a deathcamp, and in doing so she wrenches the creation story of Genesis from its original setting, rewriting it, one might say, to include the destruction of the Holocaust. Against the ancient fratricide, the primal act of violence, the initial cause for family grief, Pagis would have us imagine a fratricide boundless in scope, a lament beyond words. The poem invites us to reexamine imaginatively, in the shadow of the Holocaust, the meaning of Jewish origin and its impact on human destiny. Pagis illuminates this venture by fashioning a minidrama in which both Adam and Eve and Cain and Abel are deprived of their pristine, if familiar, roles. We continue to enact our fate on the landscape of history, but we do so on an altered stage in a setting not foreseen by the initial script. It is a new version of Scripture, a kind of counter-Genesis, a melancholy variation, documented by history, on the future of parents and children in the world of the Holocaust.

All poetry encourages us to view the human and natural scene with a fresh eye, uncontaminated by the clichés of customary speech. But Holocaust poetry, because it emerges from an experience so contaminated itself, exerts a special pressure on the poet's imagination, plunging that faculty into a desperate search for metaphors to forge analogies with the incomparable, or an equally urgent venture to sketch the bleak landscape that remains in the absence of such analogies. The poet, as Pagis writes in "Footprints," is required to learn "the declensions and ascensions / of silence," and translate them into sound, to deromanticize reality so that a heart "blue from excessive winter" would imprint itself on the reader's mind as a plausible image of Holocaust truth. Gradually, the diligent reader of these poems develops a sense of the idiom required to encompass atrocity and of the struggle to find a verbal shape and form to mirror the unspeakable. If the results numb more than they inspire, leave us frozen rather than inflamed with admiration, this should be no cause for surprise or dismay. When art addresses a theme like annihilation, what response can we expect other then a chilling gratitude? Fervent acclaim must be reserved for warmer vistas.

But a final word might be said in behalf of the word itself. If the pain of atrocity has left a lasting scar on the human spirit, language has not flagged in its search for a style to justify the hunt for wounded metaphors. Poetry may not redeem the past or erase its scars, but it has played an unusual role in helping the poet to face the harsh realities of the Holocaust. Sutzkever testifies that "living poetically" saved his life in the Vilna ghetto. During the postwar years, Celan, too, vowed that the creative effort sustained him even as he strove with the darkness of his theme, though in the end this did not prevent his suicide. When Radnóti's body was reclaimed from a mass grave after the war, his last poems were found in a pocket of the coat he was wearing when he was shot. On the verge of death, his verse may have been his final solace.

But Holocaust poetry should not be mistaken for a renewal of the spirit or used as a reason for redressing the cruelty of the doom it reflects. Its true legacy is a tribute to the resilience of language and the ability of the artistic imagination to meet a chaotic challenge and with sheer inventive skill change it into durable, if often difficult and unfamiliar, poetic forms.

25. Abraham Sutzkever

Abraham Sutzkever was born in 1913 in Smorogon, a city southwest of Vilna in what is today Lithuania. Because they feared that the local Jews might spy for the Germans, during World War I the Russians expelled more than 1 million Jews from the area. Sutzkever and his family finally settled in Omsk in western Siberia, where his father died at the age of thirty. In 1920, Sutzkever's mother returned with her three children to a suburb of Vilna, where they lived in humble circumstances for the next twenty years.

A sickly child, Sutzkever attended schools intermittently, embarking early on a program of self-education. As a young boy, he wrote poetry in Hebrew and Yiddish, studied Yiddish intensely, and became a student of secular Jewish culture. He read widely in Russian and Polish poetry, and even familiarized himself with translations of Edgar Allan Poe. He joined the group of painters and poets known as "Young Vilna" (Chaim Grade was also a member); was published in *In zikh,* the journal of the New York Introspectivists; and had his first book of poems issued in Warsaw.

With the outbreak of World War II in 1939, everything changed. Sutzkever married his childhood sweetheart. After the Germans and Russians divided Poland, Vilna, renamed Vilnius, became part of the Soviet-occupied zone and the capital of an independent Lithuania. But in May 1940, Lithuania was absorbed into the Soviet Union; a year later, in June 1941, it was occupied when the Germans invaded Russia. More than 100,000 Jews from Vilna and its environs were murdered in the woods of Ponar, a Vilna suburb, and buried in mass graves. The remaining 20,000 Jews were crowded into two small ghettos. Their ranks were periodically thinned through executions until the last ghetto was emptied in September 1943 and its occupants were deported or shot. This was the milieu that inspired many of Sutzkever's ghetto poems.

At first, Sutzkever concealed himself in a crawl space in his mother's apartment. Then, for the next few months, he joined labor details or hid elsewhere. In September 1941, he was captured by Lithuanian police and taken to be shot. In a charade similar to an experience of Dostoevsky's, Sutzkever was reprieved without warning when the Lithuanians deliberately fired over his head and returned him to the ghetto. But like Dostoevsky, he never forgot the moment and later described it in vivid detail.

His mother's hiding place was found in the ghetto, and she disappeared. His wife, Freydke, gave birth to a son, but the Germans poisoned the infant; the poem "For My Child," included here, alludes to this painful episode. Amid this misery, Sutzkever continued to write; he later insisted that during this time, poetry *became* his life: without it, he could not have managed to survive.

Together with his wife and other friends, Sutzkever joined the United Partisan Organization, a Jewish resistance group; in September 1943, they left the ghetto and walked more than sixty miles to join a Soviet partisan unit in the forests. After enduring indescribable hardships in the woods, Sutzkever and his wife were sent on a perilous journey to an airstrip, where a plane took them to Moscow. Some of Sutzkever's poems had come to the attention of Ilya Ehrenburg, a noted Soviet journalist and writer (and, like Sutzkever, Jewish), and partly through his intervention, the Soviet leadership decided that rescuing Sutzkever would be a propaganda coup. He and his wife reached Moscow in March 1944; but Vilna was liberated in July, and they returned to their home.

Sutzkever said later that he felt a mission to chronicle those "years of destruction" in his poetry. He and his wife arrived in Palestine in 1947. In 1948, he founded the Yiddish literary quarterly *Di goldene keyt (The Golden Chain)*, which he still edits. Sutzkever lives in Israel and remains the world's outstanding Holocaust poet writing in Yiddish, though like the other poets in this collection (with the possible exception of Nelly Sachs), in his long career he has addressed himself to many other themes.

Given Sutzkever's ordeal in Vilna, and the personal as well as communal losses he suffered there, it is not surprising that as a poet he remains beset by divided allegiances that leave nothing in his vision perfectly pure. The vital nature imagery that infuses so much of his verse is often tainted by a memory that cannot escape what he calls "the dark scream of the past." In "Self-Portrait," the poet wanders across a landscape whose "just ravaged whiteness" is "shrouded with soot," a landscape "lit up by a putrescent moon." This is not, of course, the tenor of all his poems, but the conjunction of natural beauty with stain helps us to understand the inner tension driving the poet. Even in a very recent work, he describes rolling "mountains into an abyss," quenching thirst with fire, draining "black honey" (reminiscent of Paul Celan's famous image, "Black milk of daybreak," in "Death Fugue") from lips that ask to be kissed.

While poets nurtured by more normal times might turn their imagination toward an infinitely expanding realm of the spirit, Sutzkever has been too honest to dismiss the legacy he inherited from the Vilna ghetto. Personal memories of that time, together with the public catastrophe of European Jewry, continue to thread through his poetry; they shadow his art, but cannot mute his voice. "I myself am my people," he declares, in a poem written as recently as 1987: "We shall both start a newborn silence / In honor of a language that has given up the ghost." The mystery of such creation from so much loss, together with the paradox of a newborn silence, is the closest thing to rebirth that Holocaust literature affords us.

War*

The same ashes will cover all of us:
The tulip—a wax candle flickering in the wind,
The swallow in its flight, sick of too many clouds,
The child who throws his ball into eternity—

And only one will remain, a poet—
A mad Shakespeare, who will sing a song, where might and wit is:
—My spirit Ariel, bring here the new fate,
And spit back the dead cities!

1939

*Referring to the outbreak of war in Europe, September 1939.

Faces in Swamps

The cycle "Faces in Swamps" was written in hiding during the first days of the Nazi occupation of Vilna. Subsequently, it was hidden in a ghetto cellar and discovered forty-nine years later in Vilnius. The manuscript contains nine poems with the following note in the poet's hand:

> Note. I wrote the nine poems of "Faces in Swamps" in the first 10 days, when the Plague marched into Vilna. Approximately between June 25 and July 5. I wrote them lying stuck in a broken chimney in my old apartment on Wilkomirska Street 14. This way I hid from the Snatchers who dragged off every Jewish male they could find.
>
> My wife carried the poems through all the horrors and tragedies. They were with her through the first provocation, were covered with blood, in prison under Schweinenberg's whip. Miraculously, my wife fled back to the ghetto with the poems, where I no longer was—I had fled in the middle of the night, during the Roundup of the Yellow Permits. When I returned, I found my wife in the hospital where she gave birth to a baby. In her labor pains, she was clutching the poems in her hands.
>
> A.S.
>
> Ghetto Vilna, May 16, 1942.

The text translated here is from the manuscript. The titles of two poems were added later.

I Faces in Swamps

. . . And overnight our thoughts grew gray. The sun
Sowed poison salt on open wounds. We choke.
White doves turned into owls. They're poking fun,
Mocking our dream that disappeared in smoke.

Why tremor, earth? Did you crack too, in trance?
Your nostrils smelled the stench of victim's flesh?

Devour us! We were cursed by overconfidence,
Devour us with our children, with our flags so fresh!

You're thirsty, earth. We, wailing pumps, will fill
With gold of our young bodies your newly opened pits.
A spiderweb of faces in a swamp will spin to kill:
Faces in a swamp—over the sunset, over huts . . .

II

Serpents of darkness: nooses choke
My breath.
Horseradish in my eyes, I toss
In a grater dungeon—
Each toss grates my skin.
Were there anything human, familiar . . .

My hand gropes: a piece of glass, the moon
Trembles imprisoned like me in the vise
Of the iron night. I grow tense:
"This was created by a human hand!"

In the glass edge I stroke the moon:
"You want?—I give you my life as a gift!"
But life is hot and the glass is cold
And it's a shame to put it to my throat . . .

III Leaves of Ash

I warm tea with your letters—
My only treasure,
Thin leaves of ash remain,
Sprinkled with glowworms
That I alone can read, can ask:
I warm tea with *your* letters,
My only treasure?

Let the wind be mute as a tombstone!
Let my shadow stand still!
One puff—
And all your healing beauty
Will stir jealousy
On all the roads.

How dear are you to me in leaves of ash,
How shining do you die in leaves of ash,
That I alone can read, can ask:
I warm tea with *your* letters,
My only treasure?

July 6, 1941

IV

Above—in a death swordplay, metal pirates
Spit whistling arrows into the heart of the moon.

Below, on a hill, among white tobacco flowers
A woman twisting on pain-and-wonder of birth.

"Who will help?"—"Hush, hush . . ." And her beloved
Weeps the glimmer of his eyes in the dust at her feet.

"My child, melody of my love, play on inside me, don't rush,
You are merely flesh and dream, and reality—is murder."

Slices of light swallow the fields. Fish in rivers scream.
The earth trembles along with the woman.

"Ghosts of death, don't dare touch, I beseech you . . ."
"Hush, hush, I am the armor against all evil."

Suddenly . . . like a piano playing among hordes of thunder,
A voice of a child slices through. And this sound—

(Whence the strength?) subjugates all fears,
And the love of the world turns the dew red.

V

Soon it will happen!
The black hoops
Grow tighter and tighter around my neck!
Impersonally, like a stone in a brook,
I shall remain lying under hooves,
Redeemed from the world.
But deep inside me—
Three ants still stray:
One,
Under the laurel of my childhood—

Will return to magicland.
The second,
Under the armor of my dream—
Will return to dreamland.
The third,
The one who carries my word—
Will have no path,
For the land of believing words
Is covered with plague.
In the valley of shadows, it will watch,
Alone and solitary,
Over my bones.

My every breath is a curse.
Every moment I am more an orphan.
I myself create my orphanhood
With fingers, I shudder to see them
Even in dark of night.

Once, through a cobblestone ghetto street
Clattered a wagon of shoes, still warm from recent feet,
A terrifying
Gift from the exterminators . . .
And among them, I recognized
My Mama's twisted shoe
With blood-stained lips on its gaping mouth.
—Mama, I run after them, Mama,
Let me be a hostage to your love,
Let me fall on my knees and kiss
The dust on your holy throbbing shoe
And put it on, a *tfillin** on my head,
When I call out your name!

But then all shoes, woven in my tears,
Looked the same as Mama's.
My stretched-out arm dropped back
As when you want to catch a dream.

Ever since that hour, my mind is a twisted shoe.
And as once upon a time to God, I wail to it
My sick prayer and wait
For new torments.
This poem too is but a howl,
A fever ripped out of its alien body.
No one to listen.
I am alone.
Alone with my thirty years.
In their pit they rot—
Those who once were called
Papa.
Mama.
Child.

Vilna Ghetto July 30, 1943

*Leather thongs wound around the forehead and the forearm by Orthodox Jews during prayer.

How?

How and with what will you fill
Your goblet on the day of Liberation?
In your joy, are you ready to feel
The dark scream of your past
Where skulls of days congeal
In a bottomless pit?

You will look for a key to fit
Your jammed locks.
Like bread you will bite the streets
And think: better the past.
And time will drill you quietly
Like a cricket caught in a fist.

And your memory will be like
An old buried city.
Your eternal gaze will crawl
Like a mole, like a mole—

Vilna Ghetto February 14, 1943

Grains of Wheat

Caves, gape open,
Split open under my ax!
Before the bullet hits me—
I bring you gifts in sacks.

Old, blue pages,
Purple traces on silver hair,
Words on parchment, created
Through thousands of years in despair.

As if protecting a baby
I run, bearing Jewish words,
I grope in every courtyard:
The spirit won't be murdered by the hordes.

I reach my arm into the bonfire
And am happy: I got it, bravo!
Mine are Amsterdam, Worms,
Livorno, Madrid, and YIVO.*

How tormented am I by a page
Carried off by the smoke and winds!
Hidden poems come and choke me:
—Hide us in your labyrinth!

And I dig and plant manuscripts,
And if by despair I am beat,
My mind recalls: Egypt,
A tale about grains of wheat.

And I tell the tale to the stars:
Once, a king at the Nile
Built a pyramid—to rule
After his death, in style.

*Jewish cultural centers. YIVO is the Jewish Scientific Institute, founded in Vilna in 1925. Sutzkever worked there for a time under the Nazis.

Let them pour into my golden coffin,
Thus an order he hurled,
Grains of wheat—a memory
For this, the earthly world.

For nine thousand years have suns
Changed in the desert their gait,
Until the grains in the pyramid
Were found after endless wait.

Nine thousand years have passed!
But when the grains were sown—
They blossomed in sunny stalks
Row after row, full grown.

Perhaps these words will endure,
And live to see the light loom—
And in the destined hour
Will unexpectedly bloom?

And like the primeval grain
That turned into a stalk—
The words will nourish,
The words will belong
To the people, in its eternal walk.

Vilna Ghetto March 1943

Stalks

Two years I longed for stalks,
Silent stalks in a familiar field.
When I struggled in the vise
That caught me
And blocked
The road,
The green road to those stalks—
But not the stalks in the familiar field.

And when my breath melted the vise—
A wind in my veins
Whistled and called:
—"Get up, son of man, the stalks are ripe.
Now your own body is like a stalk."
And as fate walks, so walked I
Through burned cities
To that call.

But when I came, weary, through the sunset,
I reached my longed-for field—
They lay there, my brothers,
Killed over the field.
And the stalks with glowing spears,
Layer upon layer, grew through
The skulls, the ribs,
And climbed higher, higher, higher,
To the sun that gathers back its light,
As if each stalk rushed to overtake
The others.
One stalk
Went wandering
Through a mouth with clenched teeth!
Two stalks crept through shoulders.
And there, a stalk searching for a way—
A hand reaching out of the earth.
And a cornflower through an eye, weeping—

What do I see now in the evening light?
I see a field with stalks, blood red.
And rushing to me closer, comes a mower
And mows the afterwar fresh bread.

Narocz Forest September 1943

Frozen Jews

Did you ever see in fields of snow
Frozen Jews, in row upon row?

Breathless they lie, marbled and blue.
Of death in their bodies, no hint and no clue.

Somewhere their spirit is frozen and saved
Like a golden fish in a frozen wave.

Not speaking. Not silent. Just *thinking* bright.
The sun too lies frozen in snow at night.

On a rosy lip, in the freeze, still glows
A smile—will not move, not budge since it froze.

Near his mother, a baby starving, at rest.
How strange: she cannot give him her breast.

The fist of a naked old man in surprise:
He cannot release his force from the ice.

So far, I have tasted all kinds of death,
None will surprise me, will catch my breath.

But now, overcome in the mid-July heat
By a frost, like madness, right in the street:

They come toward me, blue bones in a row—
Frozen Jews over plains of snow.

My skin is covered with a marble veil.
My words slow down, my light that is frail.

My motions freeze, like the old man's surprise,
Who cannot release his force from the ice.

Moscow July 10, 1944

Resurrection*

I searched for the Shofar of Messiah
In specks of grass, in scorched cities,
To awaken my friends. And thus spake
My soul of bones:
See, I glow
Inside you,
Why look for me outside?

And in my great
Forged rage,
I ripped my spirit from my body
Like a sharp horn
Of a living animal
And began to blow:
Tekiya,
Shevorim.†
Come to life, the world is now free.
Leave your not-being in the graves
And leap out with blessing.
See how pure
The stars are rocking for your sake!
But the earth—like a river—
Flowed away with grass and stone,
And human words I heard:
—We don't want, go away, your earth is foul!
—From the punishment of living we were once freed!
—We don't need your time,
Your blind limping time,
And not the stars—
Our non-light glimmers brighter!
—Reality, that's *us,*

*Refers to Sutzkever's return to Vilna after its "liberation."

†Words that accompany the blowing of the Shofar, or Ram's Horn, during Rosh Hashanah and Yom Kippur.

Vanish, cursed dream!
Gambled away, played out is your war.

Only one, with a voice unheard
Like the blooming of a forest, called to me,
Yearning: Redeem me, destined one—

—Who are you, that your command should be heard?

And grass language answered me: God.
I once lived in your word.

Moscow 1945

Black Thorns

On my mother's house
Thorns grow—
Yesterday's mad, piercing gazes!
And I—
In their thorniness I dwell.*
I seek my meaning
In black thorns.
I feel my mother's spirit
Hanging on the thorns—
The black thorns are now my Psalms.
At dusk,
When only dews know no tears,
I climb up to them,
Aching with devotion,
And my lips—clouds over words,
Prattle up a homey moon.

To him
Who planted the black thorns
I pray:
Plant me too like them,
I want to live here,
This is good, is good.

I undress.
Start dancing,
Dancing,
Dancing,
Till the thorns flower with my blood.

I want to live here,
This is good, is good.

Vilna May 1945

*Allusion to Ezekiel 16:6: "In your blood shall you live."

In the Cell

How could it be otherwise: the darkness is out to choke me!
Somewhere in my sight leaden mice gnaw away.
I toss and turn, finally sink among walls:
Is there something human here or familiar?

Groping I find a piece of glass where reflected
the trapped moon twitches in transparent pincers.
Who cares if I'm growing more feverish:
this splinter's been chipped by someone's hand.

I stroke the sharp lunar edge and ask:
"Do you want me offered up as a gift?"
But blood is hot, the glass—cold,
and it's a shame to take the sliver to my throat.

Vilna End of June 1941

For My Child

Was it from some hunger
or from greater love—
but your mother is a witness to this:
I wanted to swallow you, my child,
when I felt your tiny body losing its heat
in my fingers
as though I were pressing
a warm glass of tea,
feeling its passage to cold.

You're no stranger, no guest,
for on this earth one does not give birth to aliens.
You reproduce yourself like a ring
and the rings fit into chains.

My child,
what else may I call you but: love.
Even without the word that is who you are,
you—seed of my every dream,
hidden third one,
who came from the world's corner
with the wonder of an unseen storm,
you who brought, rushed two together
to create you and rejoice:—

Why have you darkened creation
with the shutting of your tiny eyes
and left me begging outside
in the snow swept world
to which you have returned?

No cradle gave you pleasure
whose rocking
conceals in itself the pulse of the stars.
Let the sun crumble like glass
since you never beheld its light.
That drop of poison extinguished your faith—

you thought
it was warm sweet milk.

I wanted to swallow you, my child,
to feel the taste
of my anticipated future.
Perhaps in my blood
you will blossom as before.

But I am not worthy to be your grave.
So I bequeath you
to the summoning snow,
the snow—my first respite,
and you will sink
like a splinter of dusk
into its quiet depths
and bear greetings from me
to the frozen grasslands ahead—

Vilna Ghetto January 18, 1943

Burnt Pearls

It is not just because my words quiver
like broken hands grasping for aid,
or that they sharpen themselves
like teeth on the prowl in darkness,
that you, written word, substitute for my world,
flare up the coals of my anger.

It is because your sounds
glint like burnt pearls
discovered in an extinguished pyre
And no one—not even I—shredded by time
can recognize the woman drenched in flame
for all that remains of her now
are these grey pearls
smouldering in the ash—

Vilna Ghetto July 28, 1943

Poem About a Herring

Right at the open limepit
a child broke into tears:
Mameh, I'm hungry, something to eat!
So his mother momentarily forgot where she was
—or she was forgotten
by Him,
God Who snatches time right from under our feet—
and she quickly opened her satchel
and gave her child this herring to eat.

As if it were some silver bounty
the young teeth
grabbed the herring with pleasure.
But quietly as though a nightingale suddenly burst into song
from far away across blue waters
a fiery string of notes
of a sudden
gave his head such a jolt.
And out of the broken circle
the naked child
slid punctured into a pit.

Frozen and grotesque
this picture holds like a frieze:
a child with a bloody herring in his mouth
on a certain summer's morning.
And I search for that herring's salt
and still can not
find its taste on my lips.

Warsaw August 1946

Self-Portrait

As if a lake were to rise up
on hind legs
and stand in terror
covered with icy scales
the city quivers
and its purple creases shiver
as I fondle
its glassy face.

Echo of shadows.
Sounds crucified.

And I walk forth.

Columns of light—
broken stalks.

I walk further.

Where to?
To find someone's breath,
a word leaping over clay lips,
a face to greet with: good morning!
You and the world must survive
and snakes no longer
slither up from our sleeves.

And I walk on.

Once hunger bewitched me like Lilith
and I fast devoured a swallow in an attic.
As I remember it now the swallow comes to trill
its winged revenge out of my pupils.
There are no longer tears in my eyes.
The feathered one
has pecked them all out
with a rabid cry.

Once hidden in a cellar
beside a corpse laid out like a sheet of paper
illumined by phosphorous snow from the ceiling—
I wrote a poem with a piece of coal
on the paper body of my neighbour.
Now there is not even a corpse—
just ravaged whiteness
shrouded with soot.

And still I walk.

A long falling snow sinks all round.
And with small lights
my house rears up
like a temple
bitten away by lightning . . .
I recognize it from a ten-year old's dream.

Behind my back a breath closes
like a lock.
And nails
drive the steely silence into my body.
And lit up by a putrescent moon
groping through the temple's snow
a shaggy figure draws up before me,
stooped like myself
his bones show beneath his rags.

Hey, wanderer, who are you?
And the shaggy figure howls back:
Who are you?

Do you know me?
And he with the same question:
Do you know me?

Soul?
The shaggy figure shuffles closer:
Soul?

But I see the lines on his face,
triumphantly I throw myself at him—
someone takes a swipe at my head
and I crash into the barrier of glass.

1951

26. Dan Pagis

Like the other poets in this anthology, Dan Pagis was much more than a "Holocaust poet," though the focus here must necessarily be on the portion of his work concerned with that theme. Born in 1930 in Radautz, Pagis shared with Aharon Appelfeld and Paul Celan the experience of growing up in the culturally diverse environment of the Bukovina region of Romania. His father emigrated to Palestine in 1934; his mother died soon after. The young Pagis was raised by grandparents. He spent three years of his adolescence in Nazi camps in Transnistria, an area where many Romanian Jews were deported during the war and that is now part of the former Soviet Union.

After the war, Pagis emigrated to Palestine, where he rejoined his father. He settled in Jerusalem, received a Ph.D. from Hebrew University, and went on to become a professor of medieval Hebrew literature and a leading scholar in the field, writing many books on the subject. He also taught at various universities in the United States.

Pagis began publishing poems at the age of nineteen; more than a half-dozen volumes of his poetry appeared before his early death in 1986. Whereas Celan continued to write in his native tongue, Pagis, like Appelfeld, chose to write in Hebrew, which he learned only after reaching Palestine. It is nothing less than astonishing that a writer who arrived in his adopted land ignorant of its language should become, in the words of Israeli critic Gershon Shaked, "the foremost living authority on the poetics of Hebrew literature in the High Middle Ages and the Renaissance."

Unlike Nelly Sachs's early postwar verse, Pagis's poems leave to his readers the challenge of filling in the Holocaust referents. Sachs's citations are often unmistakable, but when Pagis writes of "convoys of smoke," we are left with a vision of decomposition unreinforced by surrounding imagery that might locate it in a precise Holocaust context. Similarly, a fragment from the poem "Footprints" resists the temptation of concrete historical or autobiographical allusion:

Frozen and burst, clotted,
scarred,
charred, choked.

Linked by a carefully modulated assonance and consonance, the epithets in the original Hebrew are suffused by a ritualistic, incantatory terror, but they are no more forgiving in their refusal to enter a familiar place or time. This labor is left to our own ingenuity—to accept the spatial and temporal disruption wrought by the Holocaust and to imagine, with the artist's help, a mental landscape displaying the skewed remains.

Poets like Pagis exalt the power of language to limn silence; they write praisesongs to the very words that remind us of the calamity they deplore. Perhaps this is the ultimate paradox of all Holocaust art.

Autobiography

I died with the first blow and was buried
among the rocks of the field.
The raven taught my parents
what to do with me.

If my family is famous,
not a little of the credit goes to me.
My brother invented murder,
my parents invented grief,
I invented silence.

Afterwards the well-known events took place.
Our inventions were perfected. One thing led to another,
orders were given. There were those who murdered in their own way,
grieved in their own way.

I won't mention names
out of consideration for the reader,
since at first the details horrify
though finally they're a bore:

you can die once, twice, even seven times,
but you can't die a thousand times.
I can.
My underground cells reach everywhere.

When Cain began to multiply on the face of the earth,
I began to multiply in the belly of the earth,
and my strength has long been greater than his.
His legions desert him and go over to me,
and even this is only half a revenge.

Europe, Late

Violins float in the sky,
and a straw hat. I beg your pardon,
what year is it?
Thirty-nine and a half, still awfully early,
you can turn off the radio.
I would like to introduce you to:
the sea breeze, the life of the party,
terribly mischievous,
whirling in a bell-skirt, slapping down
the worried newspapers: tango! tango!
And the park hums to itself:
    I kiss your dainty hand, madame,
    your hand as soft and elegant
    as a white suede glove. You'll see, madame,
    that everything will be all right,
    just heavenly—you wait and see.
    No it could never happen here,
    don't worry so—you'll see-it could

Written in Pencil in the Sealed Railway-Car

here in this carload
i am eve
with abel my son
if you see my other son
cain son of man
tell him that i

The Roll Call

He stands, stamps a little in his boots,
rubs his hands. He's cold in the morning breeze:
a diligent angel, who worked hard for his promotions.
Suddenly he thinks he's made a mistake: all eyes,
he counts again in the open notebook
all the bodies waiting for him in the square,
camp within camp: only I
am not there, am not there, am a mistake,
turn off my eyes, quickly, erase my shadow.
I shall not want. The sum will be all right
without me: here forever.

Testimony

No no: they definitely were
human beings: uniforms, boots.
How to explain? They were created
in the image.

I was a shade.
A different creator made me.

And he in his mercy left nothing of me that would die.
And I fled to him, floated up weightless, blue,
forgiving—I would even say: apologizing—
smoke to omnipotent smoke
that has no face or image.

Instructions for Crossing the Border

Imaginary man, go. Here is your passport.
You are not allowed to remember.
You have to match the description:
your eyes are already blue.
Don't escape with the sparks
inside the smokestack:
you are a man, you sit in the train.
Sit comfortably.
You've got a decent coat now,
a repaired body, a new name
ready in your throat.
Go. You are not allowed to forget.

Draft of a Reparations Agreement

All right, gentlemen who cry blue murder as always,
nagging miracle-makers,
quiet!
Everything will be returned to its place,
paragraph after paragraph.
The scream back into the throat.
The gold teeth back to the gums.
The terror.
The smoke back to the tin chimney and further on and inside
back to the hollow of the bones,
and already you will be covered with skin and sinews and you will live,
look, you will have your lives back,
sit in the living room, read the evening paper.
Here you are. Nothing is too late.
As to the yellow star:
it will be torn from your chest
immediately
and will emigrate
to the sky.

Footprints

"From heaven to the heaven of heavens to the heaven of night" Yannai

Against my will
I was continued by this cloud: restless, gray,
trying to forget in the horizon, which always receded

Hail falling hard,
like the chatter of teeth:
refugee pellets pushing eagerly
into their own destruction

In another sector
clouds not yet identified.
Searchlights that set up
giant crosses of light
for the victim.
Unloading of cattle-cars.

Afterwards the letters fly up,
after the flying letters mud
hurries, snuffs, covers for a time

It's true, I was a mistake, I was forgotten
in the sealed car, my body tied up
in the sack of life

Here's the pocket where I found bread,
sweet crumbs, all from the same world

Maybe there's a window here—if you don't mind,
look near that body, maybe you can open up
a bit. That reminds me
(pardon me) of the joke about the two Jews
in the train, they were traveling to

Say something more; talk.
Can I pass from my body and onwards—

*

From heaven to the heaven of heavens to the heaven of night
long convoys of smoke

The new seraphim who haven't yet understood,
prisoners of hope, astray in the empty freedom,
suspicious as always: how to exploit
this sudden vacuum, maybe
the double citizenship will help,
the old passport,
maybe the cloud? what's new in the cloud,
here too of course
they take bribes. And between us: the biggest bills
are still nicely hidden away, sewn
between the soles—
but the shoes have been piled up below:
a great gaping heap

Convoys of smoke. Sometimes
someone breaks away,
recognizes me for some reason, calls my name.
And I put on a pleasant face, try to remember:
who else
who

Without any right to remember, I remember
a man screaming in a corner, bayonets rising
to fulfill their role
in him

Without any right to remember. What else
was there? Already I'm not afraid
that I might say

without any connection at all:
there was a heart, blue from excessive winter,
and a lamp, round, blue, kind-hearted.
But the kerosene disappears with the blood, the flame flickers—

Yes, before I forget:
the rain stole across some border, so did I,
on forbidden escape-routes, with forbidden hope,
we both passed the mouth of the pits

Maybe now
I'm looking in that rain
for the scarlet thread

Where to begin?
I don't even know how to ask.
Too many tongues are mixed in my mouth. But
at the crossing of these winds,
very diligent, I immerse myself
in the laws of heavenly grammar: I am learning

the declensions and ascensions of
silence.

Who has given you the right to jest?
What is above you you already know.
You meant to ask about what is within you,
what is abysmally through you.
How is it that you did not see?

But I didn't know I was alive.
From the heaven of heavens to the heaven of night
angels rushed, sometimes one of them
would look back, see me, shrug his shoulders,
continue from my body and onwards

*

Frozen and burst, clotted,
scarred,
charred, choked.

If it has been ordained that I pull out of here,
I'll try to descend rung by rung,
I hold on to each one, carefully—
but there is no end to the ladder, and already
no time. All I can still do is fall
into the world

And on my way back
my eyes hint to me:
you have been, what more did you want to see?
Close us and see:
you are the darkness, you are the sign.

And my throat says to me:
if you are still alive, give me an opening, I
must praise.

And my upside-down head is faithful to me,
and my hands hold me tight:
I am falling falling
from heaven to the heaven of heavens to the heaven of night

*

Well then: a world.
The gray is reconciled by the blue.
In the gate of this cloud, already a turquoise
innocence, perhaps light green. Already sleep.
Heavens renew themselves, try out their wings, see me

and run for their lives. I no longer wonder.
The gate bursts open:

a lake
void   void   pure of reflections

Over there,
in that arched blue, on the edge of the air,
I once lived. My window was fragile.
Maybe what remained of me
were little gliders that hadn't grown up:
they still repeat themselves in still-clouds, glide,
slice the moment
    (not to remember now, not to remember)
And before I arrive
    (now to stretch out to the end, to stretch out)
already awake, spread to the tips of my wings,
against my will I feel that, very near,
inside, imprisoned by hopes, there flickers
this ball of the earth,
scarred, covered with footprints.

Ready for Parting

Ready for parting, as if my back were turned,
I see my dead come toward me, transparent and breathing.
I do not consent:
one walk around the square, one rain,
and I am another, with imperfect rims, like clouds.
Gray in the passing town, passing and glad,
among transitory streetlamps,
wearing my strangeness like a coat, I am free to stand
with the people who stand at the opening of a moment
in a chance doorway, anonymous as raindrops
and, being strangers, near and flowing one into another.

Ready for parting, waiting a while
for the signs of my life which appear in the chipped plaster
and look out from the grimy windowpane. A surprise of roses.
Bursting out and already future, twisted into its veins—
a blossoming to every wind. Perhaps
not in my own time into myself and from myself and onward
from gate within gate I will go out into the jungle of rain,
free to pass on like one who has tried his strength
I will go out
from the space in between as if from the walls of denial.

27. Paul Celan

Paul Ancel (or Antschel) was born in Czernowitz in the Bukovina region of Romania in 1920. From his surname he formed the anagram Celan in 1947, and subsequently wrote as a poet whose origins were screened but not nullified by his nominally altered identity. That identity had complex geographical and cultural origins. A few years before Celan's birth, Czernowitz had been part of the Austro-Hungarian Empire. During World War II, it endured Soviet and German occupation (the latter accepted by Hitler's Romanian allies); following the conflict, it was permanently annexed to the former Soviet Union.

Celan grew up in a Jewish household, where his parents spoke German. He was exposed to Yiddish in the famous Jewish community of Czernowitz and to Hebrew from the religious education required by his Zionist father. Romanian was the language of his secular schooling. He published his first youthful poems in that language, including his most famous one, "Death Fugue," which appeared in May 1947 in a Romanian translation, under the title "Tango of Death," before it was printed in German. He also translated some of Shakespeare's sonnets, and poems by Emily Dickinson and Marianne Moore. From Russian he translated the poets Aleksandr Blok, Sergey Yesenin, and—a major influence on his own verse—Osip Mandelstam, who died in deportation, a victim of Stalin's purges. Celan gained fluency in French through a year of medical school in Tours in 1938 and a course of study in Romance philology after his return. He translated Rimbaud, Valéry, Apollinaire, and other French poets.

In June 1942, Jews from the Czernowitz ghetto, including Celan's parents, were sent on an infamous forced march to a camp in Transnistria on the Bug River in eastern Romania, where a few months later they were shot by the SS. Celan managed to escape deportation by hiding with friends. He spent nearly two years at forced labor under the Germans before fleeing to the Soviet army. He reentered Czernowitz after Russian troops liberated and annexed it in February 1944. In 1945, he moved to Bucharest and in 1947, to Vienna. In July 1948, he settled in Paris, where he studied German literature and was a lecturer in that subject at an école normale supérieure. During the next twenty years, Celan published many volumes of poetry, gaining renown as one of the leading

German-language poets and a major interpreter in verse of the bleak effects of the Holocaust on language and on our vision of human experience.

The bleak effects of the Holocaust ordeal on Celan himself, especially the murder of his parents, were more muted, but in later years he suffered a number of mental crises, culminating in suicide: he was found drowned in the Seine in April 1970. He was forty-nine years old.

In spite of the long years of self-imposed exile in France, German remained Celan's mother tongue. But it was his mother's tongue, too, as well as that of his mother's (and father's) murderers, and this paradox was a main source of affliction early in Celan's poetic career. His language, as he himself confessed, had to "pass through the thousand darknesses of death-bringing speech" before it could be purified for use as poetry. And still it retained some of its original taint.

Critic George Steiner once implied that Celan wrote German as if it were a foreign language, and insofar as this observation contains a germ of truth, it reflects Celan's efforts to atomize a language that was spoken by men and women who in the 1940s were determined to atomize their Jewish victims. Like Nelly Sachs, the other great poet of the Holocaust who wrote in German, Celan was driven by a desire to purge and renew his language while simultaneously recording its corruption and demise. The word, as he conveys in poem after poem, was one of the nonhuman victims of the catastrophe of European Jewry. But unlike Sachs, Celan—especially in his later work—was driven into an abyss of fragmentation from which it was difficult to emerge.

Celan and Sachs explore similar realms, as the opening lines of the following poems suggest, though allusion is far more condensed in Celan's work. Sachs's poem "But in the night" begins:

But in the night,
when dreams pull away
walls and ceilings with a breath of air,
the trek to the dead begins.
You search for them under the stardust—

Compare this with "Relocation among the substances":

Relocation among the substances:
go to yourself, take part
where earthlight
is missing

The final verses of Sachs's poem, however, indicate that the trek to the dead begins rather than ends her journey:

Thus dawn comes
strewn with the red seed of the sun

and night has cried itself out
into the day—

Celan, however, closes with a cryptic and bitter challenge:

there are two suns, do you hear,
two,
not one—
so what?

We are left with a universe still drifting toward division—not, as in Sachs's vista, toward possible unity or, at least, reconciliation.

Both Celan and Sachs use the technique of joining and uncoupling words in an attempt to portray the unsettled reality of their poetic vision. The German language lends itself to this sort of strategy far more easily than does English. Celan also shares with Sachs a fondness for linking images of darkness and light in a way that make them inseparable—not alternative views but reflections of each other. A word such as "seelenverfinstert" ("soul-eclipsed") fuses two images into a compressed allusion that forces the reader to use what Celan in another poem calls a "variable key" to "unlock the house in which / drifts the snow of that left unspoken." He insisted to one of his translators that his poems are not hermetic; they are open to anyone who would study and read and reread and restudy. Those who still wrestle with the compact enigmas of his later poems may feel that he exaggerated their clarity. Few would dispute, however, that more than any other writer on this subject, Celan requires of his readers patience and persistence to vanquish the exasperation of his obscurities.

Much in Celan remains "unspoken," but again this is also true of the legacy he inherited, as well as of thousands of other survivors. He was too honest to pretend otherwise. If the dialogue with hindsight and foresight that his poems provoke mirrors hidden terrain more often than vision, this is consistent with the intention of a poet who, in one of his boldest images, "seminated the night." The offspring of such a frightful union, to use his own language once more, may be "glimpsed and avoided," leading, as one of his translators has proposed, to a "sympathetic unknowing"—a modest but practical goal for those who have witnessed the constant frustration of enlightenment values in our brutal and troubled century.

Death Fugue

Black milk of daybreak we drink it at sundown
we drink it at noon in the morning we drink it at night
we drink and we drink it
we dig a grave in the breezes there one lies unconfined
A man lives in the house he plays with the serpents he writes
he writes when dusk falls to Germany your golden hair Margarete
he writes it and steps out of doors and the stars are flashing he whistles his pack out
he whistles his Jews out in earth has them dig for a grave
he commands us strike up for the dance

Black milk of daybreak we drink you at night
we drink in the morning at noon we drink you at sundown
we drink and we drink you
A man lives in the house he plays with the serpents he writes
he writes when dusk falls to Germany your golden hair Margarete
your ashen hair Shulamith we dig a grave in the breezes there one lies unconfined.

He calls out jab deeper into the earth you lot you others sing now and play
he grabs at the iron in his belt he waves it his eyes are blue
jab deeper you lot with your spades you others play on for the dance

Black milk of daybreak we drink you at night
we drink you at noon in the morning we drink you at sundown
we drink you and we drink you
a man lives in the house your golden hair Margarete
your ashen hair Shulamith he plays with the serpents

He calls out more sweetly play death death is a master from Germany
he calls out more darkly now stroke your strings then as smoke you will rise into air
then a grave you will have in the clouds there one lies unconfined

Black milk of daybreak we drink you at night
we drink you at noon death is a master from Germany

we drink you at sundown and in the morning we drink and we drink you
death is a master from Germany his eyes are blue
he strikes you with leaden bullets his aim is true
a man lives in the house your golden hair Margarete
he sets his pack on to us he grants us a grave in the air
he plays with the serpents and daydreams death is a master from Germany

your golden hair Margarete
your ashen hair Shulamith

Aspen Tree

Aspen tree your leaves glance white into the dark.
My mother's hair was never white.

Dandelion, so green is the Ukraine.
My yellow-haired mother did not come home.

Rain cloud, above the well do you hover?
My quiet mother weeps for everyone.

Round star, you wind the golden loop.
My mother's heart was ripped by lead.

Oaken door, who lifted you off your hinges?
My gentle mother cannot return.

With a Variable Key

With a variable key
you unlock the house in which
drifts the snow of that left unspoken.
Always what key you choose
depends on the blood that spurts
from your eye or your mouth or your ear.

You vary the key, you vary the word
that is free to drift with the flakes.
What snowball will form round the word
depends on the wind that rebuffs you.

Nocturnally Pouting

For Hannah and Hermann Lenz

Nocturnally pouting
the lips of flowers,
criss-crossed and linked
the shafts of the spruces,
turned grey the moss, the stone shaken,
roused for unending flight
the jackdaws over the glacier:

this is the region where
those we've caught up with rest:

they will not name the hour,
they will not count the flakes
nor follow the stream to the weir.

They stand apart in the world,
each one close up to his night,
each one close up to his death,
surly, bare-headed, hoar-frosted
with all that is near, all that's far.

They discharge the guilt that adhered to their origin,
they discharge it upon a word
that wrongly subsists, like summer.

A word—you know:
a corpse.

Let us wash it,
let us comb it,
let us turn its eye
towards heaven.

Tenebrae

We are near, Lord,
near and at hand.

Handled already, Lord,
clawed and clawing as though
the body of each of us were
your body, Lord.

Pray, Lord,
pray to us,
we are near.

Wind-awry we went there,
went there to bend
over hollow and ditch.

To be watered we went there, Lord.

It was blood, it was
what you shed, Lord.

It gleamed.

It cast your image into our eyes, Lord.
Our eyes and our mouths are so open and empty, Lord.
We have drunk, Lord.
The blood and the image that was in the blood, Lord.

Pray, Lord.
We are near.

There Was Earth Inside Them

There was earth inside them, and
they dug.

They dug and they dug, so their day
went by for them, their night. And they did not praise God.
who, so they heard, wanted all this,
who, so they heard, knew all this.

They dug and heard nothing more;
they did not grow wise, invented no song,
thought up for themselves no language.
They dug.

There came a stillness, and there came a storm,
and all the oceans came.
I dig, you dig, and the worm digs too,
and that singing out there says: They dig.

O one, o none, o no one, o you:
Where did the way lead when it led nowhere?
O you dig and I dig, and I dig towards you,
and on our finger the ring awakes.

Psalm

No one moulds us again out of earth and clay,
no one conjures our dust.
No one.

Praised be your name, no one.
For your sake
we shall flower
Towards
you.

A nothing
we were, are, shall
remain, flowering;
the nothing-, the
no one's rose.

With our pistil soul-bright
with our stamen heaven-ravaged
our corolla red
with the crimson word which we sang
over, o over
the thorn.

Alchemical

Silence, cooked like gold, in
charred
hands.

Great, grey
sisterly shape
near like all that is lost.

All the names, all those
names
burnt with the rest. So much
ash to be blessed. So much
land won
above
the weightless, so weightless
rings
of souls.

Great, grey one. Cinder-
less

You, then.
You with the pale
bit-open bud,
you in the wine-flood.

(Us too, don't you think,
this clock dismissed?

Good,
good, how your word died past us here.)

Silence, cooked like gold, in
charred, charred
hands.
Fingers, insubstantial as smoke. Like crests, crest of air
around—

Great, grey one. Wake-
less.
Re-
gal one.

Radix, Matrix

As one speaks to stone, like
you,
from the chasm, from
a home become a
sister to me, hurled
towards me, you,
you that long ago
you in the nothingness of a night,
you in the multi-night encountered, you
multi-you—:

At that time, when I was not there,
at that time when you
paced the ploughed field, alone:

Who,
who was it, that
lineage, the murdered, that looms
black into the sky:
rod and bulb—?

(Root.
Abraham's root. Jesse's root. No one's
root—O
ours.)

Yes,
as one speaks to stone, as you
with my hands grope into there,
and into nothing, such
is what is here:
this fertile
soil too gapes,
this
going down
is one of the
crests growing wild.

Tabernacle Window

The eye, dark:
as tabernacle window. It gathers,
what was world, remains world: the migrant
East, the
hovering ones, the
human beings-and-Jews,
the people of clouds, magnetically
with heart-fingers, you
it attracts, Earth:
you are coming, coming,
we shall dwell at last, dwell, something

—a breath? a name?—

moves about over orphaned ground,
light as a dancer, cloddish,
the angel's
wing, heavy with what's invisible, on
the foot rubbed sore, trimmed
down by the head, with
the black hail that
fell there too, at Vitebsk,*

—and those who sowed it, they
write it away with
a mimetic anti-tank claw!—

moves, moves about,
searches,
searches below,
searches above, far, searches
with eyes, fetches
Alpha Centauri down, and Arcturus,† fetches

*City in White Russia, whose remaining Jewish residents were liquidated by the Germans in October and November 1941.

†Bright stars.

the ray as well, from the graves,
goes to the ghetto and Eden, gathers
the constellation which they,
humankind, need for dwelling, here,
among humankind,

pacing,
musters the letters and the mortal-
immortal soul of letters,
goes to Aleph and Yod* and goes farther,

builds it, the shield of David, and lets
it flare up, once,

lets it go out—there he stands,
invisible, stands
beside Alpha and Aleph, beside Yod
and the others, beside
everyone: in
you,

Beth,†—that is
the house where the table stands with

the light and the Light.

*First and tenth letters of the Hebrew alphabet.

†Second letter of the Hebrew alphabet. Also, in a variant form, the Hebrew word for "house."

. . . And No Kind Of

. . . And no kind of
peace.

Grey nights, foreknown to be cool.
Stimulus dollops, otter-like,
over consciousness gravel
on their way to
little memory bubbles.

Grey-within-grey of substance.

A half-pain, a second one, with no
lasting trace, half-way
here. A half-desire.
Things in motion, things occupied.

Cameo
of compulsive repetition.

From Things Lost

From things lost you were cast,
perfect the mask,

along the fold
of your eyelid
in the fold of my own
close to you,

the spoor, the spoor
strew it with gray,
final, deathly.

Sound-Dead Sister-Shell

Sound-dead sister-shell,
let the dwarf-sounds in,
they have been examined:
together they muffle up the great heart
and bear it off on their shoulders to
every distress, every distress.

In the Corner of Time

In the corner of time
the alder revealed
swears to itself in stillness,

on the back of the earth, breadth of a handspan,
squats the lung
shot through,

at the edge of fields the winged hour
plucks the grain of snow
from its own eye of stone.

Streamers of light infect me,
Flaws in the crown flicker.

From the Beam

From the beam
come in like the night,
the last sail
billows,

on board
your scream
enshrined,
you were there, you are below,

you are down below,

I go, I go with my
fingers
towards you,
to you down there, with fingers,

the arm-stalks multiply,

the beacon broods
instead of the one-
starred heaven,

with the drop-keel
I get a reading from you.

28. Miklós Radnóti

Miklós Radnóti was born in Budapest in 1909. His life seems to have been encased by disaster: his mother and twin brother died during childbirth, leaving him with a burden of painful memory that invaded many of his poems; and he was killed at the age of thirty-five, a victim of the Holocaust. Growing up between the two world wars, he watched the gradual encroachment of dictatorial and antisemitic regimes in many of the countries of Eastern Europe. History and autobiography thus combine to cast an uncanny aura of foreboding over much of his work.

After graduating from high school in Budapest, Radnóti studied briefly in Czechoslovakia, and then returned to Hungary to pursue a doctorate in French and Hungarian literature. Because of the restrictions against Jews attending universities, he was unable to enroll in the University of Budapest. He earned his Ph.D. in 1934 from a university in the smaller city of Seged.

Discouraged by burgeoning anti-Jewish sentiment in Hungary, which was moving swiftly toward the political right, Radnóti reacted to his increasing alienation from the centers of national culture by converting to Catholicism. Zsuzsanna Ozsváth and Frederick Turner, his editors, insist that his commitment was more to what they call an "aesthetic Catholicism" than to Catholic ritual or faith. They say that his work "makes little use of the themes and concepts of the New Testament," but draws for the most part on "the moods, ideas, and characters of the Old Testament." None of this, however, helped Radnóti when the Hungarian government, as an ally of Germany, passed stringent new antisemitic laws that regarded converts like Radnóti as Jews. He was recruited for forced labor three times: for three months in 1940, for ten months in 1942 and 1943, and in 1944, when he was shipped to Yugoslavia with thousands of others to aid in road building for the Germans.

Already exhausted by his first two stints at forced labor, Radnóti was in a weakened condition when he and the others were taken by forced march from a camp near Bor in Yugoslavia in September 1944 back toward Hungary. It was what later came to be known as a death march. Of the 3,600 laborers who left Bor, only 800 reached the Hungarian border alive. Radnóti was among the survivors, but by then he was too feeble to continue. Hungarians officers separated him and twenty-one other emaciated men from the group. They tried to leave them, first at a hospital and then at a school housing refugees, but neither place had any room for Jews. Early in November 1944, Radnóti and the others were forced to dig a ditch near the town of Abda. They were then shot and buried in the mass grave. When Radnóti's body was later discovered and exhumed, his last poems were found in the pocket of his coat. One, dated October 31, 1944, eerily predicts his own execution.

Radnóti's poetry, several volumes of which were published during his lifetime, mingles a sense of personal joy with premonitions of catastrophe. In much of his work, such traditional subjects of lyric verse as nature, love, and even normal death are threatened by a lurking enemy whose presence poisons the purity of the poet's vision. This shadowy doom infects the tone and imagery of some lines from "Foamy Sky," written a few months before the outbreak of World War II:

Sometimes the year looks round and shrieks,
looks round and faints away.
What kind of autumn lies in wait,
what winter dulled with agony to grey?

Nature itself cannot escape the taint: "Foam gushes forth upon the moon. / A dark green venom streaks the sky." Against this onslaught, Radnóti had only one fragment to shore against his potential ruin and the ruin of his age: the form and integrity of art. It would be a mistake, I think, to speak of the redemptive power of the word, since the language of art did little to forestall the mass murder that consumed Radnóti and so many millions of others; nor does it do anything today to compensate for that disaster. Yet Radnóti shared with Nelly Sachs, Abraham Sutzkever, and the other poets in this section a belief in the unique power of language to survive and to record and preserve through art the grim reality that imperiled the men and women of their generation. It also gave them a sense of identity in a world that left them few solaces to embrace. Radnóti spoke for many of them in these lines from "Root," one of his last poems:

Root is what I am, rootpoet
here at home among the worms,
finding here the poem's terms.

Holocaust poetry may be the ultimate paradox, and the ultimate challenge, since it asks us to preserve the fragile balance of the "poem's terms" while the chaos of history mocks its claims to order.

War Diary

1 Monday Night

These times, these times, when terror will finger the heart,
when the world hangs upon distant contingencies,
childhood's an older and older memory, kept
green by the aging trees.

The mornings suspect, the nights horrible omens,
half your life you've spent between a war and a war;
and now the regime slants its bayonets at you,
keeping its shining score.

Yet still at times there rises the country of dreams,
your poems' homeland, where liberty, volatile, runs
through the fields, and waking at dawn your body bears
its fugitive fragrance.

You work occasionally, half sit, half cower
there at your desk. As if you lived in a soft mire,
your hand, gemmed with its plume, is leaden and slow, it
darkens, it loses its fire.

So time and tide turn over into a new war,
hungry clouds eat up the gentle blue of the sky,
and as it glooms over your young wife holds you close,
in fear begins to cry.

2 Tuesday Night

I'm getting some sleep at last;
slowly I go back to work again.
They prepare against me—gas, bomb, plane—
still I shall not cry out or complain.
So I live hard, like roadbuilders in the cold hills:
who, when their flimsy huts,

makeshift and quite decrepit,
collapse about them, build them again;
sleep meantime on a fragrant bed
of woodshavings, and awake refreshed
and dip their faces in the dawnbright, in the plain
clear rush of mountain streams.

I live up in the crow's nest:
see the sky lower.
Just as the ship's watch, in the storm rain,
by the lightning's glare
will sing out, guessing that he spies the shore,
so I too think I see the land, and I call
A Soul!
my voice white, white as the whitelight's flare.

And my voice is carried far
and is kindled up again
by the cold night wind and the cold night star.

3 Tired Afternoon

Dying, a wasp flies in through the open window,
my wife, asleep, speaks quickly in her dreams,
the clouds turn brownish, but a delicate wind
blows to ruffles their whitening seams.

What is there left to say? Winter comes, and war comes;
I will lie broken, out of sight of men,
in the mouth and in the eye the wormed earth will lie,
roots will transfix my body then.

Give me thy peace, O rocking afternoon,
I'll lay me down and labor when I wake,
thy sunglow hangs already on the vine,
and night begins the hills to overtake.

They slew a cloud, its blood bespatters heaven;
under the leaves like embers from an oven
glow yellow berries redolent of wine.

4 Towards Nightfall

Aslide on the fluid sky, the cadent sun;
evening walks early along the darkened lane.

The edged moon spied it, but noted it in vain:
soft mistlets drizzle down.

The hedge awakes, drags at the tired wanderer;
the dusk, slowly swirling, turns among the boughs
and, as these lines build up, murmuringly soughs:
lines that brace each other.

A scared squirrel patters in my sleeping-place
and runs five iambs and half of another:
wall to window, one brown second altogether,
vanishes without trace.

And now with him too this fleeting peace has fled;
all through the fields the worms beslither the ground,
gnawing and gnawing away without a sound
the endless rows of dead.

1935–36

Guard and Protect Me

Now nightly in my dreams the wind is blowing,
the starry snow-white sails now flap and belly,
preparing for a long and distant journey.

Just so I slowly measure out this poem,
like one who says farewell, starts his life over,
who'll scrawl his poems henceforth with a stick
upon the flying sands of Africa.

But everywhere there's weeping, horrible,
even in Africa: the fearful Child
sucks at the breast, as blue as mulberry,
of the old wetnurse Time unreconciled.

And what's the word worth here between two wars?
Scholar of words, the rare and arduous,
what worth am I?—when bombs are everywhere
in hands most lunatic and fatuous?

Flame shudders through our heavens, and he who reads
those skylit signs falls stricken to the ground.
And presently a white pain hems me in
as at ebb-tide the salt hems in the sound.

Guard and protect me, salt and whitening pain,
you snow-white consciousness, abide with me:
let not the brownly-burning smoke of fear
soil or besoot my word's white purity!

1937

In a Troubled Hour

High in the wind and sun was my dwelling-place,
motherland, now you chain in the valley of
shadow your broken son; no comfort
now are the heavenly games of evening.

Over the cliffs the skyscape is shining; I
dwell in the depths, and stones are my company,
speechless; should I then be as they are?
Why do you write? Is it death? Who asks you?—

asks of your life a reckoning, asks of this
fragment of poem how it remains but a
fragment? Know this: unmourned, unburied,
I shall lie graveless, no vale shall rock me.

Winds shall disperse my leavings; but listen, the
cliff shall re-echo—today, or tomorrow—the
song I am singing; boys and girls are
growing up now who will hear its meaning.

January 10, 1939

Like Death

Silence and darkness fall with the weight of a shroud;
faintly crackles the frost; the river aches as it films,
tinkles and sets; where the path runs through the woods a mirror
pierces its shore.

How long, winter, how long? Bones of lovers, ancient,
beautiful, buried in earthcold, shiver and crack.
In the womb of his cave he's moaning, the tousled bear;
and the fawn cries out,

cries inconsolably; over a leaden sky
shades of the clouds are swept by the darkening chill;
gleam-glimmers the moon, and the snow-colored monster flies,
shakes the rustling trees.

The play of the frost is slow, like death it's serious:
a fragile flower of ice tinkles there on the pane;
you'd think it was only a cobweb of lace dropped there,
a sweatcob of cold:

Just so the poem itself steps out before you,
taps its foot softly, flies up at once and then falls
like death, like death; the rustling stillness of winter
follows in silence.

February 27, 1940

Floral Song

Petals on your mouth are falling
from apple boughs through the bright air,
one by one the last flakes twirling
fall upon your eyes and hair;

all day upon your mouth I'm gazing
while branches sink above your eyes;
on that soft glow the light is grazing
which, when it's kissed, awakes and dies

and vanishes; you close your eyes;
shadow plays upon your lashes,
transient petal where it lies,
and baseless darkness falls like ashes,

falls, but do not be afraid:
the night's a silver serenade,
the tree of heaven is unfurled;
the moon stares on a crippled world.

Nagyvarad, Military Hospital August 25, 1942

The Fifth Eclogue (fragment)

*In memory of György Bálint**

Cold, how shuddering cold, my friend, was the breath of this poem,
with what dread did I fear its words; and today again I have fled them
scribbling half-lines.
And always of some other thing, *of some*
other thing,
I would write, but in vain! The night, this night with its hidden
prodigies
summons me: write about him.
And I startle, and rise, but the voice
already has ceased, like the dead among the Ukrainian corn.
Missing.
Nor has the autumn brought news of you.
Now in the forest
again rustles the shelterless omen of winter; the clouds draw on,
heavy with snow, and slowly come to a halt in the sky.
And are you alive?
Not even I can know that, nor can I
rage when they throw up their hands and bury their face in despair.
They can know nothing.
But are you alive? Only wounded, perhaps,
are you wading the leafdrift, the fragrant mold of the forest? or
are you
no more but the fragrance?
Already the fields flutter with snow.
Missing,—the word
stiffens and chills in the thud of the heart
and there in the ribs' cage the twisted anguish awakens,
and now my memory shivers, delivers your words from the past
in a pain so sharp I can feel the touch of your physical being
as that of the dead—
yet I can't write about you, I just can't!

November 21, 1943

*A translator and writer, György Bálint was one of Radnóti's best friends. He died as a Jewish labor serviceman in the Ukraine.

O Ancient Prisons*

O peace of ancient prisons, beautiful
  outdated sufferings, the poet's death,
images noble and heroical,
  which find their audience in measured breath—
how far away you are. Who dares to act
  slides into empty void. Fog drizzles down.
Reality is like an urn that's cracked
  and cannot hold its shape; and very soon
its rotten shards will shatter like a storm.
  What is his fate who, while he breathes, will so
speak of what *is* in measure and in form,
  and only thus he teaches how to know?

  He would teach more. But all things fall apart.
  He sits and gazes, helpless at his heart.

March 27, 1944

*On March 19, 1944, the Germans invaded Hungary. This is the first of the poems Radnóti composed after the invasion.

Dreamscape

*In memory of Clemens Brentano**

While the nightfall's soot drops in its flue
and the sky's whimsy wilts from its dusky flowers
the night is braiding in an abyssal blue
its silent wreaths of stars.

And the sky glows where the moon's head bleeds
and over the lake the glittering ringlets spread,
across the yellow landscape rush the numberless shades,
climb the hills' watershed:

where they swing into a woodland dance
scaring the nestlings there as they stamp and pass,
and the eyes of the swaying leaves stare in a trance
at the fish that slap at the glass.

And now the dreamy land leaps away,
swimming on monstrous wings over the further air,
and the terrified bird drifts in the cloudy sky
driven by this nightmare;

and my heart finds loneliness sweet as a sign,
and death, my sister, is there.

October 27, 1943–May 16, 1944

*Clemens Brentano (1778–1842) was one of the great German Romantic poets.

Fragment*

In such an age I dwelt on earth
when men had fallen so beneath their nature
that they, unbidden, for their lust would kill,
and foaming stagger in the tangles of confusion,
possessed by tainted creeds, bewildered by delusion.

In such an age I dwelt on earth
when spies were honored, and the murderer,
the traitor, and the thief were held as heroes—
but he whose zeal, it seemed, lacked voice and violence
would be abhorred as if he bore a pestilence.

In such an age I dwelt on earth
when he who would protest perforce must hide,
and gnaw his fist in bitterness and shame—
that nation in its madness laughed and thought it good
to taste such fatal horrors, drunk with filth and blood.

In such an age I dwelt on earth
when to the child the mother was accursed,
the woman who miscarried would rejoice,
the live man envied the entombed and wormy dead,
while on his table poison thickened, foamed, and spread.

. . . .

In such an age I dwelt on earth
when the dumb poet must wait and hold his peace,
hope for the day when he might find a Voice—
for none could here pronounce the dark, demanded, verse
but that Isaiah, master of the fitting curse.

. . . .

May 19, 1944

*On May 18, 1944, Radnóti received his draft card. He wrote this poem the night before he left for the barracks. A few days later, he was deported to Yugoslavia.

Root*

Root, now, gushes with its power,
rain to drink and earth to grow,
and its dream is white as snow.

Earthed, it heaves above the earthly,
crafty in its clamberings,
arm clamped like a cable's strings.

On its wrists pale worms are sleeping,
and its ankles worms caress;
world is but wormeatenness.

Root, though, for the world cares nothing,
thrives and labors there below,
labors for the leafthick bough;

marvels at the bough it nurses,
liquors succulent and sweet,
feeds celestially sweet.

Root is what I am, rootpoet
here at home among the worms,
finding here the poem's terms.

I the root was once the flower,
under these dim tons my bower,
comes the shearing of the thread,
deathsaw wailing overhead.

Lager Heidenau: in the mountains above Zagubica
August 8, 1944

*One of Radnóti's last ten poems, all of which he wrote down in his address book. This book was found in the pocket of his raincoat when his body was exhumed twenty months after his execution.

Forced March*

Crazy. He stumbles, flops, gets up,   and trudges on again.
He moves his ankles and his knees   like one wandering pain,
then sallies forth, as if a wing   lifted him where he went,
and when the ditch invites him in,   he dare not give consent,
and if you were to ask why not?   perhaps his answer is
a woman waits, a death more wise,   more beautiful than this.
Poor fool, the true believer: for weeks, above the rooves,
but for the scorching whirlwind,   nothing lives or moves:
the housewall's lying on its back,   the prunetree's smashed and bare;
even at home, when dark comes on,   the night is furred with fear.
Ah, if I could believe it: that not only do I bear
what's worth the keeping in my heart,   but home is really there;
if it might be!—as once it was,   on a veranda old and cool,
where the sweet bee of peace would buzz,   prune marmalade would chill,
late summer's stillness sunbathe   in gardens half-asleep,
fruit sway among the branches,   stark naked in the deep,
Fanni waiting at the fence   blonde by its rusty red,
and shadows would write slowly out   all the slow morning said—
but still it might yet happen!   The moon's so round today!
Friend, don't walk on. Give me a shout   and I'll be on my way.

September 1944

*This poem was found in Radnóti's address book.

Razglednicas*

1

Rolling from Bulgaria the brutal cannonade
slams at the ranges, to hesitate and fade;
men and beasts and carts and thoughts are jammed into one,
neighing the road rears up, the maned sky will run.
And you're the only constant in the changing and the mess:
you shine on eternal beneath my consciousness;
mute as an angel wondering at the catastrophe,
or the beetle of burial from his hole in a dead tree.

In the mountains† August 30, 1944

2

At nine kilometers: the pall of burning
hayrick, homestead, farm.
At the field's edge: the peasants, silent, smoking
pipes against the fear of harm.
Here: a lake ruffled only by the step
of a tiny shepherdess,
where a white cloud is what the ruffled sheep
drink in their lowliness.

Cservenka‡ October 6, 1944

*The Serbian word for "picture postcard." This poem was found in Radnóti's address book.

†Thirty-six hundred men left Bor, in Yugoslavia; under the harshest conditions, they were marched toward Hungary.

‡The Germans slaughtered about 1,000 Jewish servicemen at Cservenka.

3

The oxen drool saliva mixed with blood.
Each one of us is urinating blood.
The squad stands about in knots, stinking, mad.
Death, hideous, is blowing overhead.

Mohács* October 24, 1944

4

I fell beside him and his corpse turned over,
tight already as a snapping string.
Shot in the neck. "And that's how you'll end too,"
I whispered to myself; "lie still; no moving.
Now patience flowers in death." Then I could hear
"Der springt noch auf,"† above, and very near.
Blood mixed with mud was drying on my ear.

Szentkirályszabadja‡ October 31, 1944

*From Mohács, the men were sent to Szentkirályszabadja and from there to Germany. Out of the 3,600, only a handful survived.

†These lines refer to Miklós Lorsi, a violinist comrade of Radnóti who was murdered at Cservenka by an SS man on a horse. Having been shot once, Lorsi collapsed. He soon stood up again, staggering. "He is still moving," called the SS man, taking aim a second time, this time successfully.

‡Marched on toward Germany, the surviving servicemen ended up in German concentration camps. Radnóti, however, was too weak to continue the march. Separated from the rest of the group with twenty-one of his comrades, he was shot at the dam near Abda on or about November 8, 1944.

29. Nelly Sachs

Nelly Sachs was born in Berlin in 1891. As the daughter of a prosperous industrialist, she received a model cultural education in music, dance, and literature. She began writing at the age of seventeen, but her romantic sonnets and conventional lyrical celebrations of nature made little impact. She published a volume of stories, *Legends and Tales* (1921), many of which are set in the Middle Ages, but this too roused small interest in an artistic milieu dominated by experiments in expressionism. Readers of her early works find no hint that she would one day become the great threnodist of the destruction of European Jewry.

As a young girl, Sachs read Selma Lagerlöf's novel *Gösta Berling,* which prompted her to begin a correspondence with the Swedish writer, who would later save her life. After the Nazis came to power, Sachs started to study Jewish and Christian mysticism, paying little heed to the threats that were mounting against German Jewry. But thanks to the intervention of Lagerlöf, in May 1940 Sachs and her mother were able to emigrate to Sweden, where the poet spent the rest of her life.

During World War II, Sachs began writing poems about the fate of the Jewish people who endured the Holocaust. In these poems, images of atrocity and renewal contend in tense verbal confrontation. Her first volume, *In the Dwellings of Death,* appeared in 1946, followed by *Eclipse of the Stars* in 1949. Subsequent volumes of poems reflect Sachs's continuing struggle to pay homage to the unique anguish of the Jews during the Holocaust while trying to build a vision of universal human suffering that might lead to reconciliation, if not redemption. The latter effort is well illustrated by several brief plays and dramatic fragments, the most famous of which is *Eli: A Mystery Play of the Sufferings of Israel.* Sachs's artistic achievement was crowned in 1966, when she was given the Nobel Prize for Literature, an award she shared with the Israeli novelist S. Y. Agnon. She died in Sweden in 1970.

As a postwar writer immersed in the doom of her people, Sachs inherited a dual legacy: an ancient tradition of suffering (such as Job's), straining the idea of comprehensible cause but still compatible with spiritual aspiration and a tragic view of existence; and a modern heritage of atrocity so far beyond the possibility of comprehensible cause that the tragic view of existence collapses beneath its weight, carrying with it the ruins of the language once used to explore it. One can imagine Sachs the poet picking through the rubble to rescue separate words that may have survived the disaster, while sadly confessing that their place in the moral structure of meaning would never be the same. Indeed, in one of her essays, she argues that we must return to individual letters of the alphabet if we are to reestablish links between words and spiritual reality. In piecing these letter fragments back into words, Sachs sometimes seems to have narrowed her artistic goal, in a strictly verbal sense, to restoring the purity of the noun.

According to a scholarly concordance of her work, forty-eight of the fifty most commonly used words in her poems are nouns. It should come as no surprise that "night" and "death" lead the list. Her favorite words—"death," "night," "love," "time," "star," "earth," "sand," "blood"—unencumbered by descriptive epithets, seek to reconstruct a reality out of the void, carving small but solid blocks of imagery to rebuild the edifice of the post-Holocaust world. The spare architecture of her vision reminds us of the price we have paid for this new, if limited, vista.

Few writers on this theme have expressed such faith in the power of words, though Abraham Sutzkever shared her faith. As a *poet* of the *Holocaust,* Sachs participates in a simultaneously creative and destructive act. As this idea seeps into our consciousness through a reading of her poems, we experience the paradox of our age, when no affirmation can escape the powerful negations of the catastrophes we have endured. By entering the universe of her art, we embrace the culture of the post-Holocaust era. Poetry *becomes* truth when we understand that our human future will never escape the memories of our recently dehumanized past. Men and women now share their private fates with the public doom of others.

Sachs credits this loss of privileged identity not to a failure in modern spiritual yearning, but to the events of modern history that have radically affected, not to say afflicted, both the content and the object of that longing. She offers us a clue to approaching her poems by describing her metaphors as her wounds. Her language is helpful, provided we remember that wounds, no matter how artistically scarred, are not to be mistaken for cures.

Sachs's later efforts to reconcile atrocity with a spiritual future may be found in such volumes as *Death Still Celebrates Life* and the series *Glowing Enigmas* (poems from these volumes are translated in *O the Chimneys* [1967] and *The Seeker and Other Poems* [1970]). Her belief in possibility born of fragile tension is well illustrated by the following short poem:

Hanging on the bush of despair
and yet enduring until the saga of blossom
fulfills its prophecy—

Magically
the hawthorn is suddenly beside itself
having come from death to life—

The silent space where the Holocaust bush of despair slowly nurtures a saga of blossom conceals the secret of Sachs's art. Like the hiatus between her stanzas here, it demands of the reader's imagination a strenuous encounter with *omission.* Only after that encounter can we make sense of the life she celebrates under the shadow of death.

O the Night of the Weeping Children!

O the night of the weeping children!
O the night of the children branded for death!
Sleep may not enter here.
Terrible nursemaids
Have usurped the place of mothers,
Have tautened their tendons with the false death,
Sow it on to the walls and into the beams—
Everywhere it is hatched in the nests of horror.
Instead of mother's milk, panic suckles those little ones.

Yesterday Mother still drew
Sleep toward them like a white moon,
There was the doll with cheeks derouged by kisses
In one arm,
The stuffed pet, already
Brought to life by love,
In the other—
Now blows the wind of dying,
Blows the shifts over the hair
That no one will comb again.

Even the Old Men's Last Breath

Even the old men's last breath
That had already grazed death
You snatched away.
The empty air
Trembling
To fill the sigh of relief
That thrusts this earth away—
You have plundered the empty air!

The old men's
Parched eyes
You pressed once more
Till you reaped the salt of despair—
All that this star owns
Of the contortions of agony,
All suffering from the dungeons of worms
Gathered in heaps—

O you thieves of genuine hours of death,
Last breaths and the eyelids' Good Night
Of one thing be sure:

The angel, it gathers
What you discarded,
From the old men's premature midnight
A wind of last breaths shall arise
And drive this unloosed star
Into its Lord's hands!

What Secret Cravings of the Blood

What secret cravings of the blood,
Dreams of madness and earth
A thousand times murdered,
Brought into being the terrible puppeteer?

Him who with foaming mouth
Dreadfully swept away
The round, the circling stage of his deed
With the ash-gray, receding horizon of fear?

O the hills of dust, which as though drawn by an evil moon
The murderers enacted:

Arms up and down,
Legs up and down
And the setting sun of Sinai's people
A red carpet under their feet.

Arms up and down,
Legs up and down
And on the ash-gray receding horizon of fear
Gigantic the constellation of death
That loomed like the clock face of ages.

You Onlookers

Whose eyes watched the killing.
As one feels a stare at one's back
You feel on your bodies
The glances of the dead.

How many dying eyes will look at you
When you pluck a violet from its hiding place?
How many hands be raised in supplication
In the twisted martyr-like branches
Of old oaks?
How much memory grows in the blood
Of the evening sun?

O the unsung cradlesongs
In the night cry of the turtledove—
Many a one might have plucked stars from the sky,
Now the old well must do it for them!

You onlookers,
You who raised no hand in murder,
But who did not shake the dust
From your longing,
You who halted there, where dust is changed
To light.

If I Only Knew

If I only knew
On what your last look rested.
Was it a stone that had drunk
So many last looks that they fell
Blindly upon its blindness?

Or was it earth,
Enough to fill a shoe,
And black already
With so much parting
And with so much killing?

Or was it your last road
That brought you a farewell from all the roads
You had walked?

A puddle, a bit of shining metal,
Perhaps the buckle of your enemy's belt,
Or some other small augury
Of heaven?

Or did this earth,
Which lets no one depart unloved,
Send you a bird-sign through the air,
Reminding your soul that it quivered
In the torment of its burnt body?

Chorus of the Rescued

We, the rescued,
From whose hollow bones death had begun to whittle his flutes,
And on whose sinews he had already stroked his bow—
Our bodies continue to lament
With their mutilated music.
We, the rescued,
The nooses wound for our necks still dangle
before us in the blue air—
Hourglasses still fill with our dripping blood.
We, the rescued,
The worms of fear still feed on us.
Our constellation is buried in dust.
We, the rescued,
Beg you:
Show us your sun, but gradually.
Lead us from star to star, step by step.
Be gentle when you teach us to live again.
Lest the song of a bird,
Or a pail being filled at the well,
Let our badly sealed pain burst forth again
and carry us away—
We beg you:
Do not show us an angry dog, not yet—
It could be, it could be
That we will dissolve into dust—
Dissolve into dust before your eyes.
For what binds our fabric together?
We whose breath vacated us,
Whose soul fled to Him out of that midnight
Long before our bodies were rescued
Into the ark of the moment.
We, the rescued,
We press your hand
We look into your eye—
But all that binds us together now is leave-taking,
The leave-taking in the dust
Binds us together with you.

Chorus of the Unborn

We the unborn
The yearning has begun to plague us
The shores of blood broaden to receive us
Like dew we sink into love
But still the shadows of time lie like questions
Over our secret.

You who love,
You who yearn,
Listen, you who are sick with parting:
We are those who begin to live in your glances,
In your hands which are searching the blue air—
We are those who smell of morning.
Already your breath is inhaling us,
Drawing us down into your sleep
Into the dreams which are our earth
Where night, our black nurse,
Lets us grow
Until we mirror ourselves in your eyes
Until we speak into your ear.

We are caught
Like butterflies by the sentries of your yearning—
Like birdsong sold to earth—
We who smell of morning,
We future lights for your sorrow.

Night, Night

Night, night,
that you may not shatter in fragments
now when time sinks with the ravenous suns
of martyrdom
in your sea-covered depths—
the moons of death
drag the falling roof of earth
into the congealed blood of your silence.

Night, night,
once you were the bride of mysteries
adorned with lilies of shadow—
In your dark glass sparkled
the mirage of all who yearn
and love had set its morning rose
to blossom before you—
You were once the oracular mouth
of dream painting and mirrored the beyond.

Night, night,
now you are the graveyard
for the terrible shipwreck of a star—
time sinks speechless in you
with its sign:
The falling stone
and the flag of smoke.

If the Prophets Broke In

If the prophets broke in
through the doors of night,
the zodiac of demon gods
wound like a ghastly wreath of flowers
round the head—
rocking the secrets of the falling and rising
skies on their shoulders—

for those who long since fled in terror—

If the prophets broke in
through the doors of night,
the course of the stars scored in their palms
glowing golden—

for those long sunk in sleep—

If the prophets broke in
through the doors of night
tearing wounds with their words
into fields of habit,
a distant crop hauled home
for the laborer

who no longer waits at evening—

If the prophets broke in
through the doors of night
and sought an ear like a homeland—

Ear of mankind
overgrown with nettles,
would you hear?
If the voice of the prophets
blew
on flutes made of murdered children's bones
and exhaled airs burnt with
martyrs' cries—

if they built a bridge of old men's dying
groans—

Ear of mankind
occupied with small sounds,
would you hear?

If the prophets
rushed in with the storm-pinions of eternity
if they broke open your acoustic duct with the words:
Which of you wants to make war against a mystery
who wants to invent the star-death?

If the prophets stood up
in the night of mankind
like lovers who seek the heart of the beloved,
night of mankind
would you have a heart to offer?

O the Homeless Colors of the Evening Sky!

O the homeless colors of the evening sky!
O the blossoms of death in the clouds
like the pale dying of the newly born!

O the riddles that the swallows
ask the mystery—
the inhuman cry of the gulls
from the day of creation—

Whence we survivors of the stars' darkening?
Whence we with the light above our heads
whose shadow death paints on us?

Time roars with our longing for home
like a seashell

and the fire in the depths of the earth
already knows of our ruin—

In the Evening Your Vision Widens

In the evening your vision widens
looks out beyond midnight—
twofold I stand before you—
green bud rising out of dried-up sepal,
in the room where we are of two worlds.
You too already extend far beyond the dead,
those who are here,
and know of what has flowered
out of the earth with its bark of enigma.

As in the womb the unborn
with the primordial light on its brow
has the rimless view
from star to star—
So ending flows to beginning
like the cry of a swan.
We are in a sickroom.
But the night belongs to the angels.

Peoples of the Earth

Peoples of the earth,
you who swathe yourselves with the force of the unknown
constellations as with rolls of thread,
you who sew and sever what is sewn,
you who enter the tangle of tongues
as into beehives,
to sting the sweetness
and be stung—

Peoples of the earth,
do not destroy the universe of words,
let not the knife of hatred lacerate
the sound born together with the first breath.

Peoples of the earth,
O that no one mean death when he says life—
and not blood when he speaks cradle—

Peoples of the earth,
leave the words at their source,
for it is they that can nudge
the horizons into the true heaven
and that, with night gaping behind
their averted side, as behind a mask,
help give birth to the stars—

Landscape of Screams

At night when dying proceeds to sever all seams
the landscape of screams
tears open the black bandage,

Above Moria, the falling off cliffs to God,
there hovers the flag of the sacrificial knife
Abraham's scream for the son of his heart,
at the great ear of the Bible it lies preserved.

O hieroglyphs of screams
engraved at the entrance gate to death.

Wounded coral of shattered throat flutes.

O, O hands with finger vines of fear,
dug into wildly rearing manes of sacrificial blood—

Screams, shut tight with the shredded mandibles of fish,
woe tendril of the smallest children
and the gulping train of breath of the very old,

slashed into seared azure with burning tails.
Cells of prisoners, of saints,
tapestried with the nightmare pattern of throats,
seething hell in the doghouse of madness
of shackled leaps—

This is the landscape of screams!
Ascension made of screams
out of the bodies grate of bones,

arrows of screams, released
from bloody quivers.

Job's scream to the four winds
and the scream concealed in Mount Olive
like a crystal-bound insect overwhelmed by impotence.

O knife of evening red, flung into the throats
where trees of sleep rear blood-licking from the ground,

where time is shed
from the skeletons in Hiroshima and Maidanek.

Ashen scream from visionary eye tortured blind—

O you bleeding eye
in the tattered eclipse of the sun
hung up to be dried by God
in the cosmos—

30. Jacob Glatstein

Jacob Glatstein (Yankev Glatshteyn) was born in 1896 in Lublin, Poland, which at the time was still part of the Russian empire. He received a thorough Jewish education, studying Bible, Talmud, and commentaries, while private tutors guided him in secular subjects. He emigrated to the United States without his family in 1914 and published his first short story that year in a Yiddish newspaper. He was associated with the In zikh (literally, within the self), or Introspectivist, movement, a group of writers whose Manifesto of 1919 urged the poet to see the world "egocentrically" because it was "the most natural and therefore *the truest and most human* mode of perception." The individuality of the poet, they believed, created the individuality of the poem. Only deep in the psyche could the poet find images and associations "trustworthy" enough to prevent language from betraying one's perception of experience.

These guidelines, which encouraged him to be "subjectively attuned," may have been partly responsible for Glatstein's turning away from his Jewish background in Poland in his poetry during the decades following World War I. But in 1934, he visited Poland for the first time since he had left it, and this, combined with the approaching threat against the Jews signaled by the rise of Nazi Germany, led him to direct his attention to contemporary Jewish themes. The catastrophe of European Jewry inspired some of his most powerful poems, and after the disaster he tried to re-create the vanished, or more precisely, the perished, world of his youth.

Glatstein also worked for many years as a journalist. For more than a decade, he contributed a column to a Yiddish weekly that included literary criticism, book reviews, and analyses of Jewish issues. In 1940, he published a long prose account of his 1934 visit to Poland, translated in 1962 as *Homecoming at Twilight,* which received the Louis Lamed Prize. He was again awarded this prize in 1956 for his collected poems, *Fun mayn gantser mi (From All My Toil).*

Glatstein's first work, *Jacob Glatshteyn* (1921), is the first volume of Yiddish poetry written exclusively in free verse. His Holocaust poems represent only a small part of a long and distinguished career, whose themes range from the role of the poet in ordering experience, and the versatility of the Yiddish language and the tragedy of its shrinking familiarity, to the attempted renewal of Jewish life in Israel following World War II. Glatstein died in New York City in 1971.

Good Night, World

April 1938

Good night, wide world,
great, stinking world.
Not you, but I slam the gate.
With the long gabardine,
with the yellow patch—burning—
with proud stride
I decide—:
I am going back to the ghetto.
Wipe out, stamp out all traces of apostasy.
I wallow in your filth.
Blessed, blessed, blessed,
hunchbacked Jewish life.
Go to hell, with your polluted cultures, world.
Though all is ravaged,
I am dust of your dust,
sad Jewish life.

Prussian pig and hate-filled Pole;
Jew-killers,* land of guzzle and gorge.
Flabby democracies, with your cold
sympathy compresses.
Good night, electro-impudent world.
Back to my kerosene, tallowed shadows,
eternal October,† minute stars,
to my warped streets and hunchbacked lanterns,
my worn-out pages of the Prophets,
my Gemaras,‡ to arduous
Talmudic debates, to lucent, exegetic Yiddish,
to Rabbinical Law, to deep-deep meaning, to duty, to what is right.
World, I walk with joy to the quiet ghetto light.

*The original Yiddish mentions "Amalek thief," referring to the tribe that is the perennial enemy of Israel.

†One of the months associated with the High Holy Days and other Jewish holidays.

‡An exposition of the Mishnah (Oral Law). Together, the Mishnah and the Gemara constitute the Talmud.

Good night. It's all yours, world. I disown
my liberation.
Take back your Jesusmarxists, choke on their arrogance.
Croak on a drop of our baptized blood.
And though He* tarries, I have hope;
day in, day out, my expectation grows.
Leaves will yet green
on our withered tree.
I don't need any solace.
I return to our cramped space.
From Wagner's pagan-music to chants of sacred humming.
I kiss you, tangled strands of Jewish life.
Within me weeps the joy of coming home.

*Referring to the coming of the Messiah.

God Is a Sad Maharal*

God is a sad Maharal.
A ray of His goodness
falls on a dark world.
He broods beside the wellspring of His wonder.
He tosses heavy stones in the water.
Listen, He is miserable and alone.
He has had too much of the Golem.
He is overpowered.
Every sigh is justified.
Every outcry flashes in unattended skies.
Eternity hurries to the crazy curtain call.
God is a sad Maharal.

God is a sad Maharal.
The sound of the ram's horn†
overtakes the Days of Awe quivering in the forest.
The heart of the Great Reprover
is bitter and broken.
Even He did not conceive of such ruin.
He is ashamed. He feels small,
like every frightened Jew.
Quietly, He'll steal into the synagogue.
He will stand in the anteroom
like a penitent wrapped in a prayer shawl.
He will rip black skies with His lament.

The winds of autumn chill.
The trees sway.
The leaves fall.
God is a sad Maharal.

*An acronymic title taken from the initials of the phrase "Our teacher, the Rabbi Low," referring to Rabbi Judah Low Ben Bezalel of Prague (ca. 1525–1609), whom legend credits with having created the Golem, the monster whose task was to serve and defend Jews.

†The ram's horn (Shofar) is sounded in the synagogue during the Days of Awe (from Rosh Hashanah to Yom Kippur).

I Have Never Been Here Before

I always thought
I had been here before.
Each year of my patched-up life
I mended the fabrics
of my decrepit, tattered world.
In memory I recognized
faces and smiles,
even my father and mother reappeared
as longed-for frescoes of the past.
I have traveled old and squalid paths,
maneuvered my sails
between the shores of history.
I have continually come across the wonder
of memory inscribing itself,
and the agitated past
quietly welling up in the present.
I thought
I had always been here.

Only these last ragged years—
shreds of hair—
inventive deaths—
are my days and nights.
My warped destiny
I have lived to see.
The frozen reverie,
burnt fields,
cartography of cemeteries,
stony silence,
emblems of vicious joy—
I don't recall them.
I have never seen them before.
I have never been here before.

Be still, dead world.
Be silent, in your ruin.
Blasted ornaments will bloom again.
We shall rebuild your foundations
out of the blood that was spilled.

And yet, the dead will still cry midnight prayers—*
each corpse, a trickling voice.
Like a tiny candle over each grave,
a cry will burn,
each one for itself.
"I am I"—
thousands of slaughtered I's
will cry in the night:
"I am dead, unrecognized,
my blood still unredeemed."

Such a wealth of gravestones—
I have never seen them before.
Day and night I shall mourn the names.

I have never been here before.

*Very pious Jews in Eastern Europe rose at midnight for prayer and study.

Smoke

Through crematorium chimneys
a Jew curls toward the God of his fathers.
As soon as the smoke is gone,
upward cluster his wife and son.

Upward, toward the heavens,
sacred smoke weeps, yearns.
God—where You are—
we all disappear.

Cloud-Jew

I

The cloud-Jew—in gray—
arrives with his transparent sadness.
An astonished slumber
settles on impassive windows.
The cloud-Jew writes Yiddish letters
on an alien sky.
I want to read them,
interpret them, explain them.
They run like tears,
threads of tears,
before I can make out their meanings.

A *shin,** blind in three eyes,
lengthens, thins out,
till it suspends into a harp
that darkens and trembles,
a trampled commandment
over the ruddy crosses
of a churchly city.

The empty synagogue dissolves.
The cloud-Jew carries it far away.
Axes and hands rip apart,
but they can't reach it.
Alone, my lips tremble,
like that harp at nightfall,
when the appointed vision
closes my eyes
with its sacred shawl.

*Very pious Jews in Eastern Europe rose at midnight for prayer and study.

*The letter shin (שׁ) is formed by three thin lines wavering upward from a horizontal line, producing a harp-like image. A dot is placed over the first or last of the projections as an indication of how the letter is to be pronounced (as "s" or "sh").

II

An overwhelming, burning, ancient sadness—
pages falling from a village ledger—
sinks deep in my mind;
bit by bit wear away
yesterday's signs
of Jewish streets, Jewish stores,
synagogues, gravestones.

Dread overtakes me
as I try to claim,
catch hold of, that cloud-Jew,
star-strewn in letters
across his Jewish sky.
The letters start to merge—
they drift together.
In fear and trembling
I read the scroll of fire.

"Even before my body
was torn apart limb by limb,
I was no longer a special seed,
only a flicker of Jewish sanctity.
That's how you should render me in your poem,
you—son of the eternal cloud-Jew."

VI

Painters of Terezín

In spite of courageous but sporadic surges of rebellion in such ghettos as Warsaw and Bialystok and the Auschwitz, Treblinka, and Sobibor death camps, Holocaust history and testimony are mainly chronicles of destruction. If we nonetheless cling to the notice of protest and avoid the tokens of ruin, we merely display the natural human tropism toward hope. Similarly, nurtured by our artistic and religious heritage to applaud the triumph of the human spirit, we have learned to regard any creative act as a form of affirmation. The drawings and paintings from Terezín (called Theresienstadt by the Germans) are no exception to this rule. Despite the prevailing gloom and despair of their themes, we hail them as exhibits of spiritual resistance, vital dissents against the German attempt to suppress the will to the human.

This celebratory mood is a naïve but understandable example of the impulse to change fleeting experience into enduring art. The men and women who sketched, drew, and painted their visions of Terezín knew the danger of their enterprise: if they were caught, they might pay with their lives, and many did. What they perhaps did not realize is that they also risked the lives of their spouses and children. Nevertheless, given the opportunity and the materials, they took the risk, hoping they could portray for future generations a way of living—and dying—that the Germans tried to hide or disguise. They managed to conceal much of their work, and that is why it has survived, but they were less successful screening themselves.

A leading group of artists was arrested in July 1944 and were charged with spreading *Greuelpropaganda* (horror propaganda). After being questioned and tortured—in some cases, family members were questioned and tortured, too—those who were still alive were shipped to Auschwitz. Eventually, most of the other artists were sent there, too. With a few exceptions—Leo Haas the major one—they all perished. Otto Ungar was alive at liberation, but was so weak that he died a few months later.

The Germans labeled this art "horror propaganda," but it deserved to be called the "horror truth" of Terezín. The artists foresaw that the challenge to the postwar world would be how to imagine the reality of the camp. The real tribute to them is not to their moral courage or spiritual defiance, but their will to pit their vision of how the Holocaust should be seen against the aim of their oppressors to shape another view. Their art refutes the efforts of *any* culture to manipulate the truth to gain or maintain power, prestige, or profit. This achievement is all the more remarkable when we study the context in which it was created.

Established in November 1941 as a ghetto town on the site of an old fortress, Terezín served a number of purposes. It was an interim camp for Czech Jews from Bohemia and Moravia, and later for Jews from Germany, Austria, Denmark, and Holland; it was a labor camp, where workshops were set up to serve the needs of the Reich; it was a temporary home for elderly Jews, "prominents" (such as Leo Baeck, chief rabbi of the Berlin Jewish community, who in fact survived the war in the camp), and decorated and disabled German and Austrian Jewish veterans of World War I; it became a "show" camp to deceive a visiting Red Cross team into believing that Jews were "well treated" in their ghetto quarters; and, most notoriously, it was a transit camp, especially for Auschwitz—during its existence, more than 86,000 prisoners were sent "to the east," most of them to their death in the gas chambers.

Conditions for those who stayed in Terezín were far from ideal. The barracks were crowded, dismal, and cold; food was always scarce; disease and hunger were rampant; and, with few exceptions, families were not allowed to live together—the men, women, and children were separated on arrival. Although the SS governed the camp, an internal administration of Jews managed its daily activities. Thus a semblance of independence was maintained, and as in other ghettos, such as Lodz, the residents hoped to survive by making their work crucial to the German war effort. The Nazis encouraged this illusion, even though the harsh state of daily existence and the ominous transports to the east seemed heralds of a grimmer doom.

At first secretly, and then with the consent of the Germans, a rich cultural life developed for many of Terezín's inhabitants, including dramatic and musical performances (sometimes of pieces composed in the ghetto), lectures and readings, and even a cabaret. These have been celebrated by later commentators as an example of the heroic human spirit asserting its strength in the face of oppression. But whether such "diversions" (after all, the average death rate in Terezín was often 100 a day, not counting those deported to the East) represented the illusion or hope of normality is a question we will probably never be able to resolve. Karel Fleischmann's *Cultural Lecture* shows a room crowded with attentive listeners, making us wonder how bad daily life in the camp could have been; *Kaffeehaus* (Figure 13), by Fritz Taussig (Fritta), presents a group of desolate, blank-eyed customers presumably listening to a trio playing nearby, but if their eyes are any sign of the condition of their ears, they are hearing nothing.

Our response to this art depends on our interpretation of its content, but also on our inner needs. We simplify, however, if we seek only consolation. The artists themselves had to deal with a dual reality, as any careful observer of Fritta's drawings will discover. Above the head of the trio in *Kaffeehaus* is a large clock, and through the window we see a guard standing near a fence topped by barbed wire. Whatever pleasure this gathering may derive from the music cannot cancel their status as victims of time and space. The clock ticks away their doom, and the fence reminds us that prisoners of death do not enjoy art as free spirits do. To allow this art to distract us from the truth would violate the design of the artists. Terezín was not a monument to the cultural vigor of an oppressed people, but a way station to the gas chambers. Had the war ended a year later, and the killing centers to the east continued operating, few Jews would have been found alive upon liberation.

Among the art from Terezín are some fine portraits and even some sensitive still lifes. Beauty and dignity were not alien to these artists, but embraced what we might call the nostalgia of memory. Viewing them today, we sense what was lost, not what has been preserved. They remind us that the integrity of self *and* of form was more fragile than anyone had imagined. Perhaps more than any other model of Holocaust documentation, the portraits compel us to look beyond the face of the subject to its *fate.* They and the bleaker landscapes and scenes in other Terezín drawings and paintings train the imagination to follow the content outside the frame into the realm of its awful effects.

What we see is never enough to signify everything that was there. Holocaust art, like its literary counterpart—and, indeed, like all testimonials to this event—is an eternally unfinished art. It remains neutral until the viewer agrees to activate the realization of its theme. For example, Fritta's *Life in the Attic* (Figure 20) summons its audience to rewrite its title; surely "misery in the attic" conveys more candidly what we really see: a crowded room beneath the eaves with elderly Jews sprawled or hunched over on the floor, a blind man groping against one wall, all of the exhausted figures waiting hopelessly for nothing. Our foreknowledge of their doom dramatizes the irony of their inverted days, when the prospect of death has replaced living as the future of their being. They await not rescue, but the end. The more we look, the more we are forced to imagine our way into this desolate human landscape.

We enter these scenes through the eyes of Fritta's blind man, groping to "see." Almost as if it is intended to be an explicit commentary on Fritta's work, Leo Haas has left us a drawing that pursues this theme. It also represents a crowd of gaunt men and women in the attic of a house in Terezín. They are leaning against walls, lying down, dozing, or simply staring. They are weary and forlorn, but most of all, they seem to be waiting. We do not know exactly for what, but Haas's title offers a clue: *Expecting the Worst* (Figure 5). From this, we deduce that they are soon to be deported to the east. The reputed mutual support that helped victims endure their grief is nowhere to be seen. Each creature is alone with his or her fate. A glaring light pierces the gloom, illuminating the wretchedness of the inhabitants and challenging us to witness the spectacle with horrified, if compassionate, eyes. There is nothing to celebrate, no dignity to proclaim, and certainly no "passive" victim to scorn. It is only an appeal to try to imagine what a life shorn of supports that *we* take for granted must have been like.

It is an art of *access*—culture as vision, not elation. The only resistance it implies is our own reluctance to enter the paralyzing pain of its contents. What could Haas have had in mind but the desire to leave a visual legacy of this pain, since he knew that its living models were soon to vanish from the scene? His title, *Expecting the Worst,* to say nothing of his austere draftsmanship, makes his intention unmistakable: look and see, and seeing, grieve. Living amid a German enemy whose purpose in Terezín was to delude the world, their prisoners, and perhaps even some of their own members into believing that the place was not really what it was—a transit camp to the gas chambers of Auschwitz—it is no wonder that the painters of Terezín returned again and again to the contrast between truth and delusion.

The artists' own lives before their arrest are a vivid example of this contrast. Housed in a special barracks together with their families, most of the painters (among whom were many talented graphic artists) were recruited by the self-administered Technical Department of the camp to work in the so-called Drawing Office, where they prepared blueprints and sketches for building projects undertaken in the ghetto. An example is Leo Haas's *Drafting/Technical Studio* (Figure 1). Much of this material was used by the SS authorities to prepare reports for transmission to their superiors in Berlin. Thus a crust of normalcy formed, made firmer by the SS demand for illustrations of Jews laboring on ghetto ventures to use as supplements to their reports. Naturally, works like Fritta's and Haas's attic paintings could not be included in these official documents; their very existence had to be hidden from the Nazis. One suspects that the psychological need to counter the pretense of their daily chores in the Drawing Office helped to fuel the secret goal of these artists to gather a more truthful portfolio of life—and death—in Terezín.

Of course, this art had its lighter moments, too, as if to offer some relief from the grave task of documenting a murderous truth. Peter Kien, one of the youngest Terezín artists, was not yet twenty-three when he was deported to the camp; he died in Auschwitz two years later, in 1944. Kien left a suite of pen-and-ink caricatures of prisoners working in the hospital (Figures 10 and 11). In the upper-right-hand corner of each drawing is a small sketch of what presumably were favorite daydreams of the subjects. Two men imagine nudes; another, a cello; still another dreams of an ocean liner; and a woman thinks of a small boy in a sailor suit pulling a toy truck on a string. Their faces are smiling, as if in their instants of yearning they reflect the nostalgia of hope and memory that governs us all. Kien playfully captures their transient moods, though we know that their longings are more likely to be frustrated than fulfilled.

The SS organized a second studio in Terezín whose personnel was ordered to create artworks for use as decoration in the homes of camp officials or for resale elsewhere. It is difficult to imagine what it must have been like for craftsmen who were forced to spend their days falsifying their lives by pretending to engage freely in their chosen vocation while secretly preparing a portfolio of death as their legacy. This dilemma coalesced in two well-known episodes from the history of the camp that betray the trap of duplicity that the Nazis set for their victims and potential enemies, knowing that both, if given a choice, would prefer to meet the best rather than expect the worst. Sadly, this heritage lingers in many responses to the Holocaust, even today. "Beguiled is not forewarned" was the SS maxim. They shaped its appeal through a devilish art of their own.

The two episodes were a visit to Terezín by a team from the International Red Cross on June 23, 1944, and a film, *The Führer Gives a City to the Jews,* made by the Germans (but directed by Kurt Gerron, a theatrical producer and inmate of the camp who was later sent to Auschwitz, where he died). As a mode of euphemism, verbal and visual deceit became an intrinsic part of the SS "killing style." From expressions like "special treatment" and "resettlement" to the fake façade on the station at the Treblinka deathcamp, the Germans learned that it was easier to murder their victims by pretending, to the last fatal moment, that what lay ahead was something other than it was. The "beautification" program in Terezín in preparation for the Red Cross visit and the film showing the camp as a comfortable haven for resettled Jews were designed to impose this strategy on a different but equally unsuspecting audience.

Whether all inspectors in the Red Cross team were fooled by German efforts to spruce up the camp remains debatable, but the Danish and Swiss members filed favorable reports. According to them, notes Johanna Branson in *Seeing Through "Paradise": Artists and the Terezín Concentration Camp* (1991), "compared to the rest of wartime Europe, Terezín was very nearly a spa." Of course, they were not allowed to roam freely through the camp or to interrogate whom they wished; they were guided to selected areas and prisoners by their "hosts." Some have argued that the Danish delegate knew that he was being duped, but feared that a negative report would endanger the 400 Danish inmates of the camp. Ironically, for whatever reason, the Germans had already decided not to deport the Danish Jews to Auschwitz; most of them returned home safely after the war.

Both the film and the visit highlight the Nazi talent for deception as a mode of truth. They also kindled the artists' desire to create a bulwark against such fraud. One sharply observed example is a watercolor by a fifteen-year-old girl, Helga Weissová Hošková. *Beautification for the Visit of the Red Cross* depicts neatly dressed prisoners hastily hanging curtains, whitewashing barracks walls, carrying baskets of flowers, sweeping the street, or sitting by a cart overflowing with loaves of bread as cars, presumably carrying the visitors, approach. *We* see what they will see, but most striking of all to our suspicious imagination is the sharp contrast between Hošková's bright colors and the stark black-and-white outlines of Fritta's and Haas's drawings. It is as if Hošková is warning us of the ease with which we can be misled into a technicolor view of the Holocaust. The "normalcy" she captures is a tribute to the seductive charm of that view.

In a commemorative address for a Polish Jewish writer delivered in the fall of 1941, nearly a year *before* the mass deportations to the Treblinka deathcamp had begun, Warsaw ghetto diarist Abraham Lewin describes Jewish existence under the Nazis with some shrewd words that might have inspired the vision of most of the artists of Terezín:

> The proportions of life and death have radically changed. Times were when life occupied the primary place, when it was the main and central concern, while death was a side phenomenon, secondary to life, its termination. Nowadays, death rules in all its majesty; while life hardly glows under a thick layer of ashes. Even this faint glow of life is feeble, miserable and weak, poor, devoid of any free breath, deprived of any spark of spiritual content. The very soul, both in the individual and in the community, seems to have starved and perished, to have dulled and atrophied. There remain only the needs of the body; and it leads merely an organic-physiological existence. . . .
>
> Yet, we wish to live on, to continue as free and creative men. This shall be our test. If, under the thick layer of ashes our life is not extinguished, this will prove the triumph of the human over the inhuman and that our will to live is mightier than the will to destruction; that we are capable of overcoming all evil forces which attempt to engulf us.

History has still to settle the question of whether the will to live is stronger than the will to destruction. With a few exceptions, the artists of Terezín lost that contest. Their work survives only by chance, not as a triumph of the human spirit. It merits our respect and admiration, but it is no proxy for their own lives and the lives of their families. No one has a right to ask so high a price for art; nor did the artists mean to make such sacrifice for their work. We are left to value the created remains, while we mourn the cruel loss of their creators.

31. Leo Haas

Leo Haas was born in 1901 in Opava, Czechoslovakia. He studied in Karlsruhe and at the Berlin Art Academy, and then worked in Vienna until he returned to Opava to pursue a career as a portrait painter and lithographer. He was associated with progressive art organizations whose purpose was to bring Czech and German artists closer together.

Soon after the Germans invaded his country, he was arrested and sent to a concentration camp in Nisko, Poland, the first camp to which Czech prisoners were dispatched before the transports to Terezín began. He was deported on December 30, 1942, to Terezín, where he was employed in the Drawing Office of the Technical Department. After a few incriminating drawings of camp life were discovered, he was arrested on July 17, 1944, and confined in the so-called Small Fortress, which was not far from the ghetto. For his "illegal" activity, he was transferred on October 26, 1944, to Auschwitz, from there to Sachsenhausen and Mauthausen, and finally to one of Mauthausen's satellite camps at Ebensee, where he was liberated shortly before the end of the war.

He returned to Prague, where he worked as a painter and cartoonist. In 1955, he moved to East Berlin in the former German Democratic Republic, where he did sketches for television and film and contributed to several foreign periodicals. In addition, he was a professor at the Art Academy in Berlin, where he died in 1983.

The Affair of the Painters of Terezín

On a June day in 1944 the painters and graphic artists Bedřich Fritta,* Otto Ungar, Felix Bloch and Leo Haas were called to the "Council of the Elders," to Mr. Zucker in the Magdeburg barracks, where the "Mayor's Office" of the Terezín ghetto was located. There he disclosed to us that we were to report the next morning to the commander of the SS. Mr. Zucker, who had certainly been sworn to secrecy, merely made the suggestion that we should pack warm underclothing and take a coat, for we might have to wait in the cold cellar of the commander's office, and "it might possibly last quite a while" before we were questioned—this was just a precaution, he said, so that we would come to no harm! Next morning we found there the architect Troller and an old man, Mr. Strass from Náchod, and I must say that the precautions of Mr. Zucker were justified. It really did last quite a while! For me and, if I am correctly informed, for Troller it lasted to 9 and 22 May 1945, and for the others all eternity.

What was it all about?

The Jewish leaders of the ghetto, who were responsible to the Gestapo for everything and could not take the least decision, except under the greatest danger, and with SS approval, had a sort of drafting office as part of the construction work in the ghetto, where some of the necessary technical drafting for building purposes was done. But it was also a place where artists from different countries (Dr. L. Heilbronn of Brno, the graphic artist Pöck from Vienna, Peter Kien, etc., in addition to those I mentioned above) were able to carry out all sorts of drawings and writing, necessary for bringing this phantom life to an end. For example, I remember that I myself drew a series of posters that were addressed to the smallest inmates of the ghetto, with the slogan, "Children, don't do that," trying to help in regard to rules of hygiene and general training of children. This team was headed by the splendid graphic artist Bedřich Fritta of Prague, who was at first protected from the transports as AK-man.[1]

Naturally, we were soon making use of this activity that was sanctioned by the SS (since we were employed by the "government") and which we

*Fritz Taussig, who left many graphic drawings of life in the Terezín ghetto.

exercised as part of our work, to actually make studies and sketches of life within the walls of the ghetto, camouflaged as official work. Especially Fritta and I myself were constantly encouraged to create this unique documentary testimony. The members of different cells would use the traditional Czech saying, "Write this down, Kisch!" Well, this encouragement was not necessary. Even before this I had not been idle when, after my arrest in Ostrava, I had been working in the construction of the first European camp for Jews (in 1939–1940) in Nisko on the San. I had brought from there a thick folder of documentary sketches made in a similar way. I was driven all the more by the shocking experiences within the ghetto walls, to be constantly on the look-out, wherever I could, sketch pad in hand.

Of course we were often in danger and had to use the greatest caution in drawing, hiding from the SS men that were spread out all over the town, keeping to attics or somewhere in a crowd. Fritta, Ungar, and often Bloch and I were so oppressed by the horrible surroundings that we devoted ourselves to our "office duties" in the day and then night after night gathered in our darkened workroom, where our sketches matured as a cycle. At that time and under these conditions (perhaps it is only today that I realize what great danger we were then working under) the cycle of 150 splendid drawings by "Fritzek" (Fritta), for instance, came into being. Today they hang in the Jewish Museum in Prague. This was also how I made my cycle of drawings of Terezín some of which are in the possession of the Czechoslovak government, but a large number of which were sent on an information tour of the United States, others are among my property and still others serve as basis for subsequent works and thus help in some attempts to depict Terezín life. At that time we came in contact with the above-mentioned Mr. Strass, a businessman in Náchod, whose "Aryan" family kept up underground connections with him. He told us that this contact was made through some courageous members of the Czech gendarmerie, who served in the ghetto and helped all they could. I myself came in contact with two brothers—I think their name was Přikryl. Strass, who was a passionate collector, especially of Czech art, often begged drawings from us that then were sent out of the ghetto by the same route that food and tobacco came in. We thought that in this way something of our documentary material would survive, even if we did not. Then we learned that successful contacts had been made with foreign countries and that our works had been sent beyond the borders of the territories ruled by the Nazis. In our enthusiasm we undoubtedly underestimated the danger that had thereby increased, and we worked even more passionately and intensively on our presentation of the ever more dismal "life" in the ghetto.

Things went on like this for a full two years. Then—I think it was in the spring of 1944—the visit of an especially important Red Cross commission from Switzerland was announced, I believe I forgot to say that we had been told that our drawings had gone to Switzerland. For conspiratorial reasons we could not—consciously—meet with our colleagues.

The preparations for this commission's visit, on which the Nazis seemed to place great value, in the precarious situation they were already in by that time, we also sketched.

Then the visit of the Red Cross was over. Among the initiate it was rumoured that the SS leaders were dissatisfied with it—dissatisfied and suspicious. Evidently the representatives of the Red Cross were not content with the streets that had been polished clean with Jewish persons' toothbrushes, and wanted a look behind the scenes of the Potemkin village. They seemed to have been well informed. There has been no success in tracing their contacts, but the fact is that a few weeks after the visit came the orders from Mr. Zucker with his troubled warning, which introduced this report.

I do not know to what circumstances it is due that the SS made no raid on our rooms, I think it was to show their mastery of the situation, and to avoid any irritating surprise. Our employment in the "government" gave us the only privilege, but one that was not to be discounted, that we were lodged in barrack rooms divided off by lath walls and so aroused the illusion that we were living in our own homes. When things had gone on in this way for a half year, we succeeded in saving a large part of the rest of our work. With feverish efforts, and with the help of a building expert, engineer Beck of Náchod, we succeeded in prying out a part of the wall in our room and to wall in the drawings. In a few hours it was finished, but I did not forget to leave some unimportant drawings lying about, so that if there was a house search by the Gestapo, I could stop the mouths of the wolves to some extent. Fritta's works were also buried by friends in a tin case in a farm yard, and were saved in this way.

After a short discussion with each other, when we had heard Zucker's report, we naturally decided that our activity as "documentarians" was the basis for our questioning, and this was shown to be quite correct.

We reported on the morning of—I think—June 17, 1944*—there is an assignment card to the police headquarters in the card-file of the Jewish Religious Communities office in Prague—at the office of the Gestapo head and were immediately herded down to the cellar.

After a long wait, we were hustled up to Rahm's† room. Besides Rahm there were present: Moes, also known as the "bird of death," also Günther whom I recognized immediately from my time at Nisko, for there he had headed the organization of the transports, together with his brother in Ostrava. But then still another person came in, whom I again recognized from his inspection trips to Nisko: he was Eichmann. I knew, of course, that this meant no good for us.

I am trying to make as dispassionate as possible a portrayal of the "Affair of the Painters from Terezín," as I found, after our return in 1945, our case was called. And here also I cannot speak of any brutalities. Eich-

*Actually, July 17, 1944.

†Karl Rahm was the German commandant of the Terezín ghetto from February 1944 until its liberation in May 1945.

mann, in opening the questioning, gave the impression rather of one who was deeply hurt that his noble intentions for the Jews could be so slanderously interpreted. And it was in this sense and this tone that he spoke. Then the questioning was continued by Günther and Rahm. Both followed Eichmann in using an oily smooth, soft tone, and carried on a sort of discussion on the history of art with us. Naturally the undertones sounded that much more dangerous. They had from two to three drawings from each one of us four painters as evidence. Each of us was in a different SS-man's charge. Günther questioned me—that I remember precisely—showing me a study of Jews searching for potato peels and saying, "How could you think up such mockery of reality and draw it?" I immediately used the same tone and explained to him that it was not, as he thought, something I had invented, but a simple study from nature, such as any proper artist is accustomed to make, a sketch of what I had happened to see by chance when I was on official duties, and which I had immediately sketched, in the same way as any painter seeks for objects to paint.

Then came the question: "Do you really think there is hunger in the ghetto, when the Red Cross did not find any at all?" The subsequent questions were aimed at making us name the people who had been our contacts with the outside world and tell whether we thought they formed a Communist cell. Since we denied this and shifted everything onto our artistic interest, the questioning was interrupted and we were taken again to the cellar. I do not know how long we lay there. After some hours, Günther and Moes appeared again, and this time the latter had a long pistol in his hand. Their tone was much more like the familiar Gestapo manner, and now their sole aim was to find out our connection with some Communist cell. We could go home immediately if we would tell all the names of those who had "led" us into this activity. Then Moes, after brandishing the revolver in a terrible way, broke off the questioning and said in a German that still rings in my ears, "Just stop it! You'll get nothing out of these fellows." (It is certainly very interesting from a psychological point of view that one often retains some not very essential fragment of conversation from situations that could be a matter of life and death.) Now we were left alone and felt we'd never get home again.

It had already got dark in the cellar when we heard a heavy truck pull up in the early evening. Soon thereafter came the typical bawling of the SS guards as they clattered into the cellar and drove us with the butts of their guns and with blows up the stairs and onto a place densely covered with trucks, in which were our wives and children, crying and yet happy to see us again. There was Fritta's wife and three-year-old boy Thomas, Mrs. Bloch, Mrs. Ungar with her five-year-old daughter, my wife Erna, then the aged František Strass from Náchod and the architect Troller who was sent away with us. As architect he had had to carry out the special demands of the camp command and in some way had not suited the gentlemen.

The whole transport went to the Small Fortress that was situated in the vicinity. There we first had to stand with our faces to the wall, the children as well. Then we men were put in one cell and the women and the two children in another, but not in the women's court.

I do not want to recount here the suffering we had to endure. This is impossible for lack of space. I believe that Ungar was the first to be sent from the Small Fortress, and I think Bloch was beaten to death on the spot, then Troller went to Auschwitz also, and I heard that he survived this, but lives somewhere overseas. The aged Strass also held out for a while in Auschwitz, I believe, and there he and his wife perished. Mrs. Ungar and her daughter survived the horror (in Auschwitz, too, I believe), while Otto Ungar died of typhus, in Buchenwald after liberation, I believe. There remained in the Fortress Fritta and I, our wives and little Thomas Fritta.

In the course of our daily labour at the "Richard" works in Litoměřice I was so beaten that as a result I had a serious phlegmon, among other things, that was cut out by our cell policeman, Dr. Pavel Wurzel, with the assistance of Julius Taussig (a furrier from Teplice), who could not go to work there because of a broken leg—without sterilization, without anesthetic, with a rusty saw. We from the "Eastern nations" were not allowed access to the sick bay or to be visited when ill. While I was going through the crisis of the wound's healing I lay hidden in old rags, where our yard commander Oberscharführer Rojko found me and drove me with blows into the bunker cell and forbade anyone to give me food. I was there about a month, then more and more fellow prisoners were brought in, among them the severely ill Fritta who had dysentery. I had been saved by the solidarity of my fellow prisoners (Jewish and non-Jewish), above all our yard trusty, the Czech General Melichar. Then, before the whispered rumours of our liquidation by Pindja and Rojko came true, we two received an indictment from the Prague Gestapo, that spoke ambiguously of "Horror propaganda and its dissemination abroad," together with a warrant for our arrest that we were to sign voluntarily, "for fear of the just anger of the German people," and which bore the remark that was familiar to us: "R.u." (return not desired).

The very next day (it was towards the end of August) we rode in cell cars attached to an express train, with a stop in Dresden for questioning, then through what was then called Breslau to Auschwitz. Fritta was very feeble and was scarcely able to move, but had to be taken to the toilet, because of the dysentery, every quarter hour. I offered to care for him and thus my handcuffs were temporarily removed. For instance, I carried him, almost on my back—he was a good head taller than I—in Dresden, through half the city to the police headquarters. In Auschwitz, where we were taken to the so-called stone Camp Auschwitz I, as political prisoners (not for racial reasons, what a nuance!), the prisoners' organization was already functioning. Fritta was taken directly to the sick bay by the architect Hanuš Major and Dr. Pavel Wurzel, who had become doctor for the sick bay in the meantime. I visited Fritta there a few times in the next eight days. He had almost completely lost consciousness and on the eighth day he died, a horrible spectacle of decay, covered all over with oedema, with complete decomposition of the blood (sepsis).

I went on to Sachsenhausen in a Sonderkommando of the Reichssicherheitshauptamt.* Together with Czech pals like Adolf Burger, Stein-Skala, Karel Gottlieb, Dr. Kaufmann-Hořice, and with the well-known Berlin publicist Peter Edel, I followed all the movements of this Himmelfahrt (heaven-bound) command up to liberation in May 1945, in Ebensee, Austria.

Meanwhile my wife Erna, although her health was completely undermined, remained among the living and returned home. She had spent the year with little Thomas Fritta in solitary confinement in the Fourth Yard of the Small Fortress, where Fritta's wife Hansi had died, miserably ill. We later adopted Thomas.

In 1955, my wife Erna died, after having barely existed for ten years, one hundred per cent disabled after her imprisonment.

Immediately after we had returned we sought out the places in Terezín where we had hidden our works. We actually found them again, unharmed. Later I was able to show these (they were published in almost all centres of publication throughout the world), with the splendid drawings by Fritta and some by Ungar, as well as some strongly incriminating drawings by a Terezín doctor, Dr. Fleischmann, who had managed to save his, too. Added to other evidence, this was strong testimony on the "final solution" engineered by Eichmann and Globke.†

I feel it my duty to accuse, until the end of my life, the fascist murderers named in my report, and to accuse the men in the background—in the name of all the victims, in my own name, and above all in the name of my friends who did not return, the painters of Terezín.

*Reich Security Main Office, Heinrich Himmler's security organization.

†Hans Globke, a Nazi official in the Interior Ministry who helped write anti-Jewish legislation. After the war, he became a member of the administration of Konrad Adenauer, the first chancellor of West Germany.

1. Aufbaukommando (work group that exempted its members from inclusion in transports).

32. Painters of Terezín

Leo Haas

1 *Drafting/Technical Studio* (with Fritta and Peter Kien) (1943).
Terezín Monument, Terezín.
Reproduced by permission of the Terezín Monument.

2 *Scenes of Life in the Ghetto* (1943–1945).
Terezín Monument, Terezín.
Reproduced by permission of the Terezín Monument.

3 *The Old and the Ill* (1943–1945).
Terezín Monument, Terezín.
Reproduced by permission of the Terezín Monument.

4 *The "Café" in Terezin* (1943–1945).
Terezín Monument, Terezín.
Reproduced by permission of the Terezín Monument.

5 *Expecting the Worst* (1943–1945).
Terezín Monument, Terezín.
Reproduced by permission of the Terezín Monument.

Leo Haas

1

2

3

4

Leo Haas

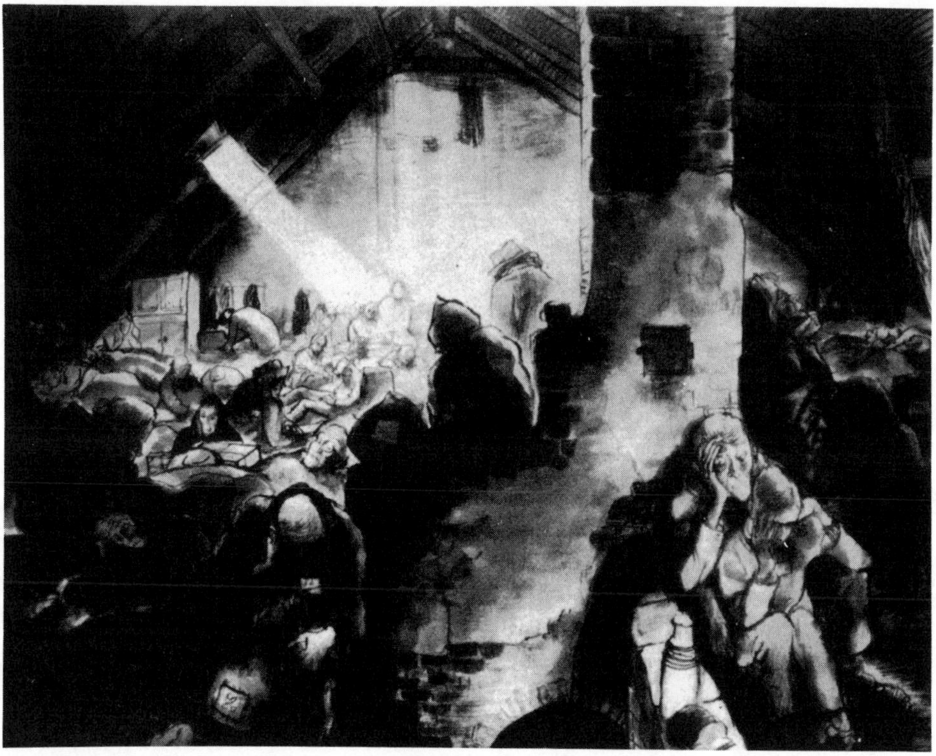

5

Karel Fleischmann

Karel Fleischmann was born in 1897 in Klatovy, Czechoslovakia. He studied medicine in Prague and became a specialist in dermatology, devoting his free time to painting and writing. He was deported on April 18, 1942, to Terezín, where he worked in health administration, but also continued to paint, draw, and write. Most of his work was preserved. On October 23, 1944, he was deported to Auschwitz, where he died.

6 *View of Terezín* (1943).
State Jewish Museum, Prague.
Reproduced by permission of the State Jewish Museum.

7 *Furniture* (1943).
State Jewish Museum, Prague.
Reproduced by permission of the State Jewish Museum.

8 *Registration for Transport* (1943).
State Jewish Museum, Prague.
Reproduced by permission of the State Jewish Museum.

Karel Fleischmann

6

7

8

Peter Kien

Peter Kien (Frantisek Petr Kien) was born in 1919 in Varnsdorf, Czechoslovakia. He studied at the Prague Art Academy and at a private graphic-art school. Together with his parents, he was deported in December 1941 to Terezín, where under Fritta he became deputy director of the camp Drawing Office. Like several other Terezín artists, he had numerous talents, devoting part of his time to music, poetry, and dramatics. But he is best remembered as a portraitist and caricaturist of important camp personalities. On October 14, 1944, he was deported with his wife and parents to Auschwitz, where they all died.

9 *Portrait of Mr. Stein* (1941–1944?).
Terezín Monument, Terezín.
Reproduced by permission of the Terezín Monument.

10 *Caricature of Dr. Stamm (with Cellist)* (1941–1944?).
Terezín Monument, Terezín.
Reproduced by permission of the Terezín Monument.

11 *Caricature of Miss Fischer (with Child)* (1941–1944?).
Terezín Monument, Terezín.
Reproduced by permission of the Terezín Monument.

Peter Kien

9

Peter Kien

11

Fritta (Fritz Taussig)

Fritz Taussig, who signed his drawings "Fritta," was born in 1906. He worked in Prague as a graphic artist and cartoonist until he was deported to Terezín in December 1941. He became director of the Technical Department, which produced official illustrations and blueprints for the Germans. Much of the unofficial, secret art was also created there. Some of Fritta's work was discovered by the Nazis in July 1944; he and other artists in the group, together with members of their families, were arrested and imprisoned in the Small Fortress not far from the ghetto, where Fritta's wife died. On October 26, 1944, Fritta and Leo Haas were deported to Auschwitz. Fritta died within a few weeks, but Haas survived and after the war adopted the Taussigs' three-year-old son.

12 *The Shops in Terezín* (1943).
State Jewish Museum, Prague.
Reproduced by permission of the State Jewish Museum.

13 *Kaffeehaus* (1943–1944).
State Jewish Museum, Prague.
Reproduced by permission of the State Jewish Museum.

14 *Life in Terezín* (1943–1944).
State Jewish Museum, Prague.
Reproduced by permission of the State Jewish Museum.

15 *Life of a Prominent* (1943–1944).
State Jewish Museum, Prague.
Reproduced by permission of the State Jewish Museum.

16 *Film and Reality* (1943–1944).
State Jewish Museum, Prague.
Reproduced by permission of the State Jewish Museum.

17 *Hospital in Cinema* (1943–1944).
State Jewish Museum, Prague.
Reproduced by permission of the State Jewish Museum.

18 *Leaving Transport* (1943–1944).
State Jewish Museum, Prague.
Reproduced by permission of the State Jewish Museum.

19 *At Kavalier (A Barracks)* (1943–1944).
State Jewish Museum, Prague.
Reproduced by permission of the State Jewish Museum.

20 *Life in the Attic* (1943–1944).
State Jewish Museum, Prague.
Reproduced by permission of the State Jewish Museum.

12

13

14

15

16

17

18

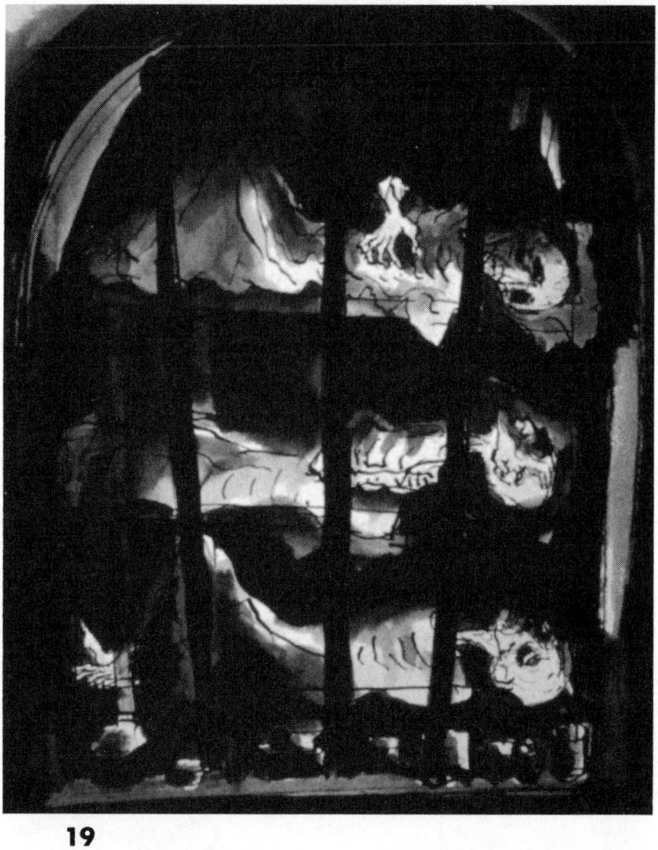

19

20